Vincent Brumfiel
Neal Golden
Mary Heins

PRE-ALGEBRA

SKILLS
PROBLEM SOLVING
APPLICATIONS

HBJ Harcourt Brace Jovanovich, Publishers
Orlando San Diego Chicago Dallas

ABOUT THE AUTHORS

VINCENT BRUMFIEL
Mathematics Instructor
Albion College
Albion, Michigan

BROTHER NEAL GOLDEN
Chairman, Department of Mathematics and Computer Science
Brother Martin High School
New Orleans, Louisiana

MARY HEINS
Mathematics Teacher
Coronado High School
El Paso, Texas

EDITORIAL ADVISORS

Kozo Nishifue
Mathematics Consultant
Division of Curriculum and Instruction
Oakland Unified School District
Oakland, California

Leo Ramirez
Mathematics Teacher
McAllen High School
McAllen, Texas

Kathy Ross-Warren
Mathematics Consultant K–12
Jefferson Parish Public Schools
Gretna, Louisiana

Pansy Rivenbark Rumley
Mathematics Teacher
D. C. Virgo Junior High School
Wilmington, North Carolina

Laurence R. Wantuck
Mathematics Supervisor K-12
Broward County Schools
Ft. Lauderdale, Florida

Richard J. Wyllie
Mathematics Department Chairman
Downers Grove South High School
Downers Grove, Illinois

Copyright © 1986 by Harcourt Brace Jovanovich, Inc.

All rights reserved. No part of this publication may be reproduced or transmitted in any form or by any means, electronic or mechanical, including photocopy, recording, or any information storage and retrieval system, without permission in writing from the publisher.

Requests for permission to make copies of any part of the work should be mailed to:
Permissions, Harcourt Brace Jovanovich, Publishers, Orlando, Florida 32887

Printed in the United States of America.

ISBN 0-15-355700-1

CONTENTS

UNIT 1 ESSENTIAL SKILLS AND APPLICATIONS

Chapter 1: Whole Numbers and Applications   1

Formulas and Whole Numbers
- 1–1 Using Formulas: Addition   2
- 1–2 Using Formulas: Subtraction   5
- 1–3 Using Formulas: Multiplication   7
- 1–4 Using Formulas: Division   10
- 1–5 Problem Solving and Applications: *Estimating with Whole Numbers*   14

Using Whole Numbers and Variables
- 1–6 Numerical Expressions   18
- 1–7 Using Variables   22

Features Consumer: 13 ● Calculator 21 ● Enrichment 29

Review/Testing Review: *Sections 1–1—1–5* 17 ● Review: *Sections 1–6—1–7* 24
Chapter Review 25 ● Chapter Test 28 ● Additional Practice 30

Chapter 2: Decimals and Applications   31

Operations with Decimals and Applications
- 2–1 Decimals and Place Value   32
- 2–2 Addition and Subtraction   35
- 2–3 Multiplication   38
- 2–4 Division   44

Decimals and Applications
- 2–5 Problem Solving and Applications: *Estimating with Decimals*   47
- 2–6 Metric Units of Measurement   51
- 2–7 Problem Solving and Applications: *Scientific Notation*   55

Features Consumer/Computer Application: 43
Calculator 50 ● Enrichment 61

Review/Testing Review: *Sections 2–1—2–3* 42 ● Review: *Sections 2–4—2–7* 57
Chapter Review 58 ● Chapter Test 60 ● Additional Practice 62

Chapter 3: Number Properties/Expressions   63

Number Properties and Applications
- 3–1 The Special Numbers 0 and 1   64
- 3–2 Simplifying Algebraic Expressions   67
- 3–3 Problem Solving and Applications: *Formulas: Area*   70

Algebraic Expressions and Applications
- 3–4 Like Terms   76
- 3–5 Combining Like Terms   79
- 3–6 Problem Solving and Applications: *Formulas: Perimeter*   81

Features	Computer 75 • Calculator 86 • Enrichment 90 • Common Errors 92	
Review/ Testing	Review: *Sections 3–1—3–3* 74 • Review: *Sections 3–4—3–6* 86 Chapter Review 87 • Chapter Test 89 Additional Practice 91 • Cumulative Review: *Chapters 1–3* 93	

Chapter 4: Number Theory — 95

Factors and Divisibility
4–1	Factors, Multiples, and Divisibility	96
4–2	Divisibility Tests	99

Properties of Prime Numbers
4–3	Prime Numbers	103
4–4	Prime Factorization	108
4–5	Least Common Multiple	111
4–6	Greatest Common Factor	115

Features	Calculator 106, 117 • Consumer 107 • Enrichment 121	
Review/ Testing	Review: *Sections 4–1—4–3* 105 • Review: *Sections 4–4—4–6* 117 Chapter Review 118 • Chapter Test 120 • Additional Practice 122	

Chapter 5: Fractions: Multiplication/Division — 123

Fractions and Applications
5–1	Writing Fractions in Lowest Terms	124
5–2	Customary Measures	126
5–3	Multiplying Fractions	129
5–4	Problem Solving and Applications: *Formulas: Temperature*	132

Mixed Numerals and Applications
5–5	Multiplying Mixed Numerals	135
5–6	Problem Solving and Applications: *Area: Parallelograms and Triangles*	140
5–7	Problem Solving and Applications: *Volume: Rectangular Prisms*	145
5–8	Dividing Fractions	148

Features	Career 138 • Consumer 139 • Enrichment 155	
Review/ Testing	Review: *Sections 5–1—5–4* 134 • Review: *Sections 5–5—5–8* 151 Chapter Review 152 • Chapter Test 154 • Additional Practice 156	

Chapter 6: Fractions: Addition/Subtraction — 157

Fractions and Mixed Numerals
6–1	Addition and Subtraction: Like Denominators	158
6–2	Addition and Subtraction: Unlike Denominators	161
6–3	Addition and Subtraction: Mixed Numerals	165

Applications of Addition/Subtraction
6–4	Problem Solving and Applications: *Time Cards*	169
6–5	Fractions and Decimals	173
6–6	Problem Solving and Applications: *Circles: Circumference and Area*	176

Features	Calculator/Computer 168 • Calculator 180 Enrichment 184 • Common Errors 186

| Review/
Testing | Review: *Sections 6–1—6–3* 167 • Review: *Sections 6–4—6–6* 180
Chapter Review 181 • Chapter Test 183
Additional Practice 185 • Cumulative Review: *Chapters 4–6* 187 | |

UNIT 2   PERCENT AND APPLICATIONS

Chapter 7:   Solving Equations   189

Equations: Subtraction and Addition

7–1	Solving Equations by Subtraction	190
7–2	Solving Equations by Addition	193
7–3	Problem Solving and Applications: *Using Equations: Addition/Subtraction*	196

Equations: Division and Multiplication

7–4	Solving Equations by Division	200
7–5	Solving Equations by Multiplication	203
7–6	Problem Solving and Applications: *Using Equations: Multiplication/Division*	206

| Features | Consumer 199 • Enrichment 213 | |
| Review/
Testing | Review: *Sections 7–1—7–3* 198 • Review: *Sections 7–4—7–6* 209
Chapter Review 210 • Chapter Test 212 • Additional Practice 214 | |

Chapter 8:   Ratio, Proportion, and Percent   215

Ratio and Proportion

8–1	Ratio	216
8–2	Proportions	219
8–3	Problem Solving and Applications: *Using Proportions*	222

Introduction to Percent

8–4	Meaning of Percent	227
8–5	Percents and Decimals	231
8–6	Finding a Percent of a Number	235
8–7	Problem Solving and Applications: *Formulas: Simple Interest*	240
8–8	Problem Solving and Applications: *Formulas: Discount*	244

| Features | Computer 226 • Calculator 243 • Enrichment 251 | |
| Review/
Testing | Review: *Sections 8–1—8–3* 225 • Review: *Sections 8–4—8–8* 247
Chapter Revew 248 • Chapter Test 250 • Additional Practice 252 | |

Chapter 9:   Percent and Applications   253

More on Percent

9–1	Finding What Percent One Number Is of Another	254
9–2	Problem Solving and Applications: *Percent of Increase and Decrease*	257
9–3	Finding a Number Given a Percent	261
9–4	Using Proportions to Solve Percent Problems	264

Graphs and Estimation

9–5	Bar Graphs	270
9–6	Circle Graphs	274
9–7	Problem Solving and Applications: *Using Estimation*	279

v

Features	Consumer 269 ● Calculator 282 Enrichment 288 ● Common Errors 290	
Review/ Testing	Review: *Sections 9–1—9–4* 267 ● Review: *Sections 9–5—9–7* 283 Chapter Review 284 ● Chapter Test 287 Additional Practice 289 ● Cumulative Review: *Chapters 7–9* 291	

UNIT 3 INTEGERS AND RATIONAL NUMBERS

Chapter 10: Rational Numbers: Addition/Subtraction — 293

Integers and Rational Numbers
- 10–1 Integers — 294
- 10–2 Rational Numbers — 297
- 10–3 Addition on the Number Line — 300

Addition and Subtraction
- 10–4 Adding Rational Numbers — 304
- 10–5 Subtracting Rational Numbers — 308
- 10–6 Properties of Addition — 312
- 10–7 Adding Polynomials — 315

Features	Consumer 303 ● Career 311 ● Calculator 314 ● Enrichment 321
Review/ Testing	Review: *Sections 10–1—10–3* 302 ● Review: *Sections 10–4—10–7* 317 Chapter Review 318 ● Chapter Test 320 ● Additional Practice 322

Chapter 11: Equations: Addition/Subtraction — 323

Addition and Subtraction Properties
- 11–1 Equations: Addition/Subtraction — 324
- 11–2 Like Terms/More Than One Equation — 326
- 11–3 Variable on Both Sides — 329

Applications of Equations
- 11–4 Problem Solving and Applications: *Words to Symbols: Addition/Subtraction* — 334
- 11–5 Inequalities on the Number Line — 338
- 11–6 Using Addition and Subtraction to Solve Inequalities — 340

Features	Calculator 331 ● Consumer 332 ● Computer 333 Calculator 342 ● Enrichment 347
Review/ Testing	Review: *Sections 11–1—11–3* 331 ● Review: *Sections 11–4—11–6* 343 Chapter Review 344 ● Chapter Test 346 ● Additional Practice 348

Chapter 12: Rational Numbers: Multiplication/Division — 349

Multiplication
- 12–1 Multiplication: Unlike Signs — 350
- 12–2 Multiplication: Like Signs — 353
- 12–3 Properties of Multiplication — 356

Division
- 12–4 Dividing Rational Numbers — 360
- 12–5 Factoring — 364

Features	Calculator 355 ● Consumer 359 ● Calculator 367 ● Enrichment 371	
Review/ Testing	Review: *Sections 12–1—12–3* 358 ● Review: *Sections 12–4—12–5* 367 Chapter Review 368 ● Chapter Test 370 ● Additional Practice 372	

Chapter 13: Equations: Multiplication/Division   373

Multiplication and Division Properties

13–1	Equations: Multiplication/Division	374
13–2	Problem Solving and Applications: *Words to Symbols: Multiplication/Division*	377
13–3	Combined Operations	381
13–4	Equations with Parentheses	383

Applications of Equations

13–5	Problem Solving and Applications: *More Than One Unknown: Angles in a Triangle*	388
13–6	Problem Solving and Applications: *More Than One Unknown: Consecutive Numbers*	390
13–7	Inequalities: Multiplication/Division	393
13–8	Problem Solving and Applications: *Using Inequalities*	396

Features	Computer 387 ● Enrichment 402 ● Common Errors 404	
Review/ Testing	Review: *Sections 13–1—13–4* 386 ● Review: *Sections 13–5—13–8* 398 Chapter Review 399 ● Chapter Test 401 Additional Practice 403 ● Cumulative Review: *Chapters 10–13* 405	

UNIT 4   REAL NUMBERS

Chapter 14: Graphing and Equations   407

The Coordinate Plane

14–1	Graphing Ordered Pairs	408
14–2	Graphing Equations	410
14–3	Problem Solving and Applications: *Broken Line Graphs*	414
14–4	Slope of a Line	417

Using Graphs

14–5	Intercepts	422
14–6	Direct and Indirect Variation	424
14–7	Solving a System of Equations by Graphing	427
14–8	Graphing Linear Inequalities	430

Features	Calculator 413 ● Computer 421 ● Enrichment 437	
Review/ Testing	Review: *Sections 14–1—14–4* 420 ● Review: *Sections 14–5—14–8* 432 Chapter Review 433 ● Chapter Test 436 ● Additional Practice 438	

Chapter 15: Special Triangles   439

Squares and Square Roots

15–1	Meaning of Square Root	440
15–2	Irrational Numbers	442
15–3	Table of Squares and Square Roots	445
15–4	Problem Solving and Applications: *Formulas: Using Square Roots*	451

Pythagorean Theorem and Special Triangles

15–5	Pythagorean Theorem and Applications	454
15–6	Tangent Ratio and Applications	458

Features Career 448 • Calculator/Computer 449 • Calculator 461
Enrichment 465 • Common Errors 467

Review/Testing Review: *Sections 15–1—15–3* 450 • Review: *Sections 15–4—15–6* 461
Chapter Review 462 • Chapter Test 464
Additional Practice 466 • Cumulative Review: *Chapters 14–15* 468

UNIT 5 OTHER APPLICATIONS

Chapter 16: Statistics and Probability 469

Statistics

16–1	Mean, Median, and Mode	470
16–2	Histograms	474

Probability

16–3	Probability	479
16–4	Probability and Tables	482
16–5	Tree Diagrams	485
16–6	Multiplying Probabilities	488

Features Career 478 • Calculator 490 • Enrichment 495

Review/Testing Review: *Sections 16–1—16–2* 477 • Review: *Sections 16–3—16–6* 491
Chapter Review 492 • Chapter Test 494 • Additional Practice 496

Chapter 17: Geometry 497

Introduction to Geometry

17–1	Introduction to Geometry	498
17–2	Angles	501
17–3	Properties of Triangles	504
17–4	Problem Solving and Applications: *Using Similar Triangles*	508
17–5	Perpendicular and Parallel Lines	514
17–6	Properties of Quadrilaterals	517

Surface Area and Volume

17–7	Problem Solving and Applications: *Surface Area and Rectangular Prisms*	520
17–8	Volume: Cylinders	524
17–9	Volume: Pyramids and Cones	527

Features Calculator 511 • Computer 512 • Enrichment 535 • Common Errors 537

Review/Testing Review: *Sections 17–1—17–4* 513 • Review: *Sections 17–5—17–9* 530
Chapter Review 531 • Chapter Test 534
Additional Practice 536 • Cumulative Review: *Chapters 16–17* 538

Glossary	540
Index	543
Answers to Selected Exercises	547

CHAPTER 1
Whole Numbers and Applications

SECTIONS
1-1 Using Formulas: Addition
1-2 Using Formulas: Subtraction
1-3 Using Formulas: Multiplication
1-4 Using Formulas: Division
1-5 Problem Solving and Applications: Estimating with Whole Numbers
1-6 Numerical Expressions
1-7 Using Variables

FEATURES
Calculator Application: *Order of Operations*
Consumer Application: *Saving $ on Housing Costs*
Enrichment: *Geometric Numbers*

1-1 Using Formulas: Addition

The following word rule tells you how to find the sticker price of a new car.

Word Rule: The **sticker price** of a new car is the sum of the base price, the destination charge (delivery charges), and the cost of the optional equipment.

A **formula** is a shorthand way of writing a word rule. In a formula, letters and mathematical symbols are used to represent the words.

Formula: $S = B + D + O$

S = sticker price; B = base price;
D = destination charge;
O = cost of optional equipment

PROCEDURE

To use formulas to solve problems:

1. Write the formula.
2. Identify the known values.
3. Replace the known values and solve.

EXAMPLE 1 The base price of a new Larkspur is $8500. The total cost of the options is $927 and the destination charge is $135.
Find the sticker price.

Solution:
1. $S = B + D + O$ Write the formula.
2. $B = 8500$; $D = 135$; $O = 927$
3. $S = 8500 + 135 + 927$
   $S = 9562$

The sticker price is **$9562**.

Self-Check

1. Find the total cost.

   Base price: $7620 Destination charge: $205 Cost of options: $2050

The table at the right shows the cost of some optional equipment. Thus, to find the total cost of the car, you must first answer the question:

What is the total cost of the options?

This is the **hidden question** in the problem. Then you can use the formula to find the sticker price.

Optional Equipment	
Air conditioning	$648
Automatic transmission	$585
Power brakes	$300
Cruise control	$205
Glass (tinted)	$ 70
Radio, AM/FM	$153
Defogger	$ 85

EXAMPLE 2 Cora Lee bought a new car with a base price of $7980. The destination charge amounted to $197. Cora chose options of air conditioning, automatic transmission and tinted glass.

What was the sticker price?

Solution:
1. $S = B + D + O$ ◄ Hidden Question: Optional Costs = ?
2. $B = 7980$; $D = 197$;
   $O = 648 + 585 + 70 = 1303$ ◄ Total cost of optional equipment
3. $S = 7980 + 197 + 1303$
   $S = 9480$

The sticker price was **$9480**.

Self-Check

2. Find the sticker price. Refer to the Table of Optional Equipment.

   Base price: $7125   Destination charge: $155
   Options: Automatic transmission, cruise control, AM/FM radio, defogger

CLASSROOM EXERCISES

Add.

1.   5
     6
   +9

2. 23
    51
   +48

3. 109
    561
   + 73

4. 5187
     609
   + 73

5. 2109
    1087
   + 679

6. $723 + 2486 + 1097$

7. $68 + 7418 + 923$

8. $75 + 6541 + 983$

Whole Numbers and Applications

WRITTEN EXERCISES

Goal: To use a formula to solve problems involving addition

For Exercises 1–7, find the sticker price for each car. (Example 1)

	Base Price	Destination Charge	Total Cost of Options
1.	$7120	$225	$2890
2.	$6800	$108	$1450
3.	$7200	$195	$1740
4.	$9050	$170	$ 948
5.	$8365	$210	$2065

6. The base price of a certain car is $8300. The total cost of optional equipment is $940 and the destination charge is $110.

7. The total cost of the options for a new car is $1875 and the destination charge is $145. The base price is $7860.

Find the sticker price. Refer to the Table of Optional Equipment on page 3. (Example 2)

	Base Price	Destination Charge	Optional Equipment
8.	$8500	$215	Air conditioning; automatic transmission; cruise control
9.	$7600	$188	Automatic transmission; power brakes; tinted glass; defogger
10.	$6950	$158	Cruise control; radio, AM/FM; automatic transmission; power brakes

Solve each problem.

11. Hector plans to buy a car with a base price of $8250 and options of air conditioning, power brakes, cruise control, and tinted glass. The destination charge is $194. What is the sticker price?

12. The base price of the car Luisa wants to buy is $9450. She chooses options of automatic transmission, power brakes, tinted glass, and defogger. The destination charge is $200. Find the sticker price.

REVIEW CAPSULE FOR SECTION 1–2

Subtract.

1. 82
   −38

2. 128
   − 9

3. 355
   − 8

4. 215 − 34

5. 403 − 78

6. 225 − 156

7. 693 − 427

The answers are on page 27.

4 Chapter 1

1-2 Using Formulas: Subtraction

The amount of an electric bill is based on the number of **kilowatt-hours (kwh)** of electricity used. You use one kilowatt-hour when you use 1 kilowatt (1000 watts) for one hour.

Word rule: The number of kilowatt-hours of electricity used equals the difference between the present reading and the previous reading.

Formula: $n = r - p$

n = number of kwh
r = present reading
p = previous reading

EXAMPLE A portion of the Miles family's electric bill is shown below. Find the number of kilowatt-hours used.

PUBLIC ELECTRIC & POWER COMPANY
025-6100

DATE FROM	READING FROM	DATE TO	READING TO	Kwh USED	AMT.
4 30	6982	5 31	7579		

ACCOUNT NUMBER   012 38 45

Solution:
1. $n = r - p$
2. Present reading, r: 7579   Previous reading, p: 6982
3. $n = 7579 - 6982$
   $n = 597$ Thus, **597 kwh** were used.

Self-Check

1. The present reading on a meter is 9123. The previous reading was 8947. How many kilowatt-hours of electricity were used?

CLASSROOM EXERCISES

Subtract.

1. 3628   2. 7142   3. 9082   4. 7111   5. 6255
   −2416   −2685   − 765   −1236   −3729

6. $3000 - 1765$ 7. $2000 - 751$ 8. $7126 - 4178$

Whole Numbers and Applications   **5**

WRITTEN EXERCISES

Goal: To use a formula to solve problems involving subtraction

For Exercises 1–8, find how many kilowatt-hours of electricity were used. (Example)

	Previous Reading	Present Reading
1.	1286	2103
2.	2335	4102
3.	5238	9442
4.	3116	3920
5.	8624	9502

6. The reading on a meter on April 5 is 4258. The reading on May 5 is 5136.

7. The reading on a meter on September 15 is 8947. The reading on October 15 is 9003.

8. The present reading on the Gibson's meter is 7705. The previous reading was 6428.

The table at the right below shows the June through October meter readings for Carlin Industries. Refer to the table for Exercises 9–12. (Example)

9. How many kilowatt-hours of electricity were used from July to August?

10. How many kilowatt-hours of electricity were used from August to September?

11. How many kilowatt-hours of electricity did Carlin Industries use from June to August?

12. If the meter reading in November is 51,011, how many kilowatt-hours of electricity were used from June to November?

Month	Meter Reading
June	9462
July	15,809
August	24,005
September	32,403
October	40,016

REVIEW CAPSULE FOR SECTION 1-3

Multiply.

1. 16 × 8
2. 24 × 9
3. 125 × 6
4. 138 × 7
5. 16 × 15
6. 35 × 23
7. 116 × 18
8. 326 × 24
9. 6 × 6 × 6
10. 1 × 1 × 1 × 0

The answers are on page 27.

1-3 Using Formulas: Multiplication

The following word rule relates distance, rate, and time.

Word Rule: Distance equals rate multiplied by time.

Formula: $d = rt$

    d — distance
    r = rate
    t = time

In the formula, rt means "r times t".
Here are several ways to indicate "r times t".

$r \times t$ rt $r \cdot t$ $(r)(t)$ These have the same meaning.

This table illustrates some units that are used in measuring distance, time, and rate (average speed).

Distance (d)	Time (t)	Rate (r)
meters	seconds	meters per second (m/sec)
kilometers	hours	kilometers per hour (km/hr)
feet	seconds	feet per second (ft/sec)
miles	hours	hour (mi/hr)
miles	minutes	minute (mi/min)

EXAMPLE 1 Light travels at the rate of about 300,000,000 meters per second. It takes 496 seconds for light from the sun to reach the earth. Find the distance from the earth to the sun.

Solution:
1. $d = rt$
2. $r = 300{,}000{,}000$ m/sec; $t = 496$ sec
3. $d = 300{,}000{,}000 \cdot 496$
   $d = 148{,}800{,}000{,}000$

The distance is about **148,800,000,000** meters.

Self-Check

1. Find d when $r = 142$ miles per hour and $t = 4$ hours.
2. Find d when $r = 250$ meters per second and $t = 8$ seconds.

Whole Numbers and Applications

You can use this formula to find the distance in feet that an object will fall in t seconds.

$$d = 16t^2$$

d = distance
t = time

In the formula, t^2 means $t \cdot t$. When a number is multiplied by itself two or more times, the number is <u>raised to a power</u>. The raised "2" is called an <u>exponent</u>. The **exponent** indicates how many times a number is multiplied by itself.

$$3^2 \leftarrow \text{Exponent}$$
$$ \leftarrow \text{Base}$$

$$3^2 = 3 \cdot 3 = 9$$
$$5^3 = 5 \cdot 5 \cdot 5 = 125$$

3^2 is read "3 squared."
5^3 is read "5 cubed."

EXAMPLE 2 Use $d = 16t^2$ to find how far an object will fall in 4 seconds.

Solution:
[1] $d = 16t^2$
[2] $t = 4$; $t^2 = 4 \cdot 4$
[3] $d = 16 \cdot 4 \cdot 4$
$d = 16 \cdot 16$
$d = 256$ The object will fall **256 feet**.

Self-Check   3. Use $d = 16t^2$ to find how far an object will fall in 3 seconds.

CLASSROOM EXERCISES

Find the distance for the given rate and time. (Example 1)

	Rate	Time	Distance
1.	5 miles per minute	3 minutes	_?_ miles
2.	80 kilometers per hour	2 hours	_?_ kilometers
3.	9 meters per second	58 seconds	_?_ meters
4.	140 feet per second	80 seconds	_?_ feet

Evaluate. (Example 2, step 2)

5. 9^2 6. 7^3 7. 4^3 8. 1^2 9. 1^3 10. 6^3

WRITTEN EXERCISES

Goal: To use a formula to solve problems involving multiplication

For Exercises 1–6, find the distance for the given rate and time. (Example 1)

	Rate	Time		Rate	Time
1.	120 feet per second	20 seconds	2.	50 miles per hour	10 hours
3.	75 kilometers per hour	5 hours	4.	25 meters per second	30 seconds

Solve each problem. (Example 1)

5. An elephant runs at a rate of 31 feet per second. How many feet will the elephant run in 15 seconds?

6. Amelia Earhart flew a plane from Newfoundland to Ireland in 15 hours at a rate of 135 miles per hour. What was the distance flown?

7. The speed of sound is about 343 meters per second. How many meters will sound travel in 6 seconds?

8. The Concorde SST flies at a speed of 1550 miles per hour. How many miles can it fly in 6 hours?

Evaluate. (Example 2, step 2)

9. 2^2 10. 2^3 11. 4^2 12. 1^4 13. 1^5 14. 5^2

15. 3^2 16. 3^3 17. 6^2 18. 10^2 19. 10^3 20. 8^2

Solve each problem. Use the formula $d = 16t^2$. (Example 2)

21. How far will an object fall in 6 seconds?

22. How far will an object fall in 12 seconds?

23. How far will an object fall in 4 seconds?

24. How far will an object fall in 8 seconds?

REVIEW CAPSULE FOR SECTION 1-4

Add.

1. $14 + 18 + 22 + 17$

2. $15 + 18 + 23 + 17 + 22$

3. $9 + 8 + 6 + 5 + 4$

4. $12 + 19 + 24 + 32 + 15 + 21$

Divide.

5. $4\overline{)36}$ 6. $6\overline{)42}$ 7. $5\overline{)65}$ 8. $7\overline{)154}$ 9. $5\overline{)235}$ 10. $4\overline{)832}$

The answers are on page 27.

Whole Numbers and Applications   9

1-4 Using Formulas: Division

Finding an average (or mean) involves finding a quotient.

Word Rule: The **average** of two or more numbers or measures is the sum of the measures divided by the number of measures.

Formula: $a = \dfrac{S}{n}$

◀ a = average
S = sum of measures
n = number of measures

EXAMPLE 1 The heights of five members of a basketball team are given below in centimeters (cm). What is the average of the players' heights?

Player	A	B	C	D	E
Height	182 cm	190 cm	178 cm	204 cm	191 cm

Solution:
1. $a = \dfrac{S}{n}$
2. $S = 182 + 190 + 178 + 204 + 191 = 945$; $n = 5$
3. $a = \dfrac{945}{5}$
   $a = 189$ The average is **189 centimeters**.

Self-Check *Find the average score for each team.*

1. Team A: 135, 172, 118, 147
2. Team B: 108, 94, 135, 151

EXAMPLE 2 The low temperatures for one week in a city are listed below in degrees Celsius. Find the mean low temperature for the week.

S	M	T	W	T	F	S
5°	3°	4°	3°	1°	2°	3°

Solution:
1. $a = \dfrac{S}{n}$
2. $S = 5 + 3 + 4 + 3 + 1 + 2 + 3 = 21$; $n = 7$
3. $a = \dfrac{21}{7}$
   $a = 3$ The mean low temperature is 3°.

◀ Mean is another word for average

Self-Check

3. Find the mean low temperature for the week.

S	M	T	W	T	F	S
6°	5°	2°	5°	4°	7°	6°

CLASSROOM EXERCISES

Divide.

1. 600 ÷ 50
2. 432 ÷ 16
3. 882 ÷ 21
4. 6300 ÷ 84
5. 7626 ÷ 62
6. 474 ÷ 79
7. 4185 ÷ 93
8. 3976 ÷ 71

Find the average. (Examples 1 and 2)

9.

Month	Rainy Days
January	2
February	3
March	7
April	20
May	8
June	2

Average number of rainy days: __?__

10.

Age at First Inaugural	
G. Washington	57
J. Adams	61
T. Jefferson	57
J. Madison	57
J. Monroe	58

Average age at first inaugural: __?__

WRITTEN EXERCISES

Goal: To use a formula to solve problems involving division

For Exercises 1–6, find the average score for each team. (Example 1)

1. Team A: 135, 172, 118, 147
2. Team B: 108, 94, 135, 151
3. Team C: 162, 145, 170, 159
4. Team D: 144, 196, 278, 110
5. Team E: 161, 163, 113, 151
6. Team F: 202, 132, 133, 129

Find the average. (Example 1)

7.

Player	Weight in kilograms
Beal	48
Davis	41
Lopez	49
Porter	45
Chang	39
Williams	48

Average weight: __?__

8.

Weekly Pay Before Taxes	
Shoe-factory Worker	$165
Clothing Worker	$165
Laundry Worker	$153
Retail-trade Worker	$149
Leather-goods Worker	$168

Average weekly pay: __?__

Whole Numbers and Applications

9. Diane's math test scores are 74, 85, 88, 86, 85, 89, 85 and 72. What is her average score?

10. Brian's pulse rates for a week were 68, 74, 77, 59, 82, 76, and 75. What was his average pulse rate?

11. Annual salaries for starting players on a professional football team are $90,000, $123,000, $225,000, $178,000, and $137,000. What is the average salary for a starter on the team?

Find the mean. (Example 2)

12. 336, 464, 304, 288, 403

13. 98, 85, 90, 92, 96, 85, 84

14. 4608, 5479, 6500, 3877

15. 18, 20, 17, 24, 24, 30, 15, 18, 20, 25, 31

16. 8916, 9214, 5684, 6382, 9144

17. 85, 612, 64, 186, 79, 189

Solve each problem.

18. The daily high temperatures for a week were 43°, 41°, 23°, 54°, 36°, 38°, 45°. What was the mean temperature?

19. A basketball player scored a total of 1875 points in 75 games. Find the mean number of points scored per game.

20. Wilson's department store received a total of 18,250 phone calls during 250 working days. What was the mean number of calls per day?

21. In Mohawk City, new traffic lights are installed when the mean number of vehicles per day using an intersection is 13,000 or more. The traffic safety division counted 102,340 vehicles using an intersection over 7 days. Will the city install a new traffic light?

REVIEW CAPSULE FOR SECTION 1-5

Round each number to the nearest 10.

1. 42
2. 28
3. 55
4. 12
5. 83
6. 96

Round each number to the nearest 100.

7. 225
8. 176
9. 389
10. 622
11. 1254
12. 795

Round each number to the nearest 1000.

13. 12,450
14. 8125
15. 18,505
16. 2050
17. 32,806
18. 13,499

The answers are on page 27.

CONSUMER APPLICATION
Saving $ on Housing Costs

On certain jobs in your home, you can save the cost of labor by doing the work yourself. The drawing below shows a plan for tiling a floor. Each of the larger squares represents a 12" × 12" square.

NOTE: The symbol " represents inches.

Exercises

1. How many 12" × 12" tiles will be needed?

2. How many 6" × 12" tiles will be needed?

3. How many 6" × 6" tiles will be needed?

4. What is the least number of 12-inch tiles needed? (HINT: you can get two 6" × 12" tiles from one 12" × 12" tile.)

5. There are 45 12-inch tiles in one box. How many boxes will be needed if only full boxes are sold?

6. The tile costs $14 per box. Adhesive costs $8 and a spreader costs $2. What is the total cost of these materials?

7. Molding will be installed around the room at the bottom of the walls. Without allowing for doorways, how many feet of molding are needed?

8. The molding cost 40¢ per foot. What will be the total cost of the molding for the room?

9. What is the total cost of all the materials for the room, excluding sales tax?

10. A carpenter will supply all the materials and do the work for $200. How much can be saved by doing the work yourself?

PROBLEM SOLVING AND APPLICATIONS

1-5 Estimating with Whole Numbers

Estimating sums, differences, products, and quotients is a skill that can be very useful in everyday life.

You use the rules for rounding to estimate answers.

PROCEDURE

To use estimation to solve a problem:

1 Write a sum, difference, product, or quotient (or a combination of these) for the problem.

2 Round each number to the nearest 10, 100, 1000, and so on as indicated.

3 Perform the indicated operations on the rounded numbers.

EXAMPLE 1 In one month, the number of vehicles that entered a National Park consisted of 556 cars, 429 vans, and 317 trucks.

Estimate the total number of vehicles to the nearest hundred.

Solution: 1 Problem   2 Rounded to the nearest 100

$$
\begin{array}{r} 556 \\ 429 \\ +317 \end{array} \longrightarrow \begin{array}{r} 600 \\ 400 \\ +300 \\ \hline \text{3}\ \ 1300 \end{array}
$$

◀ **Estimate**

About **1300 vehicles** entered the park in one month.

Self-Check

1. On a car trip, the Rulio family traveled 788 kilometers the first week, 913 kilometers the second week, and 632 kilometers the third week. Estimate the number of kilometers traveled to the nearest hundred.

2. Last year the Rulio family traveled 2576 kilometers. Estimate how many more kilometers they traveled last year than this year.

Example 2 shows how to estimate a product.

EXAMPLE 2 On each of 42 trips to a national park in July, a sightseeing bus was almost filled. The bus carries 28 passengers. Estimate the number of passengers carried by the bus in July.

Solution: ☐1 Problem   ☐2 Rounded to the nearest 10

$$\begin{array}{r}42\\\times 28\\\hline\end{array} \longrightarrow \begin{array}{r}40\\\times 30\\\hline 1200\end{array}$$

☐3 1200 ◀ **Estimate**

The bus carried about **1200 passengers** in July.

Self-Check 3. A campsite with 34 spaces for rental was almost completely filled for 27 days. Estimate the total number of daily rentals.

In problems involving division, round the dividend to the nearest convenient number that is evenly divisible by the rounded divisor.

You can use this formula to find how long it will take to travel a certain distance at a given average speed (rate).

$$t = \frac{d}{r}$$ ◀ t = time
d = distance
r = rate

EXAMPLE 3 Estimate how long it will take to drive 381 miles at an average speed of 48 miles per hour.

Solution: ☐1 Problem   ☐2 Rounded Numbers

$$\frac{381}{48} \longrightarrow \frac{400}{50}$$

☐3 $\frac{400}{50} = 8$ ◀ **Estimate**

It will take about **8 hours**.

Self-Check 4. William travels 635 kilometers at an average speed of 83 kilometers per hour. Estimate the number of hours he traveled.

5. Amtrak's Silver Meteor travels the 1325 miles from Miami to New York City at an average rate of 53 miles per hour. Estimate the number of hours the trip takes.

Whole Numbers and Applications

CLASSROOM EXERCISES

Choose the best estimate. Choose a, b, or c.
(Examples 1, 2, and 3, steps 1 and 2)

1. 112 + 631	a. 100 + 600	b. 200 + 700	c. 100 + 700
2. 2655 − 1878	a. 2700 − 1800	b. 3000 − 2000	c. 2700 − 1900
3. 72 × 48	a. 70 × 40	b. 80 × 40	c. 70 × 50
4. 774 ÷ 68	a. 770 ÷ 70	b. 760 ÷ 60	c. 800 ÷ 60

WRITTEN EXERCISES

Goal: To use estimation to solve word problems

For Exercises 1–8, choose the best estimate. Choose a, b, or c.
(Examples 1, 2, and 3)

1. 738 + 593	a. 1300	b. 1200	c. 1400
2. 4630 + 2397	a. 6000	b. 7000	c. 8000
3. 963 − 857	a. 100	b. 50	c. 120
4. 4103 − 2106	a. 2200	b. 2100	c. 2000
5. 39 × 48	a. 2000	b. 1500	c. 1200
6. 705 × 18	a. 7000	b. 14,000	c. 16,000
7. 328 ÷ 27	a. 15	b. 11	c. 14
8. 2847 ÷ 72	a. 65	b. 25	c. 40

9. Mt. Everest has an elevation of 8848 meters above sea level. The elevation of Pike's Peak is 4301 meters above sea level. Estimate, to the nearest thousand meters, the difference in their elevations.

10. On a park reserve, 18 kilograms of grain are fed to the animals each day. If there are 357 kilograms of grain on hand, estimate how many days the feed will last (nearest day).

11. In an amusement park amphitheater, there are 87 rows of seats. Each row has 31 seats. Estimate the number of seats in the amphitheater.

12. A hiking club bought T-shirts for 57 members. The total cost was $299.25. Estimate the cost of each shirt to the nearest dollar.

REVIEW: SECTIONS 1-1—1-5

Find the sticker price for each car. (Section 1-1)

1. Base Price: $6480
   Destination Charge: $125
   Total Cost of Options: $1048

2. Base Price: $12,350
   Destination Charge: $192
   Total Cost of Options: $1250

Find how many kilowatt-hours of electricity were used. (Section 1-2)

3. Previous Reading: 3819
   Present Reading: 4903

4. Previous Reading: 9426
   Present Reading: 11,054

Find the distance. (Section 1-3)

5. A cheetah can run 31 meters per second. How far can it run in 6 seconds?

6. Using the formula $d = 16t^2$, find how far an object will fall in 7 seconds.

Find the average score. (Section 1-4)

7. Blue Team: 65, 42, 73, 56

8. Red Team: 84, 87, 93, 79, 82

Estimate the answer. (Section 1-5)

9. A speed skater skated the first lap in 44 seconds, the second lap in 46 seconds, and the third lap in 42 seconds. Estimate the total time to the nearest ten seconds.

10. A basketball player scored 1401 points in 68 games. Estimate, to the nearest ten, the average number of points scored per game.

REVIEW CAPSULE FOR SECTION 1-6

Add.

1. 356
   59
   258
   +635

2. 1246
   2389
   5642
   + 384

3. $6 + 8 + 12$

4. $115 + 36 + 84$

5. $325 + 42 + 1456$

Subtract.

6. 56
   −49

7. 2040
   − 528

8. $932 - 62$

9. $5004 - 525$

Multiply.

10. 5×9

11. 25×13

12. $14 \cdot 16$

13. $156(42)$

14. $(45)(1461)$

Divide.

15. $6\overline{)42}$

16. $12\overline{)48}$

17. $62\overline{)434}$

18. $4130 \div 35$

19. $5025 \div 15$

The answers are on page 27.

1-6 Numerical Expressions

Each of the following is a *numerical expression*.

$$19 + 8 - 12$$
$$8 + 3 \times 2$$
$$(44 \times 6) \div 2$$
$$30 \div (3 \times 5)$$

Numerical expressions containing parentheses

As these four examples show, a **numerical expression** contains <u>at least one</u> of the operations of addition, subtraction, multiplication, and division.

To **evaluate** (find the value of) a numerical expression, it is important to know which operation should be done first. For example,

if you add first: $8 + 3 \times 2 = 11 \times 2 = $ **22**

if you multiply first: $8 + 3 \times 2 = 8 + 6 = $ **14**

Both cannot be correct.

To be sure that each numerical expression has only one value, follow the rules for order of operations.

ORDER OF OPERATIONS

1 When the *only operations* are addition and subtraction, perform the operations in order from left to right.

2 When the *only operations* are multiplication and division, perform the operations in order from left to right.

3 When multiplication or division is involved *along with* addition and subtraction, perform the multiplication and division before the addition and subtraction.

4 When numerical expressions contain *parentheses*, perform the operations inside the parentheses first.

The Examples in the table on the following page show how to apply the rules for order of operations.

Example	Procedure	Computation
1a. $19 + 8 - 12$	[1] Add. (Rule 1) [2] Subtract.	$19 + 8 - 12 = 27 - 12$ $= 15$
1b. $36 - 14 + 7$	[1] Subtract. (Rule 1) [2] Add.	$36 - 14 + 7 = 22 + 7$ $= 29$
2a. $48 \div 12 \times 7$	[1] Divide. (Rule 2) [2] Multiply.	$48 \div 12 \times 7 = 4 \times 7$ $= 28$
2b. $9 \times 4 \div 2$	[1] Multiply. (Rule 2) [2] Divide.	$9 \times 4 \div 2 = 36 \div 2$ $= 18$
3a. $12 \div 2 + 4$	[1] Divide. (Rule 3) [2] Add.	$12 \div 2 + 4 = 6 + 4$ $= 10$
3b. $21 - 6 \times 3$	[1] Multiply. (Rule 3) [2] Subtract.	$21 - 6 \times 3 = 21 - 18$ $= 3$
3c. $24 - 18 \div 3 \times 4$	[1] Divide. (Rule 3) [2] Multiply. (Rule 3) [3] Subtract.	$24 - 18 \div 3 \times 4 = 24 - 6 \times 4$ $24 - 6 \times 4 = 24 - 24$ $= 0$
4a. $10 \div (2 + 3)$	[1] Add. (Rule 4) [2] Divide.	$10 \div (2 + 3) = 10 \div 5$ $= 2$
4b. $5 + (3 - 1) + 4$	[1] Subtract. (Rule 4) [2] Add.	$5 + (3 - 1) + 4 = 5 + 2 + 4$ $= 11$
4c. $(5 + 3) - (1 + 4)$	[1] Add. (Rule 4) [2] Subtract.	$(5 + 3) - (1 + 4) = 8 - 5$ $= 3$

CLASSROOM EXERCISES

Name the rule that was applied in each case. (Examples 1–4, step 1)

1. $15 + 7 - 6 = 22 - 6$
2. $72 \div 36 \times 6 = 2 \times 6$
3. $4 + 9 \div 3 = 4 + 3$
4. $26 - 12 \times 2 = 26 - 24$
5. $16 - (2 + 8) = 16 - 10$
6. $5 \times 8 \div 2 = 40 \div 2$
7. $12 - 3 + 5 = 9 + 5$
8. $(8 + 4) \div 3 = 12 \div 3$
9. $3 \times 12 \div 4 = 36 \div 4$
10. $(8 - 4) \div 2 = 4 \div 2$
11. $14 - 3 \times 4 = 14 - 12$
12. $12 \div 2 \times 8 = 6 \times 8$
13. $6 + 8 \times 9 = 6 + 72$
14. $20 + 1 - 8 = 21 - 8$
15. $6 \times 8 \div 4 = 48 \div 4$

Whole Numbers and Applications

WRITTEN EXERCISES

Goal: To use order of operations in evaluating numerical expressions

For Exercises 1–8, name the operation that should be performed first. (Examples 1, 2, and 3, step 1)

1. $7 + 5 - 2$ **2.** $10 - 3 + 6$ **3.** $5 \times 4 + 7$ **4.** $4 \times 6 \div 3$
5. $14 \div 2 + 5$ **6.** $3 + 24 \div 8$ **7.** $9 \times 4 \div 3$ **8.** $14 \times 6 \div 8 - 2$

For Exercises 9–80, evaluate each numerical expression. (Example 1)

9. $5 + 6 - 3$ **10.** $15 - 3 + 6$ **11.** $29 + 6 - 8$ **12.** $23 - 9 + 14$
13. $19 - 7 + 16$ **14.** $8 + 12 - 15$ **15.** $16 + 22 - 30$ **16.** $45 - 41 + 36$

(Example 2)

17. $12 \div 3 \cdot 6$ **18.** $4 \cdot 6 \div 12$ **19.** $24 \div 4 \cdot 9$ **20.** $4 \cdot 15 \div 3$
21. $28 \div 7 \cdot 25$ **22.** $5 \cdot 6 \div 15$ **23.** $7 \cdot 8 \div 14$ **24.** $32 \div 4 \cdot 7$

(Example 3)

25. $12 - 4 \cdot 2$ **26.** $15 \div 3 + 7$ **27.** $8 \cdot 5 + 7$ **28.** $16 - 20 \div 5$
29. $45 - 2 + 51 \div 3$ **30.** $29 - 42 \div 2 + 15$ **31.** $33 - 9 + 21 \div 7$ **32.** $23 + 18 \div 6 + 52$

(Example 4)

33. $(4 + 8) \cdot 3$ **34.** $5 + (3 \cdot 2) - 8$ **35.** $(27 - 3) \cdot 4$ **36.** $(42 - 3) \cdot (4 + 6)$
37. $56 + 4 \div (10 - 6)$ **38.** $19 + (5 \div 5) - 4$ **39.** $14 + (3 \cdot 12)$ **40.** $(33 \div 3) + 5 - 11$

MIXED PRACTICE

41. $15 \div 3 \cdot 7$ **42.** $13 - 4 + 12$ **43.** $6 \cdot 8 \div 16$ **44.** $3 \div 3 \cdot (9 - 5)$
45. $30 - 9 \cdot 3 + 16$ **46.** $3 + 9 - 8$ **47.** $(54 - 8) \cdot 2 \div 4$ **48.** $12 \cdot 5 + 6 \div 6$
49. $(16 - 9) + 11$ **50.** $16 \div 4 \cdot 3 - 6$ **51.** $32 \div 4 \cdot 7$ **52.** $16 - (8 - 4) + 12$
53. $3 \cdot 4 \cdot 2 \div 6$ **54.** $8 + 9 - 11$ **55.** $9 + 7 \cdot 8 + 25$ **56.** $3 + 9 \cdot 12$
57. $6 + 7 + 2 \cdot 4$ **58.** $14 \div 2 \cdot 3 - 5$ **59.** $36 \div 3 \div 6 \div 2$ **60.** $15 \cdot 4 - 18 \div 2$
61. $5 \div 5 \cdot 7 - 5$ **62.** $37 - 7 \cdot 4 \div 2$ **63.** $(4 \cdot 8) \div 4 \cdot 3$ **64.** $6 \cdot 3 + 2 \cdot 9$
65. $5 \cdot 2 + 4 \cdot 7$ **66.** $(5 \cdot 6) \div 3 \cdot 2$ **67.** $21 \div 3 \cdot 7$ **68.** $43 - 6 \div 3 + 4$
69. $(21 - 15) + 4$ **70.** $20 \div (12 - 2)$ **71.** $(8 + 7) - (5 + 2)$ **72.** $(47 + 5) \div 4$
73. $17 + (8 - 5) - 6$ **74.** $25 \div (12 - 7)$ **75.** $(26 - 9) + (8 - 6)$ **76.** $95 - (42 + 6) - 7$
77. $(8 + 5) - (9 + 4)$ **78.** $66 \div (2 \cdot 3)$ **79.** $52 + (10 - 7) - 30$ **80.** $(15 - 9) + (21 - 8)$

MORE CHALLENGING EXERCISES

Copy each problem. Replace the ● with +, −, ×, or ÷ so that the numeral expression equals the given answer.

81. 6 ● 7 ● 3 = 27
82. 9 ● 9 ● 4 = 4
83. (9 ● 9) ● 4 = 72
84. 50 ● (40 ● 10) = 0
85. 4 ● 4 ● 4 = 64
86. 9 ● 9 ● 9 = 729
87. 8 ● 9 ● 4 ● 9 ● 6 = 73
88. 14 ● 5 ● 7 ● 3 = 13
89. 8 ● 2 ● 9 ● 3 = 43
90. (5 ● 2) ● (20 ● 4) = 15
91. (12 ● 3) ● (21 ● 7) = 45
92. 8 ● (40 ● 10) = 32
93. (6 ● 9) ● 14 = 40
94. (8 ● 6) ● (7 ● 7) = 2
95. 18 ● (35 ● 7) = 13
96. (28 ● 14) ● 16 = 32
97. (9 ● 9) ● (33 ● 11) = 0
98. 29 ● (42 ● 7) = 35

ORDER OF OPERATIONS

You can use the [M+] (memory) and [MR] (memory recall) keys to solve problems involving order of operations.

EXAMPLE: Solve: **a.** 80 − 4 × 6 **b.** 98 × 76 − 189 ÷ 7

Solutions:

a. 4 [×] 6 [=] [M+] 8 0 [−] [MR] [=] → M 56.

b. 1 8 9 [÷] 7 [=] [M+] 9 8 [×] 7 6 [=] [−] [MR] [=] → M 7421.

EXERCISES *Use a calculator to evaluate each numerical expression.*

1. 65 − 7 · 8 ÷ 4
2. 14 × 6 ÷ 12 − 3 × 2
3. 80 ÷ 10 − 5 + 6 ÷ 3
4. 11 − 15 ÷ 3 + 15 × 4
5. 20 − 12 ÷ 4 + 9 · 2
6. 110 ÷ 5 − 3 · 7 + 6
7. 75 − 25 · 2 + 10
8. 85 × 41 − 210 ÷ 3
9. 18 + 45 ÷ 9 − 50 ÷ 5
10. 22 × 3 + 18 ÷ 6 − 30 ÷ 15

REVIEW CAPSULE FOR SECTION 1-7

Perform the indicated operations.

1. 6 + 4 + 8
2. 12 + 9 + 15
3. 24 − 9 − 6
4. 35 − 12 − 8
5. 9 × 8 × 7
6. 13 × 12 × 5
7. 6 × 5 + 4
8. 14 + 3 × 9
9. 45 ÷ 5 ÷ 3
10. 88 ÷ 2 ÷ 2
11. 32 ÷ 2 ÷ 4
12. 54 ÷ 3 ÷ 2

The answers are on page 27.

Whole Numbers and Applications

1-7 Using Variables

For the square parking lot at the right, let n represent the length of each side. Since n can represent any length, n is a *variable*. A **variable** is a letter representing one or more numbers.

The sum of the lengths of the sides of a square is the **perimeter**. The perimeter of the square parking lot can be written as follows.

$n + n + n + n$ or $4 \times n$ or $4 \cdot n$ or $4(n)$ or $4n$

Each of the expressions for perimeter is an *algebraic expression*. An **algebraic expression** contains *at least one variable*. Here are some additional examples.

$$19 + c$$
$$z \div 14$$
$$5w \div (r - 3)$$
$$3 \cdot q - (t + 5)$$

◀ Algebraic expressions

To **evaluate** (find the value of) an algebraic expression, follow these steps.

PROCEDURE

Steps for Evaluating an Algebraic Expression

1. Replace each variable with a given value.
2. Perform the operations according to the rules for order of operations.

EXAMPLE 1 Evaluate $4n - 3$ when $n = 8$.

Solution:
1. Replace the variable. $\quad 4n - 3 = 4(8) - 3$
2. Multiply first. (Rule 3) $\qquad\qquad\quad = 32 - 3$
   Then subtract. $\qquad\qquad\qquad\qquad\;\; = \mathbf{29}$

Self-Check *Evaluate each expression when $t = 9$.*

1. $12t$ **2.** $t - 4$ **3.** $81 \div t - 8$ **4.** $1 + 2t$

An algebraic expression may contain more than one variable.

EXAMPLE 2 Evaluate $(2x + 3) - 5y$ when $x = 12$ and $y = 4$.

Solution:
1. Replace the variables. $\quad (2x + 3) - 5y = (2 \cdot 12 + 3) - 5(4)$
2. Evaluate inside parenthesis. $\quad\quad\quad\quad\quad = (24 + 3) - 5(4)$
   (Rule 4) $\quad\quad\quad\quad\quad\quad\quad\quad\quad\quad\quad\quad = 27 - 5(4)$
3. Multiply first. (Rule 3) $\quad\quad\quad\quad\quad\quad\quad = 27 - 20$
   Then subtract. $\quad\quad\quad\quad\quad\quad\quad\quad\quad = 7$

Self-Check *Evaluate each expression when $w = 12$ and $r = 6$.*

5. $(2w - 3) + 5r$ **6.** $(3w \div 9) + r$ **7.** $(3w \div 12 + 4r) \div 9$

CLASSROOM EXERCISES

Write the expression after each variable is replaced.
(Examples 1 and 2, Step 1)

1. $15p + 32$ when $p = 5$
2. $29 - 3k$ when $k = 8$
3. $6q + 5r$ when $q = 4$ and $r = 7$
4. $36a - 27b$ when $a = 13$ and $b = 10$
5. $24c + 18d - 30$ when $c = 19$ and $d = 15$
6. $27m + 14 - 11q$ when $m = 15$ and $q = 10$
7. $2xy - y$ when $x = 3$ and $y = 4$
8. $x \cdot x + 5xy$ when $x = 5$ and $y = 6$

Evaluate each expression when the variable equals 2. (Example 1)

9. $8 - 3n$ **10.** $4 + x$ **11.** $3y \div 2$
12. $7 - y + 2$ **13.** $x + x$ **14.** $3a + a$
15. $11 - 2y + 2$ **16.** $6n \div 3$ **17.** $5 - n + 1$
18. $6 - 2n + 2$ **19.** $3h - 2 + h$ **20.** $4p + 6 - p$

Evaluate each expression when $x = 4$ and $y = 10$. (Example 2)

21. $x + (2y \div 5)$ **22.** $(3x - y) + 7$ **23.** $(7y \div 5) + 2x$
24. $(4x + 8) - 2y$ **25.** $(12 \div x) + (90 \div y)$ **26.** $7x + (10y - 15)$
27. $15y - (10 + 5x)$ **28.** $xy - (x + y)$ **29.** $xy + (y - x)$
30. $2xy - y$ **31.** $x \cdot x + 5xy$ **32.** $x \cdot x - y$

Whole Numbers and Applications

WRITTEN EXERCISES

Goal: To evaluate (find the value of) an algebraic expression for given replacements of the variable or variables

Evaluate each expression when the variable is equal to 3.
(Example 1)

1. $2a - 1$
2. $4x + 2$
3. $2 + 2y$
4. $1 + 4c$
5. $2 + n - 1$
6. $5 - k - 1$
7. $2 + 6 \div x$
8. $12 \div p \cdot 4$
9. $8 - a + 2$
10. $15 - x + 7$
11. $n + 2 \cdot 3$
12. $x + 3 \cdot 2$

For Exercises 9–32, evaluate each expression. (Example 2)

13. $p + 8q$ when $p = 12$ and $q = 3$
14. $r + 12s$ when $r = 3$ and $s = 6$
15. $12c - 3d$ when $c = 5$ and $d = 3$
16. $15x - 8y$ when $x = 8$ and $y = 15$
17. $2v + 3 \cdot x$ when $v = 5$ and $x = 12$
18. $16 \div p - 6 \cdot r$ when $p = 2$ and $r = 1$
19. $8j - (12 + 3k)$ when $j = 6$ and $k = 8$
20. $3h - (2k + 11)$ when $h = 23$ and $k = 14$

MIXED PRACTICE

21. $8p - 19$ when $p = 6$
22. $16x + 3y$ when $x = 2$ and $y = 13$
23. $16 + 12s$ when $s = 7$
24. $12h - 25m$ when $h = 12$ and $m = 3$
25. $x \cdot x + 3x - 12$ when $x = 6$
26. $32r - 92$ when $r = 4$
27. $15 \div a - (b - 1)$ when $a = 3$ and $b = 2$
28. $5 \cdot x - (y + 1)$ when $x = 4$ and $y = 2$
29. $16 + 78 \div q$ when $q = 3$
30. $(14m - m) \cdot n$ when $m = 5$ and $n = 10$
31. $3 \cdot s \div (5 - t)$ when $s = 15$ and $t = 2$
32. $15 + 8z$ when $z = 12$
33. $3f - 8g + g^2$ when $f = 21$ and $g = 5$
34. $q \div (7 + b) \cdot 9$ when $q = 105$ and $b = 8$

REVIEW: SECTIONS 1–6 – 1–7

Evaluate each expression. (Section 1–6)

1. $8 - 6 + 4$
2. $7 - 5 + 12$
3. $5 \cdot 4 \div 2$
4. $8 \cdot 12 \div 6$
5. $3 + 8 \cdot 5$
6. $15 \div (3 + 2)$
7. $25 \div (9 - 4)$
8. $7 - 5 \cdot 6 \div 10$

Evaluate each expression. (Section 1–7)

9. $12 - 4c$ when $c = 2$
10. $3r + 15$ when $r = 9$
11. $16y - 5z$ when $y = 3$ and $z = 7$
12. $17 - p + 3q$ when $p = 11$ and $q = 16$
13. $k + 3 - 4m$ when $k = 21$ and $m = 6$
14. $a \cdot b \div 3$ when $a = 7$ and $b = 12$
15. $3 \cdot a - (2b - 1)$ when $a = 9$ and $b = 3$
16. $7m - (3n - 2)$ when $m = 5$ and $n = 4$

24 Chapter 1

CHAPTER REVIEW

PART 1: VOCABULARY

For Exercises 1–8, choose from the box at the right below the word(s) that best corresponds to each description.

1. A letter representing one or more numbers __?__ (p. 22)
2. The sum of measures divided by the number of measures __?__ (p. 10)
3. A shorthand way of writing a word rule __?__ (p. 2)
4. An expression that includes at least one of the operations of addition, subtraction, multiplication, or division __?__ (p. 18)
5. Unit for measuring the amount of electricity used __?__ (p. 5)
6. The sum of the base price, the destination charge, and the cost of optional equipment __?__ (p. 2)
7. To __?__ means to find the value of. (p. 18)
8. An expression that contains at least one variable __?__ (p. 22)

> evaluate
> average
> sticker price
> formula
> variable
> algebraic expression
> numerical expression
> kilowatt-hour

PART 2: SKILLS

For Exercises 9–12, find the sticker price for each car. (Section 1-1)

	Base Price	Destination Charge	Total Cost of Options		Base Price	Destination Charge	Total Cost of Options
9.	$4560	$135	$687	10.	$4516	$165	$1102
11.	$8032	$189	$883	12.	$10,080	$105	$752

For Exercises 13–16, find how many kilowatt-hours of electricity were used. (Section 1-2)

	Previous Reading	Present Reading
13.	1405	1624
14.	6325	8006
15.	3456	4123
16.	12,623	15,010

Whole Numbers and Applications

For Exercises 17–20, find the distance for the given rate and time. (Section 1-3)

	Rate	Time		Rate	Time
17.	40 miles per hour	4 hours	**18.**	81 kilometers per hour	3 hours
19.	6 meters per second	15 seconds	**20.**	75 feet per second	7 seconds

Evaluate. (Section 1-3)

21. 1^2 **22.** 6^2 **23.** 1^3 **24.** 4^2 **25.** 7^2 **26.** 12^2

For Exercises 27–32, use $d = 16t^2$ to find d for the given value of t. (Section 1-3)

27. $t = 2$ **28.** $t = 8$ **29.** $t = 5$ **30.** $t = 1$ **31.** $t = 15$ **32.** $t = 11$

For Exercises 33–36, find the mean temperature. (Section 1-4)

33. 8°, 12°, 10°, 16°, 24°   **34.** 41°, 52°, 38°, 50°, 49°, 40°

35. 78°, 81°, 92°, 87°, 76°, 84°   **36.** 16°, 15°, 24°, 30°, 17°, 24°, 28°

Choose the best estimate. Choose a, b, or c. (Section 1-5)

		a.	**b.**	**c.**
37.	$114 + 381$	$100 + 300$	$200 + 400$	$100 + 400$
38.	$262 - 229$	$260 - 230$	$260 - 220$	$270 - 220$
39.	24×38	20×30	20×40	30×40
40.	$644 \div 38$	$700 \div 40$	$640 \div 40$	$640 \div 30$

Evaluate each expression. (Section 1-6)

41. $13 + 5 - 6$ **42.** $32 - 4 + 16$ **43.** $15 \div 3 \cdot 4$ **44.** $6 \cdot 9 \div 3$

45. $5 \cdot (12 + 6)$ **46.** $20 \div (4 + 6)$ **47.** $42 - 18 \cdot 6 \div 9$ **48.** $16 \div 2 + 2 \cdot 3$

Evaluate each expression. (Section 1-7)

49. $19 + 3y$ when $y = 4$ **50.** $4t - 17$ when $t = 6$

51. $12a - 13b$ when $a = 5$ and $b = 3$ **52.** $3p - q + 2p$ when $p = 15$ and $q = 24$

53. $(h - 3) \cdot (k \div 4)$ when $h = 7$ and $k = 8$ **54.** $(6r + 2) \cdot s - 5$ when $r = 3$ and $s = 2$

55. $s \cdot s - 3t$ when $s = 2$ and $t = 1$ **56.** $(8 - k) + r^2$ when $k = 4$ and $r = 7$

57. $(q + 32) \div 5t$ when $q = 8$ and $t = 2$ **58.** $ac(a - c)$ when $a = 9$ and $c = 4$

PART 3: APPLICATIONS

Solve each problem.

59. The total cost of the options for a new car is $1506 and the destination charge is $185. If the base price is $8,475, find the sticker price. (Section 1–1)

60. The present reading on the Hoelthe's meter is 45,006. The previous reading was 43,100. How many kilowatt-hours of electricity were used? (Section 1–2)

61. In 1927, Charles Lindberg flew the Spirit of St. Louis from New York to Paris in about 33 hours at a rate of 107 miles per hour. Find the distance he traveled. (Section 1–3)

62. Sandra Palmer scored 78, 74, 71, and 73 to win the U.S. Open Golf Tournament. What was her average score? (Section 1–4)

63. A food wholesaler has 35 cases of peas in one warehouse, 71 cases in a second warehouse, and 29 cases in a third warehouse. Estimate the total number of cases (nearest ten). (Section 1–5)

64. A car travels 608 miles on a full tank of fuel. The fuel tank holds 18 gallons when full. Estimate how far the car can travel on one gallon of fuel. (Section 1–5)

ANSWERS TO REVIEW CAPSULES

Page 4   **1.** 44   **2.** 119   **3.** 347   **4.** 181   **5.** 325   **6.** 69   **7.** 266

Page 6   **1.** 128   **2.** 216   **3.** 750   **4.** 966   **5.** 240   **6.** 805   **7.** 2088   **8.** 7824   **9.** 216   **10.** 0

Page 9   **1.** 71   **2.** 95   **3.** 32   **4.** 123   **5.** 9   **6.** 7   **7.** 13   **8.** 22   **9.** 47   **10.** 208

Page 12 **1.** 40   **2.** 30   **3.** 60   **4.** 10   **5.** 80   **6.** 100   **7.** 200   **8.** 200   **9.** 400   **10.** 600   **11.** 1300   **12.** 800   **13.** 12,000   **14.** 8000   **15.** 19,000   **16.** 2000   **17.** 33,000   **18.** 13,000

Page 17 **1.** 1308   **2.** 9661   **3.** 26   **4.** 235   **5.** 1823   **6.** 7   **7.** 1512   **8.** 870   **9.** 4479   **10.** 45   **11.** 325   **12.** 224   **13.** 6552   **14.** 65,745   **15.** 7   **16.** 4   **17.** 7   **18.** 118   **19.** 335

Page 21 **1.** 18   **2.** 36   **3.** 9   **4.** 15   **5.** 504   **6.** 780   **7.** 34   **8.** 41   **9.** 3   **10.** 22   **11.** 4   **12.** 9

CHAPTER TEST

Complete.

1. Base price of new car: $7830
   Destination charge: $ 150
   Total cost of options $1120
   Sticker price: ___?___

2. Previous meter reading: 4825
   Present meter reading: 5604
   Total number of
   kilowatt–hours used: ___?___

3. Rate: 45 meters per second
   Time: 32 seconds
   Distance: ___?___

4. Rate: 125 kilometers per hour
   Time: 6 hours
   Distance: ___?___

Solve each problem.

5. Using the formula $d = 16t^2$, find the distance an object falls in 11 seconds.

6. Marvin's test scores are 75, 82, 95, 85, 63, and 92. What is his average score?

Choose the best estimate. Choose a, b, or c.

7. $824 + 773$ **a.** $800 + 700$ **b.** $800 + 800$ **c.** $820 + 800$

8. $1422 - 629$ **a.** $1400 - 600$ **b.** $1400 - 620$ **c.** $1420 - 600$

9. 38×87 **a.** 30×80 **b.** 40×90 **c.** 40×100

10. $601 \div 27$ **a.** $600 \div 20$ **b.** $610 \div 20$ **c.** $600 \div 30$

Evaluate each expression.

11. $18 + 6 - 4$
12. $36 \div 4 \cdot 9$
13. $(4 + 16) \div 2 - 3$
14. $17 - 4 \cdot 3 + 8$
15. $x + 3y$ when $x = 6$ and $y = 4$
16. $3s - 4t$ when $s = 8$ and $t = 2$
17. $5p \div 3 + 5q$ when $p = 9$ and $q = 4$
18. $7 \cdot (8a - 6) \div b$ when $a = 2$ and $b = 5$

Solve each problem.

19. The total cost of options for Ted's new car is $842 and the destination charge is $106. If the base price is $4580, find the sticker price.

20. Andrea scored 264 points in 12 basketball games. Find the average number of points she scored per game.

28   Chapter 1

ENRICHMENT

Geometric Numbers

Interesting relationships exist between geometric shapes and numbers. The first four **triangular numbers** and the first four **square numbers** are shown below.

Triangular Numbers

1   3   6   10

Square Numbers

1   4   9   16

EXERCISES

1. Draw the next three triangular numbers after 10.

2. The sums at the right show that the first counting number is the first triangular number, the sum of the first two counting numbers is the second triangular number, and the sum of the first three counting numbers is the third triangular number.

   $1 = 1$
   $1 + 2 = 3$
   $1 + 2 + 3 = 6$

   Write the next six triangular numbers as sums of consecutive counting numbers.

3. Draw the next three square numbers after 16.

4. The diagram at the right shows that the first odd counting number equals the first square number, and the sum of the first two odd counting numbers equals the second square number.

   1      $1 + 3 = 4$

   Write the next four square numbers as sums of consecutive odd counting numbers.

ADDITIONAL PRACTICE

SKILLS

Find the sticker price for each car. (Pages 2–4)

	Base Price	Destination Charge	Total Cost of Options
1.	$8340	$330	$1140
2.	$6800	$175	$2103

Find how many kilowatt-hours of electricity were used. (Pages 5–6)

	Previous Reading	Present Reading
3.	1254	2781
4.	3114	4001

Find the distance for the given rate and time. (Pages 7–8)

	Rate	Time
5.	95 feet per second	15 seconds
6.	42 meters per second	21 seconds

Use the formula $d = 16t^2$ to find d for each value of t. (Pages 7–8)

7. $t = 6$ seconds
8. $t = 4$ seconds
9. $t = 12$ seconds
10. $t = 14$ seconds

Find the mean. (Pages 10–12)

11. 64, 86, 70, 92
12. 17, 14, 16, 19, 4, 3, 8, 11, 16
13. 632, 436, 108, 421, 341, 306
14. 97, 421, 68, 309, 155

Choose the best estimate. Choose a, b, or c. (Pages 14–16)

15. $198 + 205 + 486$ **a.** 900   **b.** 800   **c.** 1000
16. $797 \div 21$ **a.** 30   **b.** 40   **c.** 25
17. 71×33 **a.** 2100   **b.** 2800   **c.** 21,000

Evaluate each expression. (Pages 18–20)

18. $6 + 4 \cdot 7$
19. $(42 - 12) \div 3 + 1$
20. $(15 + 12) \cdot 8 - 1$
21. $100 \div 5 + 20$

(Pages 22–24)

22. $8x - 12$ when $x = 6$
23. $9 \cdot (x + y) + y$ when $x = 2$ and $y = 3$

APPLICATIONS

24. Roger had scores of 63, 84, 75, 92, 68, and 92 on six math tests. Find his average score. (Pages 10–12)
25. A cheetah was timed running 29 meters per second. How far would it run in 7 seconds? (Pages 7–9)

CHAPTER

2 Decimals and Applications

SECTIONS
2-1 Decimals and Place Value
2-2 Addition and Subtraction
2-3 Multiplication
2-4 Division
2-5 Problem Solving and Applications: Estimating with Decimals
2-6 Metric Units of Measurement
2-7 Problem Solving and Applications: Scientific Notation

FEATURES
Calculator Application: *Estimation*
Consumer/Computer Application: *Unit Price*
Enrichment: *Using Scientific Notation*

2-1 Decimals and Place Value

In our decimal system, the value of each decimal place is *ten* times the value of the place to its right.

PROCEDURE

To give the value of any digit in a number:

1. Identify the place name.
2. Refer to the digit and the place name to give the value.

TABLE

	Hundred Thousands	Ten Thousands	Thousands	Hundreds	Tens	Ones	.	Tenths	Hundredths	Thousandths	Ten-Thousandths	Hundred-Thousandths
a.	2	7	9	4	6	3	.	5	2	8	1	2
b.					9	3	.	7	5			
c.				1	7	2	.	3				
d.						3	.	0	0	7		
e.						0	.	0	0	0	0	7

EXAMPLE 1 Give the value of the "7" for each number in the table.

	1 Place Name	2 Value	
Solutions:	a. Ten Thousands	a. 70,000	Read: "seventy thousand".
	b. Tenths	b. 0.7	Read: "seven tenths".
	c. Tens	c. 70	Read: "seventy".
	d. Thousandths	d. 0.007	Read: "seven thousandths".
	e. Hundred-Thousandths	e. 0.00007	Read: "seven hundred thousandths".

Self-Check *Give the value of the underlined digit in each number.*

1. 5̲8.4 2. 6.0̲3 3. 19.4̲ 4. 30̲1.82

32 Chapter 2

The symbols > and < are used to compare numbers.

> means **is greater than.**

< means **is less than.**

> **PROCEDURE**
>
> **To compare decimals:**
> 1. Write the decimals one under the other. Line up the decimal points.
> 2. Starting from left to right, identify and compare the first digits that are <u>not</u> alike. Annex final zeros when necessary.
> 3. Compare the decimals in the <u>same order</u> as the unlike digits in step 2.

NOTE: Annexing final zeros does not change the value of a decimal.

EXAMPLE 2 Replace the __?__ with > or <.

a. 5.39 __?__ 5.42   b. 0.031 __?__ 0.03

Solutions:

	a.	b.
1. Write the decimals one under the other. Align the decimal points.	5.39 5.42	0.031 0.03
2. Starting from the left, identify the first digits that are <u>not</u> alike. Compare the digits.	5.39 5.42 3 < 4	0.031 0.030 1 > 0
3. Compare the decimals.	**5.39 < 5.42**	**0.031 > 0.03**

(Annex one zero.)

Self-Check *Replace the __?__ with > or <.*

5. 10.83 __?__ 10.8   6. 0.1 __?__ 0.11   7. 61.4 __?__ 60.3

CLASSROOM EXERCISES

Give the <u>place name</u> for the "8" in each number.
(Example 1, step 1)

1. 18   2. 843   3. 8762   4. 9.3248   5. 9.058

Replace the __?__ with > or <. (Example 2)

6. 3.2 __?__ 3.5   7. 6.5 __?__ 6.05   8. 0.01 __?__ 0.1   9. 0.4 __?__ 0.38

WRITTEN EXERCISES

Goals: To give the value of any digit in a number
To use place value to compare decimals and to apply this skill to solving word problems

For Exercises 1-12, give the value of the underlined digit. (Example 1)

1. 13.<u>5</u>
2. 13.0<u>5</u>
3. 1<u>5</u>.03
4. 13.176<u>5</u>
5. <u>2</u>3.14
6. 21.63<u>2</u>
7. <u>2</u>56.8
8. 100.025<u>1</u>
9. <u>2</u>001.5
10. 1<u>7</u>5.4
11. 2.51<u>7</u>
12. 46.03871

For Exercises 13-17, give the value of the indicated digit in the number 2461.375. (Example 1)

13. 4
14. 1
15. 3
16. 6
17. 7
18. 5

For Exercises 19-30, replace the __?__ with > or <. (Example 2)

19. 6.51 __?__ 6.48
20. 0.02 __?__ 0.027
21. 7.01 __?__ 7.89
22. 1.63 __?__ 1.632
23. 0.319 __?__ 0.32
24. 0.72 __?__ 0.7
25. 1.007 __?__ 0.07
26. 3.05 __?__ 3.005
27. 0.628 __?__ 0.625
28. 0.421 __?__ 0.422
29. 0.914 __?__ 0.92
30. 0.00010 __?__ 0.00015

APPLICATIONS: USING PLACE VALUE

31. Luis has a batting average of 0.325. Jane has a batting average of 0.335. Who has the higher batting average?
32. Juanita needs a piece of wire 1.46 meters long for her science project. She buys 1.5 meters of wire. Does she have enough wire for her project?
33. Which is less, 22.00124 or 22.00134?

REVIEW CAPSULE FOR SECTION 2-2

Evaluate each expression. (Pages 22-24)

1. $x + y$ when $x = 15$ and $y = 24$
2. $a + b$ when $a = 280$ and $b = 124$
3. $m - n$ when $m = 2400$ and $n = 1368$
4. $p - q$ when $p = 1564$ and $q = 89$
5. $a + b + c$ when $a = 97$, $b = 259$ and $c = 803$
6. $w + x + y$ when $w = 3162$, $x = 480$ and $y = 1004$

The answers are on page 60.

2-2 Addition and Subtraction

In the 1968 Olympic Games, Jan Henne of the United States won the 100 meter women's freestyle race in 60 seconds. In the 1984 Olympics, Carrie Steinseifer and Nancy Hogshead of the United States tied for the gold medal with winning times of 55.92 seconds. To find the difference in times, you use subtraction with decimals.

> **PROCEDURE**
>
> **To add or subtract with decimals:**
> 1. Line up the decimal points one under the other.
> 2. Annex final zeros when necessary.
> 3. Add or subtract as with whole numbers.

EXAMPLE 1 Find the difference between the winning times in the 100 meter women's freestyle race in the 1984 and 1968 Olympic Games.

Solution:
1968 winning time in seconds: 60.00 ← **Annex two zeros.**
1984 winning time in seconds: −55.92
Difference: 4.08

Self-Check

1. Find the difference between 87.3 and 55.42.

Variables in an algebraic expression can be replaced with decimals.

EXAMPLE 2 Evaluate $x + y + z$ when $x = 18.2$, $y = 11.94$ and $z = 253.07$.

Solution: $x + y + z = 18.2 + 11.94 + 253.07$
$= 283.21$

```
  18.20
  11.94
+253.07
 283.21
```

◀ **Annex a zero to help line up the decimal points.**

Self-Check

2. Evaluate $t + r + s$ when $t = 9.15$, $r = 0.003$, and $s = 216.5$.

Decimals and Applications

CLASSROOM EXERCISES

Add or subtract as indicated. (Example 1)

1.	0.01	2.	1.6	3.	6.26	4.	22.44	5.	7.8
	0.22		2.3		5.8		8.72		182.1
	+0.35		+5.8		+7.88		17.435		3.24
							+ 8.04		+ 21.16

6. $0.08 + 1.07 + 0.03$
7. $0.903 + 1.07 + 100.4 + 12.89$
8. $3.38 + 1.75 + 12.8 + 3.192$
9. $4.105 + 3.2 + 108.241 + 18.16$

10.	0.86	11.	11.67	12.	265.2	13.	5.24	14.	125.6
	−0.43		− 9.35		−149.8		−3.197		− 15.385

15. $0.79 - 0.39$
16. $11.06 - 8.633$
17. $94.3 - 88.416$
18. $36.5 - 0.341$

Evaluate each expression. (Example 2)

19. $r - s$ when $r = 15.7$ and $s = 8.9$
20. $m - n$ when $m = 180.6$ and $n = 32.89$
21. $x + y + z$ when $x = 4.8$, $y = 6.5$ and $z = 12.4$
22. $a + b + c$ when $a = 45.6$, $b = 36.24$ and $c = 8.701$

WRITTEN EXERCISES

Goals: To add and subtract with decimals and to apply these skills to solving word problems

To apply addition and subtraction of decimals to evaluating algebraic expressions

Add or subtract as indicated. (Example 1)

1.	4.75	2.	6.26	3.	26.1	4.	0.33	5.	126.2
	11.04		12.58		3.95		21.56		73.08
	+ 9.13		+ 3.91		+ 8.492		38.2		1.38
							+ 7.351		+ 42.261

6. $4.65 + 3.27 + 15.98$
7. $5.93 + 12.41 + 6 + 3.4$
8. $74.656 + 8.9 + 5.013 + 713.85$
9. $57.09 + 878.015 + 6.02 + 5.535$

10.	16.78	11.	4.968	12.	7.475	13.	1.7	14.	21.195
	− 9.43		−2.375		−0.3551		−0.87		− 7.2325

15. $18.4 - 7.9$
16. $132.4 - 86.51$
17. $9 - 0.048$
18. $11.06 - 8.633$

Evaluate each expression. (Example 2)

19. $x - y$ when $x = 15.05$ and $y = 7.88$
20. $a - b$ when $a = 114.5$ and $b = 38.56$
21. $c - d$ when $c = 164$ and $d = 28.56$
22. $r - s$ when $r = 8.04$ and $s = 0.1564$
23. $x + y + z$ when $x = 5.8$, $y = 12.2$ and $z = 7.3$
24. $r + s + t$ when $r = 0.32$, $s = 1.5$ and $t = 3.16$
25. $a + b + c + d$ when $a = 35.4$, $b = 18.5$, $c = 1.23$, and $d = 16.08$
26. $m + n + p + q$ when $m = 125.8$, $n = 26.5$, $p = 32.06$, and $q = 54$

APPLICATIONS: USING DECIMALS

Solve each problem.

27. In the 800-meter relay, Nathan ran the first leg in 29.2 seconds, Ted ran the second leg in 28.6 seconds, Bill ran the third leg in 30.2 seconds, and Roberto ran the fourth leg in 28.0 seconds. Find their total time.

28. The weather bureau reported that for the five times it rained in July, 1.12 inches, 2 inches, 1.5 inches, 0.46 inches, and 0.3 inches of rain fell. What was the total rainfall for July?

29. Latasha is 1.67 meters tall and Julie is 1.58 meters tall. How much taller is Latasha than Julie?

REVIEW CAPSULE FOR SECTION 2-3

Evaluate each expression. (Pages 22-23)

1. $3x$ when $x = 14$
2. $4y$ when $y = 127$
3. $4001p$ when $p = 6$
4. $2xy$ when $x = 506$ and $y = 20$
5. $12ab$ when $a = 16$ and $b = 84$
6. $a \cdot b$ when $a = 7$ and $b = 504$

Round to the nearest cent.

7. $14.759
8. $8.962
9. $25.015
10. $37.104
11. $61.706

The answers are on page 60.

2-3 Multiplication

You use **one kilowatt-hour (kwh)** of electricity when you use 1 kilowatt (1000 watts) for one hour.

The tables below show the average number of kilowatt-hours for operating certain appliances for one month.

Appliance	Kilowatt-Hours per Month
Microwave oven	15.8
Range with oven	97.6
Refrigerator (frostless)	94.7
Refrigerator (defrost)	152.4

Appliance	Kilowatt-Hours per Month
Water heater	400.0
Radio	7.5
Television (black-white)	29.6
Television (color)	55.0

PROCEDURE

To multiply with decimals:

1 Multiply as with whole numbers.

2 Count the number of decimal places in the numbers being multiplied (factors). Starting at the right and moving left, count the same number of decimal places in the product. Insert the decimal point.

EXAMPLE 1 At a cost of $0.097 per kilowatt-hour, find the average cost of operating a refrigerator (defrost) for one month. Round the answer to the nearest cent.

Solution: Multiply the number of kilowatt-hours (refer to the table) and the cost per kilowatt-hour.

Decimal Places

Average cost = 152.4 (0.097) → 152.4 ← 1
　　　　　　= 14.7828　　　　　×0.097 ← 3
　　　　　　= 14.78　　Nearest　10668
　　　　　　　　　　　　cent　　13716
The average monthly cost is **$14.78**.　14.7828 ← 1 + 3 = 4

Self-Check　1. At a cost of $0.097 per kilowatt-hour, find the cost of operating a radio for one month. Round the answer to the nearest cent.

When multiplying, you may have to insert zeros in a product.

EXAMPLE 2 At a cost of $0.029 per hour, how much will it cost to operate a television (black–white) for 0.5 hour? Round the answer to the nearest cent.

Solution: Cost = 0.5 (0.029)
= 0.0145
= 0.01 ◀ **Nearest cent**

It will cost about **$0.01.**

```
           Decimal Places
   0.5  ←      1
 ×0.029 ←      3
   ────
    45
    10
  ──────
  0.0145  ◀ Insert 1 zero to make
            4 decimal places.
```

Self-Check 2. At a cost of $0.035 per hour, how much will it cost to operate a black–white television set for 0.5 hour? Round the answer to the nearest cent.

CLASSROOM EXERCISES

Copy each product. Then insert the decimal point. (Example 1)

1. 25
 ×0.8
 ───
 200

2. 2.5
 ×0.8
 ───
 200

3. 2.5
 × 8
 ───
 200

4. 0.25
 × 8
 ───
 200

5. 0.25
 × 0.8
 ───
 200

(Example 2)

6. 0.15
 ×0.06
 ────
 90

7. 0.015
 × 0.06
 ─────
 90

8. 0.15
 ×0.006
 ─────
 90

9. 0.015
 ×0.006
 ─────
 90

10. 0.15
 ×0.060
 ─────
 900

Multiply. (Example 1)

11. 45
 ×0.6

12. 37
 ×0.3

13. 121
 ×0.13

14. 1.4
 ×0.24

15. 32.5
 ×0.56

(Example 2)

16. 0.12
 × 0.7

17. 0.35
 ×0.04

18. 0.24
 ×0.09

19. 0.356
 ×0.008

20. 0.156
 ×0.014

Decimals and Applications 39

WRITTEN EXERCISES

Goals: To multiply with decimals
To apply the skill of multiplying with decimals to solving consumer–related problems

Multiply. (Example 1)

1. 12 ×0.8
2. 12.5 × 0.6
3. 135 ×0.72
4. 400 ×0.37
5. 210.6 ×0.065

6. (12.8)(0.023)
7. (47.6)(0.59)
8. (9.08)(1.5)
9. (8.71)(0.067)
10. (3.5)(0.014)
11. (42.6)(0.024)
12. (125.8)(0.046)
13. (1.042)(0.084)

(Example 2)

14. 0.8 ×0.06
15. 0.6 ×0.03
16. 0.54 ×0.021
17. 0.25 ×0.032
18. 0.35 ×0.0015

19. (0.61)(0.04)
20. (0.23)(0.038)
21. (0.8)(0.057)
22. (0.13)(0.052)
23. (0.35)(0.005)
24. (0.024)(0.0016)
25. (0.56)(0.013)
26. (0.68)(0.0018)

APPLICATIONS: USING MULTIPLICATION WITH DECIMALS

For Exercises 27–33, refer to the table on page 38. Find the monthly cost for each appliance at each rate. Round each answer to the nearest cent. (Example 1)

Appliance	Cost at $0.056 per kwh	Cost at $0.35 per kwh	Cost at $0.028 per kwh	Cost at $0.072 per kwh
27. Color television	?	?	?	?
28. Water heater	?	?	?	?
29. Range with oven	?	?	?	?
30. Microwave oven	?	?	?	?
31. Radio	?	?	?	?
32. Refrigerator (frostless)	?	?	?	?
33. Refrigerator (defrost)	?	?	?	?

For Exercises 34–37, find the cost of operating each appliance at the given rate and time. Round each answer to the nearest cent. (Example 2)

Appliance	Cost per hour	Time Used		Appliance	Cost per hour	Time Used
34. Electric blanket	$0.008	6 hours	35.	Iron	$0.067	0.25 hours
36. Vacuum cleaner	$0.077	0.5 hours	37.	Fan	$0.015	3.5 hours

Solve. Round each answer to the nearest cent.

38. A frostless refrigerator uses an average of 94.7 kilowatt–hours per month. At a cost of $0.065 per kilowatt–hour, find the cost of operating a frostless refrigerator for 2 months.

39. A microwave oven uses an average of 15.8 kilowatt–hours per month. At a cost of $0.034 per kilowatt–hour, find the cost of operating a microwave oven for one year.

40. The cost of operating a washing machine is $0.022 per load. The cost of operating a dryer is $0.035 per load. How much will it cost to wash and dry four loads?

41. In Exercise 40, how much will it cost to wash and dry four loads of clothes per week for 52 weeks?

MORE CHALLENGING EXERCISES

Solve. Round each answer to the nearest cent.

42. A frostless refrigerator uses an average of 94.7 kilowatt–hours per month. A manual–defrost refrigerator uses an average of 152.4 kilowatt–hours per month. At a cost of $0.057 per kilowatt–hour, find how much more it will cost to operate the manual–defrost refrigerator for 6 months?

43. The average cost of operating a coffee maker is $0.018 per pot. The average cost of operating a toaster is $0.003 per slice. Find the total cost of operating the coffee maker and the toaster to make 1 pot of coffee and 4 slices of toast each day for 365 days.

Decimals and Applications

REVIEW: SECTIONS 2–1–2–3

Complete. (Section 2–1)

1. The place name of the 5 in 22.35 is __?__.
2. The place name of the 6 in 1465.3 is __?__.

For Exercises 3–6, replace the __?__ with > or <. (Section 2–1)

3. 3.28 __?__ 3.31
4. 0.008 __?__ 0.0082
5. 5.003 __?__ 6.003
6. 4.506 __?__ 4.5055

Evaluate each expression. (Section 2–2)

7. $x + y + z$ when $x = 4.2$, $y = 3.04$ and $z = 12.7$
8. $a + b$ when $a = 132.5$ and $b = 60.2$

Solve each problem.

9. The normal annual rainfall in Houston, Texas is 48.19 inches. The normal annual rainfall in Nashville, Tennessee is 46 inches. On the average, how much more rain falls in Houston? (Section 2–2)

10. A color television uses an average of 55 kilowatt-hours per month. At a cost of $0.063 per kilowatt-hour, find the cost of operating the television for 30 days. (Section 2–3)

11. At a cost of $0.077 per hour, how much will it cost to operate a vacuum cleaner for 0.75 hour? (Section 2–3)

12. Last month, 0.02 inch, 1.38 inches, 0.08 inch and 0.57 inch of rain fell in a certain town. What was the total rainfall? (Section 2–2)

REVIEW CAPSULE FOR SECTION 2–4

Evaluate each expression. (Pages 22–24)

1. $x \div 4$ when $x = 24$
2. $90 \div y$ when $y = 6$
3. $248 \div x$ when $x = 4$
4. $2432 \div p$ when $p = 8$
5. $q \div 3$ when $q = 7623$
6. $a \div b$ when $a = 117$ and $b = 3$
7. $m \div n$ when $m = 756$ and $n = 7$
8. $x \div y$ when $x = 72$ and $y = 12$
9. $p \div q$ when $p = 1922$ and $q = 62$
10. $m \div n$ when $m = 3485$ and $n = 17$

The answers are on page 60.

CONSUMER/COMPUTER APPLICATIONS
Unit Price

You can save money if you use unit price to compare the costs of products. **Unit price** is the price per gram, per ounce, and so on.

The following word rule and the formula tell how to find unit price.

Word Rule: Unit price equals the *price of an item* divided by the *number of units*.

Formula: $U = p \div n$, or $U = \dfrac{p}{n}$

U = unit price
P = price of an item
n = number of units

The following is a program which will compute and print the unit price of an item, given the price of the item and the number of units.

Program:

```
100 REM P = TOTAL PRICE
110 REM N = NUMBER OF UNITS
120 REM U = UNIT PRICE OF AN ITEM
130 PRINT "WHAT IS THE TOTAL PRICE";
140 INPUT P
150 PRINT "HOW MANY UNITS";
160 INPUT N
170 LET U = P/N
180 PRINT "THE UNIT PRICE = $"; U
190 PRINT "ANY MORE CALCULATIONS (1 = YES, 0 = NO)";
200 INPUT Z
210 IF Z = 1 THEN 130
220 END
```

REM statements are used to add explanatory comments to a program.

⟵ $U = P \div N$

⟵ $Z = 1$ or $Z = 0$

Exercises

Use the program to compute U to the nearest tenth of a cent.

1. $p = \$1.09$; $n = 24$
2. $p = \$1.99$; $n = 46$
3. $p = \$0.99$; $n = 9$
4. $p = \$3.10$; $n = 15$
5. $p = \$0.85$; $n = 50$
6. $p = \$1.59$; $n = 36$
7. $p = \$1.88$; $n = 25$
8. $p = \$4.51$; $n = 18$
9. $p = \$2.85$; $n = 19$

2-4 Division

The **fuel economy** of a car refers to how many miles it can travel on one gallon of gasoline. To find fuel economy, you use division skills.

Fuel Economy (Miles per gallon) = Number of Miles ÷ Number of Gallons

> **PROCEDURE**
>
> **To divide a decimal by a whole number:**
>
> 1. Place the decimal point in the quotient directly above the decimal point in the dividend.
> 2. Divide as with whole numbers.

EXAMPLE 1 A certain car travels 248.4 miles on a full tank of fuel. The tank holds 12 gallons when full. How far does it travel on one gallon?

Solution:
1. Place the decimal point in the quotient. → 12)‾248.4

2.
```
    20.
12)248.4
    24
     8
```
Since 8 is not divisible by 12, write a 0 above the 8. →
```
    20.7
12)248.4
    24
     84
     84
```

The car travels **20.7 miles** on one gallon of fuel.

Self-Check

1. A car travels 366.8 miles on a full tank of fuel. The tank holds 14 gallons when full. How far does it travel on one gallon?

When dividing by a decimal, the first step is to multiply the divisor <u>and</u> the dividend by the <u>same</u> number in order to obtain a whole number divisor.

> **PROCEDURE**
>
> **To divide by a decimal:**
>
> 1. Multiply both the divisor and the dividend by 10, or by 100, or by 1000, and so on, in order to obtain a whole number divisor.
> 2. Divide.

44 Chapter 2

EXAMPLE 2 Divide: $0.224\overline{)5.6}$ Check the answer.

Solution: ① Since the divisor is 224 <u>thousandths</u>, multiply the divisor and the dividend by 1000.

$0.224\overline{)5.600}$ ◀ Annex two zeros.

②
```
         25.
224 ) 5600.
      448
      ————
      1120
      1120
```

Check:
```
    0.224
  ×    25
  ——————
    1120
     448
  ——————
   5.600
```
◀ Same as the original dividend.

Self-Check *Divide. Check each answer.*

2. $0.12\overline{)4.2}$ **3.** $0.9\overline{)657}$ **4.** $0.348\overline{)1009.2}$

CLASSROOM EXERCISES

Divide. (Example 1)

1. $6\overline{)20.88}$ **2.** $3\overline{)7.23}$ **3.** $5\overline{)3.580}$ **4.** $7\overline{)43.96}$

Multiply the divisor and the dividend by 10, or by 100, or by 1000, to make the divisor a whole number. (Example 2, step 1)

5. $0.32\overline{)8.96}$ **6.** $0.12\overline{)2.4}$ **7.** $0.036\overline{)7.2}$ **8.** $0.64\overline{)22.4}$

Divide. (Example 2)

9. $0.4\overline{)3.76}$ **10.** $0.68\overline{)30.6}$ **11.** $0.004\overline{)21.88}$ **12.** $0.7\overline{)238}$

WRITTEN EXERCISES

Goals: To divide a decimal by a whole number
To divide a decimal by a decimal
To apply these skills to solving word problems

Divide. (Example 1)

1. $6\overline{)17.04}$ **2.** $4\overline{)22.12}$ **3.** $6\overline{)58.86}$ **4.** $7\overline{)29.05}$
5. $9\overline{)0.585}$ **6.** $4\overline{)0.232}$ **7.** $15\overline{)121.05}$ **8.** $17\overline{)596.7}$
9. $42.63 \div 29$ **10.** $75.84 \div 12$ **11.** $2.912 \div 52$ **12.** $8.112 \div 78$

Decimals and Applications

Divide. Check each answer. (Example 2)

13. $0.8\overline{)4.96}$
14. $0.9\overline{)1.431}$
15. $0.06\overline{)5.34}$
16. $0.05\overline{).915}$
17. $1.2\overline{).42}$
18. $4.3\overline{)88.15}$
19. $0.38\overline{)2682.8}$
20. $0.81\overline{)50.706}$
21. $14.68 \div 0.004$
22. $2.5608 \div 0.008$
23. $94.82 \div 0.011$
24. $511.84 \div 0.056$

APPLICATIONS: USING DIVISION WITH DECIMALS

25. The fuel tank of a certain luxury car holds 24 gallons when full. The car travels 400.8 miles on a full tank. How many miles does it travel on one gallon?

26. A sub-compact car travels 535.6 miles on a full tank of fuel. A full tank holds 13 gallons. How far does it travel on one gallon?

27. A certain motorcycle travels 135.2 miles on a full tank of fuel. The tank holds 1.6 gallons when full. How far does it travel on one gallon?

28. The fuel tank in a certain sports car holds 14.2 gallons when full. The car travels 262.7 miles on a full tank. How far will it travel on one gallon?

MORE CHALLENGING EXERCISES

Determine which car has the greater fuel economy.

29. One compact car travels 500 miles on 12.5 gallons of fuel. Another compact car travels 462 miles on 13.2 gallons of fuel.

30. One sports car travels 789 kilometers on 52.6 liters of fuel. Another sports car travels 843.9 kilometers on 58.2 liters of fuel.

REVIEW CAPSULE FOR SECTION 2–5

Round to the nearest whole number.

1. 4.3
2. 8.7
3. 0.9
4. 16.82
5. 11.38

Round to the nearest tenth.

6. 3.27
7. 1.32
8. 17.84
9. 11.409
10. 0.987

Round to the nearest hundredth.

11. 5.863
12. 0.396
13. 14.649
14. 0.005
15. 19.442

The answers are on page 60.

PROBLEM SOLVING AND APPLICATIONS

2-5 Estimating with Decimals

At the local supermarket, Carlos bought items for $0.43, $1.69, $0.79, $2.29, $0.67, and $3.82. To estimate the total cost, Carlos rounded the cost of each item to the nearest dollar.

You can round decimals to the nearest whole number, to the nearest tenth, to the nearest hundredth, and so on. You can use the rules for rounding to estimate answers.

PROCEDURE

To use estimation to solve a problem:

[1] Write a sum, difference, product, or quotient (or a combination of these) for the problem.

[2] Round each number to the nearest whole number, or to the nearest tenth, or to the nearest hundredth, and so on.

[3] Perform the indicated operations on the rounded numbers.

EXAMPLE 1 Estimate Carlos' total bill at the supermarket. Round to the nearest dollar.

Solution: [1] Problem [2] Round to the nearest dollar.

0.43	⟶	0.00
1.69	⟶	2.00
0.79	⟶	1.00
2.29	⟶	2.00
0.67	⟶	1.00
3.82	⟶	+4.00

[3]   10.00   **Estimate: $10.00**

```
SOUP     ** 0.43
CATSUP   ** 1.69
CORN     ** 0.79
MILK     ** 2.29
FRUIT    ** 0.67
CHICKEN  ** 3.82
TOTAL
         THANK YOU
```

Self-Check 1. Emily bought school supplies for $3.69, $0.72, $1.37, $4.19, $0.35 and $2.55. Estimate the total cost to the nearest dollar.

Decimals and Applications

When estimating the answer to a problem that involves division of decimals or whole numbers, round the dividend and the divisor to the nearest whole numbers that divide evenly.

EXAMPLE 2 Mae is paid $0.045 for each envelope she addresses. She received $14.85 for five hours work. Estimate how many envelopes Mae addressed in the five hours.

Solution:

1. Problem

$0.045 \overline{)14.85}$ ⟶

2. Round to convenient numbers.

$0.05 \overline{)15.00}$

3. $1500 \div 5 = 300$

Mae addressed about **300** envelopes.

Self-Check 2. Karl is paid $0.042 for each pound of beans he picks. He received $2.10 for one bag. Estimate how many pounds of beans were in the bag.

CLASSROOM EXERCISES

Choose the best estimate. Choose a, b, or c.
(Examples 1 and 2, steps 1 and 2)

		a.	b.	c.
1.	$3.70 + $5.20	$3.00 + $5.00	$4.00 + $6.00	$4.00 + $5.00
2.	$14.80 + $6.75	$15.00 + $7.00	$14.00 + $7.00	$14.00 + $6.00
3.	$10.20 − $7.80	$10 − $8	$10 − $9	$11 − $7
4.	$24.60 − $19.90	$24 − $18	$25 − $19	$25 − $20
5.	3.8 (10.1)	3 · 10	4 · 10	3 · 11
6.	10.3 (38.9)	10 · 39	11 · 39	10 · 38
7.	60.3 ÷ 29.8	65 ÷ 20	60 ÷ 30	60 ÷ 20
8.	19.9 ÷ 1.8	20 ÷ 2	20 ÷ 1	18 ÷ 2
9.	2.1(29.6)	3 · 30	2 · 30	2 · 20
10.	81.8 ÷ 19.6	80 ÷ 20	90 ÷ 20	80 ÷ 15
11.	148.5 ÷ 52.2	100 ÷ 50	200 ÷ 50	150 ÷ 50
12.	38.6(3.9)	40 · 4	30 · 4	30 · 3

48 Chapter 2

WRITTEN EXERCISES

Goal: To use estimation to solve word problems involving decimals

Choose the best estimate. Choose a, b, or c. (Examples 1 and 2)

1. $16.70 + $3.90 + $21.20	a. $40	b. $42	c. $45
2. $9.10 + $8.60 + $11.90	a. $30	b. $35	c. $28
3. $11.22 − $4.85	a. $4	b. $6	c. $7
4. $22.70 − $19.80	a. $5	b. $4	c. $3
5. 2.7 (3.9)	a. 6	b. 8	c. 12
6. 11.08 (1.02)	a. 11	b. 15	c. 14
7. 17.5 ÷ 6.3	a. 6	b. 3	c. 5
8. 14.88 ÷ 3.01	a. 5	b. 3	c. 2

APPLICATIONS: USING ESTIMATION WITH DECIMALS

9. At Games Galore, Tim paid $2.79 for ping pong balls, $6.39 for dominoes, and $2.40 for a book of rules for games. Estimate the total cost to the nearest dollar.

10. The band that played for the Homecoming Dance was paid $84 per hour. The band played for 3.75 hours. Estimate to the nearest dollar how much the band was paid for the evening.

11. The regular price for a set of golf clubs was $178.95. The sale price was $113.25. Estimate the amount saved to the nearest dollar.

12. Claude was paid $6.25 for each page of manuscript he typed. His total pay amounted to $300. About how many pages did he type?

13. For a pep rally in the school gym, pennants were used to decorate the walls. Each pennant cost $2.19 and the total bill was $100.74. About how many pennants were used?

14. The French Club sold corsages to raise money for a trip. They charged $5.20 for each corsage and sold 289 corsages. Estimate, to the nearest dollar, the total receipts for the sale of corsages.

Decimals and Applications

ESTIMATION

You can use the rules for rounding and estimation and the calculator to solve problems.

EXAMPLE   54.6 ÷ 10.4

Solution   Use paper and pencil to estimate the answer: 50 ÷ 10 = 5

Use the calculator to find the exact answer.

5 4 . 6 ÷ 1 0 . 4 =   5.25

Since the estimate is 5, 5.25 is a reasonable answer.

EXERCISES   First estimate each answer. Then use a calculator to find the exact answer.

Problem	Estimated Answer	Exact Answer
1. 29 + 141 + 38 + 73	?	?
2. 5418 − 1799	?	?
3. 199 × 61	?	?
4. 39.8 × 20.3	?	?
5. 3498.6 ÷ 51	?	?
6. 90.09 ÷ 9.1	?	?

REVIEW CAPSULE FOR SECTION 2-6

Evaluate each expression for $m = 10$. (Pages 22–24)

1. $3.4m$ **2.** $15m$ **3.** $0.5m$ **4.** $4.06m$ **5.** $65.2m$ **6.** $80m$
7. $130 \div m$ **8.** $12.4 \div m$ **9.** $6 \div m$ **10.** $0.8 \div m$ **11.** $3000 \div m$ **12.** $9.6 \div m$

Evaluate each expression for $p = 100$. (Pages 22–24)

13. $2.8p$ **14.** $74p$ **15.** $0.6p$ **16.** $0.013p$ **17.** $20.4p$ **18.** $7.56p$
19. $2000 \div p$ **20.** $200 \div p$ **21.** $62 \div p$ **22.** $428.6 \div p$ **23.** $45000 \div p$ **24.** $3.5 \div p$

Evaluate each expression for $t = 1000$. (Pages 22–24)

25. $6.2t$ **26.** $8.5t$ **27.** $40.7t$ **28.** $0.3t$ **29.** $0.07t$ **30.** $38t$
31. $4000 \div t$ **32.** $405 \div t$ **33.** $32 \div t$ **34.** $78{,}000 \div t$ **35.** $3840 \div t$ **36.** $6 \div t$

The answers are on page 60.

2-6 Metric Units of Measurement

The base unit of length in the metric system is the **meter**. The other most commonly used units of length are the **centimeter**, the **millimeter**, and the **kilometer**.

The following table shows how other metric units of length are related to the meter. It also shows how each prefix is related to place value.

TABLE 1

Units of Length	Prefix	Meaning of Prefix
1 millimeter (mm) = **0.001** meter (m)	milli–	thousandths
1 centimeter (cm) = **0.01** meter	centi–	hundredths
1 decimeter (dm) = **0.1** meter	deci–	tenths
1 dekameter (dam) = **10** meters	deka–	tens
1 hectometer (hm) = **100** meters	hecto–	hundreds
1 kilometer (km) = **1000** meters	kilo–	thousands

NOTE: The value of each metric unit is 10 times the value of the next smaller unit.

PROCEDURE

1. To change to a smaller metric unit, multiply by 10, or by 100, or by 1000, and so on.
2. To change to a larger metric unit, divide by 10 or by 100, or by 1000, and so on.

EXAMPLE 1 Complete.

  a. 75 km = __?__ m   **b.** 255 mm = __?__ m

  Larger to smaller: Multiply.

Solutions: **a.** 75 km = 75 · 1000 = 75,000 m

km	hm	dam	m
10	10	10	

  Smaller to larger: Divide.

  b. 255 mm = 255 ÷ 1000 = 0.255 m

m	dm	cm	mm
10	10	10	

Decimals and Applications

Self-Check Complete.

1. 25 km = __?__ m   2. 15 mm = __?__ cm   3. 7.9 m = __?__ cm

The prefixes used for other metric units are the same as those used for units of length.

The most commonly used units of mass in the metric system are the **milligram**, the **gram**, and the **kilogram**.

TABLE 2 Units of Mass

kilogram (kg)	hectogram (hg)	dekagram (dag)	gram (g)	decigram (dg)	centigram (cg)	milligram (mg)

NOTE: A metric ton (t) equals 1000 kilograms.

EXAMPLE 2 Karen's doctor wants her to take one iron capsule per day. Each capsule contains 63 milligrams of iron.

  a. How many milligrams is this per week?
  b. How many grams is this per week?

Solution: a. 63 · 7 = 441 mg per week.

  Smaller to larger: Divide.

  b. 441 mg = 441 ÷ 1000 = 0.441 g

  Karen takes **0.441 g** of iron per week.

Self-Check Complete.

4. 36 g = __?__ cg   5. 350 mg = __?__ g   6. 5.31 g = __?__ dg

The most commonly used units of capacity are the **kiloliter**, the **liter**, and the **milliliter**.

TABLE 3 Units of Capacity

kiloliter (kL)	hectoliter (hL)	dekaliter (daL)	liter (L)	deciliter (dL)	centiliter (cL)	milliliter (mL)

Chapter 2

EXAMPLE 3   The gasoline tank of a car holds 45.6 liters. How many kiloliters is this?

Smaller to larger: Divide.

Solution: 45.6 L = 45.6 ÷ 1000 = 0.0456 kL

The tank holds **0.0456 kL**.

| kL | hL | daL | L |

10   10   10

Self-Check   *Complete.*

7. 22 mL = __?__ L   **8.** 37 L = __?__ mL   **9.** 792 kL = __?__ L

CLASSROOM EXERCISES

Complete. Use the words "multiply" or "divide."
(Tables 1, 2, and 3)

1. To change from millimeters to centimeters, __?__ by 10.

2. To change from meters to millimeters, __?__ by 1000.

3. To change from grams to centigrams, __?__ by 100.

4. To change from grams to kilograms, __?__ by 1000.

5. To change from milliliters to liters, __?__ by 1000.

6. To change from kiloliters to liters, __?__ by 1000.

7. To change from meters to decimeters, multiply by __?__.

8. To change from decimeters to centimeters, multiply by __?__.

9. To change from grams to kilograms, divide by __?__.

10. To change from centigrams to grams, divide by __?__.

WRITTEN EXERCISES

Goal: To determine equivalent metric units of length, mass, and capacity

Complete. (Example 1)

1. 3 cm = __?__ mm

2. 7.2 m = __?__ cm

3. 54 km = __?__ m

4. 65,000 cm = __?__ km

	km	m	cm
5.	3.5	?	?
7.	?	256	?

	m	cm	mm
6.	?	8.4	?
8.	?	?	32.5

Decimals and Applications   53

Complete. (Example 2)

9. 5100 g = _?_ kg **10.** 21.8 g = _?_ mg **11.** 0.9 t = _?_ kg **12.** 0.32 kg = _?_ g

	kg	g	mg
13.	3.5	?	?
15.	?	720	?

	t	kg	g
14.	0.42	?	?
16.	?	?	34,600

Complete. (Example 2)

17. 1 kL = _?_ L **18.** 3250 mL = _?_ L **19.** 0.45 L = _?_ mL **20.** 1.785 kL = _?_ L
21. 23 L = _?_ mL **22.** 400 mL = _?_ L **23.** 0.62 kL = _?_ L **24.** 43,000 mL = _?_ kL

APPLICATIONS: USING METRIC MEASURES

25. The jogging track in Central Park is 4.2 kilometers long. How many meters is this?

26. A recent estimate states that 41,000 metric tons of gold still exist in the earth. How many kilograms is this?

27. A quart of milk is about 946 milliliters. How many liters is this?

28. The polar continent Antarctica is covered by an ice dome that is almost 4 kilometers high. Express this height in meters.

29. One milliliter of water weighs (has a mass of) one gram. Express the weight of 1 mL of water in terms of milligrams.

30. A weight-lifting record of 255 kilograms was set in the 1976 Olympics. Express 255 kilograms in terms of grams.

REVIEW CAPSULE FOR SECTION 2-7

Replace the ? with the correct exponent. (Pages 7–9)

1. $10 \cdot 10 = 10^?$ **2.** $10 \cdot 10 \cdot 10 \cdot 10 \cdot 10 = 10^?$ **3.** $10 \cdot 10 \cdot 10 = 10^?$
4. $10 \cdot 10 \cdot 10 \cdot 10 \cdot 10 \cdot 10 = 10^?$ **5.** $10 \cdot 10 \cdot 10 \cdot 10 \cdot 10 \cdot 10 \cdot 10 = 10^?$ **6.** $10 \cdot 10 \cdot 10 \cdot 10 = 10^?$

Evaluate each expression. (Pages 7–9)

7. 10^3 **8.** 10^5 **9.** 10^6 **10.** 10^1 **11.** 10^4 **12.** 10^7

The answers are on page 60.

PROBLEM SOLVING AND APPLICATIONS

2-7 Scientific Notation

The distance from the earth to the moon is approximately 360,000,000 meters. Very large numbers such as this are often expressed in **scientific notation.**

$$360,000,000 = 3.6 \times 10^8$$

Number between 1 and 10 — Power of 10 — Scientific Notation

PROCEDURE

To write a number in scientific notation:

1. Count the number of places the decimal point must be moved in order to arrive at a number between 1 and 10.
2. Express it as a product in this form:
(Number between 1 and 10) × (Power of 10)

EXAMPLE 1 The distance from the sun to the earth is about 150,000,000 kilometers. Express this number in scientific notation.

Solution:
1. Count the number of places the decimal point must be moved to arrive at a number between 1 and 10.   150,000,000   8 places
2. Write the number as a product. 1.5×10^8

Number between 1 and 10 — Power of 10

Self-Check

1. The Puerto Rico Trench is the deepest known point in the Atlantic Ocean. It is about 2600 meters below sea level. Express this number in scientific notation.

You can also write very small numbers in scientific notation.

A number written in scientific notation can be expressed as a decimal.

Decimals and Applications **55**

> **PROCEDURE**
>
> **To express a number written in scientific notation as a decimal numeral:**
> 1. Evaluate the power of 10.
> 2. Multiply.

EXAMPLE 2 The diameter of the planet Mars is approximately 4.214×10^3 miles. Express the diameter as a decimal numeral.

Solution:
1. $4.214 \times 10^3 = 4.214 \times 1000$ $10^3 = 10 \cdot 10 \cdot 10 = 1000$
2. $ = 4214$

Self-Check 2. The diameter of the sun is 8.64×10^5 miles. Express the diameter as a decimal numeral.

NOTE: $10^1 = 10$ and $10^0 = 1$.

CLASSROOM EXERCISES

Complete. (Example 1, step 2)

1. $39 = 3.9 \times 10^?$
2. $395 = 3.95 \times 10^?$
3. $3950 = 3.950 \times 10^?$
4. $573 = 5.73 \times 10^?$
5. $57{,}300 = 5.73 \times 10^?$
6. $57{,}300{,}000 = 5.73 \times 10^?$

Write each number in scientific notation. (Example 1)

7. 40
8. 200,000
9. 400
10. 2,000,000
11. 4000
12. 200,000,000

Express as a decimal numeral. (Example 2)

13. 1.6×10^2
14. 3.8625×10^4
15. 3.8×10^4

WRITTEN EXERCISES

Goal: To express a number in scientific notation

To express a number written in scientific notation as a number in decimal notation

Express each number in scientific notation. (Example 1)

1. 3400
2. 8000
3. 90,000
4. 64,000
5. 1,050,000
6. 4,000,000
7. 875
8. 24,100
9. 80,000,000
10. 302
11. 76,000
12. 40,300,000

Express as a decimal numeral. (Example 2)

13. 3.2×10^3
14. 5.8×10^6
15. 2.51×10^5
16. 6.05×10^9
17. 9.123×10^6
18. 1.104×10^3
19. 5.2×10^7
20. 4.0×10^5
21. 1.42×10^1

22. The Space Shuttle's orbit is about 22,240 miles above the equator. Express this number in scientific notation.

23. The assembled Space Shuttle weighs about 2,000,000 kilograms at liftoff. Express this number in scientific notation.

24. The three main engines of the Space Shuttle orbiter burn about 242,000 liters of fuel per minute. Express this number in scientific notation.

REVIEW: SECTIONS 2-4 — 2-7

Solve each problem. (Section 2-4)

1. Arlita has a piece of fabric 2.8 meters long. She wants to cut it into 4 equal pieces. How long should each piece be?

2. Harold can travel 27.6 miles in 1.5 hours on his bicycle. Find how far he can travel in one hour.

Choose the best estimate. Choose a, b, or c. (Section 2-5)

3. $3.70 + $8.20 + $0.90   **a.** $11   **b.** $13   **c.** $14
4. $22.85 − $13.20   **a.** $10   **b.** $11   **c.** $8
5. 6.2 (4.8)   **a.** 31   **b.** 27   **c.** 30
6. 21.04 ÷ 6.83   **a.** 3   **b.** 5   **c.** 2

Complete. (Section 2-6)

7. 4.3 m = __?__ km
8. 0.4 kg = __?__ g
9. 320.4 L = __?__ kL

Complete. (Section 2-7)

10. $3200 = 3.2 \times 10^?$
11. $940{,}000 = 9.4 \times 10^?$
12. $8{,}250{,}000 = \underline{\ ?\ } \times 10^6$
13. $45{,}000{,}000 = \underline{\ ?\ } \times 10^7$

CHAPTER REVIEW

PART 1: VOCABULARY

In Exercises 1–12, choose from the box at the right below the word(s) that best corresponds to each description.

1. A unit of capacity in the metric system __?__ (p. 52)
2. A prefix meaning thousandths __?__ (p. 51)
3. A prefix meaning tens __?__ (p. 51)
4. One thousand watts __?__ (p. 38)
5. The base unit of length in the metric system __?__ (p. 51)
6. A prefix meaning thousands __?__ (p. 51)
7. One thousand kilograms __?__ (p. 52)
8. A prefix meaning hundredths __?__ (p. 51)
9. A way of expressing very large numbers __?__ (p. 55)
10. A prefix meaning tenths __?__ (p. 51)
11. The base unit of mass in the metric system __?__ (p. 52)
12. A prefix meaning hundreds __?__ (p. 51)

milli–
centi–
deci–
deka–
hecto–
kilo–
kilowatt
meter
gram
liter
metric ton
scientific notation

PART 2: SKILLS

Give the value of each underlined digit. (Section 2-1)

13. 25.<u>4</u>   14. <u>4</u>8.12   15. 15.1<u>4</u>6   16. <u>5</u>140.27   17. 9.02<u>5</u>   18. 2.5471<u>3</u>

For Exercises 19–24, replace the __?__ with > or <. (Section 2-1)

19. 2.46 __?__ 2.39   20. 5.03 __?__ 5.11   21. 1.76 __?__ 1.761
22. 0.628 __?__ 0.627   23. 0.03 __?__ 0.009   24. 2.007 __?__ 2.018

Add or subtract as indicated. (Section 2-2)

25. 2.02
     5.68
    +3.14

26.   1.9
      57.21
       3.025
    +168.37

27.   6.46
     −4.791

28. 371.082
    − 99.095

29. $0.025 + 1.76 + 21.81$
30. $0.64 - 0.58$
31. $91.23 + 4.8 + 0.062 + 9.7$
32. $84.3 - 39.561$

Evaluate each expression. (Section 2-2)

33. $x + y + z$ when $x = 8.9$, $y = 31.4$ and $z = 0.51$ 34. $a - b$ when $a = 215.3$ and $b = 48.52$

58   Chapter 2

For Exercises 35–36, find the monthly cost for each appliance at the given rate. Round each answer to the nearest cent. (Section 2-3)

Appliance	Kilowatt–hours per month	Cost at $0.098 per kwh	Cost at $0.075 per kwh	Cost at $0.047 per kwh
35. Refrigerator	152.4	?	?	?
36. Color television	55.0	?	?	?

Divide. (Section 2-4)

37. $4\overline{)23.56}$ **38.** $25\overline{)308.75}$ **39.** $341.6 \div 2.8$ **40.** $298.08 \div 0.048$

Choose the best estimate. Choose a, b, or c. (Section 2-5)

41. $18.90 + $4.30 + $35.80   **a.** 57   **b.** 62   **c.** 59
42. $45.70 − $13.35   **a.** 33   **b.** 31   **c.** 30
43. (14.8)(5.3)   **a.** 200   **b.** 75   **c.** 50
44. $19.75 \div 4.05$ **a.** 6   **b.** 4   **c.** 5

Complete. (Section 2-6)

45. 2 cm = __?__ mm   **46.** 1500 m = __?__ km   **47.** 4.51 g = __?__ mg
48. 8350 mL = __?__ L   **49.** 0.55 kL = __?__ L   **50.** 64,000 mL = __?__ kL

Express each number in scientific notation. (Section 2-7)

51. 250   **52.** 34,500   **53.** 5,000,000   **54.** 6,030,000,000

PART 3: APPLICATIONS

55. During a Women's Nordic Skiing Event, Andrea recorded runs of 5.52 minutes, 7.87 minutes, and 6.14 minutes. Estimate to the nearest minute the total time for the runs. Choose a, b, or c. (Section 2-5)

  a. 18   **b.** 22   **c.** 20

56. In 1933, the World's Land Speed Record for 1 mile was 272.109 miles per hour. In 1963, the speed record set by Craig Breedlove was 407.45 miles per hour. By how many miles per hour did the speed record increase? (Section 2-2)

57. A washing machine uses about 208.08 kilowatt–hours per month. At a cost of $0.12 per kilowatt-hour, find the cost of operating a washing machine for 12 months. (Section 2-3)

58. A car travels 343.35 miles on a full tank of fuel. The tank holds 10.5 gallons when full. How far does the car travel on one gallon? (Section 2-4)

CHAPTER TEST

Complete.

1. The place name of the 2 in 49.02 is __?__.
2. The place name of the 7 in 573.1 is __?__.

For Exercises 3–5, replace the __?__ with > or <.

3. 4.15 __?__ 4.09
4. 1.001 __?__ 1.015
5. 25.03 __?__ 25.033

6. 51.9 + 0.52 + 4.83 + 49.671 = __?__
7. 253.41 − 61.58 = __?__
8. 2.01 + 34.18 + 0.83 = __?__
9. 41.2 − 3.85 = __?__
10. Evaluate $x + y + z$ when $x = 1.07$, $y = 0.28$, and $z = 53.4$
11. Evaluate $a − b$ when $a = 512.01$ and $b = 35.64$

12. 63×0.9
13. $21 \overline{)68.46}$
14. $3.88 \div 0.4$
15. $(0.048)(0.059)$

Choose the best estimate. Choose a, b, or c.

16. $4.15 + $11.75 + $8.90   a. 26   b. 23   c. 25
17. (15.015)(9.7)   a. 135   b. 150   c. 160

Complete.

18. 51.5 cm = __?__ m
19. 2.6 kg = __?__ g
20. 9500 mL = __?__ L
21. $4100 = 4.1 \times 10^?$
22. $570,000 = 5.7 \times 10^?$
23. $8,200,000 = 8.2 \times 10^?$

24. Last month, 0.02 inch, 1.38 inches, 0.08 inch, and 0.57 inch of rain fell. What was the total rainfall?

25. A car travels 352.5 miles on a full tank (12.5 gallons) of fuel. How far does it travel on one gallon?

ANSWERS TO REVIEW CAPSULES

Page 34   **1.** 39   **2.** 404   **3.** 1032   **4.** 1475   **5.** 1159   **6.** 4646

Page 37   **1.** 42   **2.** 508   **3.** 24,006   **4.** 20,240   **5.** 16,128   **6.** 3528   **7.** $14.76   **8.** $8.96   **9.** $25.02   **10.** $37.10   **11.** $61.71

Page 42   **1.** 6   **2.** 15   **3.** 62   **4.** 304   **5.** 2541   **6.** 39   **7.** 108   **8.** 6   **9.** 31   **10.** 205

Page 46   **1.** 4   **2.** 9   **3.** 1   **4.** 17   **5.** 11   **6.** 3.3   **7.** 1.3   **8.** 17.8   **9.** 11.4   **10.** 1.0   **11.** 5.86   **12.** 0.40   **13.** 14.65   **14.** 0.01   **15.** 19.44

Page 50   **1.** 34   **2.** 150   **3.** 5   **4.** 40.6   **5.** 652   **6.** 800   **7.** 13   **8.** 1.24   **9.** 0.6   **10.** 0.08   **11.** 300   **12.** 0.96   **13.** 280   **14.** 7400   **15.** 60   **16.** 1.3   **17.** 2040   **18.** 756   **19.** 20   **20.** 2   **21.** 0.62   **22.** 4.286   **23.** 450   **24.** 0.035   **25.** 6200   **26.** 8500   **27.** 40,700   **28.** 300   **29.** 70   **30.** 38,000   **31.** 4   **32.** 0.405   **33.** 0.032   **34.** 78   **35.** 3.840   **36.** 0.006

Page 54   **1.** 2   **2.** 5   **3.** 3   **4.** 7   **5.** 8   **6.** 4   **7.** 1000   **8.** 100,000   **9.** 1,000,000   **10.** 10   **11.** 10,000   **12.** 10,000,000

ENRICHMENT

Using Scientific Notation

Space ships of the future called galactic cruisers may be able to travel at the speed of light. Light travels at the rate of 1.86×10^5 miles per second. This can also be expressed as 2.976×10^5 kilometers per second.

You can use scientific notation to multiply and divide with large numbers such as 1.86×10^5.

EXAMPLE How long would it take a galactic cruiser to travel 3.91×10^8 miles from Earth to Jupiter? Round your answer to the nearest second.

Solution: Use the distance formula.

[1] $t = \dfrac{d}{r}$ $d = 3.91 \times 10^8$ $r = 1.86 \times 10^5$ [2] $t = \dfrac{3.91 \times 10^8}{1.86 \times 10^5} = \dfrac{3.91}{1.86} \times \dfrac{10^8}{10^5}$

[3] Use a calculator to find $3.91 \div 1.86$.

   3 . 9 1 [÷] 1 . 8 6 [=] $\boxed{2.1021505}$

[4] Find $\dfrac{10^8}{10^5}$. $\dfrac{10^8}{10^5} = \dfrac{10^5 \times 10^3}{10^5} = \dfrac{\overset{1}{10^5} \times 10^3}{\underset{1}{10^5}} = 10^3$ $\dfrac{10^5}{10^5} = 1$

[5] Thus, $\dfrac{3.91 \times 10^8}{1.86 \times 10^5} = 2.1021505 \times 10^3$, or **2102.1505**

It would take **2102 seconds** (Nearest second)

EXERCISES

How long would it take a galactic cruiser to travel from the Earth to each given location? Round your answer to the nearest second.

1. Sun: 1.488×10^8 kilometers
2. Pluto: 3.68×10^9 miles
3. Today, space ships travel at a speed of about 2.4×10^4 miles per hour.
   a. At this rate, how many hours would it take to travel 3.91×10^8 miles to Jupiter? Round your answer to the nearest hour.
   b. How many days would it take?
4. The closest star (other than the Sun) to the Earth is 4×10^{13} kilometers away. To the nearest year, how long would it take a galactic cruiser to travel to this star? (HINT: First find the number of seconds it would take.)

Decimals and Applications 61

ADDITIONAL PRACTICE

SKILLS

Give the value of the underlined digit. (Pages 32–34)

1. 423.<u>8</u>   **2.** <u>3</u>071   **3.** 5<u>6</u>2.04   **4.** 4.0<u>7</u>4   **5.** 3.164<u>3</u>   **6.** 0.9371<u>8</u>

Replace each __?__ with > or <. (Pages 32–34)

7. 0.347 __?__ 0.359   **8.** 14.2 __?__ 14.199   **9.** 0.84 __?__ 0.8   **10.** 1.883 __?__ 0.9

11. Evaluate $a + b + c$ when $a = 37.4$, $b = 38.79$ and $c = 6.803$. (Pages 35–37)

12. Evaluate $x - y$ when $x = 37.4$ and $y = 9.1$. (Pages 35–37)

Multiply. (Pages 38–41)

13. $14(0.09)$ **14.** $(0.34)(0.81)$ **15.** $(0.012)(0.076)$ **16.** $(97.6)(8.4)$

Divide. (Pages 44–46)

17. $79.2 \div 6$ **18.** $2.8 \div 0.8$ **19.** $429 \div 0.65$ **20.** $10.104 \div 0.006$

Choose the best estimate. Choose a, b, or c. (Pages 47–49)

21. $10.3(59.8)$ **a.** 5980   **b.** 600   **c.** 458
22. $13.21 + 6.9$ **a.** 20   **b.** 19   **c.** 21
23. $74.9 \div 24.8$ **a.** 3.5   **b.** 3   **c.** 4
24. $120.21 - 19.8$ **a.** 100   **b.** 102   **c.** 99

Complete. (Pages 51–54)

25. 43 km = __?__ cm   **26.** 1384 g = __?__ kg   **27.** 14 L = __?__ mL

Express each number in scientific notation. (Pages 55–57)

28. 4300   **29.** 77,000   **30.** 1,300,000   **31.** 130   **32.** 90,000,000

Express as a decimal numeral. (Pages 55–57)

33. 3.1×10^2 **34.** 4.2×10^5 **35.** 3.5×10^7 **36.** 4.004×10^4

APPLICATIONS

37. At a cost of $0.007 per hour, how much will it cost to operate a 60-watt light bulb for 1.5 hours? (Pages 38–41)

38. A certain compact travels 497 miles on a full tank of fuel. A full tank holds 14.2 gallons. How far does it travel on one gallon? (Pages 44–46)

CHAPTER 3
Number Properties/Expressions

SECTIONS
3-1 The Special Numbers 0 and 1
3-2 Simplifying Algebraic Expressions
3-3 Problem Solving and Applications: Formulas: Area
3-4 Like Terms
3-5 Combining Like Terms
3-6 Problem Solving and Applications: Formulas: Perimeter

FEATURES
Calculator Application: *Using Formulas*

Computer Application: *Evaluating Expressions*

Enrichment: *Number Properties*

Common Errors

3-1 The Special Numbers 0 and 1

The numbers 0 and 1 have special properties.
The sum of any number and 0 is the number.
The product of any number and 0 is 0.
The product of any number (except 0) and 1 is the number.

Properties of 0 and 1

Addition Property of Zero

Arithmetic	Algebra
$8 + 0 = 0 + 8 = 8$	$a + 0 = 0 + a = a$

Multiplication Property of Zero

Arithmetic	Algebra
$(79.3)(0) = (0)(79.3) = 0$	$a \cdot 0 = 0 \cdot a = 0$

Multiplication Property of One

Arithmetic	Algebra
$257 \cdot 1 = 1 \cdot 257 = 257$	$a \cdot 1 = 1 \cdot a = a$

Expressions such as $8 + 0 = 8$ and $a \cdot 1 = a$ are **mathematical sentences.**

EXAMPLE 1 Write the value of the variable that makes each sentence true. Name the property that gives the reason for your choice.

	Sentence	Solution	Property
a.	$(4.81)(1) = r$	$r = 4.81$	Multiplication Property of One
b.	$p + 18.9 = 18.9$	$p = 0$	Addition Property of Zero
c.	$210 \cdot q = 0$	$q = 0$	Multiplication Property of Zero

Self-Check *Write the value of the variable that makes each sentence true.*

1. $t + 16 = 16$ **2.** $(115.6)(s) = 0$ **3.** $1 \cdot 58 = r$

Since $a \cdot 1 = a$, $a \div a = 1$, $a \neq 0$.   ≠ means "is not equal to."

That is, **any number divided by itself equals** 1.
This is true for all numbers except 0.

Arithmetic		Algebra
$10 \div 10 = 1$ $\dfrac{18}{18} = 1$		$\dfrac{a}{a} = 1$, $a \neq 0$

64 Chapter 3

EXAMPLE 2 Find the value of t that makes each sentence true.

a. $2\left(\dfrac{9}{9}\right) = t$ b. $\dfrac{23}{23}(t) = 45$ c. $6.5\left(\dfrac{t}{19}\right) = 6.5$

Solutions: a. $t = 2$ b. $t = 45$ c. $t = 19$

Self-Check Find the value of r that makes each sentence true.

4. $r = \left(\dfrac{5}{5}\right)(21)$ 5. $7\left(\dfrac{r}{15}\right) = 7$ 6. $\dfrac{10}{10}(r) = 21$

CLASSROOM EXERCISES

Name the property that makes each sentence true. (Example 1)

1. $4 + 0 = 4$ 2. $1.3(1) = 1.3$ 3. $m(0) = 0$ 4. $3y = 0 + 3y$

For Exercises 5–20, find the value of the variable that makes each sentence true. (Example 1)

5. $12 + x = 12$ 6. $(0.8)y = 0$ 7. $(1)(c) = 7$ 8. $m + 0 = 2.7$
9. $3.6(p) = 3.6$ 10. $84(n) = 0$ 11. $0.7(0) = t$ 12. $124 + s = 124$

(Example 2)

13. $\dfrac{5}{5}(m) = 24$ 14. $\dfrac{18}{18}(11) = x$ 15. $\dfrac{3}{3}(0.3) = k$ 16. $q = 1.5\left(\dfrac{8}{8}\right)$
17. $\left(\dfrac{x}{7}\right)(5) = 5$ 18. $16\left(\dfrac{91}{w}\right) = 16$ 19. $(2.5)\left(\dfrac{31}{31}\right) = a$ 20. $\left(\dfrac{5}{5}\right)(0.2) = b$

WRITTEN EXERCISES

Goal: To apply the properties of 0 and 1

For Exercises 1–44, find the value of the variable that makes each sentence true. (Example 1)

1. $6 + n = 6$ 2. $t + 11 = 11$ 3. $0.45 = 0 + x$ 4. $0 + y = 0.8$
5. $12y = 0$ 6. $0 = 3c$ 7. $(m)(3.9) = 0$ 8. $0 = (n)(14.1)$
9. $225n = 225$ 10. $46 = 46t$ 11. $(1)(4.5) = k$ 12. $2.75 = 2.75m$

(Example 2)

13. $x\left(\dfrac{3}{3}\right) = 13$ 14. $15 = y\left(\dfrac{10}{10}\right)$ 15. $z\left(\dfrac{7}{7}\right) = 19$ 16. $28 = k\left(\dfrac{15}{15}\right)$
17. $21.3\left(\dfrac{4}{4}\right) = n$ 18. $y = 0.8\left(\dfrac{7}{7}\right)$ 19. $\dfrac{8}{8}(p) = 23$ 20. $\left(\dfrac{2}{2}\right)39 = t$

Number Properties/Expressions

21. $16 = 16\left(\dfrac{3}{c}\right)$ **22.** $\left(\dfrac{z}{10}\right)21 = 21$ **23.** $9 = 9\left(\dfrac{5}{m}\right)$ **24.** $\left(\dfrac{t}{11}\right)34 = 34$

MIXED PRACTICE

25. $8.3n = 8.3$ **26.** $4.9 = (t)(4.9)$ **27.** $0.6 + r = 0.6$ **28.** $7.3 = x + 7.3$
29. $256t = 0$ **30.** $18 + 0 = v$ **31.** $0.9c = 0.9$ **32.** $93s = 0$
33. $16\left(\dfrac{8}{8}\right) = n$ **34.** $n = \left(\dfrac{9}{9}\right)(46)$ **35.** $\left(\dfrac{17}{17}\right)0 = x$ **36.** $\left(\dfrac{19}{19}\right)r = 75$
37. $11 = 11\left(\dfrac{n}{3}\right)$ **38.** $8 = 8\left(\dfrac{2}{x}\right)$ **39.** $\dfrac{n}{7}(21) = 21$ **40.** $\dfrac{x}{8}(35) = 35$
41. $n\left(\dfrac{7}{7}\right) = 3$ **42.** $12 = \dfrac{5}{5}(p)$ **43.** $\left(\dfrac{17}{17}\right)x = 13$ **44.** $n = \left(\dfrac{14}{14}\right)(8)$

MORE CHALLENGING EXERCISES

Write the value of the variable that makes each sentence true.

45. $25 \div 5b + b(0) = 1$ **46.** $(11 - 2 \cdot 5)(3x + x \cdot 0) = 12$
47. $(5 \cdot z + 5 \cdot 0)(4 + z \cdot 0) = 0$ **48.** $(18 \cdot k - 18 \cdot 0)(5 + k \cdot 0) = 90$
49. $12 + n\left(\dfrac{3}{3}\right) - 42(0) = 18$ **50.** $(14 - 3 \cdot 4)\left(\dfrac{5}{5} \cdot x\right) = 24$
51. $\left(\dfrac{r}{36} \cdot 14\right) + 39(0) = 14$ **52.** $45 \div 3a + \left(\dfrac{a}{3}\right)(0) = 3$
53. $\dfrac{5n}{5} + (5n \cdot 0) = 5$ **54.** $(5 \cdot x + 2)\left(3 + \dfrac{8}{8}\right) = 48$

REVIEW CAPSULE FOR SECTION 3-2

Multiply. (Pages 18–21)

1. a. 4×5
   b. 5×4

2. a. 3×8
   b. 8×3

3. a. 6×12
   b. 12×6

4. a. $2 \times 3 \times 5$
   b. $2 \times 5 \times 3$

5. a. $4 \times (3 \times 2)$
   b. $(4 \times 3) \times 2$

6. a. $6 \times (9 \times 7)$
   b. $(9 \times 7) \times 6$

7. a. $(7 \times 10) \times 12$
   b. $(10 \times 12) \times 7$

8. a. $5 \times 7 \times 5$
   b. $5 \times 5 \times 7$

9. a. $10 \times (11 \times 5)$
   b. $11 \times (10 \times 5)$

10. a. $(3 \times 6)(4 \times 7)$
    b. $(6 \times 4)(3 \times 7)$

11. a. $(5 \times 9)(2 \times 2)$
    b. $(2 \times 2)(5 \times 9)$

12. a. $(3 \times 4)(5 \times 6)$
    b. $(3 \times 6)(5 \times 4)$

The answers are on page 89.

3-2 Simplifying Algebraic Expressions

Multiplying two numbers in either order gives the same product. This is the **Commutative Property of Multiplication.**

$312 \cdot 9 = 2808$

$9 \cdot 312 = 2808$

When you multiply three or more numbers, you may group them in any way. This is the **Associative Property of Multiplication.**

$27(4 \cdot 2) = 27 \cdot 8 = 216$

$(27 \cdot 4) \cdot 2 = (108) \cdot 2 = 216$

Properties of Multiplication

> **Commutative Property of Multiplication**
>
> *Any two numbers can be multiplied in either order.*
>
> **Arithmetic** **Algebra**
> $4 \cdot 10 = 10 \cdot 4$ $a \cdot b = b \cdot a$
>
> **Associative Property of Multiplication**
>
> *The way three or more numbers are grouped for multiplication does not affect the product.*
>
> **Arithmetic** **Algebra**
> $(16 \cdot 3) \cdot 4 = 16 \cdot (3 \cdot 4)$ $(a \cdot b) \cdot c = a \cdot (b \cdot c)$

EXAMPLE 1 Write the value of the variable that makes each sentence true. Name the property that gives the reason for your choice.

	Sentence	Solution	Property
a.	$3 \cdot 8 = 8(c)$	$c = 3$	Commutative Property of Multiplication
b.	$(14 \cdot 2) \cdot 5 = 14 \cdot (2 \cdot n)$	$n = 5$	Associative Property of Multiplication
c.	$(d \cdot 7) \cdot 9 = 8 \cdot (7 \cdot 9)$	$d = 8$	Associative Property of Multiplication

Self-Check   *Write the value of the variable that makes each sentence true.*

1. $(2 \cdot 77) \cdot 50 = k \cdot (77 \cdot 50)$ **2.** $(6.9)(d) = 7(6.9)$

The Commutative and Associative Properties of Multiplication can be used to simplify algebraic expressions. Recall that there are several ways to indicate multiplication.

$5m \cdot p$ $(5m)(p)$ $5m(p)$ $(5m)p$ $5mp$

Number Properties/Expressions

EXAMPLE 2 Multiply: $(5e)(8g)$

Solution: $(5e)(8g)$ means $5 \cdot e \cdot 8 \cdot g$.

1. Change the order. Write the numbers first. $5 \cdot e \cdot 8 \cdot g = 5 \cdot 8 \cdot e \cdot g$
2. Group the numbers and the letters. $= (5 \cdot 8) \cdot (e \cdot g)$
   Multiply. $= 40eg$

◀ Comm. and Assoc. Properties

Self-Check *Multiply.*

3. $3(10x)$ **4.** $(15.5x)(4y)$ **5.** $(3c)(11z)$ **6.** $5x(2.2y)$

You can also simplify algebraic expressions involving exponents.

EXAMPLE 3 Multiply: $(3a)(2b)(7b^2)$

Solution: $(3a)(2b)(7b^2) = 3 \cdot a \cdot 2 \cdot b \cdot 7 \cdot b \cdot b$ ◀ $b^2 = b \cdot b$

$= 3 \cdot 2 \cdot 7 \cdot a \cdot b \cdot b \cdot b$

$= 42 \cdot a \cdot b^3$

$= \mathbf{42ab^3}$

Self-Check *Multiply.*

7. $(10c)(5cn)$ **8.** $(2a)(3b)(5a^2)$ **9.** $(8x)(2y)(3y^2)$

CLASSROOM EXERCISES

Find the value of the variable that makes each sentence true. (Example 1)

1. $8 \cdot x = 5 \cdot 8$ **2.** $7 \cdot 10 = b \cdot 7$ **3.** $20(m) = 30 \cdot 20$
4. $7 \cdot 13 = (t)7$ **5.** $2.5(3.8) = 3.8(p)$ **6.** $3.8(q) = 4(3.8)$
7. $0.5(1.2y) = (0.5 \cdot 1.2)7$ **8.** $3(4 \cdot 5) = (3 \cdot m)5$ **9.** $(20 \cdot t) \cdot 5 = 20(3 \cdot 5)$

Multiply. (Example 2)

10. $10(5y)$ **11.** $4(12t)$ **12.** $20(0.1t)$ **13.** $(1.1)(5k)$
14. $(6x)(3y)$ **15.** $(2a)(4b)$ **16.** $(100r)(0.3s)$ **17.** $(2r)(4s)$

(Example 3)

18. $(8a)(7a)$ **19.** $(2x)(3xy)$ **20.** $(3m^2)(8)(2m)$ **21.** $(3pq)(qr)(7p)$
22. $(2b^2)(3bc)(4c)$ **23.** $(6fg)(5gh)(3f)$ **24.** $(2rt)(3s^2)(rst)$ **25.** $(7n)(kn)(7kn)$

68   Chapter 3

WRITTEN EXERCISES

Goal: To simplify algebraic expressions that involve multiplication

Find the value of the variable that makes each sentence true. (Example 1)

1. $7 \cdot 6 = 6(a)$
2. $x(9) = 9 \cdot 4$
3. $(11 \cdot 2)7 = 11(n \cdot 7)$
4. $13(r \cdot 6) = (13 \cdot 4)6$
5. $3.4n = (6.2)(3.4)$
6. $(8.9)(6.2) = 6.2a$
7. $(4 \cdot 3.8)k = 4(3.8 \cdot 5)$
8. $3(t \cdot 7) = 7(15 \cdot 3)$
9. $(8.4)(1.5)x = 3(1.5)(8.4)$

Multiply. (Example 2)

10. $5(7x)$
11. $6(4y)$
12. $7(6a)$
13. $9(10b)$
14. $(8t)7$
15. $(1.2n)7$
16. $(2.3k)4$
17. $(12r)8$
18. $(9a)(7b)$
19. $(8s)(12t)$
20. $(1.3x)(3y)$
21. $(6p)(1.4q)$
22. $(5r)(20s)$
23. $(3.2a)(6b)$
24. $(56p)(2r)$
25. $(1.9r)(3.2s)(t)$

(Example 3)

26. $(13x)(3x)$
27. $(6y)(14y)$
28. $(12mn)(3n)$
29. $(3k)(12kt)$
30. $(9x)(4xy)$
31. $(7m)(6m^2n)$
32. $(10p)(0.2np)$
33. $(15st)(0.4t^2)$
34. $(3x^2)(8xy)$
35. $(20mn^2)(4mn)$
36. $(5k)(2kt)(3t^2)$
37. $(9dp)(3a^2p)(3d)$

MIXED PRACTICE

38. $(2x)(9y)$
39. $(6m)(12n)$
40. $7(12t)$
41. $11(27r)$
42. $(0.5x^2)(18xy)$
43. $(12m)(0.5mn^2)$
44. $(15w)0.5$
45. $(28a)(0.2)$
46. $(2r)(3s)(st)$
47. $(4a)(3b)(2c)$
48. $(rs)(2st)(3r^2)$
49. $(2mn)(m^2p)(3p)$

MORE CHALLENGING EXERCISES

Multiply. Then evaluate the expression when $x = 2$ and $y = 1$.

50. $(x^2)(xy)(3y)$
51. $(2xy)(4x)(3y)$
52. $x \cdot x \cdot y \cdot y^2 \cdot x \cdot y$
53. $y \cdot y \cdot x \cdot x \cdot y^3$
54. $(2x)(3xy)(4y^2)$
55. $(5y)(2x^2y)(9xy^3)$

REVIEW CAPSULE FOR SECTION 3–3

Evaluate each expression. (Pages 7–9)

1. xy when $x = 7$ and $y = 4$
2. ab when $a = 19$ and $b = 21$
3. mx when $m = 6.2$ and $x = 3.8$
4. rs when $r = 3.6$ and $s = 1.2$
5. $n \cdot n$ when $n = 6$
6. $a \cdot a$ when $a = 13.4$
7. x^2 when $x = 15$
8. s^2 when $s = 8.04$

The answers are on page 89.

PROBLEM SOLVING AND APPLICATIONS

3-3 Formulas: Area

Each side of this figure is 1 centimeter long. The area of the figure is 1 · 1 or **1 square centimeter** (1 cm²). **Area** is the number of square units needed to cover a surface.

A rectangle has four right angles.

Opposite sides of a rectangle are equal.

This word rule and formula tell how to find the area of a rectangle.

Word Rule: To find the area of a rectangle, multiply the length and the width.

Formula: $A = lw$

A = area
l = length; w = width

Small areas are measured in square centimeters (cm²) or square inches (in²). Larger areas are measured in square meters (m²), square miles (mi²), square yards (yd²), or square feet (ft²).

EXAMPLE 1 A soccer field is 119 meters long and 91 meters wide. Find the area.

Solution:
1. Write the formula. $A = lw$
2. Identify known values. $l = 119$ m; $w = 91$ m
3. Replace the variables. Solve. $A = 119(91)$
   $A = 10,829$ The area is **10,829 m²**.

Self-Check 1. A desk is 1.25 meters long and 1 meter wide. Area = __?__

A **square** is a rectangle with four equal sides. You can use this word rule and formula to find the area of a square.

Word Rule: The area of a square equals the product of the lengths of any two sides.

Formula: $A = s^2$

A = area
s = length of a side

Square

70 Chapter 3

EXAMPLE 2 Find the area of this square storm-warning flag.

Solution:
1. Formula: $A = s^2$
2. Known value: $s = 48$
3. Solve. $A = (48)(48)$
$A = 2304$

The area of the flag is **2304 in²**.

Self-Check 2. Each side of a square rug is 3 yards long. Find the area of the rug.

CLASSROOM EXERCISES

Find the area of each rectangle. (Example 1)

1. $l = 10$ cm; $w = 5$ cm
2. $l = 5$ m; $w = 3$ m
3. $l = 10$ km; $w = 0.5$ km
4. $l = 13$ ft; $w = 10$ ft
5. $l = 7$ in; $w = 3$ in
6. $l = 5$ yd; $w = 2$ yd

Find the area of each square. (Example 2)

7. $s = 8$ cm
8. $s = 12$ km
9. $s = 0.5$ m
10. $s = 1.2$ m
11. $s = 15$ in
12. $s = 12$ yd
13. $s = 25$ ft
14. $s = 8$ mi

WRITTEN EXERCISES

Goals: To find the area of a rectangle and of a square
To apply the skill of finding area to solving word problems

Find the area of each rectangle. (Example 1)

METRIC MEASURES

1. **Gym Floor** — 42 m by 27 m
2. **Dollar Bill** — 6.5 cm by 15.5 cm
3. **Swimming Pool** — 3 m by 4 m

Number Properties/Expressions **71**

Find the area of each rectangle. (Example 1)

	Length	Width	Area
4.	4 cm	3 cm	?
6.	5 m	4.6 m	?
8.	8 km	6 km	?

	Length	Width	Area
5.	2 m	1.5 m	?
7.	13 mm	2.8 mm	?
9.	7.5 m	2.8 m	?

CUSTOMARY MEASURES

10. Photograph — 7 in by 5 in

11. Garden — 14 yd by 22 yd

12. Desk Top — 5 ft by 3 ft

	Length	Width	Area
13.	13 in	8 in	?
15.	8 yd	6 yd	?
17.	3 mi	1 mi	?

	Length	Width	Area
14.	6 ft	4 ft	?
16.	32 in	28 in	?
18.	12 ft	18 ft	?

Find the area of each square. (Example 2)

METRIC MEASURES

19. Mirror — 31 m by 31 m

20. Stamp — 20.2 mm by 20.2 mm

21. Rug — 1.1 m by 1.1 m

	Length of Side	Area
22.	4 cm	?
24.	6.2 m	?

	Length of Side	Area
23.	10 m	?
25.	13.8 mm	?

CUSTOMARY MEASURES

26. Road Sign — 15 ft × 15 ft

27. City Square — 2 mi × 2 mi

28. Baseball Diamond — 30 yd × 30 yd

	Length of Side	Area		Length of Side	Area
29.	3 ft	?	30.	16 in	?
31.	16 yd	?	32.	23 ft	?
33.	4 mi	?	34.	5 yd	?

MIXED PRACTICE

35. A movie screen is 125 centimeters long and 100 centimeters wide. Find the area of the screen.

36. Each base of the twin towers in New York City's World Trade Center is a square. Each base measures 62.7 meters on a side. Find the area of one base.

37. A rectangular swimming pool is 164 feet long and 80 feet wide. Find the area of the pool.

38. The top of a rectangular table is 1 meter long and 0.8 meter wide. Find the area of the table top.

39. The length of each side of a square box is 15 inches. Find the area of the top of the box.

40. The length of a tennis court is 21.95 meters and the width is 11.4 meters. Find the area.

41. A farmer's field is 83.5 meters long and 250.4 meters wide. What is the area of the field?

42. A piece of carpet is 12 yards wide and 12 yards long. Find the area of the carpet.

43. A room has the shape of a rectangle. It is 3.1 meters long and 2.9 meters wide. Find the area.

44. A standard sheet of plywood is 48 inches wide and 96 inches long. What is the area of the sheet?

Number Properties/Expressions

REVIEW: SECTIONS 3-1 – 3-3

Name the property illustrated in each sentence. (Section 3-1)

1. $2.7(0) = 0$
2. $x + 0 = x$
3. $(4.52)(1) = 4.52$

Find the value of the variable that makes each sentence true. (Section 3-1)

4. $3.8 + 0 = t$
5. $\left(\frac{12}{12}\right)r = 32$
6. $5x = 0$

Name the property illustrated in each sentence. (Section 3-2)

7. $2(10 \cdot 8) = (2 \cdot 10)8$
8. $3 \cdot x = x \cdot 3$
9. $(26 \cdot 4)5 = 26(4 \cdot 5)$

Find the value of the variable that makes each sentence true. (Section 3-2)

10. $4.2(3.5 \cdot x) = (4.2 \cdot 3.5)8$
11. $(8.7)(g) = 6(8.7)$
12. $(15 \cdot 21)y = 15(21 \cdot 9)$

Multiply. (Section 3-3)

13. $6(4y)$
14. $(8t)(3)$
15. $(3a)(4ab)$

16. Each side of a square parking lot is 75.3 meters long. Find the area of the parking lot. (Section 3-3)

17. A rectangular drawing board is 36 inches long and 28 inches wide. Find the area of the drawing board. (Section 3-3)

18. A television screen is 62 centimeters long and 47 centimeters wide. What is the area of the screen? (Section 3-3)

19. Eric and Beth put together a square jig-saw puzzle. The puzzle is 18 inches on each side. Find the area of the puzzle. (Section 3-3)

REVIEW CAPSULE FOR SECTION 3-4

Add or subtract as indicated. (Pages 35-37)

1. $3.2 + 1.8$
2. $4.2 + 0.5$
3. $9.7 + 3$
4. $5 + 8.9$
5. $10.6 - 4.9$
6. $13.4 - 8.5$
7. $7 - 0.6$
8. $5 - 3.1$
9. $31.1 - 2.2$
10. $11 - 3.2$
11. $16 + 2.6$
12. $4.3 + 5.7$

The answers are on page 89.

COMPUTER APPLICATIONS
Evaluating Expressions

These are the symbols for arithmetic operations in BASIC.

Exponentiate means "raise to a power."

Operation	Symbol	Arithmetic	BASIC
Add.	+	9 + 7	9 + 7
Subtract.	−	8 − 3	8 − 3
Multiply.	*	6 × 5	6 * 5
Divide.	/	7 ÷ 2	7 / 2
Exponentiate.	↑, or ^	3^2	3 ↑ 2

When two or more operations appear in a BASIC expression or formula, the computer follows this order of operations.

1. Exponentiation (raising to a power) is done first.
2. Next multiplication and division are done from left to right.
3. Then addition and subtraction are done left to right.
4. When there are parentheses, do the operations inside the parentheses first, beginning with the innermost parentheses. Thus, the order listed in 1–3 is changed.

The computer will evaluate an expression if you type it as a direct command, such as **PRINT.**

EXAMPLE Type each direct command into a computer.
     a. PRINT 20 − 3 * 6   **b.** PRINT 5 ↑ 2 * (4 + 2)

Output: **a.** 2   **b.** 150

Exercises

Use a computer to evaluate each expression.

1. PRINT 4 + 5
2. PRINT 2 * 8 + 3
3. PRINT 3*(9 − 4)
4. PRINT 30 − 5 * 6
5. PRINT 5*11/2
6. PRINT 40/2*3
7. PRINT (14 + 7)/3
8. PRINT 14/(9 − 5)
9. PRINT 50/5*2
10. PRINT 50/(5*2)
11. PRINT 6 ↑ 2
12. PRINT (8 − 1) ↑ 2
13. PRINT 100 − 4 ↑ 2
14. PRINT 2 ↑ 3
15. PRINT (7 + 5)*9

3-4 Like Terms

A number such as 257 can be represented by this algebraic expression.

$$2h + 5t + 7o$$

Hundreds	+	Tens	+	Ones
2h	+	5t	+	7o

This expression contains three terms: $2h$, $5t$, and $7o$. Since the terms have different variables, h, t, and o, the terms are unlike.

Like terms have the *same variable* and the *same exponent* for each variable.

> Two or more **like terms** have
> 1. the *same* variables
> 2. the *same powers* of these variables.
>
> Like: $5a$ and $7a$
> Like: $2x^2$ and $9x^2$
> Unlike: $5a$ and $3a^2$

EXAMPLE 1 For each term in List 1 select one or more like terms from List 2.

List 1
1. $0.5d$
2. $6rs$
3. $2x^2$
4. $3xy$

List 2
a. $4rs$ b. $30x^2$
c. $1.5r$ d. $0.9xy$
e. $2s$ f. $18d$
g. $11d$ h. $1.8xy$

Solutions: 1. f, g   2. a   3. b   4. d, h

Self-Check Which pairs show like terms? Answer <u>Yes</u> or <u>No</u>.

1. $3x$ and $1.5x$
2. $6rt$ and $21rt$
3. $9r^2$ and $5s^2$
4. $0.5d$ and $1.5cd$

You can add or subtract like terms. This is called **combining like terms**. In the expression $5x^2$, 5 is the **numerical coefficient** of x^2.

> **PROCEDURE**
>
> To add or subtract like terms:
>
> 1 Add or subtract the numerical coefficients.
> 2 Write the same variable(s) and exponent(s).

EXAMPLE 2 Combine like terms: $5x + 3.1x$

Solution:
1. Add the numerical coefficients. $5 + 3.1 = 8.1$
2. Use the same variable(s) and exponent(s). $5x + 3.1x = \mathbf{8.1x}$

Self-Check *Combine like terms.*

5. $3y + 7y$ **6.** $4rs + 9rs$ **7.** $6.1z^2 + 3.4z^2$

NOTE: $13n^2 - n^2$ is the same as $13n^2 - 1n^2$.

EXAMPLE 3 Combine like terms: $13n^2 - n^2$

Solution:
1. Subtract the numerical coefficients. $13 - 1 = 12$
2. Use the same variable(s) and exponent(s). $13n^2 - n^2 = \mathbf{12n^2}$

Self-Check *Combine like terms.*

8. $15p - 9p$ **9.** $13r - r$ **10.** $5.9t^2 - 1.8t^2$

CLASSROOM EXERCISES

Identify each pair of terms as like terms or unlike terms. Give a reason for each answer. (Example 1)

1. $3x$ and $4x$ **2.** $6m$ and $3r$ **3.** $2.4y$ and $3.5y^2$ **4.** $8b$ and $5b$
5. $7xy$ and xy **6.** $4c^2$ and $2.2c^2d$ **7.** $4.2b$ and $2ab$ **8.** $5a^2$ and $6a^2$

Add or subtract the numerical coefficients in each expression. (Step 1, Examples 2 and 3)

9. $7m + 15m$ **10.** $13r - 5r$ **11.** $0.9t - 0.4t$ **12.** $4.8x + 1.3x$
13. $5k - 4k$ **14.** $26m^2 + 17m^2$ **15.** $32z^2 - 19z^2$ **16.** $80ab + 23ab$

For Exercises 17–32, combine like terms. (Example 2)

17. $8x + 3x$ **18.** $6a + 2a$ **19.** $0.3d + 0.6d$ **20.** $1.2t + 2.5t$
21. $13r^2 + 8r^2$ **22.** $9m^2 + 22m^2$ **23.** $15ad + 19ad$ **24.** $bc + 99bc$

(Example 3)

25. $9y - 4y$ **26.** $12x - x$ **27.** $40b - 15b$ **28.** $19g - 6g$
29. $4.5m^2 - 1.2m^2$ **30.** $5.9t^2 - 1.3t^2$ **31.** $35st - 15st$ **32.** $80jk - 79jk$

Number Properties/Expressions

WRITTEN EXERCISES

Goals: To identify like terms
To combine like terms

For each expression in Exercises 1–10, select one or more like terms from A–T.

1. $5x$	**2.** $3a$	**A.** $4y^2$	**B.** $12a$	**C.** $3rs$	**D.** $5y$
3. $2y$	**4.** x^2	**E.** $2x^2$	**F.** r^2t	**G.** $2mn$	**H.** $3x$
5. $0.7m$	**6.** y^2	**I.** $9ab$	**J.** $4r$	**K.** $2x^2y$	**L.** $0.7b$
7. rs	**8.** $7t^2$	**M.** $0.4t^2$	**N.** $6xy$	**O.** $3.2m$	**P.** $3xy^2$
9. $1.8a^2$	**10.** xy	**Q.** $3a^2$	**R.** $8y$	**S.** $8x^2$	**T.** $3.5xy$

For Exercises 11–34, combine like terms. (Example 2)

11. $7a + 9a$ **12.** $6m + 8m$ **13.** $12b + b$ **14.** $r + 9r$
15. $35y + 4y$ **16.** $41n + 9n$ **17.** $2.4x + 1.5x$ **18.** $3.2y + 2.5y$
19. $6m^2 + 6m^2$ **20.** $7b^2 + 9b^2$ **21.** $8xy + 6xy$ **22.** $8.4ab + 1.6ab$

(Example 3)

23. $12s - 5s$ **24.** $13t - 7t$ **25.** $6a - a$ **26.** $9x - x$
27. $7.8t - 1.9t$ **28.** $4.9b - 0.4b$ **29.** $7a - 3.2a$ **30.** $9s - 2.5s$
31. $4x^2 - 2x^2$ **32.** $5.7m^2 - 2.9m^2$ **33.** $6ab - 3ab$ **34.** $1.4xy - 0.9xy$

MIXED PRACTICE

Combine like terms.

35. $16n - 14n$ **36.** $24a^2 - 9a^2$ **37.** $22st + 23st$ **38.** $19r^2 - 8r^2$
39. $5.3f^2 + 2.9f^2$ **40.** $6cp - 5cp$ **41.** $3.83m + 12.19m$ **42.** $122.8p^2 - 23.6p^2$

REVIEW CAPSULE FOR SECTION 3–5

Add. (Pages 35–37)

1. a. $12 + 9$
 b. $9 + 12$

2. a. $4.6 + 9.8$
 b. $9.8 + 4.6$

3. a. $12 + 7.8$
 b. $7.8 + 12$

4. a. $12.6 + 5.9 + 7.4$
 b. $7.4 + 5.9 + 12.6$

5. a. $26 + 18 + 35$
 b. $18 + 26 + 35$

6. a. $9 + 13 + 17$
 b. $9 + 17 + 13$

The answers are on page 89.

3-5 Combining Like Terms

Properties of addition are similar to those of multiplication.

Properties of Addition

> **Commutative Property of Addition**
>
> Any two numbers can be added in either order.
>
> | Arithmetic | Algebra |
> |---|---|
> | $3.5 + 1.9 = 1.9 + 3.5$ | $a + b = b + a$ |
>
> **Associative Property of Addition**
>
> The way numbers are grouped for addition does not affect the sum.
>
> | Arithmetic | Algebra |
> |---|---|
> | $(7 + 8) + 15 = 7 + (8 + 15)$ | $(a + b) + c = a + (b + c)$ |

EXAMPLE 1 Write the value of the variable that makes each sentence true. Name the property or properties that give the reason for your choice.

	Sentence	Solution	Property
a.	$(6 + 11) + w = 6 + (11 + 8)$	$w = 8$	Associative Property of Addition
b.	$5.1 + 7.6 = 7.6 + p$	$p = 5.1$	Commutative Property of Addition
c.	$16 + (9 + 4) = (16 + n) + 9$	$n = 4$	Commutative and Associative Properties of Addition

Self-Check Write the value of the variable that makes each sentence true.

1. $16.9 + q = 11.5 + 16.9$
2. $3 + (1 + 5) = (t + 1) + 5$

The Commutative and Associative Properties can be used to simplify algebraic expressions.

EXAMPLE 2 Combine like terms: $15rx + 92 + 36rx$

Solution:
1. Change the order. Write the like terms first. $15rx + 36rx + 92$ ◄ **Commutative Property**

2. Group like terms and add. $(15rx + 36rx) + 92$ ◄ **Associative Property**

$51rx + 92$

Number Properties/Expressions

Self-Check Combine like terms.
  3. $4a + 7 + 13a$ 4. $6t + 4t + t$ 5. $1.8x + 13 + 2.1x$

CLASSROOM EXERCISES

Write the value of the variable that makes each sentence true. Name the property or properties that give the reason for your choice. (Example 1)

1. $7 + (9 + 11) = (9 + 7) + r$
2. $(c + 100) + 25 = 50 + (100 + 25)$
3. $(3 + 4) + x = 5 + (3 + 4)$
4. $(3.2 + 1.5) + 4.8 = (b + 1.5) + 3.2$

Combine like terms. (Example 2)

5. $36r + 10 + 16r$
6. $12b + 5 + 7b$
7. $14st + 25 + 28st$
8. $3x + 2x + 4x$
9. $0.6t + 1.8 + 0.9t$
10. $1.4s + 25 + 0.9s$
11. $71n^2 + 9 + 16n^2$
12. $4a^2 + 3a^2 + 5a^2$

WRITTEN EXERCISES

Goal: To simplify algebraic expressions that involve addition of more than two terms

Write the value of the variable that makes each sentence true. (Example 1)

1. $2.7 + 11.9 = 11.9 + r$
2. $4.5 + q = 3.5 + 4.5$
3. $(7 + 9) + x = 7 + (4 + 9)$
4. $(12 + 15) + 8 = 12 + (b + 15)$
5. $(41 + x) + 90 = (41 + 90) + 6$
6. $(m + 14) + 25 = 15 + (14 + 25)$

Combine like terms. (Example 2)

7. $3a + 2a + 4$
8. $8x + 3x + 9$
9. $12a + 15 + 17a$
10. $23t + 9 + 18t$
11. $0.9m + 4.2 + 3m$
12. $8.7r + 5 + 2.8r$
13. $21k + 7 + 34k$
14. $33n + 16 + 39n$
15. $3x + 6x + 9x$
16. $4a + 2a + 8a$
17. $1.6t + 3.8t + 4.1t$
18. $8.5p + 0.4p + 9.1p$
19. $7a^2 + 4 + 6a^2$
20. $5x^2 + 6 + 12x^2$
21. $9xy + 4 + xy$
22. $1.2mn + 3.8mn + 2.9$
23. $3p + 1.9 + 0.9p$
24. $3x^2 + 4x^2 + 9x^2$
25. $7p + 8p + 16$
26. $3.2 + 8.5t + 4.9t$

MORE CHALLENGING EXERCISES

Combine like terms.

27. $24r + 19r + r + 29r + 18r$
28. $43a^2 + 6a^2 + 9 + 8a^2 + 15$
29. $77 + 114p + 98p + 113 + 42p$
30. $112.1z^2 + 31.6z^2 + 0.9 + 101.8z^2$

REVIEW CAPSULE FOR SECTION 3-6

Evaluate each expression. (Pages 22-24)

1. $x + y + z$ when $x = 4$, $y = 3$, and $z = 5$
2. $2x + 2y$ when $x = 3$ and $y = 8$
3. $m + r + s$ when $m = 8.4$, $r = 12$, and $s = 9.5$
4. $4a$ when $a = 9.7$
5. $2a + 2b$ when $a = 5.4$ and $b = 4.2$
6. $4x$ when $x = 112$

The answers are on page 89.

PROBLEM SOLVING AND APPLICATIONS

3-6 Formulas: Perimeter

Rectangle

The **perimeter** of a figure is the distance around it. To find perimeter, add the lengths of the sides.

The following word rule and formula tell how to find the perimeter of any rectangle.

$P = l + l + w + w$
$P = 2l + 2w$

Word Rule: The perimeter of a rectangle equals twice the length plus twice the width.

Formula: $P = 2l + 2w$

P = perimeter
l = length
w = width

Perimeter is expressed in linear units such as kilometers (km), meters (m), centimeters (cm), miles (mi), yards (yd), feet (ft), and so on.

EXAMPLE 1 The sign at the right is 25 centimeters wide and 43.8 centimeters long. Find the perimeter.

Solution:

1. Write the formula. $P = 2l + 2w$
2. Identify known values. $l = 43.8$ cm; $w = 25$ cm
3. Replace the known values and solve. $P = 2(43.8) + 2(25)$
                               $P = 87.6 + 50$
                               $P = \mathbf{137.6}$

The perimeter is **137.6 centimeters**.

Self-Check   *Find the perimeter of each rectangle.*

1. $l = 24$ m; $w = 17$ m
2. $l = 22$ ft; $w = 15$ ft

Number Properties/Expressions   **81**

This word rule tells how to find the perimeter of any triangle.

Word Rule: The perimeter of a triangle equals the sum of the lengths of the sides.

Formula: $P = a + b + c$

P = perimeter
a, b, c = lengths of the sides

EXAMPLE 2 The course for a boat race is triangular in shape. One leg of the race is 50 miles long. The second leg is 46 miles long and the third leg is 62 miles long. Find the total length of the race.

Solution: Since the course has the shape of a triangle, use the formula for the perimeter of a triangle.

1. Formula: $P = a + b + c$
2. Known values: $a = 50;\ b = 46;\ c = 62$
3. $P = 50 + 46 + 62$

$P = 158$ The total length of the race is **158 miles.**

Self-Check *Find the perimeter of each triangle.*

3. $a = 4$ m; $b = 7$ m; $c = 9$ m   **4.** $a = 2$ yd; $b = 5$ yd; $c = 6$ yd

Since a square is a rectangle with four equal sides, you can write a formula for the perimeter of a square by adding like terms.

$P = s + s + s + s$

Formula: $P = 4s$

P = perimeter
s = length of a side

CLASSROOM EXERCISES

Find the perimeter of each rectangle. (Example 1)

1. $l = 7$ cm; $w = 4$ cm   **2.** $l = 12.4$ m; $w = 8.9$ m   **3.** $l = 15$ m; $w = 12$ m
4. $l = 3$ ft; $w = 1$ ft   **5.** $l = 8$ in; $w = 4$ in   **6.** $l = 6$ yd; $w = 4$ yd

Find the perimeter of each triangle. (Example 2)

7. $a = 6$ cm; $b = 9$ cm, $c = 4$ cm
8. $a = 2.5$ m; $b = 5.1$ m; $c = 3$ m
9. $a = 12$ cm; $b = 13.8$ cm; $c = 8$ cm
10. $a = 4$ ft; $b = 5$ ft; $c = 2$ ft
11. $a = 4$ in; $b = 10$ in; $c = 8$ in
12. $a = 22$ yd; $b = 18$ yd; $c = 13$ yd

WRITTEN EXERCISES

Goals: To find the perimeter of a rectangle and of a triangle
To apply the skill of finding perimeter to solving word problems

Find the perimeter of each rectangle. (Example 1)

METRIC MEASURES

1. Newspaper — 28 cm by 37 cm
2. Backgammon Board — 40 cm by 40 cm
3. Door — 2.4 m by 1.2 m

	Length	Width	Perimeter
4.	8 cm	5 cm	?
6.	3.4 cm	2.1 cm	?
8.	4.2 m	4.2 m	?

	Length	Width	Perimeter
5.	25.1 m	9.8 m	?
7.	3.2 cm	1.9 cm	?
9.	16 mm	16 mm	?

CUSTOMARY MEASURES

10. Book Cover — 9 in by 7 in
11. Table Top — 6 ft by 2 ft
12. Checker Board — 15 in by 15 in

Number Properties/Expressions **83**

	Length	Width	Perimeter
13.	14 in	7 in	?
15.	6 yd	6 yd	?
17.	5 ft	5 ft	?

	Length	Width	Perimeter
14.	2 ft	3 ft	?
16.	3 in	4 in	?
18.	12 yd	8 yd	?

Find the perimeter of each triangle. (Example 2)

METRIC MEASURES

19. Table Top — 0.7 m, 1 m, 1.2 m

20. Pennant — 40 cm, 40 cm, 16 cm

21. Sail — 8.6 m, 10.5 m, 6 m

	a	b	c	Perimeter
22.	8 cm	12 cm	6 cm	?
24.	4 cm	4 cm	4 cm	?
26.	4.2 m	3.5 m	6.4 m	?

	a	b	c	Perimeter
23.	3.2 m	4.1 m	2.5 m	?
25.	12 cm	12 cm	8 cm	?
27.	3 m	3 m	3 m	?

CUSTOMARY MEASURES

28. Flag — 5 yd, 5 yd, 3 yd

29. Sail — 12 ft, 15 ft, 9 ft

30. Gable of a Roof — 12 ft, 12 ft, 20 ft

	a	b	c	Perimeter
31.	3 ft	4 ft	5 ft	?
33.	5 yd	5 yd	5 yd	?
35.	22 in	11 in	16 in	?

	a	b	c	Perimeter
32.	9 in	11 in	7 in	?
34.	4 ft	4 ft	6 ft	?
36.	2 in	3 in	4 in	?

84 Chapter 3

MIXED PRACTICE

37. The rectangular building in which spacecraft are assembled at Cape Canaveral in Florida is 654 feet long and 474 feet wide. Find its perimeter.

38. The base of the Great Pyramid of Egypt is a square. Each side of the square is 236.4 meters long. Find the perimeter of the base.

39. A triangular lot has sides having lengths of 32 yards, 45 yards and 27 yards. Find the perimeter of the lot.

40. A football field has a rectangular shape. It is 110 meters long and 49 meters wide. Find the perimeter of the field.

MORE CHALLENGING EXERCISES

Find the perimeter of each figure.

41. Square with sides $2b$.

42. Triangle with two sides a and base c.

43. Equilateral triangle with sides a.

44. Hexagon with all sides b.

45. Pentagon with sides c, b, b, a, a.

46. Octagon with all sides a.

47. The length of a rectangle is $4x$ and the width is $2x$. Find the perimeter of the rectangle.

48. The lengths of the sides of a triangle are $4s$, $6s$, and $7s$. Find the perimeter of the triangle.

Number Properties/Expressions

REVIEW: SECTIONS 3-4 — 3-6

Which pairs show like terms? Answer Yes or No. (Section 3-4)

1. $4x$ and $5x$
2. $2.5c^2d$ and $7.1cd^2$
3. $2xy$ and $1.5xy$

Combine like terms. (Section 3-4)

4. $2x + 3x$
5. $6.2b + b$
6. $4.7ef - 2.8ef$
7. $9n^2 - n^2$

Write the value of the variable that makes each sentence true. (Section 3-5)

8. $3.4 + 5 = 5 + a$
9. $(7 + c) + 9 = 7 + (4 + 9)$
10. $9 + (m + 6) = (9 + 6) + 3$

Combine like terms. (Section 3-5)

11. $2b + 6b + 4b$
12. $9g + 5g + 17$
13. $7.2t + 10 + 4.1t$
14. $5x^2 + 2.5 + 3.2x^2$

Solve each problem. (Section 3-6)

15. A boat race is run over a triangular course. The first leg is 72 miles long; the second leg is 30 miles long; and the third is 78 miles long. Find the perimeter. (Section 3-6)

16. A rectangular pasture needs to be fenced. The pasture is 2.5 kilometers wide and 6.7 kilometers long. Find the perimeter of the pasture. (Section 3-6)

USING FORMULAS

You can use a calculator and the following formula to find the sum of a given number of counting numbers. (The counting numbers are 1, 2, 3, 4, 5, · · ·)

$$S = \frac{n(n + 1)}{2}$$

S represents the sum.
n represents how many numbers are to be added.

EXAMPLE Find the sum of the first 230 counting numbers.

Solution $S = \dfrac{n(n + 1)}{2}$ Replace *n* with 230. Then $n + 1 = 231$.

$$S = \frac{230(230 + 1)}{2} = \frac{230(231)}{2}$$

2 3 0 [×] 2 3 1 [÷] 2 [=]   `26565.`

EXERCISES Find the sum of the first *n* counting numbers for the given value of *n*.

1. 17
2. 23
3. 68
4. 493
5. 605
6. 1308

CHAPTER REVIEW

PART 1: VOCABULARY

For Exercises 1–11, choose from the box at the right the word(s) or numerical expression(s) that best corresponds to each description.

1. The number of square units needed to cover a surface __?__ (p. 70)
2. Addition Property of Zero __?__ (p. 64)
3. Commutative Property of Addition __?__ (p. 79)
4. Multiplication Property of One __?__ (p. 64)
5. A rectangle with four equal sides __?__ (p. 70)
6. Associative Property of Addition __?__ (p. 79)
7. Multiplication Property of Zero __?__ (p. 64)
8. Terms that have the same variable and the same exponents for each variable __?__ (p. 76)
9. Commutative Property of Multiplication __?__ (p. 67)
10. Associative Property of Multiplication __?__ (p. 67)
11. The distance around a figure __?__ (p. 81)

$7 + 0 = 7$
$25 \cdot 0 = 0$
$(8)(1) = 8$
$5 \cdot 12 = 12 \cdot 5$
$4 + 3 = 3 + 4$
$3 \cdot (6 \cdot 8) = (3 \cdot 6) \cdot 8$
$(2 + 6) + 9 = 2 + (6 + 9)$
perimeter
like terms
square
area

PART 2: SKILLS

Name the property that makes each sentence true.
(Sections 3–1 – 3–2)

12. $2.4(1) = 2.4$
13. $(46)(0) = 0$
14. $7x = 0 + 7x$
15. $(8)(y) = (y)(8)$
16. $(23 \cdot 9)4 = 23(9 \cdot 4)$
17. $5(16 \cdot 32) = (5 \cdot 16)32$

Find the value of the variable that makes each sentence true.
(Sections 3–1 – 3–2)

18. $0 = 5e$
19. $1.9 = 0 + y$
20. $3.2 \cdot t = 3.2$
21. $6 \cdot x = 3 \cdot 6$
22. $4(q \cdot 7) = 7(9 \cdot 4)$
23. $(2 \cdot 8.5)k = 4(8.5 \cdot 2)$

Multiply. (Section 3–2)

24. $5(4x)$
25. $(4.2m)6$
26. $(2x)(12y)$
27. $(10r)(1.7t)$
28. $(12x)(3x)$
29. $(5a)(9a)$
30. $(2x)(5xy)$
31. $(17pq)(4p)$
32. $(15cd)(2.4c^2)$
33. $(20ab^2)(5ab)$
34. $(14e^2f)(3ef)$
35. $(9x^2y^2)(6y)$

Number Properties/Expressions

Find the area of each rectangle or square. (Section 3-3)

	Length	Width	Area
36.	27 m	16 m	?
37.	8.4 cm	8.4 cm	?
38.	9.2 mm	4.8 mm	?
39.	21 m	21 m	?

Identify each pair of terms as like terms or unlike terms. (Section 3-4)

40. $1.4z$ and $2x$
41. a and $7a$
42. $25s$ and $1.3s$
43. $2c$ and cx
44. $16xy$ and $8xy$
45. z and z^2

Combine like terms. (Section 3-4)

46. $18x + 12x$
47. $7n + 6n$
48. $2.7y + 6.8y$
49. $a + 10a$
50. $9s - 4s$
51. $18t - 5.9t$
52. $20x - x$
53. $8.4n^2 - 5.6n^2$

Find the value of the variable that makes each sentence true. (Section 3-5)

54. $3.1 + 12.4 = 12.4 + m$
55. $(5 + 8) + y = 5 + (8 + 3)$

Combine like terms. (Section 3-5)

56. $4x + 7x + 2x$
57. $5a + 1.7a + 3$
58. $6y + 2 + 3y$
59. $4ab + 5 + 9ab$
60. $13c^2 + 12 + c^2$
61. $7t + 2t + 8$
62. $1 + u + 5u$
63. $2x^2 + x^2 + 6x^2$

Find the perimeter of each rectangle. (Section 3-6)

64. $l = 6$ cm; $w = 5$ cm
65. $l = 10$ in; $w = 32$ in
66. $l = 8.7$ m; $w = 12.5$ m

Find the perimeter of each triangle. (Section 3-6)

67. $a = 14.2$ cm; $b = 9.7$ cm; $c = 5.6$ cm
68. $a = 24$ in; $b = 55$ in; $c = 38$ in

PART 3: APPLICATIONS

69. A rectangular wing on a single-engine airplane is 3 feet wide and 36 feet long. Find the area of the wing. (Section 3-3)

70. The sides of a square Oriental rug are 2.4 meters long. Find the area of the rug. (Section 3-3)

71. A triangular courtyard has sides measuring 50.4 meters, 47.2 meters and 39 meters. Find the perimeter of the courtyard. (Section 3-6)

72. A rectangular picture window is 6 feet long and 9 feet wide. Find the perimeter of the picture window. (Section 3-6)

CHAPTER TEST

Find the value of the variable that makes each sentence true.

1. $x + 0 = 5.2$
2. $3(6 \cdot y) = (3 \cdot 6)7$
3. $(48)(a) = 48$
4. $5 \cdot b = 9 \cdot 5$

Multiply.

5. $4.1\,(3.6x)$
6. $(13s)(5t)$
7. $(4.5x)(8x)$
8. $(7cd)(8cd^2)$

Find the area of each rectangle or square.

9. length: 3.4 m; width: 3.4 m
10. length: 9 ft; width: 10 ft
11. length: 41 in; width: 8 in
12. length: 2.3 cm; width: 2.3 cm

Combine like terms.

13. $3x + 12x$
14. $5.2y - y$
15. $4a + 3 + 10a$
16. $7m^2 - 3.9m^2$

Find the perimeter of each rectangle.

17. $l = 14$ m; $w = 6.2$ m
18. $l = 25$ in; $w = 19$ in
19. $l = 8.9$ cm; $w = 4.6$ cm

Find the perimeter of each triangle.

20. $a = 2$ m; $b = 6.5$ m; $c = 7.3$ m
21. $a = 48$ ft; $b = 27$ ft; $c = 36$ ft
22. $a = 45$ in; $b = 60$ in; $c = 75$ in
23. $a = 9.8$ cm; $b = 5.4$ cm; $c = 6.7$ cm

24. Maria Gala plans to wallpaper her bedroom. In finding the area of the walls she must allow for the area of the doorway. A rectangular doorway is 3 feet wide and 7 feet long. Find the area of the doorway.

25. A rectangular sign is 0.75 meters wide and 3.4 meters long. Find the perimeter of the sign.

ANSWERS TO REVIEW CAPSULES

Page 66   1. 20; 20   2. 24; 24   3. 72; 72   4. 30; 30   5. 24; 24   6. 378; 378   7. 840; 840   8. 175; 175   9. 550; 550   10. 504; 504   11. 180; 180   12. 360; 360

Page 69   1. 28   2. 399   3. 23.56   4. 4.32   5. 36   6. 179.56   7. 225   8. 64.6416

Page 74   1. 5.0   2. 4.7   3. 12.7   4. 13.9   5. 5.7   6. 4.9   7. 6.4   8. 1.9   9. 28.9   10. 7.8   11. 18.6   12. 10.0

Page 78   1. 21; 21   2. 14.4; 14.4   3. 19.8; 19.8   4. 25.9; 25.9   5. 79; 79   6. 39; 39

Page 81   1. 12   2. 22   3. 29.9   4. 38.8   5. 19.2   6. 448

Number Properties/Expressions

ENRICHMENT

Number Properties

Number properties can be used to simplify numerical computations.

EXAMPLES Use number properties to evaluate each expression.

Solutions:

a. $(93 + 8) + (7 + 72) = (93 + 7) + (8 + 72)$ ◄ By the Commutative and Associative Properties for Addition
$= 100 + 80 = \mathbf{180}$

b. $(6 \cdot 25) \cdot (7 \cdot 4) = (6 \cdot 7) \cdot (25 \cdot 4)$ ◄ By the Commutative and Associative Properties for Multiplication
$= 42 \cdot 100 = \mathbf{4200}$

c. $997 \cdot 42 + 3 \cdot 42 = 997x + 3x$ ◄ Think of $997 \cdot 42$ as $997x$. Think of $3 \cdot 42$ as $3x$.

$= 1000x$ ◄ Now replace x with 42.

$= 1000 \cdot 42 = \mathbf{42{,}000}$

d. $17 \cdot 54 - 7 \cdot 54 = 17y - 7y$ ◄ Think of $17 \cdot 54$ as $17y$. Think of $7 \cdot 54$ as $7y$.

$= 10y$ ◄ Replace y with 54.

$= 10 \cdot 54 = \mathbf{540}$

EXERCISES

Use number properties to evaluate each expression.

1. $(64 + 78) + (36 + 12)$
2. $(450 + 19) + (50 + 1)$
3. $(50 \cdot 42) \cdot (2 \cdot 5)$
4. $(250 \cdot 371) \cdot (4 \cdot 2)$
5. $51 \cdot 371 + 49 \cdot 371$
6. $822 \cdot 12 + 178 \cdot 12$
7. $109 \cdot 9 - 9 \cdot 91$
8. $1364 \cdot 9 - 364 \cdot 9$
9. $(10 \cdot 4 + 5) + (90 \cdot 4 + 95)$
10. $(37 \cdot 12 + 32) + (63 \cdot 12 + 54)$
11. $88 \cdot 161 + 11 \cdot 161 + 161$
12. $960 \cdot 19 + 39 \cdot 19 + 19$

ADDITIONAL PRACTICE

SKILLS *Find the value of the variable that makes each sentence true.* (Pages 64–66)

1. $72 + x = 72$
2. $0 = 4v$
3. $7.1x = 7.1$
4. $43 = y\left(\dfrac{12}{12}\right)$
5. $\left(\dfrac{6}{x}\right)134 = 134$

(Pages 67–69)

6. $17x = 33(17)$
7. $(5 \cdot 4)y = 5 \cdot (4 \cdot 13)$
8. $3(t \cdot 6) = (3 \cdot 6) \cdot 12$

Multiply. (Pages 67–69)

9. $4(9x)$
10. $(3xy)(2xy)$
11. $(10p)(2p^2)$
12. $(4.3a)(2.1c)$
13. $(4x^2y^2)(5xy)$

Find the area of each rectangle or square. (Pages 70–73)

14. Rectangle: $l = 13$ cm; $w = 8$ cm
15. Square: $s = 9.2$ cm
16. Square: $s = 27$ m
17. Rectangle: $l = 12.4$ m; $w = 11.1$ m

Combine like terms. (Pages 76–78)

18. $6x + 8x$
19. $4cd - cd$
20. $13.2x^2y - 4.7x^2y$
21. $16.8x^2y^2 + 19.9x^2y^2$

(Pages 79–80)

22. $10c + 12 + c$
23. $4xy + 3 + xy$
24. $12ab^2 + 4ab^2 + 6$
25. $9x + 4 + 3.8x$
26. $4.1xy + 11.9xy + 2$
27. $9.6m^2n + 13 + 8.7m^2n$

Find the perimeter of each figure. (Pages 81–85)

28. Rectangle: $l = 15$ in; $w = 6$ in
29. Square: $s = 52$ yd
30. Triangle: $a = 13.5$ m; $b = 13.5$ m; $c = 13.5$ m
31. Triangle: $a = 17$ cm; $b = 25$ cm; $c = 30$ cm

APPLICATIONS

32. A desk top has a length of 1.5 meters and a width of 0.75 meter. Find the perimeter and the area of the desk top. (Pages 70–73; 81–85)

33. Each side of a triangular lot is 17 yards long. Each side of a square lot is 13 yards long. Which lot has the greater perimeter? How much greater? (Pages 81–85)

34. A boat race is run over a triangular course. The first leg is 82 kilometers long, the second leg is 27 kilometers long, and the third leg is 91 kilometers long. Find the perimeter of the course. (Pages 81–85)

35. A rectangular room is 5.3 yards long and 4.4 yards wide. Find the number of square yards of carpet needed to carpet the room. (Pages 70–73)

COMMON ERRORS

Each of these problems contains a common error.

a. Find the correct answer.

b. Find the error.

Evaluate each expression.

1. $3^2 = 3 \cdot 2$
   $= 6$

2. $6 + 4 \cdot (5 - 3) = 6 + 4 \cdot 2$
   $= 10 \cdot 2 = 20$

3. $3 + 6 \cdot 2 - 4 = 9 \cdot 2 - 4$
   $= 18 - 4$
   $= 14$

4. $xy + 5$ when $x = 6$ and $y = 4$
   $xy + 5 = 6(4) + 5$
   $= 64 + 5 = 69$

Complete.

5. $3.24 + 1.5 + 16.65 = \underline{\ ?\ }$

   3.24
   1.5
   $\underline{+16.65}$
   20.04

6. $2.5(0.45) = \underline{\ ?\ }$

   2.5
   $\underline{\times 0.45}$
   125
   $\underline{100}$
   11.25

7. $4 \text{ m } \underline{\ ?\ } \text{ mm}$
   $4 \text{ m} = (4 \div 1000) \text{ mm}$
   $= 0.004 \text{ mm}$

8. $(4a)(7a) = \underline{\ ?\ }$
   $(4a)(7a) = 4 \cdot 7 \cdot a$
   $= 28a$

Combine like terms.

9. $2b + 7b = \underline{\ ?\ }$
   $2 + 7 = 9$

   Thus, $2b + 7b = 9b^2$.

10. $12y + 6 + 3y = \underline{\ ?\ }$
    $12 + 6 + 3 = 21$

    Thus, $12y + 6 + 3y = 21y$.

11. Find the perimeter of this rectangle.
    l: 10 cm; w: 6 cm

    $P = l + w$
    $P = 10 + 6$
    $P = 16 \text{ cm}$

12. Find the area of a square with a side 8 feet in length.

    $A = s^2$
    $A = 8^2$
    $A = 16 \text{ feet}$

Chapter 3

CUMULATIVE REVIEW: CHAPTERS 1-3

Choose the correct answer.
Choose a, b, c, or d.

NEW CAR	
BASE PRICE	$9475
OPTIONAL EQUIPMENT	$ 655
DESTINATION CHARGE	$ 160

1. Find the sticker price of the car.
   a. $10,300
   b. $10,280
   c. $10,290
   d. $10,190

2. The reading on a meter on June 10 is 8146. The reading on May 10 was 7359. How many kilowatt-hours of electricity were used?
   a. 887
   b. 797
   c. 793
   d. 787

3. An airplane is flying at a rate of 358 kilometers per hour. How many kilometers will the plane fly in 12 hours?
   a. 4286
   b. 4196
   c. 4296
   d. 29.8

4. The formula $d = 16t^2$ gives the distance in feet that an object will fall in t seconds. How many feet will the object fall in 7 seconds?
   a. 784
   b. 224
   c. 12,544
   d. 512

5. Erica's science test scores are 68, 74, 82, 96, 62, and 74. Find the average score.
   a. 76
   b. 78
   c. 75
   d. 74

6. Choose the best estimate: 179×299
   a. 51,000
   b. 54,000
   c. 49,300
   d. 53,280

7. Evaluate: $6 + 4 \times 17$
   a. 170
   b. 74
   c. 160
   d. 64

8. Evaluate: $(20 + 60) \div 5 + 5$
   a. 37
   b. 26
   c. 8
   d. 21

9. Evaluate $13x - 11y$ when $x = 14$ and $y = 12$.
   a. 40
   b. 50
   c. 30
   d. 60

10. Evaluate $(18 + 72) \div p$ when $p = 3$.
    a. 32
    b. 42
    c. 30
    d. 40

11. Which decimal has the greatest value?
    a. 0.511
    b. 0.509
    c. 0.4978
    d. 0.5059

12. Subtract: $96.05 - 3.48$
    a. 92.57
    b. 92.67
    c. 93.57
    d. 93.53

13. Evaluate $x + y + z$ when $x = 7.8$, $y = 14.7$, and $z = 6.9$.
    a. 28.4   b. 29.4   c. 2.4   d. 19.4

14. Multiply: 2.84×0.008
    a. 2.272   b. 0.02272   c. 22.72   d. 0.2272

15. Divide: $.81 \overline{) 51.03}$
    a. 630   b. .63   c. 63   d. 6.3

16. Harold's subcompact car travels 518.4 miles on 12 gallons of gas. How many miles does it travel on one gallon?
    a. 432   b. 44.2   c. 6220.8   d. 43.2

17. Choose the best estimate: $99.8 \div 24.9$
    a. 4.3   b. 5   c. 3.9   d. 4

18. A bottle holds 1250 milliliters of water. How many liters is this?
    a. 12.50   b. 125   c. 1.25   d. 0.1250

19. Saturn is 887,000,000 miles from the sun. Express this number in scientific notation.
    a. 88.7×10^7 b. 887×10^6 c. 0.887×10^9 d. 8.87×10^8

20. Multiply: $(4x)(3x)$
    a. $7x^2$ b. $7x$ c. $12x$ d. $12x^2$

21. Multiply: $(3x)(4xy)(3x^2)$
    a. $108x^4y$ b. $36x^3y$ c. $36x^4y$ d. $108x^3y$

22. The length of each side of a square parking lot is 72 meters. Find the area of the parking lot.
    a. 288 m²   b. 5084 m²   c. 358 m²   d. 5184 m²

23. Combine like terms: $4.1xy - 2.5xy$
    a. $2.6xy$ b. $1.6x^2y^2$ c. $2.6x^2y^2$ d. $1.6xy$

24. Combine like terms: $1.4pt + 8.1 + 6.7pt + 9.2$
    a. $8.1p^2t^2 + 17.3$ b. $7.1p^2t^2 + 17.3$ c. $8.1pt + 17.3$ d. $7.1pt / 17.3$

25. Write the value of the variable that makes this sentence true.
    $$(4.6 + 9.4) + 13 = 13 + (9.4 + x)$$
    a. 0   b. 13   c. 9.4   d. 4.6

26. The cover of a book is 10 inches long and 8 inches wide. Find the perimeter of the cover.
    a. 80 in   b. 36 in   c. 800 in   d. 26 in

94 Cumulative Review: Chapters 1–3

CHAPTER 4: Number Theory

SECTIONS
- 4-1 Factors, Multiples, and Divisibility
- 4-2 Divisibility Tests
- 4-3 Prime Numbers
- 4-4 Prime Factorization
- 4-5 Least Common Multiple
- 4-6 Greatest Common Factor

FEATURES
Calculator Application: *Prime Numbers and Exponents*

Calculator Application: *Prime Numbers*

Consumer Application: *Painting and Estimation*

Enrichment: *Factor Trees*

4-1 Factors, Multiples, and Divisibility

> A number **is divisible** by another number if the remainder is 0.

$$4\overline{\smash{)}24} \quad \underline{24} \quad 0 \quad = 6$$

◀ 24 is divisible by 4.

$$6\overline{\smash{)}32} \quad \underline{30} \quad 2 \quad = 5$$

◀ 32 is not divisible by 6.

Since 24 is divisible by 4, 4 is a **factor** of 24. Here are <u>all</u> the factors of 24.

$$1, 2, 3, 4, 6, 8, 12, 24$$ ◀ Factors of 24

NOTE: The only factors that we consider are counting numbers.

Counting numbers: 1, 2, 3, 4, 5, . . .   ◀ These are also called <u>natural</u> numbers.

The number 24 is a **multiple** of each of its factors.

$$1 \cdot 24 = 24 \qquad 2 \cdot 12 = 24 \qquad 3 \cdot 8 = 24 \qquad 4 \cdot 6 = 24$$

The table below shows several related statements using the words <u>is divisible by</u>, <u>is a factor of</u>, and <u>is a multiple of</u>.

TABLE: Related Statements

Statement 1	Statement 2	Statement 3
24 is divisible by 8.	8 is a factor of 24.	24 is a multiple of 8.
9 is divisible by 3.	3 is a factor of 9.	9 is a multiple of 3.
7 is divisible by 1.	1 is a factor of 7.	7 is a multiple of 1.

EXAMPLE 1 Write <u>True</u> or <u>False</u> for each statement. Give a reason for each answer.

 a. 50 <u>is a multiple of</u> 5.
 b. 28 <u>is divisible by</u> 9.
 c. 9 <u>is a factor of</u> 27.
 d. 17 <u>is a multiple of</u> 17.

Solutions:
 a. True, since $5 \cdot 10 = 50$.
 b. False, since $28 \div 9 = 3$ R 1.
 c. True, since $27 \div 9 = 3$.
 d. True, since $1 \cdot 17 = 17$.

Self-Check   Write <u>True</u> or <u>False</u> for each statement. Give the reason.

1. 36 is a multiple of 5.   **2.** 4 is a factor of 28.

> **PROCEDURE**
>
> **To determine the factors of a counting number, n:**
>
> **1** Starting with $1 \cdot n$, list all the ways of writing n as the product of two numbers. Stop when the factors begin to repeat.
>
> **2** Use the list to write the factors of n. Start with the least.

EXAMPLE 2 List all the factors of 30.

Solution: **1** List all the pairs of numbers whose product is 30.

$1 \cdot 30 \qquad 2 \cdot 15 \qquad 3 \cdot 10 \qquad 5 \cdot 6$ ◀ **Stop!** $6 \cdot 5$ has the same factors as $5 \cdot 6$.

2 Write the factors in order.
Start with the least. Factors of 30: **1, 2, 3, 5, 6, 10, 15, 30**

Self-Check List all the factors of each number.

3. 12   **4.** 48   **5.** 3   **6.** 11   **7.** 76   **8.** 105

CLASSROOM EXERCISES

Write a true statement about each pair of numbers. Use the words is divisible by. (Definition)

1. 40 and 20   **2.** 4 and 4   **3.** 2 and 6   **4.** 6 and 12
5. 5 and 40   **6.** 35 and 5   **7.** 7 and 7   **8.** 7 and 56

Write a true statement about each pair of numbers. Use the words is a factor of. (Table)

9. 55 and 11   **10.** 1 and 8   **11.** 8 and 40   **12.** 12 and 24
13. 7 and 1   **14.** 39 and 13   **15.** 33 and 33   **16.** 47 and 47

Write a true statement about each pair of numbers. Use the words is a multiple of. (Table)

17. 24 and 8   **18.** 13 and 1   **19.** 6 and 6   **20.** 16 and 4
21. 1 and 9   **22.** 5 and 30   **23.** 3 and 15   **24.** 11 and 11

Show all the ways of writing each number as a product of two numbers. (Example 2, Step 1)

25. 13   **26.** 9   **27.** 18   **28.** 28   **29.** 30   **30.** 80

Number Theory **97**

WRITTEN EXERCISES

Goal: To identify the factors of a number

Write True or False for each sentence. Write a reason for each answer. (Example 1)

1. 6 is a factor of 18.
2. 6 is a factor of 36.
3. 21 is divisible by 8.
4. 58 is divisible by 17.
5. 29 is a multiple of 29.
6. 28 is a multiple of 7.
7. 75 is a multiple of 5.
8. 132 is divisible by 12.
9. 143 is divisible by 13.

List all the factors of each number. (Example 2)

10. 8
11. 13
12. 15
13. 23
14. 27
15. 31
16. 40
17. 60
18. 24
19. 46
20. 72
21. 49

MIXED PRACTICE

For Exercises 22–26, write Yes or No to answer the questions in a, b, and c. For d in each row, list the factors of n.

n	Is n Divisible by This Number?	Is This Number a Factor of n?	Is n a Multiple of This Number?	List of Factors of n
22. 72	a. 24 ?	b. 18 ?	c. 48 ?	d. ?
23. 15	a. 30 ?	b. 30 ?	c. 30 ?	d. ?
24. 77	a. 27 ?	b. 11 ?	c. 7 ?	d. ?
25. 56	a. 8 ?	b. 7 ?	c. 4 ?	d. ?
26. 60	a. 12 ?	b. 9 ?	c. 15 ?	d. ?

REVIEW CAPSULE FOR SECTION 4–2

For each number:
a. *Write the remainder when the number is divided by 3.*
b. *Write the remainder when the number is divided by 9.*

1. 17
2. 12
3. 18
4. 11
5. 31
6. 44

For each number:
a. *Write the remainder when the number is divided by 4.*
b. *Write the remainder when the number is divided by 5.*

7. 21
8. 18
9. 39
10. 108
11. 226
12. 198

The answers are on page 120.

4-2 Divisibility Tests

For a family reunion dinner, the Gills would like the same number of persons seated at each table and all the tables completely filled. For the 157 expected guests, the caterer used the test for divisibility by 9 to determine whether each table could seat exactly nine persons.

The tests for divisibility involve the **digits** 0, 1, 2, 3, 4, 5, 6, 7, 8, and 9 and the set of <u>whole numbers</u>.

Whole numbers: 0, 1, 2, 3, 4, 5, . . .

Whole Numbers:
0 + Counting Numbers

TESTS FOR DIVISIBILITY

A whole number is divisible by:
2 if it ends in 0, 2, 4, 6, or 8.
3 if the sum of its digits is divisible by 3.
4 if its last two digits are divisible by 4.
5 if it ends in 0 or 5.
9 if the sum of its digits is divisible by 9.
10 if it ends in 0.

Any whole number that does <u>not</u> meet one of these tests is not divisible by the number shown. Recall that the following statements are related (see page 96).

38 <u>is divisible by</u> 2.   38 <u>is a multiple of</u> 2.   2 <u>is a factor of</u> 38.

EXAMPLE 1 Write <u>True</u> or <u>False</u> for each statement. Give the reason.

Solutions:

	Statement	True or False? Why?
a.	456 is a multiple of 2.	True. 456 ends in 6.
b.	522 is divisible by 4.	False. 22 is not divisible by 4.
c.	5 is a factor of 723.	False. 723 does not end in 0 or 5.
d.	10 is a factor of 520.	True. 520 ends in 0.

Number Theory   99

Self-Check *Which of these numbers are divisible by 2? by 4? by 5? by 10?*

1. 48   **2.** 412   **3.** 260   **4.** 1005   **5.** 1340

> A number that is divisible by 2 is an **even number**.
> A number that is <u>not</u> divisible by 2 is an **odd number**.

EXAMPLE 2 Write <u>True</u> or <u>False</u> for each statement. Give the reason.

Solutions:

	Statements	Digit Sum	True or False? Why?
a.	113 is divisible by 3.	$1+1+3=5$	**False.** 5 is not divisible by 3.
b.	3 is a factor of 546.	$5+4+6=15$	**True.** 15 is divisible by 3.
c.	468 is divisible by 9.	$4+6+8=18$	**True.** 18 is divisible by 9.

Self-Check *Which of these numbers are divisible by 3? by 9?*

6. 58,014   **7.** 768   **8.** 2346   **9.** 40,693

The tests for divisibility can be used to solve problems.

EXAMPLE 3 At a reunion of the Gill Family, 157 persons were present.

a. Can the guests be seated so that there are <u>exactly</u> 9 persons at <u>each</u> table?

b. If not, how many more persons are needed to exactly fill the tables?

Solutions:

a. Determine whether 157 is divisible by 9. $1+5+7=13$ ◀ **13 is not divisible by 9.**

Since 13 is not divisible by 9, 157 is not divisible by 9.
Therefore, 157 persons **cannot be seated** with exactly 9 persons at each table.

b. Find the nearest whole number greater than 13 that is divisible by 9. $13+5=18$ ◀ **18 is divisible by 9.**

Thus, **5 more persons** would be needed to fill the tables with exactly 9 persons per table.

Self-Check *Audrey has 148 books. She plans to build shelves holding 10 books each.*

10. Will the 148 books completely fill the shelves?

11. If not, how many more books will be needed to do this?

100   Chapter 4

CLASSROOM EXERCISES

For Exercises 1-18, write True or False for each statement. Write a reason for each answer. (Example 1)

1. 71 is divisible by 2.
2. 375 is divisible by 5.
3. 116 is a multiple of 4.
4. 622 is a multiple of 4.
5. 385 is divisible by 10.
6. 321 is divisible by 2.
7. 5 is a factor of 4130.
8. 10 is a factor of 4130.

(Example 2)

9. 126 is divisible by 3.
10. 3 is a factor of 372.
11. 72,631 is a multiple of 9.
12. 576 is divisible by 9.
13. 648 is a multiple of 9.
14. 9 is a factor of 71,173.
15. 9 is a factor of 3429.
16. 1403 is divisible by 3.
17. 2463 is a multiple of 3.
18. 3 is a factor of 81,273.
19. 9 is a factor of 9.
20. 3 is a factor of 81.

WRITTEN EXERCISES

Goals: To test a whole number for divisibility by 2, 3, 4, 5, 9, or 10
To apply these tests to solving word problems

For Exercises 1-32, write True or False for each statement. Write a reason for each answer. (Example 1)

1. 428 is divisible by 2.
2. 57 is divisible by 5.
3. 95 is divisible by 5.
4. 936 is a multiple of 4.
5. 717 is a multiple of 4.
6. 9020 is divisible by 10.
7. 428 is divisible by 4.
8. 2 is a factor of 313.
9. 10 is a factor of 926.
10. 7032 is divisible by 2.

(Example 2)

11. 825 is divisible by 3.
12. 657 is divisible by 3.
13. 3 is a factor of 4321.
14. 3 is a factor of 15,423.
15. 90,426 is a multiple of 3.
16. 72,646 is a multiple of 3.
17. 74,613 is divisible by 9.
18. 35,226 is divisible by 9.
19. 671,346 is a multiple of 9.
20. 1,301,206 is a multiple of 9.

Number Theory

MIXED PRACTICE

21. 52,374 is divisible by 10.
22. 648 is a multiple of 9.
23. 62,235 is a multiple of 3.
24. 5 is a factor of 323.
25. 4 is a factor of 7044.
26. 324 is divisible by 2.
27. 67,237 is divisible by 2.
28. 7,324,675 is a multiple of 3.
29. 5 is a factor of 675.
30. 7150 is divisible by 10.
31. 7630 is a multiple of 9.
32. 4 is a factor of 655.

APPLICATIONS: USING THE RULES FOR DIVISIBILITY

Solve. (Example 3)

33. Sheila built shelves for her collection of 982 books. Each shelf will hold 10 books.
    a. Will the 982 books completely fill the shelves?
    b. If not, how many more books are needed to do this?

34. Mr. Garcia wishes to give $70,622 in equal amounts to four different charities.
    a. Can $70,622 be divided into four equal whole-dollar amounts?
    b. If not, how many dollars must be added so that this can be done?

35. A farmer has 3788 kilograms of potatoes. He plans to sell the potatoes in bags that hold 5 kilograms each.
    a. Will all the bags be completely filled?
    b. If not, how many additional kilograms are needed to completely fill the bags?

36. To make money, a bicycle club with 9 members plans to sell 326 boxes of cookies.
    a. Can each member sell the same number of boxes?
    b. If not, how many more boxes are needed for this to be possible?

REVIEW CAPSULE FOR SECTION 4-3

Complete. (Pages 7-9)

1. In the numerical expression 3^2, 3 is called the base and the raised 2 is called the __?__.

2. In Exercise 1, the raised 2 in 3^2 indicates how many times to multiply 3 by __?__.

Evaluate. (Pages 7-9)

3. 3^2 4. 2^3 5. 4^2 6. 2^4 7. 1^2 8. 3^3

The answers are on page 120.

102 Chapter 4

4-3 Prime Numbers

Each number listed below is a *prime number*.

2, 3, 5, 7, 11, 13, 17, 19, 23

> A **prime number** is a counting number greater than 1 that has exactly two counting-number factors, the number itself and 1.
>
> $3 = 3 \cdot 1$
> $59 = 59 \cdot 1$

Numbers that are <u>not</u> prime are *composite numbers*.

> A **composite number** is a counting number greater than 1 that is not a prime number.
>
> $32 = 2 \cdot 16$
> $75 = 3 \cdot 5 \cdot 5$

The numbers 0 and 1 are <u>neither prime nor composite</u>.

PROCEDURE

To determine the smallest prime-number factor of a given number:

1. Starting with 2, write the first several prime numbers in order.
2. Test whether each prime is a factor of the given number. If none of these primes is a factor, the smallest prime factor is the number itself.

EXAMPLE 1 What is the smallest prime number that is a factor of 119?

Solution:

1. Write several prime numbers in order.   2, 3, 5, 7, 11, 13, 17, 19, 23, . . .

2. **Think**

Is 119 divisible by 2?	No. 119 is an odd number.
Is 119 divisible by 3?	No. $1 + 1 + 9 = 11$. Since 11 is not divisible by 3, 119 is not divisible by 3.
Is 119 divisible by 5?	No. 119 does not end in 0 or 5.
Is 119 divisible by 7?	$\begin{array}{r} 17 \\ 7\overline{)119} \\ \underline{7} \\ 49 \\ \underline{49} \end{array}$ ▶ Since the remainder is 0, 119 is divisible by 7.

Thus, the smallest prime factor of 119 is **7**.

Number Theory

Self-Check *What is the smallest prime factor of each number?*

1. 51 **2.** 550 **3.** 62 **4.** 365 **5.** 31

Exponents are often useful in expressing a product of prime numbers.

> **PROCEDURE**
>
> **To use exponents to express a product of prime numbers:**
>
> 1 Write each prime number once as a factor.
> Write the factors in order from least to greatest.
>
> 2 For each prime number, write the exponent that indicates how many times it is a factor.

EXAMPLE 2 Use exponents to rewrite this expression: $3 \cdot 7 \cdot 2 \cdot 5 \cdot 2 \cdot 2 \cdot 13 \cdot 5 \cdot 3$

Solution:

1 Starting with the smallest, list each prime factor once. $2 \cdot 3 \cdot 5 \cdot 7 \cdot 13$

2 Use exponents to indicate how many times each prime number is a factor. $2^3 \cdot 3^2 \cdot 5^2 \cdot 7 \cdot 13$

Self-Check *Use exponents to rewrite each expression.*

6. $5 \cdot 5 \cdot 3 \cdot 7 \cdot 3 \cdot 7 \cdot 11$ **7.** $3 \cdot 5 \cdot 3 \cdot 5 \cdot 2 \cdot 19 \cdot 5$

NOTE: $2^3 \cdot 3^2 \cdot 5^2 \cdot 7 \cdot 13$ is read: "two cubed times three squared times five squared times seven times 13."

CLASSROOM EXERCISES

Identify each number as prime or composite. (Definitions)

1. 17 **2.** 55 **3.** 12 **4.** 41 **5.** 59 **6.** 29

Write the smallest prime factor of each number. (Example 1)

7. 10 **8.** 121 **9.** 9 **10.** 80 **11.** 77 **12.** 162

Use exponents to write each product. (Example 2)

13. $2 \cdot 5 \cdot 3 \cdot 5$
14. $3 \cdot 7 \cdot 2 \cdot 2 \cdot 3 \cdot 7 \cdot 7$
15. $2 \cdot 11 \cdot 3 \cdot 2 \cdot 2 \cdot 3 \cdot 7 \cdot 11$
16. $7 \cdot 5 \cdot 5 \cdot 5 \cdot 3 \cdot 7$
17. $2 \cdot 11 \cdot 11 \cdot 2 \cdot 2 \cdot 2$
18. $17 \cdot 17 \cdot 23 \cdot 2 \cdot 3 \cdot 23 \cdot 23$
19. $3 \cdot 3 \cdot 3 \cdot 3 \cdot 7 \cdot 4 \cdot 7 \cdot 7 \cdot 4$
20. $11 \cdot 11 \cdot 11 \cdot 3 \cdot 5 \cdot 5 \cdot 5 \cdot 11$

Chapter 4

WRITTEN EXERCISES

Goal: To rewrite a product of prime factors with exponents

Find the smallest prime factor of each number. (Example 1)

1. 6	2. 8	3. 15	4. 21	5. 25	6. 91
7. 175	8. 696	9. 187	10. 343	11. 169	12. 84
13. 29	14. 23	15. 219	16. 1573	17. 1859	18. 2187

Use exponents to rewrite each product. (Example 2)

19. $2 \cdot 2 \cdot 3 \cdot 3 \cdot 2$
20. $3 \cdot 3 \cdot 5 \cdot 5 \cdot 3 \cdot 7 \cdot 7$
21. $2 \cdot 2 \cdot 3 \cdot 3 \cdot 2 \cdot 5 \cdot 11 \cdot 11$
22. $7 \cdot 7 \cdot 11 \cdot 2 \cdot 2 \cdot 7 \cdot 7$
23. $11 \cdot 11 \cdot 13 \cdot 13 \cdot 2 \cdot 2 \cdot 11$
24. $17 \cdot 11 \cdot 17 \cdot 13 \cdot 17 \cdot 2 \cdot 2 \cdot 2$
25. $2 \cdot 3 \cdot 3 \cdot 2 \cdot 2 \cdot 3 \cdot 5 \cdot 7 \cdot 5$
26. $3 \cdot 3 \cdot 5 \cdot 3 \cdot 5 \cdot 7 \cdot 7 \cdot 3 \cdot 5$
27. $2 \cdot 2 \cdot 3 \cdot 2 \cdot 2 \cdot 2 \cdot 3 \cdot 2 \cdot 3$
28. $2 \cdot 3 \cdot 2 \cdot 2 \cdot 2 \cdot 3 \cdot 2 \cdot 3 \cdot 7$

REVIEW: SECTIONS 4-1 – 4-3

List all the factors of each number. (Section 4-1)

| 1. 33 | 2. 10 | 3. 40 | 4. 26 | 5. 15 | 6. 24 |
| 7. 20 | 8. 12 | 9. 7 | 10. 13 | 11. 30 | 12. 32 |

Write True or False for each statement. Give a reason for each answer. (Section 4-2)

13. 3 is a factor of 31,206.
14. 72,429 is divisible by 3.
15. 31,867 is divisible by 9.
16. 4 is a multiple of 8024.
17. 644 is a factor of 4.
18. 2 is a factor of 1379.
19. 5 is a factor of 2160.
20. 13,370 is a multiple of 10.

Find the smallest prime factor of each number. (Section 4-3)

21. 6012
22. 37,055
23. 1263
24. 539

Use exponents to rewrite each product. (Section 4-3)

25. $3 \cdot 3 \cdot 3 \cdot 5 \cdot 5$
26. $2 \cdot 2 \cdot 2 \cdot 3 \cdot 7 \cdot 7$
27. $2 \cdot 2 \cdot 3 \cdot 3 \cdot 11 \cdot 11 \cdot 11$
28. $2 \cdot 3 \cdot 3 \cdot 3 \cdot 5 \cdot 5 \cdot 5 \cdot 7 \cdot 7$
29. $3 \cdot 3 \cdot 3 \cdot 5 \cdot 5 \cdot 5$
30. $2 \cdot 2 \cdot 3 \cdot 3 \cdot 4 \cdot 4 \cdot 5 \cdot 5$

Number Theory

Solve. (Section 4–3)

31. A grocer has 1347 apples to pack in boxes. Each box holds 10 apples.
  a. Will all the boxes be completely filled?
  b. If not, how many more apples are needed to fill the boxes?

32. Maria wishes to spend $71 to buy gifts for three friends.
  a. Can the $71 be divided into three equal whole-dollar amounts?
  b. If not, how many more dollars are needed in order to do this?

PRIME NUMBERS AND EXPONENTS

Some prime numbers can be found by using this formula.

$$P = 2^n - 1$$

EXAMPLE: Use the formula above to find P when $n = 5$.

Solution: Use a calculator. On some calculators, you can keep pressing the $\boxed{\times}$ key to evaluate powers.

2 [×] [=] [=] [=] [=] [−] 1 [=]   **31.**

EXERCISES Use the formula $P = 2^n - 1$ to find P for each value of n.

1. 3   **2.** 7   **3.** 10   **4.** 11   **5.** 15   **6.** 17

REVIEW CAPSULE FOR SECTION 4–4

Which of these numbers are divisible by 2? (Pages 99–102)

1. 105   **2.** 328   **3.** 400   **4.** 63   **5.** 225   **6.** 630

Which of these numbers are divisible by 3? (Pages 99–102)

7. 603   **8.** 128   **9.** 516   **10.** 71   **11.** 543   **12.** 89

Which of these numbers are divisible by 5? (Pages 99–102)

13. 125   **14.** 374   **15.** 600   **16.** 95   **17.** 101   **18.** 558

Find each quotient.

19. $7\overline{)161}$ **20.** $3\overline{)102}$ **21.** $5\overline{)1015}$ **22.** $17\overline{)323}$ **23.** $13\overline{)221}$

The answers are on page 120.

CONSUMER APPLICATION
Painting and Estimation

To estimate the amount of paint needed to paint a room, you must first answer these questions.

What is the combined area of the walls and ceiling?

How much area must be subtracted for the doors and windows?

These are the **hidden questions** in the problem.

EXAMPLE A rectangular-shaped room is 4.6 meters long and 3.4 meters wide. It is 2.6 meters high. The combined area of windows and doors is 7.1 square meters. One liter of paint will cover 9 square meters. How many liters of paint are needed to paint the room?

SOLUTION

1. Find the area of the walls. Multiply the perimeter of the room by the height of the room.
   Perimeter: $2(4.6) + 2(3.4) = 16$ m
   Area of walls: $16 \times 2.6 = 41.6$ m^2
2. Area of ceiling: $4.6 \times 3.4 = 15.64$ m^2
3. Combined area of walls and ceiling: $41.6 + 15.64 = 57.24$ m^2
4. Subtract the area of doors and windows: $57.24 - 7.1 = 50.14$ m^2
5. Divide the area to be painted by the area covered by one liter of paint.
   $50.14 \div 9 = 5.57\overline{1}$ or **6 liters**.   ◀ **Round up to the next whole liter.**

Exercises

Find the number of liters of paint needed to paint each room. Assume that one liter of paint covers 9 square meters.

Room	Length	Width	Height	Combined Area Windows–Doors
1. Kitchen	4.1 m	3.7 m	2.5 m	7.1 m^2
2. Bedroom	5.4 m	4.8 m	2.7 m	10.8 m^2
3. Den	3.2 m	3.6 m	2.7 m	4.5 m^2
4. Family Room	6.7 m	4.2 m	2.8 m	5.3 m^2

4-4 Prime Factorization

Every composite number can be expressed as a product of prime factors.

$$75 = 3 \cdot 25 \quad \longleftarrow \text{ 25 is a composite number.}$$
$$= 3 \cdot 5 \cdot 5 \quad \longleftarrow \text{ Each factor is a prime number.}$$

The product, $3 \cdot 5 \cdot 5$, is called the **prime factorization** of 75. In a prime factorization, the factors are usually written in order from least to greatest.

The prime numbers less than 24 are listed below. Refer to them in the Examples and Exercises.

$$2, 3, 5, 7, 11, 13, 17, 19, 23$$

EXAMPLE 1 Write the prime factorization of 105.

Solution:

[1] Is 105 divisible by 2? No; 105 is an odd number.

[2] Is 105 divisible by 3? Yes, by the rule for divisibility by 3. So divide.

$$3 \underline{|105}$$
$$35$$

← Not a prime number   The quotient is **35**.

[3] Is 35 divisible by 3? No

[4] Is 35 divisible by 5? Yes, by the rule for divisibility by 5. So divide.

$$3 \underline{|105}$$
$$5 \underline{|35}$$
$$7$$

← Prime number   The quotient is **7**.

Since 7 is a prime number, all the prime factors of 105 have been determined.

[5] Write the prime factorization of 105.

$$105 = 3 \cdot 5 \cdot 7$$

Self-Check *Write the prime factorization of each number.*

1. 77   **2.** 33   **3.** 30   **4.** 210

When the same prime number appears more than once in a prime factorization, exponents are often used to indicate the number of times each prime factor occurs.

EXAMPLE 2 Write the prime factorization of 126.

Solution:

1. Is 126 divisible by 2? Yes. Divide. The quotient is 63.

   2 ⌊126
       63 ◀ Not a prime number

2. Is 63 divisible by 2? No.

3. Is 63 divisible by 3? Yes. Divide. The quotient is 21.

   2 ⌊126
   3 ⌊63
       21 ◀ Not a prime number

4. Is 21 divisible by 3? Yes. Divide. The quotient is 7. Since 7 is a prime number, all the prime factors of 126 have been found.

   2 ⌊126
   3 ⌊63
   3 ⌊21
       7 ◀ Prime number

5. Write the prime factorization: $126 = 2 \cdot 3 \cdot 3 \cdot 7$, or $2 \cdot 3^2 \cdot 7$.

Self-Check *Write the prime factorization of each number.*

5. 20   **6.** 18   **7.** 100   **8.** 242

CLASSROOM EXERCISES

Write the prime factorization of each number.
(Step 5, Examples 1 and 2)

1. 455
5 ⌊455
7 ⌊91
   13

2. 261
3 ⌊261
3 ⌊87
   29

3. 150
2 ⌊150
3 ⌊75
5 ⌊25
   5

4. 204
2 ⌊204
2 ⌊102
3 ⌊51
   17

5. 492
2 ⌊492
2 ⌊246
3 ⌊123
   41

6. 315
3 ⌊315
3 ⌊105
5 ⌊35
   7

7. 198
? ⌊198
? ⌊99
? ⌊33
   11

8. 220
? ⌊220
? ⌊110
? ⌊55
   11

9. 261
? ⌊261
? ⌊87
   29

10. 396
? ⌊396
? ⌊198
? ⌊99
? ⌊33
   11

11. 450
? ⌊450
? ⌊225
? ⌊75
? ⌊25
   5

12. 312
? ⌊312
? ⌊156
? ⌊78
? ⌊39
   13

(Example 1)

13. 35   **14.** 70   **15.** 15   **16.** 165   **17.** 130   **18.** 182

(Example 2)

19. 36   **20.** 135   **21.** 140   **22.** 525   **23.** 594   **24.** 375

WRITTEN EXERCISES

Goal: To write the prime factorization of a composite number

Write the prime factorization of each number. (Example 1)

| 1. 6 | 2. 55 | 3. 34 | 4. 39 | 5. 55 | 6. 85 |
| 7. 110 | 8. 143 | 9. 85 | 10. 38 | 11. 91 | 12. 231 |

(Example 2)

13. 56	14. 54	15. 50	16. 48	17. 78	18. 105
19. 84	20. 60	21. 128	22. 168	23. 64	24. 125
25. 378	26. 364	27. 324	28. 288	29. 540	30. 420

MIXED PRACTICE

31. 180	32. 399	33. 119	34. 115	35. 468	36. 133
37. 187	38. 504	39. 512	40. 729	41. 484	42. 418
43. 1768	44. 1071	45. 2431	46. 221	47. 323	48. 1547
49. 3185	50. 4998	51. 437	52. 391	53. 1150	54. 1425
55. 4655	56. 546	57. 42,588	58. 250,965	59. 12,675	60. 17,850

REVIEW CAPSULE FOR SECTION 4-5

For Exercises 1–5, list the first six multiples of each number. (Pages 96–98)

1. 2
2. 3
3. 4
4. 5
5. 6

6. List any numbers that are in the answers to <u>both</u> Exercises 1 and 2.
7. List any numbers that are in the answers to <u>both</u> Exercises 3 and 5.
8. List any numbers that are in the answers to <u>both</u> Exercises 3 and 6.

Use exponents to rewrite each expression. (Pages 103–105)

9. $2 \cdot 2 \cdot 5 \cdot 3 \cdot 3 \cdot 3$
10. $2 \cdot 2 \cdot 5 \cdot 5 \cdot 2 \cdot 5$
11. $2 \cdot 2 \cdot 2 \cdot 3 \cdot 3 \cdot 5 \cdot 5 \cdot 2 \cdot 3$
12. $5 \cdot 11 \cdot 11 \cdot 5 \cdot 5 \cdot 11 \cdot 17 \cdot 5$
13. $2 \cdot 3 \cdot 4 \cdot 3 \cdot 2 \cdot 2 \cdot 3 \cdot 4$
14. $6 \cdot 9 \cdot 9 \cdot 6 \cdot 6 \cdot 9 \cdot 12 \cdot 12$

The answers are on page 120.

4-5 Least Common Multiple

Brian and his sister Sara have part-time jobs after school. Brian works every third day and Sara works every fourth day. If both start to work on the same day, what is the first working day after this when they *both* will be working at their part-time jobs?

Sara solved this problem by using *common multiples* of 3 and 4.

12 is a **common multiple** of 3 and 4.

24 is a **common multiple** of 3 and 4.

Multiples of 3: 3, 6, 9, **12**, 15, 18, 21, **24**, 27, . . .
Multiples of 4: 4, 8, **12**, 16, 20, **24**, 28, 32, 36, . . .

Answer: Since 12 is the smallest common multiple of 3 and 4, they would both be working again at their part-time jobs on the **twelfth day.**

The number 12 is the *least common multiple* of 3 and 4.

> The **least common multiple (LCM)** of two or more counting numbers is the smallest counting number that is divisible by the given numbers.
>
> LCM of 2 and 3: **6.**

You can use prime factors to find the LCM of two, three, or more counting numbers.

> **PROCEDURE**
> **To find the LCM of two or more counting numbers:**
> **1** Write the prime factorization of each number.
> **2** Write a product using each prime factor only once.
> **3** For each prime factor, write the highest exponent used for that factor in any of the prime factorizations of the numbers.
> **4** Multiply these factors.

Number Theory

EXAMPLE 1 Find the least common multiple (LCM).

    a 4 and 6   **b.** 12 and 18

Solutions:

a.

1. Prime factorization: $4 = 2 \cdot 2 = 2^2$
   Prime factorization: $6 = 2 \cdot 3$
2. Write a product. Use each prime factor once. $2 \cdot 3$
3. For each prime factor, write the highest exponent used in any of the prime factorizations. $2^2 \cdot 3$
4. Multiply. $4 \cdot 3 = 12$ ◀ **LCM of 4 and 6**

b.

1. $12 = 2 \cdot 2 \cdot 3 = 2^2 \cdot 3$
   $18 = 2 \cdot 3 \cdot 3 = 2 \cdot 3^2$
2. $2 \cdot 3$
3. $2^2 \cdot 3^2$
4. $4 \cdot 9 = 36$ ◀ **LCM of 12 and 18**

Self-Check   *Write the LCM of each pair of numbers.*

    1. 4 and 8   **2.** 12 and 14   **3.** 9 and 21

NOTE: The LCM of two numbers is always equal to, or larger than, either of the numbers.

Follow the same procedure to find the LCM of three or more counting numbers.

EXAMPLE 2 Find the least common multiple of 8, 36, and 45.

Solution:
1. Prime factorization: $8 = 2 \cdot 2 \cdot 2 = 2^3$
   Prime factorization: $36 = 2 \cdot 2 \cdot 3 \cdot 3 = 2^2 \cdot 3^2$
   Prime factorization: $45 = 3 \cdot 3 \cdot 5 = 3^2 \cdot 5$
2. Product that includes each prime factor once. $2 \cdot 3 \cdot 5$
3. Product that shows the highest exponent for each factor. $2^3 \cdot 3^2 \cdot 5$
4. Multiply. $2^3 \cdot 3^2 \cdot 5 = 8 \cdot 9 \cdot 5 = 360$

    LCM of 8, 36, and 45: **360**

Self-Check   *Write the LCM of each group of numbers.*

    4. 4, 6, and 12   **5.** 6, 18, and 27   **6.** 15, 20, and 25

CLASSROOM EXERCISES

The prime factorizations of two or three numbers are given. Write the prime factorization of their least common multiple.
(Step 3, Examples 1 and 2)

1. $2 \cdot 3^2$
   $2^2 \cdot 3$
2. $2^2 \cdot 3^2 \cdot 5$
   $2^3 \cdot 3$
3. $2^2 \cdot 3^3 \cdot 5^2$
   $2^3 \cdot 3^2 \cdot 7$
4. $11 \cdot 17$
   11^2
5. $2^3 \cdot 3^2 \cdot 5^3$
   $2^2 \cdot 3^3 \cdot 5$
6. $2 \cdot 3^2 \cdot 7$
   2^4
7. $2 \cdot 3 \cdot 5$
   $3 \cdot 5$
   $2^3 \cdot 3^2$
8. $2 \cdot 3 \cdot 5^2$
   $2^2 \cdot 3^3$
   5

Find the least common multiple. (Example 1)

9. 4 and 10
10. 3 and 27
11. 3 and 5
12. 2 and 3
13. 7 and 27
14. 6 and 24
15. 27 and 18
16. 4 and 18
17. 3 and 7
18. 2 and 11
19. 15 and 25
20. 6 and 14
21. 120 and 16
22. 135 and 54
23. 240 and 16
24. 216 and 45

(Example 2)

25. 3, 5, and 15
26. 9, 5, and 14
27. 3, 15, and 27
28. 6, 27, and 10
29. 3, 9, and 27
30. 5, 10, and 15
31. 3, 15, and 25
32. 2, 27, and 18
33. 7, 14, and 21
34. 15, 27, and 25
35. 9, 6, and 4
36. 2, 16, and 24

WRITTEN EXERCISES

Goals: To find the least common multiple of two or more numbers

To apply the skill of finding the least common multiple to solving word problems

For Exercises 1-48, write the least common multiple. (Example 1)

1. 15 and 75
2. 8 and 40
3. 6 and 20
4. 45 and 28
5. 3 and 11
6. 5 and 13
7. 28 and 40
8. 36 and 405
9. 12 and 56
10. 45 and 27
11. 5 and 27
12. 45 and 7
13. 49 and 42
14. 18 and 168
15. 42 and 96
16. 120 and 150

(Example 2)

17. 8, 28, and 21
18. 20, 30, and 50
19. 36, 32, and 12
20. 50, 35, and 28
21. 10, 15, and 24
22. 54, 45, and 36
23. 15, 10, and 16
24. 15, 35, and 49
25. 135, 225, and 405
26. 27, 45, and 100
27. 24, 30, and 42
28. 16, 20, and 48

Number Theory

MIXED PRACTICE

29. 5, 25, and 40
30. 16 and 40
31. 6, 36, and 30
32. 16, 6, and 30
33. 42, 100, and 14
34. 6 and 20
35. 54 and 8
36. 24, 28, and 16
37. 11 and 17
38. 9, 27, and 81
39. 11 and 29
40. 8, 16, and 128
41. 30, 24, and 20
42. 9, 20, and 48
43. 30 and 126
44. 108 and 90
45. 48, 12, and 36
46. 72 and 45
47. 21 and 70
48. 18, 72, and 105

APPLICATIONS: USING LEAST COMMON MULTIPLES

49. Jill, Jean, and Alex have part-time jobs after school. Jill works every third day, Jean every fourth day, and Alex every other day. If all three start to work on the same day, on what working day after that will all three again be working at their part-time jobs?

50. In a grocery store, produce is delivered every other business day, milk every third business day, and paper goods every twentieth business day. All three of these items are delivered on Tuesday. How many business days will pass before all three are again delivered on the same day?

51. In a warehouse, cartons 36, 40, and 24 inches tall are being stacked next to each other in three separate piles. What is the shortest height possible for all three piles to have the same height?

52. May earns $16 per hour and Joan earns $18 per hour. On Thursday, both girls worked a whole number of hours and they both earned the same amount. What is the least amount they could have earned?

REVIEW CAPSULE FOR SECTION 4-6

List all the factors of each number. (Pages 96–98)

1. 8
2. 10
3. 15
4. 24
5. 36
6. 42

Write the prime factorization of each number. (Pages 108–110)

7. 6
8. 16
9. 20
10. 32
11. 45
12. 50

The answers are on page 120.

4-6 Greatest Common Factor

Martin made a list of the factors of 12 and 18. He noticed that 12 and 18 have some of the same factors. These are called **common factors.**

Factors of 12: ①, ②, ③, 4, ⑥, 12
Factors of 18: ①, ②, ③, ⑥, 9, 18

Common factors: **1, 2, 3, 6**

The greatest common factor (GCF) is 6.

> The **greatest common factor (GCF)** of two or more counting numbers is the greatest number that is a factor of each number.

The table below shows how to find the GCF by listing the factors of the numbers.

TABLE

Numbers	Factors	Common Factors	GCF
16 24	①, ②, ④, ⑧, 16 ①, ②, 3, ④, 6, ⑧, 12, 24	1, 2, 4, 8	8
9 15	①, ③, 9 ①, ③, 5, 15	1, 3	3
10 20	①, ②, ⑤, ⑩ ①, ②, 4, ⑤, ⑩, 20	1, 2, 5, 10	10

Another way to find the GCF is to write the prime factorization of each number.

PROCEDURE

To determine the greatest common factor (GCF) of two or more numbers:

1 Write the prime factorization of each number. Identify the common factors.

2 Write the common factors as a product.

Number Theory

EXAMPLE Find the GCF of 24 and 36.

Solution:
1. $24 = ②\cdot ②\cdot 2 \cdot ③$
   $36 = ②\cdot ②\cdot ③\cdot 3$
   Common factors: 2, 2, and 3
2. GCF: $2 \cdot 2 \cdot 3 = 12$

Self-Check *Find the GCF of each pair of numbers.*

1. 4 and 10   2. 16 and 28   3. 15 and 48

CLASSROOM EXERCISES

Complete. (Table)

1. Factors of 6: 1, 2, 3, 6
   Factors of 8: 1, 2, 4, 8
   Common factors: 1, 2
   GCF: __?__

2. Factors of 12: 1, 2, 3, 4, 6, 12
   Factors of 16: 1, 2, 4, 8, 16
   Common factors: 1, 2, 4
   GCF: __?__

(Example)

3. $16 = 2 \cdot 2 \cdot 2 \cdot 2$
   $30 = 2 \cdot 3 \cdot 5$
   GCF: __?__

4. $27 = 3 \cdot 3 \cdot 3$
   $36 = 2 \cdot 2 \cdot 3 \cdot 3$
   GCF: __?__

5. $8 = 2 \cdot 2 \cdot 2$
   $18 = $ __?__
   GCF: __?__

WRITTEN EXERCISES

Goal: To find the greatest common factor (GCF) of two or more numbers

Complete. Find the GCF of each pair of numbers.
(Table and Example)

1. $6 = 2 \times$ __?__
   $10 = 2 \times$ __?__
   GCF: __?__

2. $30 = 2 \times 3 \times$ __?__
   $45 = 3 \times 3 \times$ __?__
   GCF: $3 \times$ __?__ $=$ __?__

3. $18 = 2 \times 3 \times$ __?__
   $24 = 2 \times 2 \times 2 \times$ __?__
   GCF: $2 \times$ __?__ $=$ __?__

Find the GCF. (Table and Example)

4. 15 and 18
5. 8 and 14
6. 20 and 30
7. 9 and 15
8. 30 and 45
9. 40 and 24
10. 36 and 24
11. 16 and 60
12. 60 and 100
13. 45 and 75
14. 9 and 25
15. 15 and 30
16. 4, 8, 10
17. 12, 18, 24
18. 6, 9, 15
19. 10, 15, 25
20. 12, 15, 21
21. 16, 20, 24
22. 22, 66, 88
23. 18, 27, 36

REVIEW: SECTIONS 4-4—4-6

Write the prime factorization of each number. (Section 4-4)

1. 48   **2.** 40   **3.** 180   **4.** 200   **5.** 125   **6.** 66
7. 75   **8.** 340   **9.** 95   **10.** 363   **11.** 288   **12.** 50

Write the least common multiple. (Section 4-5)

13. 8 and 40   **14.** 18, 20, and 24   **15.** 12, 14, and 16   **16.** 50 and 60
17. 18, 36, and 210   **18.** 42 and 70   **19.** 13 and 19   **20.** 14, 20, and 35

Solve. (Section 4-5)

21. Tim and Michelle have part-time jobs after school. Tim works every third day and Michelle works every fourth day. If they both work on Wednesday, how many days will pass before they both will again be working on the same day? (Section 4-5)

22. In a meat market, pork is delivered every third business day, lamb every eighth business day, and beef every other business day. All three of these items are delivered on Monday. How many business days will pass before all three items are again delivered on the same day? (Section 4-5)

Find the GCF of each pair of numbers. (Section 4-6)

23. 18 and 72   **24.** 40 and 24   **25.** 12 and 108   **26.** 22 and 77
27. 63 and 15   **28.** 20 and 25   **29.** 28 and 36   **30.** 24 and 42

PRIME NUMBERS

Some prime numbers can be determined by evaluating the formula $n^2 - n + 41$ for values of n from 1 through 40.

EXAMPLE What prime number is produced by evaluating $n^2 - n + 41$ for $n = 29$?

Solution $n^2 - n + 41 = 29^2 - 29 + 41$

Use a calculator with a $\boxed{x^2}$ key.

$2\ 9\ \boxed{x^2}\ \boxed{-}\ 2\ 9\ \boxed{+}\ \boxed{=}$

Display: $853.$

EXERCISES Determine the prime number produced when $n^2 - n + 41$ is evaluated for each value of n.

1. 10   **2.** 17   **3.** 26   **4.** 35   **5.** 38   **6.** 40

Number Theory

CHAPTER REVIEW

PART 1: VOCABULARY

For Exercises 1–12, choose from the box at the right the word(s) that best correspond to each description.

1. The name of the symbols 0, 1, 2, 3, 4, 5, 6, 7, 8, and 9 __?__ (p. 99)	divisible by
2. A counting number greater than 1 that has exactly two factors __?__ (p. 103)	factor
3. The smallest counting number divisible by two or more counting numbers __?__ (p. 111)	counting numbers
4. If the remainder is zero when a first number is divided by a second, the first number is said to be __?__ the second. (p. 96)	multiple
5. Counting numbers greater than 1 that are not prime __?__ (p. 103)	whole numbers
6. Expressing a counting number as a product of primes __?__ (p. 108)	digits
7. A number that is divisible by 2 __?__ (p. 100)	even number
8. Since 24 is divisible by 4, 4 is a __?__ of 24. (p. 96)	odd number

Word box: divisible by, factor, counting numbers, multiple, whole numbers, digits, even number, odd number, prime number, composite numbers, prime factorization, least common multiple

9. The numbers 1, 2, 3, 4, . . . __?__ (p. 96)

10. The counting numbers and zero __?__ (p. 99)

11. A counting number is a __?__ of each of its factors. (p. 96)

12. A number not divisible by 2 __?__ (p. 100)

PART 2: SKILLS

Write the factors of each number. (Section 4–1)

13. 14 **14.** 24 **15.** 36 **16.** 40 **17.** 18 **18.** 32
19. 10 **20.** 6   **21.** 17 **22.** 13 **23.** 72 **24.** 80

Write True or False for each statement. Give a reason based on the rules on page 99. (Section 4-2)

25. 435 is divisible by 5.

26. 9723 is divisible by 3.

27. 9 is a factor of 103,421.

28. 6420 is a multiple of 10.

29. 531 is divisible by 2.

30. 1024 is divisible by 4.

Use exponents to rewrite each product. (Section 4–3)

31. $2 \cdot 3 \cdot 5 \cdot 3 \cdot 3 \cdot 7 \cdot 5 \cdot 11 \cdot 3$ **32.** $3 \cdot 5 \cdot 2 \cdot 2 \cdot 3 \cdot 2 \cdot 5$
33. $5 \cdot 27 \cdot 3 \cdot 5 \cdot 27 \cdot 5 \cdot 5$ **34.** $2 \cdot 5 \cdot 11 \cdot 11 \cdot 11 \cdot 2 \cdot 2 \cdot 2 \cdot 5 \cdot 7 \cdot 7$

Write the prime factorization of each number. (Section 4–4)

35. 105   **36.** 44   **37.** 300   **38.** 1210   **39.** 88   **40.** 240

The prime factorizations of two or three numbers are given below. Write the prime factorization of their least common multiple. (Section 4–5)

41. $2 \cdot 3^2 \cdot 5^3 \cdot 13$; $2 \cdot 3 \cdot 5 \cdot 13^2$ **42.** $2^5 \cdot 3 \cdot 7^3 \cdot 11$; $2^3 \cdot 5^2 \cdot 11^4$
43. $2 \cdot 3 \cdot 5^3$; $3^4 \cdot 5 \cdot 7$; $2^4 \cdot 5^3 \cdot 11$ **44.** $2^4 \cdot 5 \cdot 17$; $2^3 \cdot 5^4 \cdot 11$; $2^2 \cdot 5^2 \cdot 11^3$

Find the least common multiple of the given numbers. (Section 4–5)

45. 8 and 30   **46.** 15 and 14   **47.** 36, 42, and 18   **48.** 15, 18, and 50

Find the GCF. (Section 4–6)

49. 35 and 49   **50.** 16 and 34   **51.** 9 and 24   **52.** 20 and 8
53. 5, 15, 25   **54.** 8, 10, 16   **55.** 9, 12, 18   **56.** 21, 35, 14

PART 3: APPLICATIONS

57. A manager of a grocery store has 121 pounds of onions that are to be put into 3 pound bags.
  a. Will all the bags be completely filled?
  b. If not, how many additional pounds of onions are needed to completely fill the bags?
  (Section 4–1)

58. Nine high school students earned $4967 one summer.
  a. Can the $4967 be divided into nine equal whole-dollar amounts?
  b. If not, how many more dollars are needed in order to do this?
  (Section 4–1)

59. In a grocery store, boxes 20 inches tall, 24 inches tall, and 45 inches tall are being stacked next to each other in three piles. What is the shortest height possible so that all three piles will have the same height?
(Section 4–5)

Number Theory 119

CHAPTER TEST

List all of the factors of each number.

1. 30   **2.** 11   **3.** 24   **4.** 40

For Exercises 5–8, write True or False for each sentence. Give a reason for each answer.

5. 47,209,131 is divisible by 3.   **6.** 130,244 is a multiple of 4.
7. 15 is a factor of 13,324.   **8.** 6,734,385 is divisible by 9.

9. What is the smallest prime factor of 847?

Use exponents to rewrite each product.

10. 2 · 2 · 3 · 3 · 2 · 5 · 2 · 3 · 5 · 11   **11.** 2 · 2 · 3 · 2 · 3 · 2 · 3 · 5 · 5
12. 2 · 2 · 2 · 3 · 3 · 3 · 3 · 5 · 5 · 2   **13.** 5 · 7 · 7 · 5 · 5 · 11 · 5 · 13 · 5

For Exercises 14–17, write the prime factorization of each number.

14. 42   **15.** 90   **16.** 225   **17.** 360

18. Find the least common multiple of 100 and 56.

19. Find the greatest common factor of 12, 63, and 98.

20. In a neighborhood grocery, milk is delivered every third business day, bread every fifth business day, and paper goods every twelfth business day. All three of these items are delivered on Monday. How many more business days will pass before all three items are delivered on the same day?

21. A rock band with three members was offered $245 to play at a prom.
  a. Could the $245 be divided evenly so that each member received a whole-dollar amount?
  b. If not, how many more dollars would the band need to earn for this to be possible?

ANSWERS TO REVIEW CAPSULES

Page 98   **1.** 2; 8 **2.** 0; 3 **3.** 0; 0 **4.** 2; 2 **5.** 1; 4 **6.** 2; 8 **7.** 1; 1 **8.** 2; 3 **9.** 3; 4 **10.** 0; 3 **11.** 2; 1 **12.** 2; 3

Page 102   **1.** exponent **2.** 3, or itself **3.** 9 **4.** 8 **5.** 16 **6.** 16 **7.** 1 **8.** 27

Page 106   **1.** no **2.** yes **3.** yes **4.** no **5.** no **6.** yes **7.** yes **8.** no **9.** yes **10.** no **11.** yes **12.** no **13.** yes **14.** no **15.** yes **16.** yes **17.** no **18.** no **19.** 23 **20.** 34 **21.** 203 **22.** 19 **23.** 17

Page 110   **1.** 2, 4, 6, 8, 10, 12 **2.** 3, 6, 9, 12, 15, 18 **3.** 4, 8, 12, 16, 20, 24 **4.** 5, 10, 15, 20, 25, 30 **5.** 6, 12, 18, 24, 30, 36 **6.** 6, 12 **7.** 12, 24 **8.** 12 **9.** $2^2 \cdot 3^3 \cdot 5$ **10.** $2^3 \cdot 5^3$ **11.** $2^4 \cdot 3^3 \cdot 5^2$ **12.** $5^4 \cdot 11^3 \cdot 17$ **13.** $2^3 \cdot 3^3 \cdot 4^2$ **14.** $6^3 \cdot 9^3 \cdot 12^2$

Page 114   **1.** 1, 2, 4, 8 **2.** 1, 2, 5, 10 **3.** 1, 3, 5, 15 **4.** 1, 2, 3, 4, 6, 8, 12, 24 **5.** 1, 2, 3, 4, 6, 9, 12, 18, 36 **6.** 1, 2, 3, 6, 7, 14, 21, 42 **7.** $2 \cdot 3$ **8.** 2^4 **9.** $2^2 \cdot 5$ **10.** 2^5 **11.** $3^2 \cdot 5$ **12.** $2 \cdot 5^2$

ENRICHMENT

Factor Trees

Another way to write the prime factorization of a composite number is to use a **factor tree**.

EXAMPLE Use a factor tree to factor 180. Then use exponents to write the prime factors.

Solution: A factor tree can begin with any pair of factors of the number.

$180 = 2 \times 2 \times 5 \times 3 \times 3$
$\quad\;\; = 2^2 \times 3^2 \times 5$

$180 = 2 \times 5 \times 2 \times 3 \times 3$
$\quad\;\; = 2^2 \times 3^2 \times 5$

EXERCISES

Copy and complete the factor trees below. Then use exponents to write the prime factorization of the given number.

1. 294
2. 1200

Use a factor tree to find the prime factorization of each number. Then use exponents to write each prime factorization.

3. 36
4. 40
5. 100
6. 80
7. 150
8. 200
9. 1000
10. 1260
11. 2025
12. 2079

ADDITIONAL PRACTICE

SKILLS

List all the factors of each number. (Pages 96–98)

1. 9 **2.** 19 **3.** 48 **4.** 60 **5.** 32 **6.** 200

Write True or False for each sentence. Write a reason for each answer. (Pages 99–102)

7. 9 is a factor of 54.
8. 97 is divisible by 7.
9. 66 is a multiple of 11.
10. 4 is a factor of 92.
11. 135 is divisible by 5.
12. 1642 is divisible by 4.
13. 1342 is a multiple of 10.
14. 9 is a factor of 47,358.

Find the smallest prime factor of each number. (Pages 103–105)

15. 37 **16.** 121 **17.** 4035 **18.** 343 **19.** 7419 **20.** 1331

Use exponents to rewrite each product. (Pages 103–105)

21. $2 \cdot 2 \cdot 3 \cdot 2 \cdot 3 \cdot 5$
22. $2 \cdot 2 \cdot 3 \cdot 2 \cdot 2 \cdot 5 \cdot 5 \cdot 3 \cdot 5$
23. $2 \cdot 11 \cdot 2 \cdot 3 \cdot 11 \cdot 2 \cdot 3 \cdot 17$
24. $17 \cdot 2 \cdot 3 \cdot 3 \cdot 2 \cdot 5 \cdot 3 \cdot 2 \cdot 5 \cdot 17$

Write the prime factorization of each number. (Pages 108–110)

25. 65 **26.** 42 **27.** 182 **28.** 360 **29.** 216 **30.** 1400

For Exercises 31–34, write the LCM and the GCF. (Pages 111–116)

31. 18 and 50   **32.** 18 and 45   **33.** 6, 45, and 35   **34.** 20, 54, and 180

APPLICATIONS

35. The members of a student council want to divide $874 evenly between 3 clubs.
   a. Can each club receive a whole-number dollar amount?
   b. If not, how many additional dollars are needed for this to be possible? (Pages 99–102)

36. Three satellites are in orbit around the earth. The satellites make one complete orbit every 60, 80, and 120 minutes, respectively. If all three satellites are directly over Los Angeles, California, how many minutes will pass before they are all over Los Angeles again? (Pages 111–114)

CHAPTER 5
Fractions: Multiplication/Division

FEATURES
Career Application: *Engineering Technician*
Consumer Application: *Saving $ on Cooling Costs*
Enrichment: *Special Areas*

SECTIONS
5-1 Writing Fractions in Lowest Terms
5-2 Customary Measures
5-3 Multiplying Fractions
5-4 Problem Solving and Applications: Formulas: Temperature
5-5 Multiplying Mixed Numerals
5-6 Problem Solving and Applications: Area: Parallelograms and Triangles
5-7 Problem Solving and Applications: Volume: Rectangular Prisms
5-8 Dividing Fractions

5-1 Writing Fractions in Lowest Terms

Six out of every 300 gallons of water used in a home are used for cleaning purposes. Six out of 300 can be written as a fraction.

$\dfrac{6}{300}$ ← Numerator
          ← Denominator

The fraction $\dfrac{6}{300}$ can be simplified; that is, it can be written in *lowest terms*.

A fraction is in **lowest terms** when its numerator and denominator have no common prime factor.

$$\dfrac{12}{35} = \dfrac{2 \cdot 2 \cdot 3}{5 \cdot 7}$$

↑ Lowest Terms

Here are three methods for writing fractions in lowest terms.

TABLE

Method 1 (GCF)	$\dfrac{6}{300} = \dfrac{6 \div 6}{300 \div 6}$ $= \dfrac{1}{50}$	Divide the numerator and denominator by the GCF. Lowest terms
Method 2 (Prime factorization)	$\dfrac{12}{20} = \dfrac{2 \cdot 2 \cdot 3}{2 \cdot 2 \cdot 5}$ $= \dfrac{\cancel{2} \cdot \cancel{2} \cdot 3}{\cancel{2} \cdot \cancel{2} \cdot 5} = \dfrac{3}{5}$	Write the prime factorizations. Lowest terms
Method 3 (Common factors)	$\dfrac{12}{18} = \dfrac{\overset{6}{\cancel{12}}}{\underset{9}{\cancel{18}}}$ $= \dfrac{6}{9}$ $= \dfrac{\overset{2}{\cancel{6}}}{\underset{3}{\cancel{9}}} = \dfrac{2}{3}$	2 is a common factor. Divide the numerator and denominator by 2. Not in lowest terms. 3 is also a common factor. Lowest terms

124 Chapter 5

CLASSROOM EXERCISES

Determine whether each fraction is in lowest terms. Answer <u>Yes</u> or <u>No</u>. (Table)

1. $\frac{8}{24}$
2. $\frac{12}{48}$
3. $\frac{9}{25}$
4. $\frac{24}{51}$
5. $\frac{15}{40}$
6. $\frac{21}{52}$

WRITTEN EXERCISES

Goal: To write a fraction in lowest terms

Write each fraction in lowest terms. (Table)

1. $\frac{6}{9}$
2. $\frac{6}{12}$
3. $\frac{16}{28}$
4. $\frac{12}{20}$
5. $\frac{2}{16}$
6. $\frac{15}{48}$
7. $\frac{21}{24}$
8. $\frac{4}{10}$
9. $\frac{6}{45}$
10. $\frac{24}{64}$
11. $\frac{100}{300}$
12. $\frac{25}{55}$
13. $\frac{9}{16}$
14. $\frac{22}{88}$
15. $\frac{16}{36}$
16. $\frac{16}{49}$
17. $\frac{57}{90}$
18. $\frac{12}{30}$

APPLICATIONS: USING FRACTIONS

19. Sheri cut a cantaloupe into 10 slices. She then ate 4 of the slices. What fractional part (lowest terms) of the cantaloupe did she eat?

20. The string section of a certain orchestra has 42 instruments. Twenty of these instruments are violins. What fractional part (lowest terms) of the string section is not violins?

MORE CHALLENGING EXERCISES

Simplify. No variable equals zero.

EXAMPLE: $\frac{3a}{6} = \frac{\overset{1}{\cancel{3}} \cdot a}{\underset{2}{\cancel{6}}} = \frac{1 \cdot a}{2} = \frac{a}{2}$

21. $\frac{5t}{10}$
22. $\frac{6p}{8}$
23. $\frac{4}{6y}$
24. $\frac{9}{6x}$
25. $\frac{6p}{9pn}$
26. $\frac{8st}{10t}$
27. $\frac{6y}{14y^2}$
28. $\frac{4a^2}{7a}$
29. $\frac{3pn}{5p^2}$
30. $\frac{8x^2y}{12xy^2}$
31. $\frac{27cde}{60def}$
32. $\frac{12rst}{8st^2}$

Fractions: Multiplication/Division

5-2 Customary Measures

This table shows some commonly used customary units.

TABLE

Length	Capacity
12 inches (in) = 1 foot (ft) 3 feet = 1 yard (yd) 36 inches = 1 yd 5280 feet = 1 mile (mi)	2 cups (c) = 1 pint (pt) 2 pints = 1 quart (qt) 4 quarts = 1 gallon (gal)
	Weight
	16 ounces (oz) = 1 pound (lb) 2000 pounds = 1 ton (T)

A measure such as 3 feet 9 inches can be written as a mixed numeral. **A mixed numeral** (shown at the right), represents the sum of a whole number and a fraction.

Mixed Numeral

$3\frac{3}{4}$

Whole Number — Fraction

EXAMPLE 1 Write a mixed numeral for each measure.

    a. 3 ft 9 in   **b.** 5 lb 2 oz

Solutions: **a.** Since 12 in = 1 ft, 9 in = $\frac{9}{12}$, or $\frac{3}{4}$ ft. ◀ Lowest terms

       Thus, 3 ft 9 in = $3\frac{3}{4}$ **ft.**

    b. Since 16 oz = 1 lb, 2 oz = $\frac{2}{16}$, or $\frac{1}{8}$ lb.

       Thus, 5 lb 2 oz = $5\frac{1}{8}$ **lb.**

Self-Check *Write a mixed numeral for each measure.*

   1. 3 lb 12 oz   **2.** 4 ft 8 in   **3.** 2 yd 2 ft

Example 2 shows how to add and subtract with customary measures.

EXAMPLE 2 Add or subtract as indicated. Express the answer as a mixed numeral.

    a.   3 ft 9 in   **b.**   4 gal 2 qt   ◀ 4 gal 2 qt =
       +2 ft 8 in   −1 gal 3 qt   3 gal + 4 qt + 2 qt =
                                                               3 gal 6 qt

126 *Chapter 5*

Solutions: **a.** 3 ft 9 in
           +2 ft 8 in
           5 ft 17 in = 5 ft + 1 ft 5 in
                     = 6 ft 5 in
                     = $6\frac{5}{12}$ ft

b. Since you cannot subtract 3 qt from 2 qt, write 4 gal 2 qt as 3 gal 6 qt.

   4 gal 2 qt   3 gal 6 qt
   −1 gal 3 qt →   −1 gal 3 qt
                     2 gal 3 qt,
                     or $2\frac{3}{4}$ **gal**

Self-Check *Add or subtract as indicated.*

4. 6 yd 2 ft
       +1 yd 2 ft

5. 8 lb 6 oz
       −3 lb 8 oz

6. 9 gal 1 qt
       −4 gal 3 qt

CLASSROOM EXERCISES

Write a mixed numeral for each measure. (Example 1)

1. 4 gal 2 qt **2.** 3 yd 1 ft **3.** 5 qt 1 pt **4.** 6 ft 9 in
5. 6 lb 8 oz **6.** 1 T 200 lb **7.** 3 qt 1 pt **8.** 3 lb 10 oz

Add or subtract as indicated. Express the answer as a mixed numeral. (Example 2)

9. 2 ft 3 in
       +4 ft 5 in

10. 5 gal 2 qt
       +2 gal 1 qt

11. 12 lb 8 oz
       + 8 lb 12 oz

12. 4 yd 2 ft
       +8 yd 2 ft

13. 9 ft 8 in
       −4 ft 10 in

14. 15 lb 12 oz
       −10 lb 4 oz

15. 7 yd 1 ft
       −3 yd 2 ft

16. 13 gal 1 qt
       − 2 gal 3 qt

WRITTEN EXERCISES

Goals To add and subtract customary measures
         To apply these skills to solving word problems

Write a mixed numeral for each measure. (Example 1)

1. 6 gal 3 qt   **2.** 3 ft 5 in   **3.** 6 yd 2 ft   **4.** 7 yd 12 in
5. 4 gal 3 qt   **6.** 24 lb 6 oz   **7.** 2 pt 1 c   **8.** 3 T 500 lb
9. 1 mi 528 ft   **10.** 1 qt 5 pt   **11.** 1 yd 5 ft **12.** 98 lb 12 oz

Fractions: Multiplication/Division **127**

Add or subtract as indicated. Write each answer as a mixed numeral.
(Example 2)

13.   1 ft 7 in
  +2 ft 4 in

14.   5 ft 9 in
  +2 ft 3 in

15.   1 yd 2 ft
  +8 yd 2 ft

16.   6 yd 1 ft
  +5 yd 1 ft

17.   8 gal 2 qt
  +3 gal 1 qt

18.   19 gal 3 qt
  +   2 qt

19.   3 lb 12 oz
  +9 lb 8 oz

20.   16 lb 14 oz
  +12 lb 15 oz

21.   3 ft 8 in
  −1 ft 5 in

22.   9 ft 6 in
  −4 ft 8 in

23.   9 yd 1 ft
  −2 yd 2 ft

24.   4 yd 2 ft
  −1 yd 1 ft

25.   10 gal 3 qt
  − 3 gal 2 qt

26.   6 gal 1 qt
  −2 gal 3 qt

27.   18 lb 4 oz
  −13 lb 12 oz

28.   98 lb
  − 16 lb 8 oz

APPLICATIONS: **USING CUSTOMARY MEASURES**

Solve. Write each answer as a mixed numeral.

29. May measured the amount of water collected from a dripping faucet.

    Monday: 1 gallon 2 quarts
    Tuesday: 1 gallon 3 quarts

    Find the total amount collected in the two days.

30. Larry wishes to replace the weather-stripping at the top and on one side of a door. The height of the door is 6 feet 11 inches and the width 2 feet 9 inches. How many feet of weather-stripping will be needed?

31. Juan is 6 feet 3 inches tall. His younger sister is 4 feet 8 inches tall. How much taller is Juan than his sister?

32. An airline passenger is allowed to bring luggage weighing 60 pounds. Coleen's luggage weighs 72 pounds 4 ounces. How many pounds over the limit is the weight of her luggage?

REVIEW CAPSULE FOR SECTION 5–3

Multiply. (Pages 7–9)

1. $8 \cdot 22$ 2. $42 \cdot 9$ 3. $10 \cdot 15$ 4. $12 \cdot 13$ 5. $16 \cdot 21$ 6. $18 \cdot 32$

Write each fraction in lowest terms. (Pages 124–125)

7. $\frac{4}{6}$ 8. $\frac{6}{14}$ 9. $\frac{12}{18}$ 10. $\frac{28}{30}$ 11. $\frac{50}{75}$ 12. $\frac{18}{81}$

The answers are on page 154.

5-3 Multiplying Fractions

Two-thirds of Grant City Park is covered with grass. The Director of City Parks plans to replace $\frac{1}{5}$ of this grass with shrubs.

Grass covers $\frac{2}{3}$ of the park. $\frac{1}{5} \cdot \frac{2}{3} = \frac{1 \cdot 2}{5 \cdot 3} = \frac{2}{15}$

15 equal parts
2 of the 15 will have shrubs.

One way of multiplying with fractions is to multiply the numerators <u>and</u> multiply the denominators. Then write the answer in lowest terms.

EXAMPLE 1   Multiply: a. $\frac{2}{3} \cdot \frac{7}{10}$ b. $\frac{3}{16} \cdot 5$

Solution: a. ⬜1 $\frac{2}{3} \cdot \frac{7}{10} = \frac{2 \cdot 7}{3 \cdot 10}$

$= \frac{14}{30}$ ◀ Not in lowest terms

⬜2 $= \frac{14 \div 2}{30 \div 2}$

$= \frac{7}{15}$ ◀ Lowest terms

b. ⬜1 $\frac{3}{16} \cdot 5 = \frac{3}{16} \cdot \frac{5}{1}$ $5 = \frac{5}{1}$

⬜2 $= \frac{15}{16}$

Self-Check *Multiply:*

1. $\frac{3}{4} \cdot \frac{3}{10}$ 2. $\frac{4}{15} \cdot 2$ 3. $\frac{1}{18} \cdot 11$ 4. $\frac{4}{5} \cdot \frac{3}{12}$

It is sometimes easier to divide a numerator <u>and</u> a denominator by a common factor <u>before</u> multiplying.

EXAMPLE 2   Multiply: a. $\frac{5}{7} \cdot \frac{4}{15}$ b. $\frac{2}{9} \cdot \frac{3}{4}$

Solutions: a. $\frac{5}{7} \cdot \frac{4}{15} = \frac{\overset{1}{\cancel{5}}}{7} \cdot \frac{4}{\underset{3}{\cancel{15}}}$ $5 \div 5 = 1$
$15 \div 5 = 3$

$= \frac{1 \cdot 4}{7 \cdot 3} = \frac{4}{21}$

b. $\frac{2}{9} \cdot \frac{3}{4} = \frac{\overset{1}{\cancel{2}}}{\underset{3}{\cancel{9}}} \cdot \frac{\overset{1}{\cancel{3}}}{\underset{2}{\cancel{4}}}$

$= \frac{1 \cdot 1}{3 \cdot 2} = \frac{1}{6}$

Fractions: Multiplication/Division

Self-Check *Multiply.*

5. $\frac{6}{7} \cdot \frac{14}{15}$
6. $\frac{7}{9} \cdot \frac{18}{21}$
7. $\frac{12}{15} \cdot \frac{3}{4}$
8. $\frac{2}{3} \cdot \frac{3}{8}$

> For the fractions $\frac{a}{b}$ and $\frac{c}{d}$, $b \neq 0$, $d \neq 0$,
> $\frac{a}{b} \cdot \frac{c}{d} = \frac{a \cdot c}{b \cdot d} = \frac{ac}{bd}.$ $\quad\quad \frac{2}{5} \cdot \frac{3}{7} = \frac{2 \cdot 3}{5 \cdot 7} = \frac{6}{35}$

CLASSROOM EXERCISES

Multiply. Write each answer in lowest terms. (Examples 1 and 2)

1. $\frac{1}{2} \cdot \frac{1}{7}$
2. $\frac{1}{4} \cdot \frac{1}{5}$
3. $\frac{3}{4} \cdot \frac{1}{4}$
4. $\frac{4}{5} \cdot \frac{1}{7}$
5. $\frac{2}{5} \cdot \frac{8}{9}$
6. $\frac{3}{4} \cdot \frac{3}{5}$
7. $\frac{5}{6} \cdot \frac{2}{7}$
8. $\frac{2}{5} \cdot \frac{1}{6}$
9. $5 \cdot \frac{3}{7}$
10. $2 \cdot \frac{1}{5}$
11. $\frac{2}{7} \cdot 3$
12. $\frac{2}{9} \cdot 4$
13. $\frac{4}{5} \cdot \frac{5}{7}$
14. $\frac{3}{4} \cdot \frac{4}{5}$
15. $\frac{3}{5} \cdot \frac{2}{3}$
16. $\frac{4}{7} \cdot \frac{3}{4}$
17. $\frac{1}{3} \cdot \frac{3}{5}$
18. $\frac{7}{8} \cdot \frac{8}{9}$
19. $\frac{1}{2} \cdot \frac{4}{5}$
20. $\frac{1}{4} \cdot \frac{8}{9}$
21. $\frac{5}{9} \cdot \frac{7}{10}$
22. $\frac{4}{5} \cdot \frac{5}{12}$
23. $\frac{9}{10} \cdot \frac{1}{3}$
24. $\frac{3}{4} \cdot \frac{2}{9}$

WRITTEN EXERCISES

Goals: To multiply with fractions
To apply the skill of multiplying with fractions to solving word problems

Multiply. Write each answer in lowest terms. (Examples 1 and 2)

1. $\frac{1}{4} \cdot \frac{1}{2}$
2. $\frac{1}{3} \cdot \frac{1}{5}$
3. $\frac{3}{5} \cdot \frac{1}{2}$
4. $\frac{3}{5} \cdot \frac{2}{7}$
5. $\frac{2}{9} \cdot \frac{4}{5}$
6. $\frac{3}{11} \cdot \frac{5}{7}$
7. $4 \cdot \frac{1}{5}$
8. $5 \cdot \frac{1}{8}$
9. $3 \cdot \frac{1}{6}$
10. $\frac{3}{10} \cdot 3$
11. $\frac{2}{15} \cdot 4$
12. $6 \cdot \frac{3}{20}$
13. $\frac{4}{5} \cdot \frac{7}{12}$
14. $\frac{10}{11} \cdot \frac{3}{5}$
15. $\frac{5}{9} \cdot \frac{3}{7}$
16. $\frac{3}{4} \cdot \frac{2}{5}$
17. $\frac{4}{5} \cdot \frac{3}{8}$
18. $\frac{5}{6} \cdot \frac{1}{5}$
19. $\frac{2}{3} \cdot \frac{3}{8}$
20. $\frac{4}{5} \cdot \frac{5}{8}$
21. $\frac{9}{10} \cdot \frac{5}{6}$
22. $\frac{2}{3} \cdot \frac{9}{10}$
23. $\frac{5}{8} \cdot \frac{4}{15}$
24. $\frac{3}{4} \cdot \frac{8}{15}$
25. $\frac{5}{8} \cdot \frac{11}{25}$
26. $\frac{5}{6} \cdot \frac{4}{5}$
27. $\frac{10}{21} \cdot \frac{14}{15}$
28. $\frac{9}{10} \cdot \frac{8}{15}$
29. $\frac{4}{3} \cdot \frac{3}{4}$
30. $\frac{1}{9} \cdot 6$
31. $2 \cdot \frac{3}{8}$
32. $\frac{3}{8} \cdot \frac{5}{7}$
33. $\frac{8}{9} \cdot \frac{9}{8}$
34. $\frac{3}{4} \cdot \frac{4}{9}$
35. $\frac{2}{3} \cdot \frac{4}{5}$
36. $\frac{3}{8} \cdot \frac{8}{9}$

APPLICATIONS: USING MULTIPLICATION WITH FRACTIONS

Solve. Write each answer in lowest terms.

37. Angelo spent $\frac{5}{6}$ of an hour practicing basketball. He spent $\frac{2}{3}$ of that time practicing lay–up shots. What fractional part of an hour did he spend practicing lay–up shots?

38. Susan can mow her front lawn in $\frac{3}{4}$ hour. Her brother can mow the lawn in $\frac{4}{5}$ that time. How long does it take Susan's brother to mow the lawn?

39. Mr. Mitchell's lawn mower will hold $\frac{9}{10}$ gallon of gasoline. The tank is $\frac{1}{3}$ full. How much gasoline is in the tank?

40. Band practice lasted 2 hours. One third of that time was spent marching. What fractional part of an hour was spent marching?

MORE CHALLENGING EXERCISES

Multiply. No variable equals zero.

EXAMPLES: **a.** $\dfrac{3}{t} \cdot \dfrac{t}{9} = \dfrac{\overset{1}{\cancel{3}} \cdot \overset{1}{\cancel{t}}}{\underset{1}{\cancel{t}} \cdot \underset{3}{\cancel{9}}} = \dfrac{1 \cdot 1}{1 \cdot 3} = \dfrac{1}{3}$ **b.** $10 \cdot \dfrac{1}{b} = \dfrac{10}{1} \cdot \dfrac{1}{b} = \dfrac{10 \cdot 1}{1 \cdot b} = \dfrac{10}{b}$

41. $\dfrac{2}{a} \cdot \dfrac{a}{6}$
42. $\dfrac{5}{x} \cdot \dfrac{x}{10}$
43. $\dfrac{a^2}{5} \cdot \dfrac{3}{a}$
44. $\dfrac{2m}{3} \cdot \dfrac{5}{6m}$
45. $\dfrac{3st^2}{5} \cdot \dfrac{5}{3t}$
46. $6 \cdot \dfrac{1}{x}$
47. $\dfrac{2}{b} \cdot 3$
48. $9 \cdot \dfrac{1}{3t}$
49. $12 \cdot \dfrac{1}{4m}$
50. $8 \cdot \dfrac{3}{6p}$

REVIEW CAPSULE FOR SECTION 5-4

Evaluate each expression. (Pages 18–21)

1. $2 \cdot 10 + 1$
2. $8 \cdot 15 + 16$
3. $9 \cdot 32 - 25$
4. $13 \cdot 25 - 47$

Evaluate each expression. (Pages 22–24)

5. $\frac{2}{3}x + 15$ when $x = 6$
6. $\frac{4}{9}y + 24$ when $y = 36$
7. $\frac{4}{5}(a + 10)$ when $a = 75$
8. $\frac{5}{6}s - 45$ when $s = 90$
9. $\frac{3}{4}(d - 8)$ when $d = 120$
10. $\frac{3}{8}m - 51$ when $m = 424$

The answers are on page 154.

Fractions: Multiplication/Division

PROBLEM SOLVING AND APPLICATIONS

5-4 Formulas: Temperature

The **Celsius (C)** scale is used to measure temperature. In the United States, temperatures are also given in degrees **Fahrenheit (°F).**

It may be helpful to memorize the information in this table.

TABLE

Temperature in °C	Meaning	Temperature in °F
100°C	Boiling point of water	212°F
37°C	Normal body temperature	98.6°F
35°C	Hot summer day	95°F
20°C	Comfortable room temperature	68°F
0°C	Freezing point of water	32°F

You can use this formula to write temperatures in °C (read: degrees Celsius) as temperatures in °F.

$$F = \frac{9}{5}C + 32$$

Given: °C
Find: Corresponding °F

EXAMPLE 1 The temperature at which copper boils is 2300°C. Express this as a Fahrenheit reading.

Solution: $F = \frac{9}{5}C + 32$ Replace C with 2300.

$F = \frac{9}{5}(2300) + 32$

$F = 9(460) + 32$

$F = 4140 + 32 = 4172$ 2300 °C corresponds to 4172 °F.

Self-Check *Express each of the following as a Fahrenheit reading.*

1. 5°C **2.** 105°C **3.** 800°C **4.** 1200°C

You can use the following formula to write temperatures in °F as temperature in °C.

$$C = \frac{5}{9}(F - 32)$$

Given: °F
Find: Corresponding °C

132 Chapter 5

When using the formula at the bottom of page 132, be sure to do the subtraction <u>inside the parentheses first</u>. Then multiply by $\frac{5}{9}$.

EXAMPLE 2 The temperature of a cup of hot tea is 176°F. What is the corresponding Celsius reading?

Solution: $C = \frac{5}{9}(F - 32)$ ◀ Replace F with 176.

$C = \frac{5}{9}(176 - 32)$

$C = \frac{5}{9} \cdot 144 \qquad C = 80$ ◀ 176°F corresponds to 80°C

Self-Check *Express each of the following as a Celsius reading.*

5. 41°F **6.** 221°F **7.** 986°F **8.** 185°F

CLASSROOM EXERCISES

Express each of the following as a Fahrenheit reading. (Example 1)

1. 50°C **2.** 25°C **3.** 15°C **4.** 90°C **5.** 200°C **6.** 1000°C

Express each of the following as a Celsius reading. (Example 2)

7. 68°F **8.** 86°F **9.** 140°F **10.** 77°F **11.** 131°F **12.** 338°F

WRITTEN EXERCISES

Goal: To use the formulas for Celsius–Fahrenheit temperatures

Express each of the following as a Fahrenheit reading. (Example 1)

1. 10°C **2.** 45°C **3.** 100°C **4.** 400°C **5.** 225°C **6.** 85°C

Express each of the following as a Celsius reading. (Example 2)

7. 50°F **8.** 257°F **9.** 113°F **10.** 59°F **11.** 464°F **12.** 932°F

13. The average high temperature on the moon's surface is about 248°F. Express this as a Celsius reading.

14. The average surface temperature of Venus is about 860°F. Express this as a Celsius reading.

15. The average surface temperature of the sun is about 5525°C. Express this as a Fahrenheit reading.

Fractions: Multiplication/Division

REVIEW: SECTIONS 5-1—5-4

Express each fraction in lowest terms. (Section 5-1)

1. $\frac{9}{12}$
2. $\frac{20}{32}$
3. $\frac{18}{21}$
4. $\frac{200}{500}$
5. $\frac{12}{15}$
6. $\frac{36}{54}$

Add or subtract as indicated. Write each answer as a mixed numeral. (Section 5-2)

7. 5 ft 4 in
   +3 ft 7 in

8. 10 lb 9 oz
   + 8 lb 12 oz

9. 7 yd 1 ft
   −3 yd 2 ft

10. 8 gal 2 qt
    −5 gal 3 qt

Multiply. Write each answer in lowest terms. (Section 5-3)

11. $\frac{1}{2} \cdot \frac{5}{6}$
12. $\frac{3}{8} \cdot \frac{8}{9}$
13. $\frac{2}{7} \cdot 3$
14. $\frac{4}{9} \cdot \frac{3}{10}$

Complete. (Section 5-4)

15. 68°F = __?__ °C
16. 50°C = __?__ °F
17. 356°F = __?__ °C
18. 500°C = __?__ °F

19. David's sailboat is 13 feet 10 inches long. Ellie's sailboat is 15 feet 5 inches long. How much longer is Ellie's boat than David's? (Section 5-2)

20. Tennis practice lasted 3 hours. One-fourth of that time was spent practicing the serve. What fractional part of an hour was spent practicing the serve? (Section 5-3)

REVIEW CAPSULE FOR SECTION 5-5

Write the remainder.

1. $2\overline{)3}$
2. $3\overline{)5}$
3. $5\overline{)8}$
4. $7\overline{)20}$
5. $8\overline{)15}$
6. $3\overline{)20}$

Multiply. Write each answer in lowest terms. (Pages 124–125)

7. $\frac{1}{3} \cdot \frac{3}{4}$
8. $\frac{4}{7} \cdot \frac{3}{8}$
9. $\frac{5}{9} \cdot \frac{7}{10}$
10. $\frac{4}{5} \cdot \frac{7}{12}$
11. $\frac{3}{10} \cdot 2$
12. $\frac{2}{15} \cdot 6$

The answers are on page 154.

5-5 Multiplying Mixed Numerals

The length of the bolt at the right can be given as $\frac{7}{4}$ inches or as $1\frac{3}{4}$ inches.

$$\frac{7}{4} = 1\frac{3}{4}$$

— Fraction
— Whole Number

Recall from page 126 that $1\frac{3}{4}$ represents the sum of a whole number and a fraction. It is called a **mixed numeral**. Any fraction whose numerator is greater than its denominator can be expressed as a mixed numeral.

$\frac{1}{4}$ $\frac{2}{4}$ $\frac{3}{4}$ $\frac{4}{4}$ $\frac{5}{4}$ $\frac{6}{4}$ $\frac{7}{4}$, or $1\frac{3}{4}$

PROCEDURE

To write a mixed numeral for a fraction whose numerator is greater than the denominator:

1 Divide the numerator by the denominator.

2 Write a fraction for the remainder.

EXAMPLE 1 Write a mixed numeral for $\frac{37}{5}$.

Solution: **1** $\frac{37}{5} = 37 \div 5$

$5\overline{)37}$ → 7 R 2, or $7\frac{2}{5}$
$\underline{35}$
2

2 $= 7\frac{2}{5}$ ◄ Mixed numeral

Self-Check *Write a mixed numeral for each fraction.*

1. $\frac{23}{7}$ 2. $\frac{41}{5}$ 3. $\frac{16}{3}$ 4. $\frac{81}{6}$

This table shows how to write a fraction for a mixed numeral.

Mixed Numeral	Procedure		
	1 Multiply the denominator and the whole number.	**2** Add this product to the numerator.	**3** Write this sum over the denominator.
$5\frac{3}{4}$	$4 \times 5 = 20$	$20 + 3 = 23$	$\frac{23}{4}$ ◄ $5\frac{3}{4} = \frac{23}{4}$

Fractions: Multiplication/Division

> **PROCEDURE**
> To multiply with mixed numerals:
> 1. Write a fraction for each mixed numeral.
> 2. Multiply the fractions.
> 3. Write the product in lowest terms.

EXAMPLE 2 Multiply. a. $4 \cdot 3\frac{1}{16}$ b. $6\frac{2}{3} \cdot 2\frac{3}{4}$

Solutions: a. ① $4 \cdot 3\frac{1}{16} = \frac{4}{1} \cdot \frac{49}{16}$ b. ① $6\frac{2}{3} \cdot 2\frac{3}{4} = \frac{20}{3} \cdot \frac{11}{4}$

② $= \frac{\overset{1}{\cancel{4}}}{1} \cdot \frac{49}{\underset{4}{\cancel{16}}}$ ← $\frac{1 \cdot 49}{1 \cdot 4}$ ② $= \frac{\overset{5}{\cancel{20}}}{3} \cdot \frac{11}{\underset{1}{\cancel{4}}}$

$= \frac{49}{4}$ $= \frac{55}{3}$

③ $= 12\frac{1}{4}$ ← Lowest terms → ③ $= 18\frac{1}{3}$

Self-Check *Multiply.*

5. $2\frac{1}{2} \cdot 4$ 6. $9 \cdot 6\frac{2}{3}$ 7. $1\frac{1}{3} \cdot 4\frac{1}{8}$ 8. $1\frac{5}{8} \cdot 3\frac{1}{2}$

CLASSROOM EXERCISES

Write a mixed numeral for each fraction. (Example 1)

1. $\frac{9}{2}$ 2. $\frac{11}{5}$ 3. $\frac{27}{5}$ 4. $\frac{59}{6}$ 5. $\frac{46}{7}$ 6. $\frac{37}{4}$ 7. $\frac{8}{5}$

Write a fraction for each mixed numeral. (Table)

8. $3\frac{1}{2}$ 9. $2\frac{1}{4}$ 10. $5\frac{3}{4}$ 11. $9\frac{3}{8}$ 12. $2\frac{2}{3}$ 13. $3\frac{7}{16}$ 14. $5\frac{3}{7}$

Multiply. Write each answer in lowest terms. (Example 2)

15. $7 \cdot 2\frac{1}{2}$ 16. $8 \cdot 6\frac{1}{3}$ 17. $2\frac{2}{3} \cdot 1\frac{1}{2}$ 18. $1\frac{5}{8} \cdot 1\frac{1}{4}$ 19. $2\frac{2}{5} \cdot 3\frac{1}{3}$

WRITTEN EXERCISES

Goals: To multiply with mixed numerals

To apply multiplication with mixed numerals to solving word problems

Write a mixed numeral for each fraction. (Example 1)

1. $\frac{10}{7}$ 2. $\frac{11}{4}$ 3. $\frac{12}{5}$ 4. $\frac{22}{3}$ 5. $\frac{15}{8}$ 6. $\frac{25}{12}$ 7. $\frac{27}{10}$

136 Chapter 5

Write a fraction for each mixed numeral. (Table)

8. $4\frac{7}{8}$ **9.** $3\frac{2}{3}$ **10.** $4\frac{3}{4}$ **11.** $7\frac{2}{5}$ **12.** $6\frac{5}{7}$ **13.** $3\frac{4}{9}$ **14.** $3\frac{7}{10}$

Multiply. Write each answer in lowest terms. (Example 2)

15. $3\frac{1}{2} \cdot 2$ **16.** $5\frac{1}{3} \cdot 4$ **17.** $4 \cdot 1\frac{1}{2}$ **18.** $6 \cdot 2\frac{1}{3}$ **19.** $2\frac{1}{4} \cdot 4$
20. $2\frac{1}{3} \cdot 9$ **21.** $1\frac{3}{4} \cdot 4\frac{1}{2}$ **22.** $2\frac{1}{4} \cdot 3\frac{1}{3}$ **23.** $4\frac{1}{2} \cdot 2\frac{3}{5}$ **24.** $5\frac{2}{3} \cdot 2\frac{1}{5}$
25. $5\frac{1}{2} \cdot 6\frac{1}{3}$ **26.** $8\frac{2}{3} \cdot 5\frac{1}{6}$ **27.** $4\frac{1}{8} \cdot 4\frac{3}{5}$ **28.** $3\frac{1}{2} \cdot 4\frac{2}{3}$ **29.** $1\frac{1}{2} \cdot 1\frac{5}{9}$
30. $1\frac{7}{8} \cdot 3\frac{2}{3}$ **31.** $3\frac{2}{3} \cdot 3\frac{2}{3}$ **32.** $4\frac{1}{2} \cdot 2\frac{1}{3}$ **33.** $10 \cdot 1\frac{5}{6}$ **34.** $6\frac{1}{2} \cdot 12$

APPLICATIONS: USING MIXED NUMERALS

Solve. Write each answer in lowest terms.

35. A sports car averages $12\frac{1}{2}$ miles to one gallon of fuel. The tank holds $9\frac{1}{3}$ gallons. How far will the car travel on a full tank of fuel?

36. Vivian works $5\frac{1}{6}$ hours every Saturday. How many hours does she work in a month if there are 4 Saturdays in the month?

37. A casserole recipe calls for $1\frac{1}{4}$ cups of chicken. How much chicken is needed to make 3 times that recipe?

38. Jules is putting carpet in his hall. The hall is $8\frac{2}{3}$ feet long and $4\frac{1}{2}$ feet wide. How many square feet of carpet will he need to cover the hall floor?

MORE CHALLENGING EXERCISES

Multiply. No variable equals zero.

EXAMPLE: $2\frac{1}{4} \cdot \frac{x}{3} = \frac{9}{4} \cdot \frac{x}{3}$

$$= \frac{\overset{3}{\cancel{9}} \cdot x}{4 \cdot \underset{1}{\cancel{3}}} = \frac{3 \cdot x}{4 \cdot 1} = \frac{3x}{4}$$

39. $1\frac{1}{2} \cdot \frac{p}{3}$ **40.** $6\frac{2}{3} \cdot \frac{y}{5}$ **41.** $2\frac{1}{4} \cdot \frac{2}{a}$ **42.** $5\frac{1}{3} \cdot \frac{9}{b}$ **43.** $2\frac{1}{3} \cdot \frac{m}{3}$
44. $\frac{3d}{2} \cdot 1\frac{2}{3}$ **45.** $\frac{5t}{6} \cdot 4\frac{1}{5}$ **46.** $3\frac{1}{3} \cdot \frac{6t}{10}$ **47.** $5\frac{1}{4} \cdot \frac{2}{7p}$ **48.** $1\frac{2}{5} \cdot \frac{3}{5x}$

Fractions: Multiplication/Division

CAREER APPLICATIONS
Engineering Technician

Engineering technicians often work with engineers and perform some of the same tasks as engineers.

In construction work, it is important to know the greatest safe load that a beam can bear. The formula below applies to a steel I–beam with the load in the middle of the beam. The load is in pounds.

$$S = \frac{1795Ad}{l}$$

S = greatest safe load (lbs)
A = cross-sectional area (in^2)
d = depth of beam (in)
l = length of beam (ft)

EXAMPLE Calculate the greatest safe load in pounds for a steel I–beam 24 feet long, 6 inches deep, and with a cross–sectional area of 28 square inches.

SOLUTION
$$S = \frac{1795Ad}{l}$$

$$S = \frac{1795(28)(6)}{24}$$

Use a calculator or pencil and paper.

1 7 9 5 [×] 2 8 [×] 6 [÷] 2 4 [=]

Display: 12565.

24 ft
$A = 28$ in^2

The greatest safe load is **12,565 pounds**.

Exercises

Calculate the greatest safe load in pounds of a steel I–beam that has the given dimensions.

1. $A = 15$ in^2
   $d = 4$ in
   $l = 12$ ft

2. $A = 8$ in^2
   $d = 4$ in
   $l = 10$ ft

3. $A = 48$ in^2
   $d = 8$ in
   $l = 20$ ft

4. $A = 21$ in^2
   $d = 6$ in
   $l = 15$ ft

5. The formula $S = \frac{1780Ad}{l}$ is used to calculate the greatest safe load in pounds of a solid steel beam with the load distributed along the beam. Find the greatest safe load of such a beam with a length of 24 feet, a depth of 6 inches, and a cross–sectional area of 30 square inches.

CONSUMER APPLICATION
Saving $ on Cooling Costs

When buying an air conditioner, it is important to select a model that is the right size. The size or **cooling capacity** of an air conditioner is measured in **British Thermal units (BTU's)**. One **BTU** is the energy needed to raise the temperature of one pound of water one degree Fahrenheit.

The graph shown at the right can be used to estimate how much cooling capacity is needed for a given room of a house or apartment. (The graph can be used for any room that is about eight feet high provided that it is not a kitchen and not directly below an attic floor.) The exposure of the room's exterior wall determines which line you should use.

EXAMPLE The room of a house has a northern exposure. Its floor area is 300 square feet. What must be the size (cooling capacity) of the air conditioner for this room?

SOLUTION
1. On the Floor Area scale, find the floor area, 300 square feet.
2. Since the room has a northern exposure, find the point on the northern exposure line directly to the right of the 300 reading.
3. From this point on the line read directly <u>down</u> on the Cooling Capacity scale to find the correct cooling capacity, **5500 BTU's per hour.**

Exercises
Estimate the size of an air conditioner needed to cool each room described. Round each answer to the nearest 500 BTU's.

1. Southern exposure; 300 square feet
2. Eastern exposure; 300 square feet
3. Southern exposure; 500 square feet
4. Northern exposure; 500 square feet
5. Western exposure; 100 square feet
6. Northern exposure; 100 square feet

REVIEW CAPSULE FOR SECTION 5-6

Evaluate each expression   (Pages 22–24)

1. $(4.5)(a)$ when $a = 2.7$
2. $(3.1)(y)$ when $y = 6$
3. $2n$ when $n = 3.5$
4. $\frac{1}{2}bc$ when $b = 8$ and $c = 5$
5. $\frac{1}{2}pq$ when $p = \frac{2}{3}$ and $q = 4$
6. mt when $m = \frac{8}{9}$ and $t = 5$

Multiply.   (Pages 38–41)

7. $3.4(8.1)$
8. $1.4(3.5)$
9. $4(2.3)$
10. $(1.6)(2.1)$
11. $12.4(3.5)$
12. $0.5(8.1)$

Multiply. Write each answer in lowest terms.   (Pages 124–125)

13. $\frac{1}{2} \cdot \frac{4}{5}$
14. $\frac{1}{4} \cdot \frac{8}{9}$
15. $\frac{2}{3} \cdot 6$
16. $\frac{3}{5} \cdot 15$
17. $\frac{4}{9} \cdot \frac{1}{4}$
18. $16 \cdot \frac{5}{6}$

The answers are on page 154.

PROBLEM SOLVING AND APPLICATIONS

5-6 Area: Parallelograms and Triangles

A rectangle can be changed into a **parallelogram** of equal area.

Thus, the formula for finding the area of a parallelogram is similar to the formula for finding the area of a rectangle.

Word Rule: The area of a parallelogram is the product of the base and the height.

Formula: $A = bh$

A = area
b = length of a base
h = corresponding height

The base of a parallelogram can be any side.
The height of a parallelogram is measured along a line at right angles to the base.

140   Chapter 5

EXAMPLE 1 An address plate on an office building has the shape of a parallelogram. The plate is 32 centimeters long and 16.5 centimeters high. Find the area.

Solution:
1. Formula: $A = bh$
2. Known values: $b = 32$ cm; $h = 16.5$ cm
3. Replace b with 32 and h with 16.5. $A = 32 \times 16.5$
4. Multiply. $A = 528.0$ The area is **528 cm²**.

Self-Check *Find the area of each parallelogram.*

1. $b = 7\frac{1}{2}$ in; $h = 3$ in
2. $b = 12.8$ cm; $h = 5.6$ cm

A parallelogram can be changed into two triangles with equal area. The area of each triangle is one-half the area of the parallelogram.

Word Rule: The area of a triangle is one-half the product of the base and the height.

Formula: $A = \frac{1}{2}bh$

A = area
b = length of a base
h = corresponding height

EXAMPLE 2 The gable of a roof has the shape of a triangle. The gable is 18 yards long and $5\frac{1}{3}$ yards high. Find the area.

Solution:
1. Formula: $A = \frac{1}{2}bh$
2. Known values: $b = 18$ yd; $h = 5\frac{1}{3}$ yd
3. Replace b with 18 and h with $5\frac{1}{3}$. Write a fraction for $5\frac{1}{3}$. $A = \frac{1}{2} \times 18 \times \frac{16}{3}$
4. Multiply. $A = \frac{1}{2} \times \frac{\cancel{18}^{9}}{1} \times \frac{16}{\cancel{3}_{1}}$

$A = \frac{48}{1} = 48$

The area is **48 yd²**.

Self-Check *Find the area of each triangle.*

3. $b = 4\frac{1}{2}$ ft; $h = 6$ ft
4. $b = 8.6$ m; $h = 10$ m

Fractions: Multiplication/Division

CLASSROOM EXERCISES

Find the area of each parallelogram. (Example 1)

CUSTOMARY MEASURES

1. $b = 8$ in; $h = 3\frac{1}{2}$ in **2.** $b = 2\frac{2}{3}$ ft; $h = 1\frac{3}{4}$ ft **3.** $b = 13$ yd; $h = 7$ yd

METRIC MEASURES

4. $b = 6.2$ cm; $h = 4$ cm **5.** $b = 13$ m; $h = 7.6$ m **6.** $b = 24.5$ cm; $h = 9.2$ cm

Find the area of each triangle. (Example 2)

CUSTOMARY MEASURES

7. $b = 4$ ft; $h = 3$ ft **8.** $b = 6\frac{1}{4}$ in; $h = 4$ in **9.** $b = 9$ yd; $h = 4\frac{2}{3}$ yd

METRIC MEASURES

10. $b = 5$ m; $h = 4$ m **11.** $b = 12.2$ cm; $h = 6$ cm **12.** $b = 10.4$ mm; $h = 8.2$ mm

WRITTEN EXERCISES

Goals: To use a formula to find the area of a parallelogram and of a triangle

To apply these formulas to solving word problems

Find the area of each parallelogram. (Example 1)

CUSTOMARY MEASURES

1. Stair Rail — 45 in, 5 in

2. Product Label — SUPER STRENGTH, $1\frac{1}{2}$ in, 3 in

3. Address Plate — 4251, $\frac{3}{4}$ ft, $2\frac{1}{3}$ ft

	Base	Height	Area
4.	6 in	$3\frac{1}{3}$ in	?
6.	$4\frac{1}{5}$ yd	$2\frac{1}{3}$ yd	?

	Base	Height	Area
5.	12 ft	8 ft	?
7.	16 in	$12\frac{1}{2}$ in	?

142 Chapter 5

METRIC MEASURES

8. Parking Lot — 7 dam, 9 dam

9. Decorative Tile — 13 cm, 17.5 cm

10. Product Label — 1.5 cm, 4 cm

	Base	Height	Area
11.	3.2 cm	4 cm	?
13.	7 m	12 m	?

	Base	Height	Area
12.	15.5 mm	18 mm	?
14.	35 cm	14 cm	?

Find the area of each triangle. (Example 2)

CUSTOMARY MEASURES

15. Table Top — 2 ft, $4\frac{1}{2}$ ft

16. Sail — $9\frac{1}{3}$ yd, 3 yd

17. Tie Back — $3\frac{5}{6}$ in, 14 in

	Base	Height	Area
18.	4 in	6 in	?
20.	$8\frac{1}{6}$ ft	4 ft	?

	Base	Height	Area
19.	10 ft	$18\frac{1}{4}$ ft	?
21.	5 yd	$3\frac{1}{9}$ yd	?

METRIC MEASURES

22. Gable of a Roof — 1.9 m, 6 m

23. Flag — 2.5 m, 4.6 m

24. Sail — 8.6 m, 6 m

Fractions: Multiplication/Division 143

Find the area of each triangle. (Example 2)

	Base	Height	Area
25.	6 cm	4.2 cm	?
27.	3.5 cm	2.6 cm	?

	Base	Height	Area
26.	3 m	2.5 m	?
28.	16 mm	14 mm	?

MIXED PRACTICE

Solve each problem.

29. The mainsail on Jim's catamaran is in the shape of a triangle. The base is $6\frac{2}{3}$ feet and the height is $20\frac{1}{4}$ feet. Find the area of the sail.

30. Ellen's garden has the shape of a parallelogram. The base is 6 meters and the height is 5.6 meters. Find the area of the garden.

31. A wing on a model rocket has the shape of a parallelogram. The base is $2\frac{1}{3}$ inches and the height is $6\frac{3}{4}$ inches. Find the area of the wing.

32. Each side of a tent is triangular in shape. The base of the triangle is 2.4 meters and the height is 3 meters. Find the area of one side of the tent.

REVIEW CAPSULE FOR SECTION 5-7

Evaluate each expression. (Pages 22–24)

1. xyz when $x = 2$, $y = 4$, and $z = 12$
2. ab when $a = 3$ and $b = 9$
3. kmn when $k = 6$, $m = 1.7$, and $n = 3.2$
4. st when $s = 0.6$ and $t = 4.9$
5. cd when $c = 8$ and $d = 3\frac{1}{3}$
6. rst when $r = 2$, $s = 3\frac{1}{2}$, and $t = 1\frac{1}{3}$

Multiply. (Pages 38–41)

7. $4(3.2)$
8. $7.1(5)$
9. $2.8(1.6)$
10. $1.9(3.5)$
11. $(5.2)(1.8)$
12. $(3.6)(0.7)$
13. $(12)(3.8)$
14. $(6)(2.9)$

Multiply. Write each answer in lowest terms. (Pages 129–131)

15. $1\frac{1}{2} \cdot 2$
16. $3 \cdot 2\frac{1}{4}$
17. $1\frac{1}{3} \cdot 4\frac{1}{2}$
18. $2\frac{2}{5} \cdot 1\frac{1}{8}$
19. $1\frac{1}{6} \cdot 2\frac{1}{7}$
20. $8\frac{1}{3} \cdot 2\frac{2}{5}$
21. $4 \cdot 2\frac{3}{4}$
22. $6\frac{1}{4} \cdot 2\frac{3}{5}$

The answers are on page 154.

PROBLEM SOLVING AND APPLICATIONS

5-7 Volume: Rectangular Prisms

A **rectangular prism** is a solid such as the shoe box at the right. The **volume** of a rectangular prism is the amount of space it contains. Volume is measured in **cubic units** such as cubic centimeters (cm³), cubic meters (m³), cubic inches (in³), cubic feet (ft³), and so on.

Word Rule: The volume of a rectangular prism is the product of the length, the width, and the height.

Formula: $V = lwh$ ◀ V = volume; l = length; w = width; h = height

EXAMPLE A shoe box has a length of 35 centimeters, a width of 15 centimeters, and a height of 10 centimeters. Find the volume of the shoe box.

Solution:
1. Formula: $V = lwh$
2. Known values: $l = 35$ cm; $w = 15$ cm; $h = 10$ cm
3. Replace l with 35, w with 15, and h with 10.
   $V = 35 \times 15 \times 10$
   $V = 5250$

The volume is **5250 cm³**.

Self-Check *Find the volume of each rectangular prism.*
1. $l = 6\frac{1}{2}$ ft; $w = 3\frac{1}{3}$ ft; $h = 2$ ft
2. $l = 12$ m; $w = 4.8$ m; $h = 5.2$ m

NOTE: The formula for the volume of a rectangular prism can also be written as follows.

$V = l \cdot w \cdot h$

$V = B \cdot h$, or

$V = Bh$ ◀ B = area of base
h = height

Fractions: Multiplication/Division

CLASSROOM EXERCISES

Find the volume of each rectangular prism. (Example)

1. $l = 9$ ft; $w = 6$ ft; $h = 2\frac{1}{2}$ ft
2. $l = 10$ in; $w = 5\frac{1}{4}$ in; $h = 3$ in
3. $l = 0.5$ cm; $w = 1$ cm; $h = 1.7$ cm
4. $l = 2.7$ m; $w = 3.1$ m; $h = 5$ m
5. $l = 8\frac{1}{2}$ yd; $w = 3$ yd; $h = 2\frac{1}{3}$ yd
6. $l = 4\frac{1}{4}$ ft; $w = 3$ ft; $h = 5\frac{1}{7}$ ft
7. $l = 16.4$ mm; $w = 9.2$ mm; $h = 5$ mm
8. $l = 2.5$ cm; $w = 1.4$ cm; $h = 4$ cm

WRITTEN EXERCISES

Goals: To use a formula to find the volume of a rectangular prism
To apply this formula to solving word problems

Find the volume of each rectangular prism. (Example)

CUSTOMARY MEASURES

1. Shed

6 ft, 4 ft, $4\frac{1}{4}$ ft

2. Truck

$3\frac{1}{2}$ yd, $2\frac{3}{4}$ yd, 1 yd

3. Refrigerator

$3\frac{3}{4}$ ft, 3 ft, 6 ft

	Length	Width	Height
4.	10 ft	5 ft	4 ft
6.	5 yd	$6\frac{1}{5}$ yd	$3\frac{1}{2}$ yd

	Length	Width	Height
5.	$16\frac{1}{2}$ in	8 in	$2\frac{1}{4}$ in
7.	12 ft	$3\frac{1}{8}$ ft	$3\frac{1}{3}$ ft

METRIC MEASURES

8. Planter

20 cm, 10 cm, 10 cm

9. Paper Weight

52 mm, 24 mm, 38 mm

10. Stereo Speaker

26 cm, 18 cm, 40 cm

	Length	Width	Height
11.	5 cm	2 cm	4 cm
13.	9 m	6 m	5.1 m

	Length	Width	Height
12.	8 m	6 m	3.9 m
14.	2.1 mm	1.9 mm	5.4 mm

MIXED PRACTICE

15. Stephanie McCloud is building a kitchen cabinet that is 3 feet long, $1\frac{3}{4}$ feet wide, and $2\frac{1}{2}$ feet high. Find the volume of the cabinet.

16. A wall oven is 100 centimeters wide, 85 centimeters deep, and 64 centimeters high. Find the volume of the oven.

17. A fish tank is 45 centimeters long, 30 centimeters wide, and 25.4 centimeters high. Find the volume.

18. A wading pool is 12 feet long, $8\frac{1}{2}$ feet wide, and $4\frac{1}{6}$ feet high. Find the volume of the wading pool.

MORE CHALLENGING EXERCISES

Find the volume of each figure.

19.

20.

21. The term "1 inch of rainfall" means a rainfall 1 inch high covering 1 acre of ground. An acre equals 43,560 square feet. Find the volume of water in cubic feet for 1 inch of rainfall. (HINT: 1 inch = $\frac{1}{12}$ foot)

22. Twenty-one people crowded into a telephone booth together. The telephone booth was $3\frac{1}{2}$ feet long, $3\frac{1}{2}$ feet wide, and $7\frac{1}{5}$ feet high. How many cubic feet of space did each person have in the booth?

REVIEW CAPSULE FOR SECTION 5-8

Write a fraction for each mixed numeral. (Pages 135–137)

1. $1\frac{1}{2}$
2. $2\frac{3}{4}$
3. $5\frac{1}{3}$
4. $4\frac{2}{5}$
5. $3\frac{1}{6}$
6. $1\frac{2}{7}$

Multiply. Write each answer in lowest terms. (Pages 129–131)

7. $\frac{1}{3} \cdot \frac{9}{1}$
8. $6 \cdot \frac{1}{2}$
9. $\frac{6}{5} \cdot \frac{10}{9}$
10. $\frac{7}{4} \cdot \frac{8}{3}$
11. $8 \cdot \frac{1}{4}$
12. $\frac{3}{2} \cdot \frac{4}{9}$

The answers are on page 154.

5-8 Dividing Fractions

When the product of two numbers is 1, the numbers are **reciprocals** of each other. Reciprocals are sometimes called **multiplicative inverses**.

TABLE

Number	Reciprocal	Reason
$\frac{5}{16}$	$\frac{16}{5}$	$\frac{5}{16} \cdot \frac{16}{5} = 1$
$\frac{1}{12}$	12	$\frac{1}{12} \cdot 12 = 1$
$\frac{10}{3}$	$\frac{3}{10}$	$\frac{10}{3} \cdot \frac{3}{10} = 1$
$3\frac{1}{2}$, or $\frac{7}{2}$	$\frac{2}{7}$	$\frac{7}{2} \cdot \frac{2}{7} = 1$

Dividing by a number gives the same result as multiplying by its reciprocal.

$$18 \div 2 = 9 \qquad 18 \cdot \frac{1}{2} = 9$$

PROCEDURE

To divide with fractions:

1. Using the reciprocal of the divisor, write the division problem as a multiplication problem.
2. Multiply.
3. Write the answer in lowest terms.

EXAMPLE 1 Divide.

a. $\frac{5}{16} \div \frac{3}{4}$

b. $\frac{1}{12} \div 9$

Solutions:

a.
1. $\frac{5}{16} \div \frac{3}{4} = \frac{5}{16} \cdot \frac{4}{3}$ ◄ The reciprocal of $\frac{3}{4}$ is $\frac{4}{3}$.
2. $= \frac{5}{\cancel{16}_{4}} \cdot \frac{\cancel{4}^{1}}{3}$
3. $= \frac{5}{12}$ ◄ Lowest terms

b.
1. $\frac{1}{12} \div 9 = \frac{1}{12} \cdot \frac{1}{9}$ ◄ The reciprocal of 9 is $\frac{1}{9}$.
2. $= \frac{1 \cdot 1}{12 \cdot 9}$
3. $= \frac{1}{108}$

Self-Check *Divide.*

1. $\frac{3}{5} \div \frac{4}{15}$
2. $\frac{7}{8} \div 21$
3. $\frac{4}{5} \div 10$
4. $\frac{2}{9} \div \frac{24}{45}$

> **PROCEDURE**
> **To divide with mixed numerals:**
> 1. Write a fraction for each mixed numeral.
> 2. Divide the fractions.
> 3. Write the answer in lowest terms.

EXAMPLE 2 Divide.

a. $2\frac{1}{4} \div 3\frac{3}{8}$ b. $24 \div 5\frac{1}{3}$

Solution:

a. ① $2\frac{1}{4} \div 3\frac{3}{8} = \frac{9}{4} \div \frac{27}{8}$

② $= \frac{9}{4} \cdot \frac{8}{27}$ ◀ The reciprocal of $\frac{27}{8}$ is $\frac{8}{27}$.

$= \frac{\cancel{9}^1}{\cancel{4}_1} \cdot \frac{\cancel{8}^2}{\cancel{27}_3}$

$= \frac{1 \cdot 2}{1 \cdot 3}$

③ $= \frac{2}{3}$

b. ① $24 \div 5\frac{1}{3} = 24 \div \frac{16}{3}$

② $= \frac{24}{1} \cdot \frac{3}{16}$

$= \frac{\cancel{24}^3}{1} \cdot \frac{3}{\cancel{16}_2}$

$= \frac{3 \cdot 3}{1 \cdot 2}$

③ $= \frac{9}{2} = 4\frac{1}{2}$

Self-Check Divide.

5. $3\frac{1}{4} \div 6\frac{1}{2}$ 6. $12 \div 2\frac{4}{7}$ 7. $5\frac{1}{3} \div 8$ 8. $4\frac{1}{2} \div 2\frac{3}{4}$

> For the fractions $\frac{a}{b}$ and $\frac{c}{d}$, $b \neq 0$, $c \neq 0$, $d \neq 0$,
>
> $\frac{a}{b} \div \frac{c}{d} = \frac{a}{b} \cdot \frac{d}{c} = \frac{ad}{bc}$ $\frac{1}{2} \div \frac{3}{5} = \frac{1}{2} \cdot \frac{5}{3} = \frac{5}{6}$

CLASSROOM EXERCISES

Write the reciprocal of each number. (Table)

1. $\frac{2}{3}$ 2. $\frac{3}{5}$ 3. $\frac{9}{7}$ 4. 2   5. 4   6. $\frac{10}{3}$ 7. $\frac{3}{2}$

Divide. Write each answer in lowest terms. (Example 1)

8. $\frac{1}{4} \div \frac{1}{3}$ 9. $\frac{3}{8} \div 3$ 10. $\frac{2}{7} \div \frac{3}{8}$ 11. $\frac{4}{7} \div \frac{2}{3}$ 12. $\frac{3}{5} \div \frac{15}{20}$

13. $\frac{7}{10} \div 14$ 14. $\frac{4}{9} \div 16$ 15. $\frac{9}{10} \div 15$ 16. $\frac{1}{7} \div 21$ 17. $\frac{3}{5} \div 12$

Fractions: Multiplication/Division

(Example 2)

18. $2\frac{1}{2} \div 3\frac{1}{3}$ 19. $9\frac{2}{3} \div 4\frac{1}{6}$ 20. $2\frac{1}{4} \div 3\frac{3}{8}$ 21. $2\frac{1}{2} \div 2\frac{1}{2}$ 22. $4\frac{3}{4} \div 3\frac{1}{4}$

23. $7\frac{1}{2} \div 15$ 24. $9\frac{3}{4} \div 13$ 25. $4 \div 1\frac{1}{3}$ 26. $3 \div 7\frac{1}{2}$ 27. $21 \div 3\frac{3}{7}$

WRITTEN EXERCISES

Goals: To divide with fractions and mixed numerals
To apply this skill to solving word problems

Write the reciprocal of each number. (Table)

1. $\frac{1}{3}$ 2. $\frac{3}{5}$ 3. $\frac{4}{7}$ 4. $\frac{5}{18}$ 5. 8 6. 4 7. 12

8. $\frac{11}{5}$ 9. $\frac{18}{11}$ 10. $\frac{21}{4}$ 11. $\frac{13}{4}$ 12. 21 13. $\frac{7}{9}$ 14. 9

Divide. Write each answer in lowest terms. (Example 1)

15. $\frac{1}{2} \div \frac{3}{4}$ 16. $\frac{1}{4} \div \frac{1}{3}$ 17. $\frac{3}{5} \div \frac{4}{5}$ 18. $\frac{5}{8} \div \frac{5}{6}$ 19. $\frac{2}{3} \div \frac{4}{5}$

20. $\frac{4}{7} \div \frac{7}{12}$ 21. $\frac{1}{3} \div \frac{5}{9}$ 22. $\frac{1}{8} \div \frac{1}{4}$ 23. $\frac{5}{9} \div \frac{11}{12}$ 24. $\frac{7}{9} \div 14$

25. $\frac{1}{6} \div 5$ 26. $\frac{1}{5} \div 6$ 27. $4 \div \frac{2}{3}$ 28. $8 \div \frac{1}{4}$ 29. $\frac{2}{3} \div 5$

(Example 2)

30. $6\frac{1}{3} \div 1\frac{1}{3}$ 31. $8\frac{1}{5} \div 2\frac{1}{2}$ 32. $3\frac{1}{4} \div 6\frac{1}{2}$ 33. $1\frac{3}{5} \div 3\frac{1}{5}$ 34. $3\frac{3}{8} \div 2\frac{1}{4}$

35. $6\frac{6}{7} \div 2\frac{1}{7}$ 36. $2 \div 1\frac{1}{4}$ 37. $6 \div 1\frac{1}{2}$ 38. $5\frac{2}{3} \div 2$ 39. $4\frac{4}{5} \div 6$

40. $4 \div 6\frac{1}{4}$ 41. $6\frac{2}{3} \div 1\frac{3}{4}$ 42. $9\frac{1}{7} \div 8\frac{1}{2}$ 43. $3\frac{7}{8} \div 4$ 44. $5\frac{1}{4} \div 6\frac{7}{8}$

MIXED PRACTICE

Divide. Write each answer in lowest terms.

45. $6 \div \frac{1}{2}$ 46. $\frac{8}{9} \div \frac{2}{3}$ 47. $3\frac{1}{2} \div 7$ 48. $4\frac{2}{3} \div 1\frac{7}{9}$ 49. $3 \div \frac{2}{5}$

50. $1\frac{1}{2} \div 3\frac{3}{8}$ 51. $\frac{2}{5} \div \frac{4}{15}$ 52. $\frac{3}{4} \div \frac{1}{8}$ 53. $1\frac{7}{10} \div 3\frac{2}{5}$ 54. $2 \div 4\frac{1}{2}$

55. $1\frac{1}{5} \div 2\frac{1}{5}$ 56. $\frac{4}{9} \div 4$ 57. $5\frac{1}{4} \div 4\frac{3}{8}$ 58. $7 \div 3\frac{2}{3}$ 59. $\frac{11}{12} \div \frac{5}{6}$

60. $\frac{7}{8} \div \frac{1}{6}$ 61. $3\frac{3}{8} \div 2\frac{1}{2}$ 62. $7\frac{1}{3} \div 4\frac{2}{3}$ 63. $\frac{3}{7} \div \frac{9}{14}$ 64. $8 \div 4\frac{1}{2}$

APPLICATIONS: USING FRACTIONS AND MIXED NUMERALS

65. A stack of magazines weighs 10 pounds. Each magazine weighs $\frac{5}{8}$ pound. How many magazines are in the stack?

66. Jason rides his bicycle 25 kilometers in $1\frac{1}{4}$ hours. How far does he ride in 1 hour?

67. How many boards $2\frac{1}{4}$ feet long can be cut from a board $20\frac{1}{4}$ feet long?

MORE CHALLENGING EXERCISES

Divide. No variable equals zero.

EXAMPLE: $\frac{x}{5} \div x = \frac{x}{5} \cdot \frac{1}{x}$

The reciprocal of x is $\frac{1}{x}$.

$$= \frac{\overset{1}{\cancel{x}} \cdot 1}{5 \cdot \underset{1}{\cancel{x}}} = \frac{1 \cdot 1}{5 \cdot 1} = \frac{1}{5}$$

68. $\frac{y}{4} \div y$
69. $\frac{p}{6} \div p$
70. $\frac{2}{5} \div \frac{n}{5}$
71. $\frac{1}{4} \div \frac{t}{4}$
72. $\frac{r}{4} \div \frac{r}{5}$
73. $\frac{x}{3} \div \frac{x}{6}$
74. $1\frac{1}{3} \div a$
75. $4\frac{1}{2} \div 3p$
76. $3\frac{1}{3} \div \frac{y}{3}$
77. $6\frac{1}{4} \div \frac{5}{t}$

REVIEW: SECTIONS 5-5 — 5-8

Multiply. Write each answer in lowest terms. (Section 5-5)

1. $1\frac{2}{5} \cdot 2\frac{1}{7}$
2. $3\frac{1}{3} \cdot 1\frac{1}{2}$
3. $4\frac{1}{6} \cdot 5\frac{3}{5}$
4. $2\frac{2}{9} \cdot 6\frac{1}{10}$

Find the area of each triangle. (Section 5-6)

5. $b = 5$ ft; $h = 6$ ft
6. $b = 8.6$ m; $h = 3.5$ m
7. $b = 4\frac{1}{8}$ in; $h = 6$ in

Find the volume of each rectangular prism. (Section 5-7)

8. $l = 6$ yd; $w = 8$ yd; $h = 2\frac{1}{8}$ yd
9. $l = 3.7$ cm; $w = 5.2$ cm; $h = 2.5$ cm

Divide. Write each answer in lowest terms. (Section 5-8)

10. $\frac{1}{2} \div \frac{1}{3}$
11. $\frac{3}{4} \div 6$
12. $3 \div 2\frac{1}{7}$
13. $1\frac{1}{2} \div 1\frac{3}{5}$

14. The base of a parallelogram is $3\frac{1}{2}$ inches. The height is $2\frac{1}{4}$ inches. Find the area. (Section 5-6)

15. A car travels 120 miles on $6\frac{2}{5}$ gallons of fuel. How far does it travel on one gallon of fuel? (Section 5-8)

CHAPTER REVIEW

PART 1: VOCABULARY

For Exercises 1–6, choose from the box at the right the word(s) that best corresponds to each description.

1. The boiling point of water is 212° on this temperature scale. __?__ (p. 132)
2. Two numbers whose product is 1 __?__ (p. 148)
3. Represents the sum of a whole number and a fraction __?__ (p. 126)
4. The freezing point of water is 0° on this temperature scale. __?__ (p. 132)
5. A numerator and denominator having no common prime factors __?__ (p. 124)
6. The amount of space that a solid contains __?__ (p. 145)

> lowest
> volume
> mixed numeral
> Celsius
> Fahrenheit
> reciprocals

PART 2: SKILLS

Write each fraction in lowest terms. (Section 5–1)

7. $\frac{3}{6}$
8. $\frac{4}{18}$
9. $\frac{6}{24}$
10. $\frac{28}{30}$
11. $\frac{14}{32}$
12. $\frac{18}{51}$

Add or subtract as indicated. Write each answer as a mixed numeral. (Section 5–2)

13. 1 ft 8 in
    +2 ft 3 in

14. 5 ft 6 in
    +9 ft 6 in

15. 4 gal 3 qt
    +5 gal 3 qt

16. 7 lb 15 oz
    +3 lb 8 oz

17. 8 yd 1 ft
    −3 yd 2 ft

18. 7 ft 5 in
    −5 ft 8 in

19. 10 gal 2 qt
    − 4 gal 1 qt

20. 57 lb
    −28 lb 7 oz

Multiply. Write each answer in lowest terms. (Section 5–3)

21. $\frac{1}{2} \cdot \frac{3}{8}$
22. $\frac{3}{4} \cdot \frac{1}{2}$
23. $\frac{4}{5} \cdot \frac{1}{2}$
24. $\frac{2}{5} \cdot \frac{3}{4}$
25. $4 \cdot \frac{1}{6}$
26. $\frac{2}{3} \cdot \frac{9}{10}$
27. $\frac{5}{6} \cdot \frac{3}{10}$
28. $\frac{4}{7} \cdot \frac{1}{8}$
29. $\frac{5}{8} \cdot \frac{4}{5}$
30. $\frac{8}{9} \cdot \frac{3}{32}$

Complete. (Section 5–4)

31. 40°C = __?__ °F
32. 104°F = __?__ °C
33. 10°C = __?__ °F

Write a mixed numeral for each fraction. (Section 5–5)

34. $\frac{5}{3}$
35. $\frac{7}{2}$
36. $\frac{25}{12}$
37. $\frac{15}{4}$
38. $\frac{13}{7}$
39. $\frac{21}{8}$

Write a fraction for each mixed numeral. (Section 5-5)

40. $3\frac{1}{5}$ **41.** $1\frac{1}{4}$ **42.** $6\frac{1}{3}$ **43.** $4\frac{3}{8}$ **44.** $7\frac{3}{4}$ **45.** $2\frac{4}{7}$

Multiply. Write each answer in lowest terms. (Section 5-5)

46. $3\frac{3}{5} \cdot 5$ **47.** $4\frac{1}{3} \cdot 9$ **48.** $6\frac{1}{2} \cdot 3$ **49.** $\frac{5}{6} \cdot 3\frac{3}{10}$ **50.** $\frac{5}{8} \cdot 2\frac{1}{5}$

51. $1\frac{1}{6} \cdot 1\frac{4}{5}$ **52.** $\frac{1}{8} \cdot 3\frac{3}{4}$ **53.** $3\frac{1}{2} \cdot 4\frac{2}{3}$ **54.** $1\frac{1}{4} \cdot 1\frac{1}{4}$ **55.** $1\frac{7}{8} \cdot 6\frac{2}{5}$

Find the area of each parallelogram. (Section 5-6)

	Base	Height	Area
56.	52.7 m	15 m	?
57.	4 in	$2\frac{1}{2}$ in	?

Find the area of each triangle. (Section 5-6)

	Base	Height	Area
58.	20 ft	$16\frac{1}{10}$ ft	?
59.	3.5 cm	2.6 cm	?

Find the volume of each rectangular prism. (Section 5-7)

	Length	Width	Height
60.	5 ft	2 ft	$5\frac{1}{4}$ ft
62.	4.6 m	1.2 m	0.6 m

	Length	Width	Height
61.	6.2 m	4.1 m	5 m
63.	$12\frac{1}{2}$ in	$8\frac{2}{5}$ in	$4\frac{1}{3}$ in

Divide. Write each answer in lowest terms. (Section 5-8)

64. $\frac{5}{6} \div \frac{2}{3}$ **65.** $3\frac{1}{3} \div 1\frac{1}{5}$ **66.** $6 \div \frac{3}{8}$ **67.** $1\frac{1}{8} \div 3$ **68.** $2\frac{1}{4} \div 3\frac{3}{8}$

69. $\frac{3}{8} \div \frac{3}{4}$ **70.** $2\frac{1}{5} \div 4$ **71.** $2 \div \frac{4}{5}$ **72.** $3\frac{1}{2} \div \frac{3}{4}$ **73.** $4\frac{1}{4} \div \frac{1}{4}$

PART 3: APPLICATIONS

74. A Lunar Rover carrying an astronaut wearing a space suit and backpack weighs 880 pounds 5 ounces. The astronaut wearing a spacesuit and backpack weighs 379 pounds 13 ounces. Find the weight of the Lunar Rover. (Section 5-2)

75. At sunset, the temperature on the moon is about 59°F. Express this as a Celsius reading. (Section 5-4)

76. The wing on a hang glider is in the shape of a triangle. The base is 5.4 meters and the height is 1.5 meters. Find the area. (Section 5-6)

77. A rectangular decompression chamber is $10\frac{1}{2}$ feet long, 5 feet wide, and $9\frac{1}{3}$ feet high. Find the volume. (Section 5-7)

Fractions: Multiplication/Division **153**

CHAPTER TEST

Write each fraction in lowest terms.

1. $\frac{7}{14}$
2. $\frac{6}{36}$
3. $\frac{15}{40}$
4. $\frac{12}{28}$

Add or subtract as indicated. Write each answer as a mixed numeral.

5. $\begin{array}{r} 3 \text{ ft } 7 \text{ in} \\ +5 \text{ ft } 2 \text{ in} \\ \hline \end{array}$

6. $\begin{array}{r} 5 \text{ gal } 1 \text{ qt} \\ -3 \text{ gal } 3 \text{ qt} \\ \hline \end{array}$

Multiply. Write each answer in lowest terms.

7. $\frac{1}{2} \cdot \frac{5}{6}$
8. $\frac{4}{5} \cdot \frac{5}{12}$
9. $2\frac{1}{3} \cdot 6$
10. $2\frac{1}{4} \cdot 3\frac{5}{6}$

Find the area of each parallelogram.

11. $b = 15$ m; $h = 8.7$ m
12. $b = 14\frac{3}{8}$ ft; $h = 3\frac{1}{5}$ ft

Find the area of each triangle.

13. $b = 10$ in; $h = 5\frac{1}{3}$ in
14. $b = 12.1$ cm; $h = 8.8$ cm

Divide. Write each answer in lowest terms.

15. $\frac{1}{2} \div \frac{1}{4}$
16. $\frac{5}{6} \div \frac{3}{4}$
17. $3 \div \frac{4}{9}$
18. $2\frac{2}{5} \div 3\frac{3}{4}$

19. On a certain day in Death Valley, California, the temperature reached 131°F. Write this as a Celsius reading.

20. A storage closet for linen is 2 meters long, 1 meter wide, and 0.8 meters high. Find the volume of the closet.

ANSWERS TO REVIEW CAPSULES

Page 128 1. 176 2. 378 3. 150 4. 156 5. 336 6. 576 7. $\frac{2}{3}$ 8. $\frac{3}{7}$ 9. $\frac{2}{3}$ 10. $\frac{14}{15}$ 11. $\frac{2}{3}$ 12. $\frac{2}{9}$

Page 131 1. 21 2. 136 3. 263 4. 278 5. 19 6. 40 7. 68 8. 30 9. 84 10. 108

Page 134 1. 1 2. 2 3. 3 4. 6 5. 7 6. 2 7. $\frac{1}{4}$ 8. $\frac{3}{14}$ 9. $\frac{7}{18}$ 10. $\frac{7}{15}$ 11. $\frac{3}{5}$ 12. $\frac{4}{5}$

Page 140 1. 12.15 2. 18.6 3. 7 4. 20 5. $1\frac{1}{3}$ 6. $4\frac{4}{9}$ 7. 27.54 8. 4.9 9. 9.2 10. 3.36 11. 43.4 12. 4.05 13. $\frac{2}{5}$ 14. $\frac{2}{9}$ 15. 4 16. 9 17. $\frac{1}{9}$ 18. $13\frac{1}{3}$

Page 144 1. 96 2. 27 3. 32.64 4. 2.94 5. $26\frac{2}{3}$ 6. $9\frac{1}{3}$ 7. 12.8 8. 35.5 9. 4.48 10. 6.65 11. 9.36 12. 2.52 13. 45.6 14. 17.4 15. 3 16. $6\frac{3}{4}$ 17. 6 18. $2\frac{7}{10}$ 19. $2\frac{1}{2}$ 20. 20 21. 11 22. $16\frac{1}{4}$

Page 147 1. $\frac{3}{2}$ 2. $\frac{11}{4}$ 3. $\frac{16}{3}$ 4. $\frac{22}{5}$ 5. $\frac{19}{6}$ 6. $\frac{9}{7}$ 7. 3 8. 3 9. $1\frac{1}{3}$ 10. $4\frac{2}{3}$ 11. 2 12. $\frac{2}{3}$

ENRICHMENT

Special Areas

You sometimes have to use more than one formula to find area.

Remember: Opposite sides of rectangles, parallelograms, and squares have the same length.

EXAMPLE Find the area of the shaded region.

Solution:

Area of Shaded Region	=	Area of Parallelogram	−	Area of Triangle
A	=	bh	−	$\frac{1}{2}bh$
A	=	$30 \cdot 20$	−	$\frac{1}{2} \cdot 8 \cdot 10$
A	=	$600 - 40$, or **560 yd^2**		

EXERCISES

Find the area of the shaded region.

1. (5 m × 5 m square with 3 m triangle on top)

2. (trapezoid: top 3 cm, bottom 4 cm, height 2 cm)

3. (triangle base 15 m, height 10 m, with 3 m × 3 m square removed)

4. (6 in × 6 in square with triangle base 6 in, height 3 in removed)

5. (rectangle 8 ft × 4 ft with triangle base 2 ft, height 4 ft removed; 6 ft marked)

6. (shape: top 6 yd, right side 2 yd, left side 4 yd)

Fractions: Multiplication/Division

ADDITIONAL PRACTICE

SKILLS

Write each fraction in lowest terms. (Pages 124–125)

1. $\frac{15}{30}$
2. $\frac{7}{21}$
3. $\frac{12}{42}$
4. $\frac{30}{45}$
5. $\frac{42}{98}$
6. $\frac{15}{18}$

Add or subtract as indicated. Write each answer as a mixed numeral. (Pages 126–128)

7. 13 lb 8 oz
   +14 lb 12 oz

8. 16 gal 2 qt
   −11 gal 3 qt

9. 4 ft 6 in
   −2 ft 9 in

10. 5 yd 2 ft
    +7 yd 2 ft

Multiply. Write each answer in lowest terms. (Pages 129–131)

11. $\frac{2}{5} \cdot \frac{5}{8}$
12. $\frac{3}{7} \cdot 2$
13. $\frac{2}{7} \cdot \frac{3}{5}$
14. $\frac{3}{8} \cdot \frac{2}{9}$
15. $\frac{2}{11} \cdot 3$
16. $\frac{4}{9} \cdot \frac{3}{10}$

Complete. (Pages 132–134)

17. 35°C = __?__ F
18. 104°F = __?__ C
19. 32°F = __?__ C

Multiply. Write each answer in lowest terms. (Pages 135–137)

20. $3\frac{1}{3} \cdot 3$
21. $2\frac{1}{4} \cdot 10$
22. $1\frac{1}{3} \cdot 2\frac{1}{2}$
23. $4\frac{1}{5} \cdot 1\frac{3}{7}$
24. $1\frac{5}{6} \cdot \frac{15}{22}$

Find the area of each parallelogram. (Pages 140–144)

25. Base: 8 cm; Height: $3\frac{1}{2}$ cm
26. Base: $2\frac{1}{2}$ ft; Height: $4\frac{1}{4}$ ft

Find the area of each triangle. (Pages 141–144)

27. Base: 4 m; Height: 6 m
28. Base: 8.4 ft; Height: 2.5 ft

Find the volume of each rectangular prism. (Pages 145–147)

29. $l = 8$ m; $w = 4\frac{1}{2}$ m; $h = 1\frac{2}{3}$ m
30. $l = 8.2$ ft; $w = 4.1$ ft; $h = 3$ ft

Divide. Write each answer in lowest terms. (Pages 148–151)

31. $\frac{2}{3} \div \frac{8}{15}$
32. $\frac{2}{3} \div \frac{5}{6}$
33. $1\frac{1}{5} \div 2\frac{2}{3}$
34. $4\frac{1}{2} \div 4$
35. $1\frac{1}{3} \div \frac{1}{4}$

APPLICATIONS

36. Mark is $6\frac{1}{2}$ feet tall. His sister is $\frac{3}{5}$ Mark's height. How much taller is Mark than his sister? (Pages 129–131)

37. A rectangular pool is 50 meters long, 25 meters wide, and 6 meters deep. Find the volume of the pool. (Pages 145–147)

CHAPTER 6
Fractions: Addition/Subtraction

SECTIONS
- **6-1** Addition and Subtraction: Like Denominators
- **6-2** Addition and Subtraction: Unlike Denominators
- **6-3** Addition and Subtraction: Mixed Numerals
- **6-4** Problem Solving and Applications: Time Cards
- **6-5** Fractions and Decimals
- **6-6** Problem Solving and Applications Circles: Circumference and Area

FEATURES
Calculator Application: *Comparing Fractions*
Calculator/Computer Application: *Evaluating Powers*
Enrichment: *Precision of Measurement*
Common Errors

6-1 Addition and Subtraction: Like Denominators

Susan is making a pocketbook. She needs $\frac{3}{8}$ yard of felt for the pocketbook and $\frac{1}{8}$ yard for the strap. To find the total number of yards, Susan adds $\frac{3}{8}$ and $\frac{1}{8}$.

$$\frac{3}{8} + \frac{1}{8} = \frac{3+1}{8} = \frac{4}{8}, \text{ or } \frac{1}{2} \quad \blacktriangleleft \text{ Lowest terms}$$

Thus, she needs $\frac{1}{2}$ **yard** of felt in all.

Like fractions, such as $\frac{3}{8}$ and $\frac{1}{8}$, have **like** denominators.

PROCEDURE

To add or subtract like fractions:

1. Add or subtract the numerators. Write the sum or difference over the <u>like</u> denominator.
2. Write the answer in lowest terms.

EXAMPLE 1 Add or subtract as indicated: **a.** $\frac{7}{12} + \frac{11}{12}$ **b.** $\frac{13}{15} - \frac{7}{15}$

Solutions: **a.** [1] $\frac{7}{12} + \frac{11}{12} = \frac{7+11}{12}$ **b.** [1] $\frac{13}{15} - \frac{7}{15} = \frac{13-7}{15}$

$= \frac{18}{12}$ ◀ Write a mixed numeral for $\frac{18}{12}$. $= \frac{6}{15}$

[2] $= 1\frac{6}{12}$, or $1\frac{1}{2}$ [2] $= \frac{2}{5}$

Self-Check *Add or subtract as indicated.*

1. $\frac{5}{12} + \frac{11}{12}$ 2. $\frac{14}{15} - \frac{8}{15}$ 3. $\frac{19}{20} - \frac{11}{20}$ 4. $\frac{9}{10} + \frac{7}{10}$

Mixed numerals such as $1\frac{1}{6}$ and $3\frac{5}{6}$ also have like denominators.

PROCEDURE

To add or subtract mixed numerals:

1. Add or subtract the whole numbers.
2. Add or subtract the fractional parts.
3. Write the answer in lowest terms.

EXAMPLE 2 Add or subtract as indicated: **a.** $4\frac{1}{5} + 3\frac{2}{5}$ **b.** $10\frac{7}{8} - 5\frac{3}{8}$

Solutions: **a.** $4\frac{1}{5} + 3\frac{2}{5} = 7\frac{1+2}{5}$ **b.** $10\frac{7}{8} - 5\frac{3}{8} = 5\frac{7-3}{8}$
$\qquad\qquad\qquad\quad = 7\frac{3}{5}$ $\qquad\qquad\qquad\qquad = 5\frac{4}{8}$, or $5\frac{1}{2}$

Self-Check *Add or subtract as indicated.*

5. $2\frac{1}{4} + 5\frac{3}{4}$ **6.** $5\frac{3}{4} - 2\frac{1}{4}$ **7.** $8\frac{9}{10} - 3\frac{7}{10}$ **8.** $5\frac{1}{5} + 2\frac{3}{5}$

> For fractions $\frac{a}{b}$ and $\frac{c}{b}$, $b \neq 0$,
> $\frac{a}{b} + \frac{c}{b} = \frac{a+c}{b}$ and $\frac{a}{b} - \frac{c}{b} = \frac{a-c}{b}$. $\frac{8}{9} - \frac{4}{9} = \frac{8-4}{9}$, or $\frac{4}{9}$

CLASSROOM EXERCISES

Add or subtract. Write each answer in lowest terms. (Example 1)

1. $\frac{1}{4} + \frac{1}{4}$ **2.** $\frac{7}{16} - \frac{3}{16}$ **3.** $\frac{9}{7} - \frac{2}{7}$ **4.** $\frac{3}{10} + \frac{3}{10}$ **5.** $\frac{11}{16} + \frac{13}{16}$

(Example 2)

6. $3\frac{1}{3} + 2\frac{1}{3}$ **7.** $4\frac{7}{8} - 2\frac{3}{8}$ **8.** $5\frac{3}{5} + 2\frac{2}{5}$ **9.** $4\frac{3}{4} - 3\frac{1}{4}$ **10.** $2\frac{7}{8} - 1\frac{1}{8}$

WRITTEN EXERCISES

Goals: To add and subtract fractions and mixed numerals (like denominators)

To apply the skill of adding and subtracting fractions and mixed numerals (like denominators) to solving word problems.

Add or subtract. Write each answer in lowest terms. (Example 1)

1. $\frac{1}{3} + \frac{1}{3}$ **2.** $\frac{1}{5} + \frac{2}{5}$ **3.** $\frac{1}{6} + \frac{1}{6}$ **4.** $\frac{1}{8} + \frac{1}{8}$ **5.** $\frac{3}{4} - \frac{1}{4}$

6. $\frac{7}{15} - \frac{4}{15}$ **7.** $\frac{11}{13} + \frac{2}{13}$ **8.** $\frac{11}{25} + \frac{12}{25}$ **9.** $\frac{7}{10} + \frac{7}{10}$ **10.** $\frac{19}{20} + \frac{11}{20}$

(Example 2)

11. $3\frac{1}{7} + 2\frac{1}{7}$ **12.** $4\frac{1}{5} + 9\frac{2}{5}$ **13.** $11\frac{7}{9} - 2\frac{2}{9}$ **14.** $13\frac{11}{12} - 9\frac{4}{12}$

15. $5\frac{3}{6} + 1\frac{1}{6}$ **16.** $4\frac{7}{8} - 2\frac{1}{8}$ **17.** $9\frac{11}{12} - 4\frac{3}{12}$ **18.** $17\frac{11}{20} + 13\frac{5}{20}$

19. $5\frac{3}{7} + 4\frac{2}{7}$ **20.** $6\frac{5}{8} + 4\frac{2}{8}$ **21.** $6\frac{9}{11} + 3\frac{3}{11}$ **22.** $4\frac{11}{13} + 9\frac{2}{13}$

Fractions: Addition/Subtraction

MIXED PRACTICE

23. $\frac{1}{10} + \frac{5}{10}$
24. $\frac{7}{8} + \frac{3}{8}$
25. $\frac{6}{25} - \frac{3}{25}$
26. $\frac{3}{14} - \frac{1}{14}$
27. $6\frac{1}{5} + 2\frac{1}{5}$
28. $4\frac{1}{7} + 3\frac{6}{7}$
29. $3\frac{3}{8} - 1\frac{1}{8}$
30. $5\frac{7}{8} - 3\frac{3}{8}$
31. $11\frac{5}{32} + 3\frac{3}{32}$
32. $2\frac{3}{20} + 5\frac{9}{20}$
33. $27\frac{7}{8} - 14\frac{5}{8}$
34. $9\frac{3}{4} - 4\frac{3}{4}$

APPLICATIONS: USING ADDITION AND SUBTRACTION

35. Stan is flying from Washington, D.C. to Seattle with a stop in San Francisco. It takes $4\frac{5}{8}$ hours for the flight to San Francisco and $1\frac{7}{8}$ hours for the flight to Seattle. What is the total time? Write the answer in lowest terms.

36. In 1978, petroleum production in the United States was $18\frac{4}{5}$ million barrels per day. By 1982, production dropped to $15\frac{3}{5}$ million barrels per day. How many millions of barrels per day did oil production drop?

MORE CHALLENGING EXERCISES

Add or subtract. Write each answer in lowest terms.

EXAMPLES: a. $\frac{3}{x} + \frac{5}{x} = \frac{3+5}{x} = \frac{8}{x}$ b. $\frac{11a}{9} - \frac{5a}{9} = \frac{11a - 5a}{9} = \frac{6a}{9} = \frac{2a}{3}$ ◀ $11a - 5a = 6a$

37. $\frac{3}{a} + \frac{5}{a}$
38. $\frac{2}{n} + \frac{7}{n}$
39. $\frac{12}{x} - \frac{3}{x}$
40. $\frac{5}{n} - \frac{2}{n}$
41. $\frac{x}{3} + \frac{x}{3}$
42. $\frac{3a}{16} + \frac{9a}{16}$
43. $\frac{4s}{9} - \frac{2s}{9}$
44. $\frac{5y}{8} - \frac{y}{8}$
45. $\frac{3m}{10} - \frac{m}{10}$
46. $\frac{5t}{7} - \frac{3t}{7}$
47. $\frac{7x}{3y} + \frac{2x}{3y}$
48. $\frac{2a}{3c} + \frac{4a}{3c}$
49. $\frac{7a}{b} - \frac{2a}{b}$
50. $\frac{5n}{2s} - \frac{n}{2s}$
51. $\frac{7a}{3f} - \frac{2a}{3f}$

REVIEW CAPSULE FOR SECTION 6-2

Write the least common multiple of each pair of numbers. (Pages 111–114)

1. 4 and 6
2. 3 and 8
3. 2 and 12
4. 2 and 3
5. 15 and 20

Write each fraction in lowest terms. (Pages 124–125)

6. $\frac{5}{10}$
7. $\frac{8}{20}$
8. $\frac{14}{18}$
9. $\frac{12}{30}$
10. $\frac{24}{56}$

The answers are on page 183.

160 Chapter 6

6-2 Addition and Subtraction: Unlike Denominators

Unlike fractions have unlike denominators. To add or subtract fractions having unlike denominators, you first write *equivalent fractions*.

Equivalent fractions name the same number.

$$\frac{4}{6} = \frac{\cancel{2} \cdot 2}{\cancel{2} \cdot 3} \qquad \frac{6}{9} = \frac{2 \cdot \cancel{3}}{3 \cdot \cancel{3}} \qquad \frac{10}{15} = \frac{2 \cdot \cancel{5}}{3 \cdot \cancel{5}}$$

$$\frac{4}{6} = \frac{2}{3} \qquad \frac{6}{9} = \frac{2}{3} \qquad \frac{10}{15} = \frac{2}{3}$$

Thus, $\frac{4}{6}$, $\frac{6}{9}$, and $\frac{10}{15}$ are **equivalent fractions**.

PROCEDURE

To write equivalent fractions, multiply or divide <u>both</u> the numerator and denominator of a fraction by the <u>same</u> number.

EXAMPLE 1 Write a fraction that is equivalent to $\frac{5}{6}$ and has a denominator of 30.

Solution: Think: $\frac{5}{6} = \frac{?}{30}$

$$\frac{5}{6} = \frac{5}{6} \cdot \frac{5}{5}$$

$$= \frac{25}{30}$$

◀ Multiply both the numerator and denominator by the same number.

Self-Check *Complete by writing an equivalent fraction as indicated.*

1. $\frac{1}{2} = \frac{?}{10}$ 2. $\frac{2}{5} = \frac{?}{20}$ 3. $\frac{7}{12} = \frac{?}{60}$ 4. $\frac{5}{16} = \frac{?}{64}$

To add or subtract fractions having unlike denominators, you write equivalent fractions having like denominators.

PROCEDURE

To add or subtract with fractions having unlike denominators:

1 Find the LCM of the denominators.
2 Write each fraction as an equivalent fraction having the least common multiple as its new denominator.
3 Add or subtract as with like fractions.
4 Write the answer in lowest terms.

Fractions: Addition/Subtraction

EXAMPLE 2 Add: $\frac{1}{4} + \frac{7}{8}$

Solution:
[1] Find the LCM of 4 and 8. $\left.\begin{array}{l}4 = 2 \cdot 2 = 2^2 \\ 8 = 2 \cdot 2 \cdot 2 = 2^3\end{array}\right\}$ LCM: $2^3 = 8$

[2] Write an equivalent fraction for $\frac{1}{4}$ having 8 as a denominator. $\frac{1}{4} + \frac{7}{8} = \frac{1}{4} \cdot \frac{2}{2} + \frac{7}{8}$

[3] Add. $= \frac{2}{8} + \frac{7}{8}$ ◀ $\frac{2+7}{8}$

$= \frac{9}{8}$, or $1\frac{1}{8}$

Self-Check *Add.*

5. $\frac{5}{6} + \frac{2}{3}$ 6. $\frac{1}{2} + \frac{7}{10}$ 7. $\frac{3}{4} + \frac{1}{5}$ 8. $\frac{3}{8} + \frac{5}{12}$

Example 3 shows how to subtract fractions having unlike denominators.

EXAMPLE 3 Subtract: $\frac{5}{8} - \frac{7}{12}$

Solution:
[1] Find the LCM of 8 and 12.

$8 = 2 \cdot 2 \cdot 2 = 2^3$
$12 = 2 \cdot 2 \cdot 3 = 2^2 \cdot 3$
LCM: $2^3 \cdot 3 = 8 \cdot 3 = 24$

[2] Write equivalent fractions with a denominator of 24 for $\frac{5}{8}$ and $\frac{7}{12}$. $\frac{5}{8} - \frac{7}{12} = \frac{5}{8} \cdot \frac{3}{3} - \frac{7}{12} \cdot \frac{2}{2}$

[3] Subtract. $= \frac{15}{24} - \frac{14}{24}$ ◀ $\frac{15-14}{24}$

[4] $= \frac{1}{24}$

Self-Check *Subtract.*

9. $\frac{7}{12} - \frac{1}{3}$ 10. $\frac{5}{6} - \frac{1}{3}$ 11. $\frac{5}{8} - \frac{3}{5}$ 12. $\frac{7}{16} - \frac{3}{8}$

CLASSROOM EXERCISES

Complete. Write an equivalent fraction as indicated. (Example 1)

1. $\frac{1}{3} = \frac{?}{9}$ 2. $\frac{2}{5} = \frac{?}{15}$ 3. $\frac{3}{4} = \frac{?}{16}$ 4. $\frac{1}{8} = \frac{?}{24}$ 5. $\frac{3}{7} = \frac{?}{35}$ 6. $\frac{4}{9} = \frac{?}{18}$

Add. Write each answer in lowest terms. (Example 2)

7. $\frac{5}{8} + \frac{1}{4}$ 8. $\frac{5}{6} + \frac{1}{3}$ 9. $\frac{1}{6} + \frac{2}{9}$ 10. $\frac{1}{4} + \frac{5}{6}$ 11. $\frac{1}{3} + \frac{3}{8}$

12. $\frac{2}{5} + \frac{1}{2}$ 13. $\frac{3}{7} + \frac{11}{14}$ 14. $\frac{1}{2} + \frac{5}{6}$ 15. $\frac{4}{5} + \frac{3}{8}$ 16. $\frac{9}{10} + \frac{7}{20}$

Subtract. Write each answer in lowest terms. (Example 3)

17. $\frac{2}{3} - \frac{1}{6}$
18. $\frac{3}{4} - \frac{1}{8}$
19. $\frac{5}{6} - \frac{3}{8}$
20. $\frac{7}{10} - \frac{4}{15}$
21. $\frac{3}{4} - \frac{1}{5}$
22. $\frac{4}{5} - \frac{1}{3}$
23. $\frac{5}{12} - \frac{1}{4}$
24. $\frac{7}{8} - \frac{1}{4}$
25. $\frac{1}{2} - \frac{3}{8}$
26. $\frac{6}{10} - \frac{3}{20}$

WRITTEN EXERCISES

Goals: To add and subtract fractions with unlike denominators

To apply the skill of adding and subtracting fractions with unlike denominators to solving word problems

Complete. Write an equivalent fraction as indicated. (Example 1)

1. $\frac{1}{3} = \frac{?}{6}$
2. $\frac{2}{5} = \frac{?}{15}$
3. $\frac{5}{8} = \frac{?}{40}$
4. $\frac{3}{4} = \frac{?}{20}$
5. $\frac{5}{6} = \frac{?}{36}$
6. $\frac{1}{9} = \frac{?}{18}$
7. $\frac{3}{7} = \frac{?}{42}$
8. $\frac{9}{10} = \frac{?}{100}$
9. $\frac{2}{3} = \frac{?}{15}$
10. $\frac{1}{2} = \frac{?}{8}$
11. $\frac{7}{12} = \frac{?}{36}$
12. $\frac{8}{15} = \frac{?}{45}$

Add. Write each answer in lowest terms. (Example 2)

13. $\frac{5}{8} + \frac{1}{4}$
14. $\frac{1}{4} + \frac{1}{3}$
15. $\frac{7}{15} + \frac{1}{3}$
16. $\frac{3}{20} + \frac{2}{5}$
17. $\frac{2}{9} + \frac{1}{6}$
18. $\frac{5}{12} + \frac{3}{8}$
19. $\frac{2}{9} + \frac{4}{15}$
20. $\frac{1}{8} + \frac{1}{6}$
21. $\frac{3}{5} + \frac{1}{3}$
22. $\frac{2}{7} + \frac{1}{5}$
23. $\frac{5}{8} + \frac{1}{3}$
24. $\frac{2}{3} + \frac{1}{10}$
25. $\frac{5}{6} + \frac{2}{3}$
26. $\frac{3}{4} + \frac{7}{12}$
27. $\frac{11}{14} + \frac{5}{21}$

Subtract. Write each answer in lowest terms. (Example 3)

28. $\frac{7}{12} - \frac{1}{2}$
29. $\frac{4}{5} - \frac{3}{10}$
30. $\frac{9}{14} - \frac{3}{7}$
31. $\frac{7}{15} - \frac{1}{3}$
32. $\frac{9}{10} - \frac{1}{4}$
33. $\frac{5}{12} - \frac{2}{9}$
34. $\frac{11}{15} - \frac{3}{10}$
35. $\frac{5}{6} - \frac{2}{9}$
36. $\frac{3}{5} - \frac{1}{3}$
37. $\frac{5}{9} - \frac{2}{5}$
38. $\frac{6}{7} - \frac{1}{4}$
39. $\frac{11}{13} - \frac{2}{3}$
40. $\frac{4}{5} - \frac{3}{4}$
41. $\frac{1}{2} - \frac{3}{8}$
42. $\frac{5}{7} - \frac{1}{3}$

MIXED PRACTICE

Add or subtract. Write each answer in lowest terms. (Examples 1–3)

43. $\frac{5}{6} + \frac{1}{2}$
44. $\frac{3}{5} - \frac{1}{6}$
45. $\frac{5}{8} - \frac{1}{10}$
46. $\frac{3}{7} + \frac{1}{4}$
47. $\frac{9}{11} - \frac{1}{4}$
48. $\frac{1}{9} + \frac{2}{3}$
49. $\frac{4}{5} + \frac{2}{9}$
50. $\frac{3}{5} - \frac{1}{10}$
51. $\frac{9}{14} + \frac{2}{7}$
52. $\frac{7}{8} - \frac{1}{6}$

Fractions: Addition/Subtraction

APPLICATIONS: UNLIKE DENOMINATORS

53. On Tuesday, the rainfall was $\frac{5}{8}$ inch. On Wednesday, the rainfall was $\frac{3}{4}$ inch. What was the total rainfall for the two days?

54. Carlton is making beef stew. The recipe calls for $\frac{3}{4}$ of a pound of beef. He has $\frac{3}{10}$ of a pound. How much more beef does he need?

55. Yolanda lives $\frac{9}{10}$ of a mile from school. Eric lives $\frac{2}{3}$ of a mile from school. How much farther from the school does Yolanda live?

56. Elena spent $\frac{5}{6}$ of an hour doing her math homework and $\frac{3}{5}$ of an hour doing her science homework. How much time did she spend altogether?

MORE CHALLENGING EXERCISES

Add or subtract. Write each answer in lowest terms. No variable equals zero.

EXAMPLE: a. $\frac{t}{6} + \frac{t}{2} = \frac{t + 3t}{6} = \frac{4t}{6} = \frac{2t}{3}$

b. $\frac{7}{f} - \frac{5}{2f} = \frac{14}{2f} - \frac{5}{2f} = \frac{14 - 5}{2f} = \frac{9}{2f}$ LCM: $2f$

57. $\frac{x}{5} + \frac{x}{25}$ **58.** $\frac{n}{6} + \frac{n}{3}$ **59.** $\frac{a}{6} - \frac{a}{9}$ **60.** $\frac{y}{2} - \frac{y}{5}$ **61.** $\frac{3x}{8} + \frac{x}{4}$

62. $\frac{5y}{8} + \frac{y}{2}$ **63.** $\frac{9}{2y} + \frac{2}{y}$ **64.** $\frac{1}{2m} + \frac{5}{m}$ **65.** $\frac{3}{x} - \frac{2}{3x}$ **66.** $\frac{7n}{3t} - \frac{n}{t}$

67. $\frac{3a}{n} + \frac{a}{2n}$ **68.** $\frac{5x}{y} + \frac{3x}{2y}$ **69.** $\frac{21y}{16z} - \frac{3y}{4z}$ **70.** $\frac{11n}{2p} - \frac{2n}{3p}$ **71.** $\frac{17s}{5t} + \frac{2s}{3t}$

REVIEW CAPSULE FOR SECTION 6-3

Write the least common multiple of each pair of numbers. (Pages 111–114)

1. 3 and 11   **2.** 6 and 20   **3.** 8 and 32   **4.** 10 and 15   **5.** 4 and 9

Write each fraction in lowest terms. (Pages 124–125)

6. $\frac{10}{20}$ **7.** $\frac{5}{30}$ **8.** $\frac{12}{26}$ **9.** $\frac{8}{24}$ **10.** $\frac{15}{40}$

The answers are on page 183.

6-3 Addition and Subtraction: Mixed Numerals

The record for the high jump at Doug's school is $6\frac{2}{3}$ feet. Doug's best jump at the last intramural meet was $6\frac{1}{4}$ feet. To find how much short of the record he is, Doug found this difference.

$6\frac{2}{3} - 6\frac{1}{4}$ ◀ **Mixed numerals with unlike denominators**

Adding and subtracting mixed numerals with unlike denominators is similar to adding and subtracting fractions having unlike denominators.

EXAMPLE 1 Add or subtract as indicated: **a.** $2\frac{5}{8} + 5\frac{3}{4}$ **b.** $6\frac{2}{3} - 6\frac{1}{4}$

Solutions:

a. LCM: 8

$2\frac{5}{8} + 5\frac{3}{4} = 2\frac{5}{8} + 5\frac{6}{8}$ ◀ $2 + 5 = 7$; $\frac{5}{8} + \frac{6}{8} = \frac{11}{8}$

$= 7\frac{11}{8}$

$= 7 + 1\frac{3}{8}$

$= \mathbf{8\frac{3}{8}}$

b. LCM: 12

$6\frac{2}{3} - 6\frac{1}{4} = 6\frac{8}{12} - 6\frac{3}{12}$ ◀ $6 - 6 = 0$; $\frac{8}{12} - \frac{3}{12} = \frac{5}{12}$

$= \frac{5}{12}$ ◀ Doug's jump was $\frac{5}{12}$ ft short of the record

Self-Check Add or subtract as indicated.

1. $8\frac{4}{5} + 5\frac{1}{3}$ **2.** $5\frac{3}{4} - 2\frac{1}{2}$ **3.** $7\frac{3}{4} - 5\frac{1}{3}$ **4.** $6\frac{7}{8} + 1\frac{1}{4}$

When subtracting with mixed numerals, it is sometimes necessary to "borrow" from the whole number.

EXAMPLE 2 Subtract: **a.** $12\frac{1}{4} - 2\frac{2}{5}$ **b.** $14 - 8\frac{5}{6}$

Solutions: **a.** LCM: 20

$12\frac{1}{4} - 2\frac{2}{5} = 12\frac{5}{20} - 2\frac{8}{20}$ ◀ Since $\frac{5}{20}$ is less than $\frac{8}{20}$, borrow 1, or $\frac{20}{20}$, from 12.

$= 11\frac{25}{20} - 2\frac{8}{20}$

$= \mathbf{9\frac{17}{20}}$

b. LCM: 6 $14 - 8\frac{5}{6} = 13\frac{6}{6} - 8\frac{5}{6}$ ◀ $14 = 13\frac{6}{6}$

$= \mathbf{5\frac{1}{6}}$

Fractions: Addition/Subtraction

Self-Check *Subtract.*

5. $5\frac{1}{4} - 3\frac{3}{4}$
6. $7\frac{1}{6} - 4\frac{2}{3}$
7. $9 - 3\frac{3}{5}$
8. $5 - 2\frac{1}{3}$

CLASSROOM EXERCISES

Complete. (Example 1)

1. $3\frac{1}{4} + 2\frac{1}{2} = 3\frac{1}{4} + 2\frac{?}{4}$
2. $6\frac{3}{10} + 2\frac{2}{5} = 6\frac{3}{10} + 2\frac{?}{10}$
3. $6\frac{4}{5} - 2\frac{1}{4} = 6\frac{?}{20} - 2\frac{5}{20}$
4. $5\frac{2}{3} - 1\frac{1}{4} = 5\frac{?}{12} - 1\frac{3}{12}$

Complete. (Example 2)

5. $7\frac{1}{6} - 2\frac{2}{3} = 6\frac{?}{6} - 2\frac{4}{6}$
6. $11\frac{1}{2} - 2\frac{5}{8} = 10\frac{?}{8} - 2\frac{5}{8}$
7. $6 - 3\frac{1}{4} = 5\frac{?}{4} - 3\frac{1}{4}$
8. $3 - 1\frac{5}{6} = 2\frac{?}{6} - 1\frac{5}{6}$

WRITTEN EXERCISES

Goals: To add and subtract with mixed numerals

To apply the skill of adding and subtracting with mixed numerals to solving word problems

Add or subtract. Write each answer in lowest terms. (Example 1)

1. $5\frac{1}{2} + 2\frac{1}{4}$
2. $2\frac{3}{8} + 1\frac{1}{2}$
3. $1\frac{1}{6} + \frac{3}{4}$
4. $8\frac{4}{5} + 5\frac{1}{3}$
5. $6\frac{3}{4} + 1\frac{2}{8}$
6. $12\frac{1}{2} + 11\frac{1}{8}$
7. $6\frac{2}{5} + \frac{3}{4}$
8. $12\frac{1}{12} + 13\frac{1}{4}$
9. $5\frac{3}{4} - 2\frac{1}{2}$
10. $5\frac{11}{12} - 4\frac{3}{8}$
11. $2\frac{5}{6} - \frac{3}{4}$
12. $7\frac{7}{10} - \frac{2}{5}$
13. $8\frac{7}{8} - 6\frac{3}{4}$
14. $12\frac{1}{2} - 3\frac{1}{3}$
15. $6\frac{8}{15} - \frac{2}{5}$

Subtract. Write each answer in lowest terms. (Example 2)

16. $5\frac{1}{4} - 1\frac{3}{4}$
17. $7\frac{1}{6} - 2\frac{1}{2}$
18. $9\frac{2}{3} - 4\frac{11}{12}$
19. $14 - 12\frac{1}{2}$
20. $6 - 4\frac{2}{3}$
21. $11\frac{1}{2} - \frac{5}{6}$
22. $2\frac{1}{6} - \frac{3}{8}$
23. $7\frac{1}{5} - 2\frac{5}{12}$
24. $3\frac{1}{4} - 1\frac{2}{3}$
25. $8 - 2\frac{9}{10}$
26. $9\frac{1}{12} - 3\frac{2}{3}$
27. $7\frac{1}{8} - 3\frac{1}{4}$
28. $5 - 3\frac{1}{3}$
29. $9 - 2\frac{5}{16}$
30. $15\frac{1}{4} - 8\frac{3}{8}$

MIXED PRACTICE

Add or subtract as indicated. Write each answer in lowest terms.

31. $7\frac{3}{5} + 2\frac{1}{3}$
32. $2\frac{1}{2} - 1\frac{1}{8}$
33. $8\frac{1}{3} + 7\frac{3}{4}$
34. $4\frac{11}{15} + 6\frac{3}{5}$
35. $8\frac{1}{8} - 2\frac{1}{4}$
36. $9\frac{1}{2} - 3\frac{1}{5}$
37. $3 - 1\frac{7}{12}$
38. $4\frac{6}{7} + 2\frac{5}{6}$
39. $20\frac{2}{3} - 8\frac{1}{4}$
40. $11\frac{3}{11} + 2\frac{1}{3}$

APPLICATIONS: USING MIXED NUMERALS

Solve. Write each answer in lowest terms.

41. The school record for running the mile is $4\frac{3}{4}$ minutes. Rick's record is $5\frac{3}{10}$ minutes. Find the difference between the two records.

42. Rose can throw the shot put $34\frac{3}{8}$ feet. Julie can throw the shot put $39\frac{1}{4}$ feet. How much farther can Julie throw the shot put than Rose?

43. Arthur is $5\frac{2}{3}$ feet tall. His record in the long jump is $10\frac{1}{2}$ feet more than his height. What is his long jump record?

REVIEW: SECTIONS 6–1–6–3

Add or subtract. Write each answer in lowest terms. (Section 6–1)

1. $\frac{1}{8} + \frac{5}{8}$ **2.** $3\frac{1}{3} + 4\frac{1}{3}$ **3.** $\frac{8}{13} - \frac{5}{13}$ **4.** $15\frac{9}{20} - 4\frac{3}{20}$ **5.** $6\frac{7}{8} + 3\frac{5}{8}$

Complete. Write an equivalent fraction as indicated. (Section 6–2)

6. $\frac{1}{2} = \frac{?}{20}$ **7.** $\frac{3}{7} = \frac{?}{28}$ **8.** $\frac{3}{4} = \frac{?}{12}$ **9.** $\frac{5}{6} = \frac{?}{30}$ **10.** $\frac{7}{8} = \frac{?}{48}$

Add or subtract. Write each answer in lowest terms. (Section 6–2)

11. $\frac{1}{2} + \frac{2}{3}$ **12.** $\frac{1}{5} + \frac{3}{10}$ **13.** $\frac{5}{6} - \frac{1}{8}$ **14.** $\frac{4}{5} - \frac{3}{10}$ **15.** $\frac{1}{2} - \frac{1}{6}$

(Section 6–3)

16. $1\frac{5}{6} + 2\frac{1}{4}$ **17.** $4\frac{2}{3} + 5\frac{1}{5}$ **18.** $5\frac{2}{3} - 1\frac{1}{6}$ **19.** $8\frac{3}{5} - 5\frac{3}{4}$ **20.** $15 - 4\frac{1}{8}$

Solve each problem. Write answers in lowest terms.

21. Tamara swam $2\frac{1}{5}$ miles on Wednesday and $1\frac{3}{5}$ miles on Friday. How many miles did she swim in all? (Section 6–1)

22. Willie spent $\frac{3}{4}$ of an hour mowing his front yard. He spent $\frac{2}{3}$ of an hour edging the driveway and front curb. How much time did he spend altogether on mowing and edging? (Section 6–2)

Fractions: Addition/Subtraction

CALCULATOR/COMPUTER APPLICATIONS
Evaluating Powers

To evaluate a formula such as

$$x = b^n - 2$$

you first have to evaluate b^n. For $b = 3$ and $n = 4$,

$$x = 3^4 - 2$$
$$= 3 \cdot 3 \cdot 3 \cdot 3 - 2$$
$$= 81 - 2 = \mathbf{79}$$

You can use a calculator to evaluate a power.

Calculator: Evaluate 15^3.

Solution: Press " = " *once* for the b^2, *twice* for the b^3, *three* times for b^4, and so on. Since the exponent in 15^3 is three, press the = key twice.

15 [×] [=] [=]   `3375.`

Computer: You can use the following program to evaluate b^n, where b and n are counting numbers. Note that this program does not use the ↑ operator of BASIC.

Program:
```
10 INPUT B,N
20 LET X = 1
30 FOR I = 1 TO N
40 LET X = X * B
50 NEXT I
60 PRINT "ANSWER IS"; X
70 END
```

Statements 30–50 represent a **FOR-NEXT** loop. This loop multiplies B times itself N times. However, before the loop begins, X is set to 1 in statement 20.

Exercises

Use a calculator or a computer to evaluate each of the following.

1. 7^2
2. 9^3
3. 8^4
4. 9^6
5. 4^5
6. 11^3
7. 11^7
8. 9^8
9. 13^7
10. 32^5
11. 45^4
12. 87^4

REVIEW CAPSULE FOR SECTION 6-4

Multiply. Write each answer in lowest terms. (Pages 135–137)

1. $\frac{3}{4} \cdot \frac{9}{10}$
2. $\frac{5}{6} \cdot \frac{8}{15}$
3. $2\frac{1}{4} \cdot 4\frac{2}{5}$
4. $1\frac{1}{3} \cdot 3\frac{1}{6}$
5. $5 \cdot 1\frac{1}{2}$
6. $9 \cdot 2\frac{2}{3}$
7. $24 \cdot 3\frac{1}{8}$
8. $150 \cdot 2\frac{3}{10}$

Complete. (Pages 161–164)

9. $15 \text{ min} = \frac{15}{60} \text{ hr} = \underline{\ ?\ } \text{ hr}$
10. $\frac{3}{4} \text{ hr} = \frac{?}{60} \text{ hr} = \underline{\ ?\ } \text{ min}$
11. $10 \text{ min} = \frac{?}{60} \text{ hr} = \frac{1}{6} \text{ hr}$
12. $\frac{1}{3} \text{ hr} = \frac{?}{60} \text{ hr} = \underline{\ ?\ } \text{ min}$

The answers are on page 183.

PROBLEM SOLVING AND APPLICATIONS

6-4 Time Cards

Rosemarie and Karl have part time jobs at a supermarket near their home. At the supermarket, time cards are used to record how many hours each employee works per week.

EXAMPLE 1 A copy of Rosemarie's time card for one week is shown at the right. Find the total number of hours she worked that week.

DAYS	IN	OUT	IN	OUT
1	—	—	5:30	8:00
2	—	—	3:00	6:15
3	—	—	4:00	7:20
4	7:00	11:30	1:00	4:00
5	9:00	12:00	1:30	5:15

Solution:

[1] Find the hours worked each day.

Day 1	Day 2	Day 3	Day 4	Day 5
$2\frac{1}{2}$ hr	$3\frac{1}{4}$ hr	$3\frac{1}{3}$ hr	$7\frac{1}{2}$ hr	$6\frac{3}{4}$ hr

[2] Find the total hours for the week. LCM: 12

$$2\frac{1}{2} + 3\frac{1}{4} + 3\frac{1}{3} + 7\frac{1}{2} + 6\frac{3}{4} = 2\frac{6}{12} + 3\frac{3}{12} + 3\frac{4}{12} + 7\frac{6}{12} + 6\frac{9}{12}$$

$$= 21\frac{28}{12} \qquad \frac{28}{12} = 2\frac{4}{12} = 2\frac{1}{3}$$

$$= 23\frac{1}{3} \text{ hours}$$

Self-Check

1. Karl worked $3\frac{1}{2}$ hours on Tuesday, $6\frac{1}{4}$ hours on Friday, and $5\frac{2}{3}$ hours on Saturday. Find the total number of hours worked.

Fractions: Addition/Subtraction

Rosemarie is paid $4.35 per hour and Karl is paid $4.15 per hour. To determine how much Rosemarie and Karl earn each week, you must first answer the question:

How many hours do they work each week?

This is the **hidden question** in the problem. Then you can find the weekly earnings.

EXAMPLE 2 Rosemarie is paid $4.35 per hour. How much did she earn for the week discussed in Example 1?

Solutions:

⒈ Find the total number of hours worked.

Total hours: $23\frac{1}{3}$ ◀ From Example 1

⒉ Weekly Earnings = (Pay per Hour) × (Total Hours)

$$= (4.35)\left(23\frac{1}{3}\right) \quad 23\frac{1}{3} = \frac{70}{3}$$

$$= (\cancel{4.35})^{1.45}\left(\frac{70}{\cancel{3}}\right)_1$$

$$= \$101.50$$

Rosemarie's weekly earnings amounted to **$101.50**.

Self-Check 2. Karl is paid $4.15 per hour. How much did he earn for the week discussed in the Self-Check that follows Example 1?

CLASSROOM EXERCISES

For Exercises 1–2:
a. *Find the number of hours worked each day.*
b. *Find the total number of hours for the week.* (Example 1)

1.

DAYS	IN	OUT	IN	OUT	DAILY TOTALS
1	—	—	1:00	5:30	?
2	9:00	12:15	1:15	5:00	?
3	8:30	12:00	1:00	4:30	?
4	8:45	12:00	1:00	4:15	?
5	9:15	12:00	—	—	?
WEEKLY TOTAL					?

2.

DAYS	IN	OUT	IN	OUT	DAILY TOTALS
1	9:00	12:30	1:30	5:00	?
2	9:00	12:20	1:00	5:15	?
3	8:45	11:30	—	—	?
4	9:00	12:45	1:45	5:00	?
5	8:40	12:00	1:00	5:30	?
WEEKLY TOTAL					?

WRITTEN EXERCISES

Goal: To solve two-step problems involving earnings by applying the technique of the "hidden question"

For Exercises 1–4:
a. *Find the total number of hours worked each day.*
b. *Find the total number of hours worked for the week.* (Example 1)

1.

Name: Sonya Wilson					
DAYS	IN	OUT	IN	OUT	DAILY TOTALS
1	8:00	12:00	1:00	4:15	?
2	8:00	12:00	1:00	4:30	?
3	8:00	12:30	1:00	5:15	?
4	8:00	12:00	1:00	4:30	?
5	8:00	12:15	1:15	4:30	?
WEEKLY TOTAL					?

2.

Name: Paul Chan					
DAYS	IN	OUT	IN	OUT	DAILY TOTALS
1	9:00	11:30	12:30	5:30	?
2	8:30	11:30	12:00	5:20	?
3	8:15	11:30	12:30	5:15	?
4	9:00	12:30	1:30	5:30	?
5	8:30	11:30	12:30	5:40	?
WEEKLY TOTAL					?

3.

Name: Ted Schultz					
DAYS	IN	OUT	IN	OUT	DAILY TOTALS
1	9:00	1:00	2:00	5:30	?
2	9:00	1:00	2:00	5:45	?
3	8:15	12:00	—	—	?
4	8:30	12:00	1:00	5:15	?
5	9:00	12:30	1:30	5:30	?
WEEKLY TOTAL					?

4.

Name: Lisa Kotlowski					
DAYS	IN	OUT	IN	OUT	DAILY TOTALS
1	9:00	12:20	1:20	5:00	?
2	—	—	1:15	5:00	?
3	8:30	1:00	2:00	4:20	?
4	8:40	1:00	2:00	5:00	?
5	9:00	12:30	1:30	5:15	?
WEEKLY TOTAL					?

For Exercises 5–8, use the information from Exercises 1–4 to solve each problem.

5. Sonya Wilson is paid $5.20 per hour (see Exercise 1). Find her total weekly earnings.

6. Paul Chan is paid $4.50 per hour (see Exercise 2). Find his total weekly earnings.

7. Ted Schultz is paid $6.20 per hour (see Exercise 3). Find his total weekly earnings.

8. Lisa Kotlowski is paid $3.78 per hour (see Exercise 4). Find her total weekly earnings.

Fractions: Addition/Subtraction

Solve each problem.

9. Tom Hastings worked $6\frac{1}{2}$ hours on Monday, $5\frac{3}{4}$ hours on Wednesday, and $4\frac{1}{2}$ hours on Friday. He is paid $5.60 per hour. Find his weekly earnings.

10. Julie Pendleton worked $3\frac{1}{2}$ hours on Tuesday, $3\frac{3}{4}$ hours on Friday, and $8\frac{1}{4}$ hours on Saturday. She is paid $4.80 per hour. Find her weekly earnings.

11. Robert Felch worked from 9:00 A.M. to 4:15 P.M. on Monday, Wednesday, and Friday, and from 8:30 A.M. to 5:15 P.M. on Saturday. He is paid $4.40 per hour. Find his weekly earnings.

12. Doris Youngblood worked from 3:00 P.M. to 9:30 P.M. on Monday and Tuesday, from 3:00 P.M. to 8:20 P.M. on Thursday, and from 3:15 P.M. to 10:00 P.M. on Friday. She is paid $6.00 per hour. Find her weekly earnings.

MORE CHALLENGING EXERCISES

13. Tim Murray works $38\frac{1}{4}$ hours each week. He is paid $4.60 per hour. After his raise he will be paid $5.20 per hour. How much more money will Tim earn each week?

14. Denise Turner is paid $165.55 per week for $38\frac{1}{2}$ hours of work. How much does Denise earn in an hour?

15. Mike Fu is paid $344.00 for 40 hours of work. He is paid $1\frac{1}{2}$ times the hourly rate for time over 40 hours. How much is Mike paid per hour for time over 40 hours?

16. Marjorie Russell was paid $344.00 for 40 hours of work. She was paid $9.30 per hour for time over 40 hours. She worked $45\frac{1}{5}$ hours last week. How much did Marjorie earn?

REVIEW CAPSULE FOR SECTION 6-5

Divide to the hundredths place. (Pages 44–46)

1. $2\overline{)1.7}$
2. $8\overline{)4.4}$
3. $4\overline{)2.6}$
4. $5\overline{)1.3}$
5. $6\overline{)4.5}$

Replace the __?__ with > or <. (Pages 32–34)

6. 0.421 __?__ 0.422
7. 0.914 __?__ 0.92
8. 0.021 __?__ 0.039
9. 0.77 __?__ 0.68
10. 0.628 __?__ 0.625
11. 0.01 __?__ 0.15
12. 0.163 __?__ 0.1603
13. 0.199 __?__ 0.1991
14. 0.004 __?__ 0.014

The answers are on page 183.

6-5 Fractions and Decimals

The San Andreas Fault extends $\frac{3}{4}$ of the length of California's coastal region.

To write a decimal for $\frac{3}{4}$, divide 3 by 4.

> **PROCEDURE**
>
> To write a decimal for a fraction, divide the numerator of the fraction by the denominator.

EXAMPLE 1 Write a decimal for $\frac{3}{4}$.

Solution: $\frac{3}{4}$ means $3 \div 4$.

```
    0.75
4)3.00
  2 8
    20
    20
     0
```
The remainder is 0.
The decimal <u>terminates</u>.

Self-Check *Write a decimal for each fraction.*

1. $\frac{1}{2}$ 2. $\frac{2}{5}$ 3. $\frac{5}{8}$ 4. $\frac{5}{2}$ 5. $\frac{9}{5}$

The fraction $\frac{3}{4} = 0.75$. The decimal 0.75 is called a **terminating decimal**. Some decimals do not terminate.

EXAMPLE 2 About $\frac{2}{3}$ of the earth's surface is covered with water. Write a decimal for $\frac{2}{3}$. Carry the division to three decimal places.

Solution:
```
   0.666
3)2.000
  1 8
    20
    18
     20
     18
      2
```

The digit, 6, repeats on and on.

$\frac{2}{3} = 0.666 \cdots$

The three dots mean that "6" repeats on and on.

Fractions: Addition/Subtraction **173**

Self-Check *Write a decimal for each fraction. Use dots to show repeating digits.*

6. $\frac{1}{3}$ **7.** $\frac{5}{6}$ **8.** $\frac{2}{15}$ **9.** $\frac{5}{9}$ **10.** $\frac{5}{18}$

The decimal 0.666 · · · is called a **repeating decimal** because the digit 6 repeats indefinitely. The decimal can be written in another way.

$$0.666 \cdots = 0.\overline{6}$$

◀ The bar indicates the repeating digit(s).

NOTE: Repeating decimals may have one, two, or several repeating digits. Also, the digits may start to repeat only after a certain point.

$\frac{5}{11} = 0.\overline{45}$ ◀ 0.454545 · · · $\frac{1}{12} = 0.08\overline{3}$ ◀ 0.083333 . . .

One way of comparing two fractions is to write a decimal for each fraction. Then compare the decimals (see Example 2 on page 33).

EXAMPLE 3 Suellen wanted to buy $\frac{3}{4}$ of a yard of material. The clerk told her there was $\frac{5}{8}$ of a yard in stock. Was there enough material in stock?

Solution: ☐1 Write a decimal for each fraction.

$\frac{3}{4} = 0.75$ $\frac{5}{8} = 0.625$ ◀ See Example 1 on page 173.

☐2 Compare the decimals. $0.75 > 0.625$

Therefore, $\frac{3}{4} > \frac{5}{8}$

Since $\frac{3}{4} > \frac{5}{8}$, there is **not enough material.**

Self-Check *Complete. Write >, <, or =.*

11. $\frac{2}{5}$ __?__ $\frac{1}{2}$ **12.** $\frac{5}{8}$ __?__ $\frac{3}{4}$ **13.** $\frac{7}{8}$ __?__ $\frac{3}{5}$ **14.** $\frac{2}{3}$ __?__ $\frac{6}{9}$

CLASSROOM EXERCISES

Replace the __?__ with = or ≠ (is not equal to). (Example 1)

1. $\frac{1}{25}$ __?__ 0.04   **2.** $\frac{4}{5}$ __?__ 0.6   **3.** $\frac{3}{20}$ __?__ 0.17   **4.** $\frac{7}{50}$ __?__ 0.14

5. $\frac{7}{5}$ __?__ 1.25   **6.** $\frac{9}{4}$ __?__ 2.25   **7.** $\frac{19}{10}$ __?__ 1.9   **8.** $\frac{27}{8}$ __?__ 3.75

State whether each of the following is a repeating or a terminating decimal. Write R for repeating and T for terminating. (Definitions)

9. 0.32   **10.** 0.10   **11.** $0.\overline{2}$ **12.** $0.5\overline{2}$ **13.** 1.5   **14.** $3.8\overline{61}$

Compare each pair of fractions. Write >, <, or =. (Example 3)

15. $\frac{1}{2}$ __?__ $\frac{3}{5}$ **16.** $\frac{1}{4}$ __?__ $\frac{3}{10}$ **17.** $\frac{1}{5}$ __?__ $\frac{1}{3}$ **18.** $\frac{3}{8}$ __?__ $\frac{3}{4}$

WRITTEN EXERCISES

Goals: To write a decimal for a fraction

To apply the skill of writing a decimal for a fraction to comparing fractions

Write a decimal for each fraction. (Example 1)

1. $\frac{11}{20}$ **2.** $\frac{7}{4}$ **3.** $\frac{19}{5}$ **4.** $\frac{9}{25}$ **5.** $\frac{3}{20}$

6. $\frac{3}{50}$ **7.** $\frac{5}{8}$ **8.** $\frac{7}{8}$ **9.** $\frac{13}{40}$ **10.** $\frac{7}{16}$

Write a decimal for each fraction. Use a bar to show any digit or digits that repeat. (Example 2)

11. $\frac{1}{3}$ **12.** $\frac{5}{9}$ **13.** $\frac{7}{9}$ **14.** $\frac{7}{15}$ **15.** $\frac{11}{15}$

16. $\frac{17}{3}$ **17.** $\frac{10}{9}$ **18.** $\frac{9}{11}$ **19.** $\frac{7}{12}$ **20.** $\frac{10}{33}$

21. $\frac{5}{6}$ **22.** $\frac{4}{15}$ **23.** $\frac{7}{3}$ **24.** $\frac{15}{11}$ **25.** $\frac{1}{6}$

Compare each pair of fractions. Write >, <, or =. (Example 3)

26. $\frac{3}{5}$ __?__ $\frac{5}{8}$ **27.** $\frac{7}{8}$ __?__ $\frac{3}{4}$ **28.** $\frac{9}{20}$ __?__ $\frac{7}{10}$ **29.** $\frac{3}{10}$ __?__ $\frac{2}{5}$ **30.** $\frac{1}{5}$ __?__ $\frac{3}{10}$

31. $\frac{2}{3}$ __?__ $\frac{5}{8}$ **32.** $\frac{5}{6}$ __?__ $\frac{5}{8}$ **33.** $\frac{7}{8}$ __?__ $\frac{9}{10}$ **34.** $\frac{18}{25}$ __?__ $\frac{7}{8}$ **35.** $\frac{2}{3}$ __?__ $\frac{17}{20}$

APPLICATIONS: COMPARING DECIMALS

36. Pearl and Jane are reading the same book. Pearl has read $\frac{7}{8}$ of the book and Jane has read $\frac{7}{12}$ of the book. Who has read fewer pages, Pearl or Jane?

37. Earl and Miyako are painting a fence. Earl has completed $\frac{3}{8}$ of the job and Miyako has completed $\frac{1}{3}$ of the job. Who has completed the greater part?

38. Mario needs $1\frac{7}{10}$ yards of chain to hang a light. He ordered $1\frac{7}{8}$ yards. Did he order enough?

39. Phyllis and Flora are running in the marathon. Phyllis has completed $\frac{11}{12}$ of the distance and Flora has completed $\frac{13}{15}$ of the distance. Who is ahead?

Fractions: Addition/Subtraction

REVIEW CAPSULE FOR SECTION 6-6

Multiply. (Pages 38–42)

1. $2 \times 3.14 \times 9.5$
2. $2 \times 3.14 \times 27.8$
3. $3.14 \times 4 \times 4$
4. $3.14 \times 10 \times 10$

Write a fraction for each mixed numeral. (Pages 135–137)

5. $3\frac{1}{7}$
6. $1\frac{3}{4}$
7. $5\frac{1}{4}$
8. $10\frac{7}{8}$
9. $3\frac{1}{10}$
10. $2\frac{1}{8}$
11. $11\frac{1}{2}$
12. $9\frac{3}{5}$

Multiply. (Pages 135–137)

13. $2 \times 49 \times \frac{22}{7}$
14. $2 \times 280 \times \frac{22}{7}$
15. $\frac{22}{7} \times 8\frac{2}{3} \times 8\frac{2}{3}$
16. $\frac{22}{7} \times 5\frac{1}{4} \times 5\frac{1}{4}$

Round to the nearest whole number. (Pages 14–16)

17. 21.3
18. 73.5
19. 34.72
20. $\frac{3}{4}$
21. $5\frac{1}{8}$
22. $63\frac{2}{5}$

The answers are on page 183.

PROBLEM SOLVING AND APPLICATIONS

6-6 Circles: Circumference and Area

The diameter and radius of a circle such as a bicycle wheel are shown at the right. In any circle, the diameter is twice the radius. That is,

$d = 2 \cdot r$, or $d = 2r$

d = diameter
r = radius

The circumference, C, of a circle, is the distance around the circle.

Start

Turning

Circumference

One Complete Turn

For all circles, the quotient $\frac{\text{circumference}}{\text{diameter}}$ is the same. It is the number π (Greek letter "pi"). The value of π is approximately 3.14 or $\frac{22}{7}$.

176 Chapter 6

PROCEDURE

To find the circumference, C, of a circle, find the product of the radius and 2π <u>or</u> the product of the diameter and π.

$$C = 2 \cdot \pi \cdot r \text{ or } C = \pi \cdot d$$

EXAMPLE 1 The radius of this bicycle wheel is 13 inches. Find the circumference. Round your answer to the nearest whole number.

Solutions:

Method 1

$C = 2 \cdot \pi \cdot r$ ◀ $r = 13$ in

$C = 2 \cdot \frac{22}{7} \cdot 13$

$C = 81\frac{5}{7}$

Method 2

$C = \pi \cdot d$

$C = \frac{22}{7} \cdot 26$ ◀ $d = 2r$
$d = 2 \cdot 13$

$C = 81\frac{5}{7}$

The circumference is about **82 inches**.

Self-Check *Find the circumference of each circle. Round each answer to the nearest whole number.*

1. $r = 3$ in; $\pi = \frac{22}{7}$
2. $r = 10$ ft; $\pi = \frac{22}{7}$
3. $d = 5$ cm; $\pi = 3.14$
4. $d = 25$ m; $\pi = 3.14$

The formula for the area of a circle can be developed from the formula for the area of a parallelogram.

$A = b \cdot h$ ◀ Area of parallelogram

$A = \frac{1}{2} \cdot C \cdot r$

$A = \frac{1}{2} \cdot 2 \cdot \pi \cdot r \cdot r$

$A = \pi \cdot r \cdot r$

PROCEDURE

To find the area of a circle, multiply π times the radius times the radius.

$$A = \pi \cdot r \cdot r \text{ or } A = \pi r^2$$ ◀ r^2 means $r \cdot r$.

Fractions: Addition/Subtraction

EXAMPLE 2 The top of a round table has a radius of 0.5 meter. Find the area. Use 3.14 for π. Round your answer to the nearest whole number.

Solution: $A = \pi \cdot r \cdot r$
$A = 3.14 \cdot 0.5 \cdot 0.5$
$A = 0.785$ The area is about **1 m²**.

Self-Check *Find the area of each circle. Round answers to the nearest whole number.*

5. $r = 10$ cm; $\pi = 3.14$ **6.** $r = 3\frac{1}{2}$ ft; $\pi = \frac{22}{7}$

CLASSROOM EXERCISES

For Exercises 1–3, find the circumference of each circle. Round answers to the nearest whole number. (Example 1)

1. $r = 14$ in; $\pi = \frac{22}{7}$ **2.** $r = 10.5$ cm; $\pi = 3.14$ **3.** $d = 30$ m; $\pi = 3.14$

For Exercises 4–6, find the area of each circle. Round answers to the nearest whole number. (Example 2)

4. $r = 15$ cm; $\pi = 3.14$ **5.** $r = 3$ yd; $\pi = \frac{22}{7}$ **6.** $r = 5.6$ m; $\pi = 3.14$

WRITTEN EXERCISES

Goal: To find the circumference and area of a circle

METRIC MEASURES   *Find the circumference. Round answers to the nearest whole number. For Exercises 1–8, use 3.14 for π.*

1. Auto Tire
$r = 30$ cm

2. Bicycle Wheel
$r = 33$ cm

3. Tennis Ball
$r = 3.2$ cm

4. Record
$r = 12.7$ cm

5. Tractor Tire
$d = 1.3$ m

6. Earth
$d = 12,766$ km

7. Basketball
$r = 12.1$ cm

8. Sun
$r = 696,000$ km

CUSTOMARY MEASURES   *Find the circumference. Round answers to nearest whole number. For Exercises 9–16 use $\frac{22}{7}$ for π.*

9. Fountain Base
$r = 7\frac{1}{2}$ ft

10. Bicycle Gear
$r = 4\frac{2}{5}$ in

11. Clock
$d = 13\frac{1}{2}$ in

12. Moon
$d = 2250$ mi

13. Douglas Fir
$r = 2$ yd

14. Mars
$d = 11{,}152$ mi

15. Softball
$d = 12$ in

16. Reel of Tape
$r = 3\frac{1}{2}$ in

METRIC MEASURES   *Find the area. Round answers to the nearest whole number. For Exercises 17–24, use 3.14 for π.* (Example 2)

17. Tray
$r = 20$ cm

18. Mirror
$r = 23$ cm

19. Rug
$r = 0.8$ m

20. Camera Lens
$r = 25$ mm

21. Plate
$r = 11.4$ cm

22. Pool Cover Top
$r = 3$ m

23. Garden
$r = 2.7$ m

24. Top of a Jar
$r = 3.5$ cm

CUSTOMARY MEASURES   *Use $\frac{22}{7}$ for π.*

25. Top of a Drum
$r = 8\frac{1}{2}$ in

26. Hassock Top
$r = 9$ in

27. Emblem
$r = 2\frac{1}{4}$ in

28. Window
$r = 8$ ft

29. Paper Doily
$r = 4$ in

30. Target
$r = 1\frac{1}{2}$ ft

31. Flower Center
$r = 2$ in

32. Clock Face
$r = 2\frac{3}{4}$ in

Fractions: Addition/Subtraction   **179**

REVIEW: SECTIONS 6-4 – 6-6

For Exercises 1-2, use the information in the time card at the right to solve each problem. (Section 6-4)

Name:		Mark Thomas		
DAYS	IN	OUT	IN	OUT
1	7:00	11:00	12:00	3:15
2	7:00	11:15	12:15	3:00
3	7:00	11:20	12:20	3:20
4	6:45	11:00	12:00	3:00
5	7:00	11:00	12:00	2:40

1. **a.** Find the number of hours Mark Thomas worked each day.
   b. Find the total number of hours he worked for the week.

2. Mark Thomas is paid $5.40 per hour. Find his total weekly earnings.

Compare each pair of fractions. Write >, <, or =. (Section 6-5)

3. $\frac{1}{4}$? $\frac{2}{5}$

4. $\frac{1}{2}$? $\frac{3}{8}$

5. $\frac{4}{10}$? $\frac{2}{5}$

6. $\frac{8}{5}$? $\frac{5}{3}$

Find the circumference and area of each circle. Round answers to the nearest whole number. (Section 6-6)

7. $r = 6$ in; $\pi = \frac{22}{7}$

8. $r = 4.2$ m; $\pi = 3.14$

9. Joan caught a fish that weighed $1\frac{1}{3}$ pounds. Tara caught a fish that weighed $1\frac{2}{5}$ pounds. Whose fish weighed more, Joan's or Tara's? (Section 6-5)

10. Cathy cut four circle posters for the bulletin board at her school. She cut two circles 7 inches in diameter and two circles 10 inches in diameter. How many square inches of paper did she use? Use $\frac{22}{7}$ for π. (Section 6-6)

COMPARING FRACTIONS

You can use a calculator to compare fractions.

EXAMPLE Which is greater, $\frac{5}{6}$ or $\frac{7}{8}$?

Solution Write a decimal for each fraction.

5 ÷ 6 = [0.8333333]   7 ÷ 8 = [0.875]

Since $0.875 > 0.8\overline{3}$, $\frac{7}{8} > \frac{5}{6}$.

EXERCISES Use a calculator to compare each pair of fractions in Exercises 31-35 on page 175.

CHAPTER REVIEW

PART 1: VOCABULARY

For Exercises 1–5, choose from the box at the right the word(s) that best correspond to each description.

1. The number 0.25 is a __?__ decimal. (p. 173)
2. The number 1.1$\overline{6}$ is a __?__ decimal. (p. 174)
3. Fractions such as $\frac{3}{5}$ and $\frac{1}{5}$ have __?__ denominators. (p. 158)
4. Fractions that name the same number are __?__. (p. 161)
5. Fractions such as $\frac{2}{3}$ and $\frac{5}{8}$ have __?__ denominators. (p. 161)

> like
> equivalent fractions
> repeating
> terminating
> unlike

PART 2: SKILLS

Add or subtract as indicated. Write each answer in lowest terms. (Section 6-1)

6. $\frac{3}{5} + \frac{1}{5}$
7. $\frac{2}{7} + \frac{4}{7}$
8. $\frac{5}{8} + \frac{2}{8}$
9. $\frac{9}{16} - \frac{1}{16}$
10. $\frac{11}{12} - \frac{5}{12}$
11. $2\frac{1}{3} + 1\frac{1}{3}$
12. $6\frac{1}{4} + 2\frac{3}{4}$
13. $6\frac{7}{8} - 2\frac{3}{8}$
14. $12\frac{9}{10} - 3\frac{7}{10}$
15. $3\frac{3}{5} - 3\frac{2}{5}$

Complete. Write an equivalent fraction as indicated. (Section 6-2)

16. $\frac{3}{5} = \frac{?}{15}$
17. $\frac{4}{9} = \frac{?}{18}$
18. $\frac{1}{7} = \frac{?}{35}$
19. $\frac{7}{12} = \frac{?}{36}$
20. $\frac{2}{3} = \frac{?}{30}$

Add or subtract as indicated. Write each answer in lowest terms. (Section 6-2)

21. $\frac{3}{5} + \frac{1}{3}$
22. $\frac{5}{8} + \frac{1}{2}$
23. $\frac{7}{16} + \frac{11}{24}$
24. $\frac{4}{7} + \frac{3}{14}$
25. $\frac{2}{9} + \frac{1}{3}$
26. $\frac{7}{12} - \frac{1}{6}$
27. $\frac{11}{15} - \frac{3}{10}$
28. $\frac{7}{9} - \frac{1}{4}$
29. $\frac{6}{7} - \frac{2}{21}$
30. $\frac{17}{24} - \frac{3}{8}$

(Section 6-3)

31. $2\frac{1}{2} + 1\frac{1}{3}$
32. $3\frac{1}{6} + 5\frac{2}{5}$
33. $9\frac{2}{7} + 6\frac{2}{3}$
34. $10\frac{3}{14} + 4\frac{1}{2}$
35. $4\frac{5}{6} + 3\frac{1}{8}$
36. $7\frac{2}{3} - 3\frac{1}{8}$
37. $10 - 4\frac{5}{9}$
38. $8\frac{2}{5} - 5\frac{3}{4}$
39. $17\frac{1}{8} - 10\frac{5}{6}$
40. $4\frac{1}{2} - 2\frac{1}{6}$

For Exercises 41–42, complete the table. (Section 6-4)

| | Hours Per Day ||||| Total Hours | Rate Per Hour | Weekly Earnings |
	1	2	3	4	5			
41.	$7\frac{1}{2}$	8	$7\frac{1}{3}$	$8\frac{1}{4}$	$7\frac{3}{4}$	?	$4.80	?
42.	$6\frac{1}{2}$	5	$8\frac{1}{3}$	$7\frac{1}{2}$	$6\frac{3}{4}$	?	$7.20	?

Fractions: Addition/Subtraction

Write a decimal for each fraction. (Section 6-5)

43. $\frac{1}{8}$ **44.** $\frac{4}{9}$ **45.** $\frac{6}{11}$ **46.** $\frac{7}{5}$ **47.** $\frac{3}{20}$

Compare each pair of fractions. Write >, <, or =. (Section 6-5)

48. $\frac{2}{5}$ __?__ $\frac{1}{4}$ **49.** $\frac{5}{6}$ __?__ $\frac{7}{8}$ **50.** $\frac{7}{12}$ __?__ $\frac{1}{2}$ **51.** $\frac{9}{4}$ __?__ $\frac{11}{5}$ **52.** $\frac{1}{3}$ __?__ $\frac{2}{9}$

Find the circumference and area of each circle. Round answers to the nearest whole number. (Section 6-6)

53. $r = 3.8$ m; $\pi = 3.14$

54. $r = 5.3$ km; $\pi = 3.14$

55. $d = 24$ ft; $\pi = \frac{22}{7}$

56. $d = 16$ in; $\pi = \frac{22}{7}$

PART 3: APPLICATIONS

57. It takes Diane $\frac{3}{5}$ of an hour to walk to school. It takes Nara $\frac{4}{5}$ of an hour to walk to school. How much longer does it take Nara to walk to school than Diane? (Section 6-1)

58. Martin bought $\frac{1}{2}$ pound of pecans and $\frac{3}{8}$ pound of walnuts. How many pounds of nuts did he buy in all? (Section 6-2)

59. Grant bicycled $8\frac{2}{5}$ miles on Saturday and $10\frac{1}{4}$ miles on Sunday. How much farther did he bicycle on Sunday than on Saturday? (Section 6-3)

60. Rafael worked $\frac{4}{5}$ of his math problems correctly. Hiram worked $\frac{7}{8}$ of his math problems correctly. Who worked more problems correctly, Rafael or Hiram? (Section 6-5)

61. The diameter of a round exercise track on a space station is 32 meters. To the nearest meter, how far must an astronaut walk to go around the track? (Section 6-6)

62. Jane painted the top of a round picnic table. The radius of the table is 21 inches. To the nearest square inch, what is the area she painted? Use $\frac{22}{7}$ for π. (Section 6-6)

CHAPTER TEST

Complete. Write an equivalent fraction as indicated.

1. $\frac{4}{5} = \frac{?}{10}$
2. $\frac{2}{9} = \frac{?}{45}$
3. $\frac{5}{12} = \frac{?}{48}$
4. $\frac{3}{4} = \frac{?}{24}$

Add or subtract as indicated. Write each answer in lowest terms.

5. $\frac{5}{12} + \frac{1}{12}$
6. $\frac{3}{5} - \frac{1}{4}$
7. $\frac{8}{9} + \frac{1}{6}$
8. $2\frac{1}{2} - 1\frac{1}{3}$
9. $6 - 2\frac{3}{7}$
10. $3\frac{2}{5} + 2\frac{3}{8}$
11. $8\frac{3}{8} + 4\frac{5}{6}$
12. $12\frac{1}{8} - 4\frac{2}{3}$

Write a decimal for each fraction.

13. $\frac{3}{5}$
14. $\frac{2}{3}$
15. $\frac{5}{6}$
16. $\frac{7}{20}$

17. Find the circumference and area of a circle with a radius of 2.3 meters. Use $\frac{22}{7}$ for π. Round your answer to the nearest whole number.

18. California produces $\frac{1}{4}$ of the cotton crop for the United States. Texas produces $\frac{9}{40}$ of the cotton crop for the United States. Which state produces more cotton, California of Texas?

19. Tanya is paid $8.40 per hour. On Monday she worked from 9:00 A.M. to 11:30 A.M.; on Thursday she worked from 10:00 A.M. to 3:15 P.M.; on Friday she worked from 9:00 A.M. to 12:20 P.M. Find her total earnings.

20. The Johnson's bought 30 feet of fence to put around a round flower bed. The diameter of the flower bed is 8 feet. To the nearest foot, how many feet of fence did they have left?

ANSWERS TO REVIEW CAPSULES

Page 160   1. 12   2. 24   3. 12   4. 6   5. 60   6. $\frac{1}{2}$ 7. $\frac{2}{5}$ 8. $\frac{7}{9}$ 9. $\frac{2}{5}$ 10. $\frac{3}{7}$

Page 164   1. 33   2. 60   3. 32   4. 30   5. 36   6. $\frac{1}{2}$ 7. $\frac{1}{6}$ 8. $\frac{6}{13}$ 9. $\frac{1}{3}$ 10. $\frac{3}{8}$

Page 169   1. $\frac{27}{40}$ 2. $\frac{4}{9}$ 3. $9\frac{9}{10}$ 4. $4\frac{2}{9}$ 5. $7\frac{1}{2}$ 6. 24   7. 75   8. 345   9. $\frac{1}{4}$ 10. $\frac{45}{60}$; 45   11. $\frac{10}{60}$ 12. $\frac{20}{60}$; 20

Page 172   1. 0.85   2. 0.55   3. 0.65   4. 0.26   5. 0.75   6. <   7. <   8. <   9. >   10. >   11. <   12. >   13. <   14. <

Page 176   1. 59.66   2. 174.584   3. 50.24   4. 314   5. $\frac{22}{7}$ 6. $\frac{7}{4}$ 7. $\frac{21}{4}$ 8. $\frac{87}{8}$ 9. $\frac{31}{10}$ 10. $\frac{17}{8}$ 11. $\frac{23}{2}$ 12. $\frac{48}{5}$ 13. 308   14. 1760   15. $236\frac{4}{63}$ 16. $86\frac{5}{8}$ 17. 21   18. 74   19. 35   20. 1   21. 5   22. 63

Fractions: Addition/Subtraction

ENRICHMENT

Precision of Measurement

In measurement, the word **precision** refers to the *smallest measurement* that can be made on the instrument used to measure. The last digit written in a measurement may be in error by as much as one-half the unit of precision. This is the **maximum possible error.**

EXAMPLE Find the maximum possible error for a measurement of 1.6 cm.

Solution: Unit of precision: 0.1 cm ◀ The last digit of 1.6 cm is in the tenths place.

Maximum possible error: $\frac{1}{2} \times 0.1 =$ **0.05**

Since all measurements are approximations, a maximum possible error of 0.05 means that a measurement can vary by ±0.05. That is, a measurement of 4.8 centimeters can be as long as (4.8 + 0.05), or 4.85 centimeters and as short as (4.8 − 0.05), or 4.75 centimeters.

This table shows other examples.

Measurement	Precision	Maximum Possible Error	Greatest Length	Shortest Length
4.7 m	0.1 m	±0.05 m	4.75 m	4.65 m
3.41 cm	0.01 cm	±0.005 cm	3.415 cm	3.405 cm
$4\frac{3}{8}$ in	$\frac{1}{8}$ in	$\pm\frac{1}{16}$ in	$4\frac{7}{16}$ in	$4\frac{5}{16}$ in

EXERCISES

Complete.

	Measurement	Precision	Maximum Possible Error	Greatest Length	Shortest Length
1.	4.3 cm	?	?	?	?
2.	61.71 cm	?	?	?	?
3.	$2\frac{1}{2}$ in	?	?	?	?
4.	$5\frac{3}{8}$ in	?	?	?	?

ADDITIONAL PRACTICE

SKILLS

Add or subtract. Write each answer in lowest terms.
(Pages 158–160)

1. $\frac{3}{10} + \frac{3}{10}$
2. $4\frac{5}{7} + 3\frac{2}{7}$
3. $\frac{11}{12} - \frac{5}{12}$
4. $27\frac{3}{4} - 18\frac{1}{4}$

(Pages 161–164)

5. $\frac{1}{2} + \frac{1}{4}$
6. $\frac{1}{3} + \frac{3}{5}$
7. $\frac{2}{7} + \frac{1}{5}$
8. $\frac{5}{8} + \frac{9}{10}$
9. $\frac{2}{3} - \frac{1}{6}$
10. $\frac{4}{5} - \frac{2}{9}$
11. $\frac{11}{14} - \frac{1}{7}$
12. $\frac{3}{4} - \frac{7}{10}$

(Pages 165–167)

13. $3\frac{1}{6} + 5\frac{1}{3}$
14. $4\frac{1}{3} + 2\frac{1}{2}$
15. $5\frac{7}{8} + 3\frac{3}{4}$
16. $17\frac{7}{8} + 4\frac{9}{10}$
17. $5\frac{4}{9} - 2\frac{1}{6}$
18. $17\frac{1}{8} - 9\frac{3}{4}$
19. $14\frac{1}{5} - 8\frac{1}{3}$
20. $16\frac{7}{10} - 8\frac{3}{4}$

Write a decimal for each fraction. Use a bar to indicate any digit or digits that repeat. (Pages 173–175)

21. $\frac{7}{10}$
22. $\frac{5}{9}$
23. $\frac{4}{15}$
24. $\frac{1}{16}$
25. $\frac{5}{12}$
26. $\frac{5}{3}$

Compare each pair of fractions. Write >, <, or =. Give a reason. (Pages 173–175)

27. $\frac{4}{5}$? $\frac{3}{4}$
28. $\frac{7}{10}$? $\frac{4}{5}$
29. $\frac{2}{5}$? $\frac{3}{8}$
30. $\frac{2}{3}$? $\frac{4}{6}$
31. $\frac{3}{8}$? $\frac{1}{3}$

Find the circumference and area of each circle. Use 3.14 for π. Round answers to the nearest whole number. (Pages 176–179)

32. $r = 40$ cm
33. $d = 25$ m
34. $r = 11$ km
35. $d = 30.4$ cm

Use $\frac{22}{7}$ for π. (Pages 176–179)

36. $r = 9\frac{1}{2}$ ft
37. $d = 4$ mi
38. $r = 3\frac{3}{4}$ in
39. $d = 18$ yds

APPLICATIONS

40. David Yaw works from 1:00 A.M. to 3:15 P.M. on Monday, Wednesday, and Friday and from 8:30 A.M. to 1:00 P.M. on Tuesday and Thursday. He is paid $4.80 per hour. Find his weekly earnings. (Pages 169–172)

41. About $\frac{1}{10}$ of the households in America have a computer. About $\frac{9}{100}$ of American households own a cordless telephone. Are there more computers or cordless telephones in American households?

Fractions: Addition/Subtraction

COMMON ERRORS

Each of these problems contains a common error.

    a. Find the correct answer.

    b. Find the error.

1. List all the factors of 18.
Factors of 18: **2, 3, 6, 9**

2. Write the prime factorization of 24.
$24 = 1 \cdot 2 \cdot 2 \cdot 2 \cdot 3 = \mathbf{1 \cdot 2^3 \cdot 3}$

3. Find the LCM of 6 and 8.
$6 = 2 \cdot 3$
$8 = 2 \cdot 2 \cdot 2$
LCM of 6 and 8: $2 \cdot 3 = \mathbf{6}$

4. Find the GCF of 4 and 10.
$4 = ②\cdot 2$
$10 = ②\cdot 5$
GCF of 4 and 10: $2 \cdot 2 = \mathbf{4}$

5. $\frac{3}{5} \cdot \frac{4}{5} = \underline{\ ?\ }$
$\frac{3}{5} \cdot \frac{4}{5} = \frac{12}{5}$, or $\mathbf{2\frac{2}{5}}$

6. $\frac{2}{3} \div \frac{4}{7} = \underline{\ ?\ }$
$\frac{2}{3} \div \frac{4}{7} = \frac{2}{3} \cdot \frac{4}{7} = \mathbf{\frac{8}{21}}$

7. $3\frac{3}{5} \div 1\frac{2}{3} = \underline{\ ?\ }$
$3\frac{3}{5} \div 1\frac{2}{3} = 3\frac{3}{5} \cdot 1\frac{3}{2} = \mathbf{3\frac{9}{10}}$

8. $2\frac{1}{3} \cdot 4\frac{1}{5} = \underline{\ ?\ }$
$2\frac{1}{3} \cdot 4\frac{1}{5} = \mathbf{8\frac{2}{15}}$

9. $\frac{2}{3} + \frac{3}{4} = \underline{\ ?\ }$
$\frac{2}{3} + \frac{3}{4} = \frac{2}{12} + \frac{3}{12} = \mathbf{\frac{5}{12}}$

10. $\frac{5}{8} - \frac{3}{5} = \underline{\ ?\ }$
$\frac{5}{8} - \frac{3}{5} = \mathbf{\frac{2}{3}}$

11. $4\frac{1}{6} + 5\frac{1}{6} = \underline{\ ?\ }$
$4\frac{1}{6} + 5\frac{1}{6} = 9\frac{2}{12}$
$= \mathbf{9\frac{1}{6}}$

12. $10\frac{1}{4} - 3\frac{3}{5} = \underline{\ ?\ }$
$10\frac{1}{4} - 3\frac{3}{5} = 10\frac{5}{20} - 3\frac{12}{20}$
$= 9\frac{15}{20} - 3\frac{12}{20} = \mathbf{6\frac{3}{20}}$

13. Find the area of this triangle.
$b = 3$ yd; $h = 1\frac{1}{2}$ yd
$A = bh$
$A = 3\left(\frac{3}{2}\right)$
$A = \frac{9}{2}$, or $\mathbf{4\frac{1}{2}\ yd^2}$

14. Find the area of this circle.
$r = 8$ cm; $\pi = 3.14$
$A = \pi r^2$
$A = 3.14(8)(2)$
$A = \mathbf{50.24\ cm^2}$

CUMULATIVE REVIEW: CHAPTERS 4–6

Choose the correct answer. Choose a, b, c, or d.

1. Which number is <u>not</u> a factor of 27?
   a. 1 b. 6 c. 3 d. 9

2. Which number is <u>not</u> divisible by 9?
   a. 315 b. 756 c. 664 d. 558

3. Which number is <u>not</u> a multiple of 10?
   a. 770 b. 660 c. 1300 d. 995

4. Find the smallest prime factor of 1925.
   a. 3 b. 2 c. 5 d. 7

5. Rewrite this product using exponents: $2 \cdot 2 \cdot 3 \cdot 2 \cdot 3 \cdot 5 \cdot 5 \cdot 2 \cdot 3$
   a. $2^3 \cdot 3^3 \cdot 5^2$ b. $2^4 \cdot 3^3 \cdot 5^2$ c. $2^4 \cdot 3^2 \cdot 5^3$ d. $2^3 \cdot 3^2 \cdot 5^2$

6. Find the prime factorization of 540.
   a. $2^2 \cdot 3^2 \cdot 5$ b. $2^2 \cdot 3^3 \cdot 5$ c. $2^3 \cdot 3^3 \cdot 5$ d. $2^2 \cdot 3^2 \cdot 5^2$

7. Find the least common multiple of 15, 30, and 50.
   a. 150 b. 60 c. 300 d. 90

8. Find the greatest common factor of 12 and 20.
   a. 4 b. 2 c. 24 d. 60

9. Write this fraction in lowest terms: $\frac{18}{54}$
   a. $\frac{9}{27}$ b. $\frac{6}{18}$ c. $\frac{1}{3}$ d. $\frac{1}{6}$

10. A basketball player made 25 out of 55 attempted shots. What fractional part of the attempted shots did the player make?
    a. $\frac{2}{5}$ b. $\frac{5}{11}$ c. $\frac{3}{5}$ d. $\frac{3}{10}$

11. Subtract. Express the answer as a mixed numeral.   17 lb 4 oz
                                                      $-$12 lb 12 oz
    a. $5\frac{1}{2}$ lb b. $5\frac{1}{8}$ lb c. $4\frac{1}{2}$ lb d. $4\frac{1}{8}$ lb

12. Multiply: $\frac{4}{15} \cdot \frac{3}{4}$
    a. $\frac{2}{5}$ b. $\frac{4}{5}$ c. $\frac{3}{5}$ d. $\frac{1}{5}$

13. On the planet Mercury, the temperature can climb to 100°C. Use the formula $F = \frac{9}{5}C + 32$ to write this temperature in degrees Fahrenheit.
    a. 124° b. 224° c. 112° d. 212°

14. Multiply: $1\frac{1}{3} \cdot 1\frac{7}{8}$
  a. $2\frac{1}{2}$
  b. $1\frac{7}{24}$
  c. $1\frac{1}{4}$
  d. $2\frac{1}{4}$

15. Sam earns $3.80 per hour. How much will Sam earn in 37 hours?
  a. $140.60
  b. $142.50
  c. $145.00
  d. $172.50

16. Find the area of the gable of a roof that has a base of 6.8 meters and a height of 4.4 meters.
  a. 74.4 m²
  b. 14.96 m²
  c. 7.44 m²
  d. 149.6 m²

17. Find the volume of a box of detergent 22 centimeters long, 10 centimeters wide, and 30 centimeters high.
  a. 6600 cm³
  b. 62 cm³
  c. 6660 cm³
  d. 124 cm³

18. Divide: $1\frac{1}{3} \div 2\frac{2}{3}$
  a. $\frac{1}{2}$
  b. $\frac{24}{9}$
  c. 2
  d. $\frac{1}{3}$

19. Maria and Andrea ride their bicycles 20 kilometers in $1\frac{1}{4}$ hours. How many kilometers could they ride in one hour?
  a. 25
  b. 16
  c. 18
  d. 20

20. Subtract: $\frac{7}{15} - \frac{3}{10}$
  a. $\frac{4}{5}$
  b. $\frac{1}{6}$
  c. $1\frac{1}{5}$
  d. $\frac{11}{30}$

21. Subtract: $4\frac{1}{6} - 2\frac{1}{2}$
  a. $2\frac{1}{4}$
  b. $2\frac{1}{3}$
  c. $1\frac{2}{3}$
  d. $1\frac{1}{2}$

22. Yoshiro worked $5\frac{1}{2}$ hours on Monday, $2\frac{3}{4}$ hours on Tuesday, and $7\frac{1}{4}$ hours on Wednesday. At $3.40 per hour, how much did Yoshiro earn for the three days?
  a. $52.70
  b. $51.40
  c. $53.80
  d. $55.20

23. Write a decimal for $\frac{7}{25}$.
  a. 0.24
  b. 0.26
  c. 0.25
  d. 0.28

24. Find the circumference of a wheel that has a diameter of 40 centimeters. Use 3.14 for π.
  a. 47.10 cm
  b. 125.60 cm
  c. 1412.50 cm
  d. 5024 cm

CHAPTER 7 Solving Equations

SECTIONS
7-1 Solving Equations by Subtraction
7-2 Solving Equations by Addition
7-3 Problem Solving and Applications: Using Equations: Addition/Subtraction
7-4 Solving Equations by Division
7-5 Solving Equations by Multiplication
7-6 Problem Solving and Applications: Using Equations: Multiplication/Division

FEATURES
Consumer Application: *Wallpaper and Estimation*
Enrichment: *Repeating Decimals*

7-1 Solving Equations by Subtraction

This table shows that subtraction "undoes" addition.

1 Choose a number.	2 Add any number to it.	3 Subtract the same number.	4 Result: The original number
12	12 + 8	12 + 8 − 8	12
3.2	3.2 + 8.1	3.2 + 8.1 − 8.1	3.2
$5\frac{1}{3}$	$5\frac{1}{3} + \frac{1}{2}$	$5\frac{1}{3} + \frac{1}{2} - \frac{1}{2}$	$5\frac{1}{3}$
n	$n + 4$	$n + 4 - 4$	n

Thus, subtraction is the **opposite,** or **inverse operation,** of addition. This idea can be applied to solving equations. An **equation** is a mathematical sentence that contains the equality symbol "=".

To **solve an equation** means to find the number(s) that can replace the variable to make the equation true. To solve an addition equation such as

$$x + 9 = 21,$$

subtract 9 from each side of the equation.

The goal in solving an equation is <u>to get the variable alone on one side of the equation</u>.

PROCEDURE

To solve the equation $x + b = c$ for x, subtract b from each side of the equation.

EXAMPLE 1 Solve and check: $x + 9 = 21$

Solution: To get x alone, subtract 9 from each side.

$x + 9 = 21$ **Check:** $x + 9 = 21$
$x + 9 - 9 = 21 - 9$ $12 + 9$
$x = 12$ 21

▶ Replace x with 12.

Self-Check *Solve and check.*

1. $x + 7 = 43$ **2.** $t + 9 = 12$ **3.** $a + 13 = 13$

Chapter 7

As Example 1 shows, you check an equation by replacing the variable with the value you obtained. If this value makes the equation true, it is a **solution** of the equation.

EXAMPLE 2 Solve and check.
 a. $12.6 = t + 1.3$
 b. $g + 1\frac{1}{4} = 2\frac{5}{8}$

Solutions:

a. $12.6 = t + 1.3$ ◀ Subtract 1.3 from both sides.
$12.6 - 1.3 = t + 1.3 - 1.3$
$11.3 = t$

b. $g + 1\frac{1}{4} = 2\frac{5}{8}$ ◀ Subtract $1\frac{1}{4}$ from both sides.
$g + 1\frac{1}{4} - 1\frac{1}{4} = 2\frac{5}{8} - 1\frac{1}{4}$
$g = 2\frac{5}{8} - 1\frac{2}{8}$
$g = 1\frac{3}{8}$

The checks are left for you.

Self-Check *Solve and check.*

4. $19.6 = n + 17.8$
5. $y + \frac{1}{4} = \frac{5}{8}$
6. $p + 1\frac{1}{2} = 2\frac{3}{4}$

You can also use the following method to solve equations.

$$\begin{array}{r} x + 9 = 21 \\ -9 = -9 \\ \hline x = 12 \end{array}$$

$$\begin{array}{r} 12.6 = t + 1.3 \\ -1.3 = -1.3 \\ \hline 11.3 = t \end{array}$$

To be sure that your answer is a solution of the equation, always use the original equation to check.

CLASSROOM EXERCISES

Evaluate each expression. (Table)

1. $8 + 9 - 9$
2. $12 + 3 - 3$
3. $n + 7 - 7$
4. $m + 5 - 5$
5. $x + \frac{1}{2} - \frac{1}{2}$
6. $a + \frac{3}{4} - \frac{3}{4}$
7. $n + 112 - 112$
8. $c + \frac{2}{3} - \frac{2}{3}$

Write the equation formed as the first step in solving each equation. (Step 1, Examples 1–2)

9. $x + 5 = 19$
10. $a + 12 = 14$
11. $n + 5.7 = 10.3$
12. $14.6 = p + 9.7$
13. $b + \frac{1}{4} = \frac{1}{2}$
14. $y + 1\frac{3}{4} = 3\frac{7}{8}$
15. $6\frac{7}{8} = y + 2\frac{3}{4}$
16. $x + 12.5 = 21.3$
17. $3.2 + r = 9$
18. $27 = 5 + a$
19. $6 + c = 9$
20. $5.6 + z = 8.2$

Solving Equations 191

WRITTEN EXERCISES

Goal: To use subtraction to solve equations

For Exercises 1–51, solve and check each equation. Show all steps. (Example 1)

1. $n + 14 = 21$
2. $x + 18 = 35$
3. $46 = y + 29$
4. $42 = r + 23$
5. $t + 19 = 43$
6. $s + 34 = 51$
7. $a + 39 = 56$
8. $b + 45 = 71$
9. $64 = q + 37$
10. $120 = t + 90$
11. $h + 48 = 97$
12. $n + 17 = 71$

(Example 2)

13. $k + 5.8 = 7.4$
14. $x + 3.4 = 7.6$
15. $5.8 = y + 2.3$
16. $6.9 = n + 4.5$
17. $t + 6.8 = 9.7$
18. $b + 12.8 = 15.1$
19. $r + 23.7 = 35.2$
20. $s + 0.43 = 0.9$
21. $0.85 = y + 0.47$
22. $0.74 = x + 0.24$
23. $m + 1.05 = 22.79$
24. $p + 4.31 = 18.19$
25. $r + \frac{3}{8} = 1\frac{5}{8}$
26. $t + \frac{7}{12} = 3\frac{11}{12}$
27. $5\frac{3}{4} = x + 3\frac{1}{2}$
28. $9\frac{1}{2} = n + 5\frac{1}{4}$
29. $c + 3\frac{3}{4} = 7\frac{1}{4}$
30. $a + 10\frac{1}{2} = 15\frac{1}{4}$
31. $c + \frac{1}{2} = \frac{5}{6}$
32. $\frac{3}{5} + p = \frac{9}{10}$
33. $8\frac{1}{3} = 4\frac{7}{9} + m$
34. $9\frac{3}{4} = y + 5\frac{1}{2}$
35. $15\frac{1}{5} + q = 17\frac{1}{3}$
36. $9\frac{1}{2} + n = 17\frac{3}{8}$

MIXED PRACTICE

37. $5.19 = a + 2.65$
38. $t + 3\frac{1}{8} = 5\frac{3}{16}$
39. $x + 4\frac{3}{4} = 8\frac{7}{12}$
40. $7.03 = g + 4.28$
41. $y + 190 = 403$
42. $s + 8417 = 9378$
43. $x + 0.39 = 1.46$
44. $4204 = k + 3787$
45. $205 = m + 198$
46. $v + 5\frac{2}{3} = 12\frac{1}{5}$
47. $4\frac{3}{16} = r + 2\frac{5}{16}$
48. $3\frac{1}{8} = s + 1\frac{7}{8}$
49. $y + 0.57 = 2.15$
50. $145 = b + 145$
51. $w + 5.23 = 5.23$

REVIEW CAPSULE FOR SECTION 7–2

Complete. (Pages 5–6; 35–37; 165–167)

1. $7 = 6 - 6 + \underline{}$
2. $1.5 - 1.5 + \underline{} = 6$
3. $9\frac{1}{2} = 1\frac{1}{4} - 1\frac{1}{4} + \underline{}$
4. $58.6 - 58.6 + \underline{} = 0.9$
5. $q = 18 - 18 + \underline{}$
6. $\frac{3}{8} - \frac{3}{8} + c = \underline{}$
7. $1.9 - 1.9 + \underline{} = f$
8. $p + 1800 - p = \underline{}$

The answers are on page 212.

7-2 Solving Equations by Addition

This table shows that addition "undoes" subtraction.

1 Choose a number.	2 Subtract any number.	3 Add the same number.	4 Result: the original number
18	$18 - 7$	$18 - 7 + 7$	18
23.5	$23.5 - 11.8$	$23.5 - 11.8 + 11.8$	23.5
$12\frac{1}{2}$	$12\frac{1}{2} - 3\frac{3}{4}$	$12\frac{1}{2} - 3\frac{3}{4} + 3\frac{3}{4}$	$12\frac{1}{2}$
n	$n - 12$	$n - 12 + 12$	n

Thus, subtraction is the inverse operation of addition. To solve a subtraction equation such as

$$r - 9 = 15$$

add 9 **to each side** of the equation.

PROCEDURE

To solve the equation $x - b = c$ for x, add b **to each side** of the equation.

EXAMPLE 1 Solve and check: $r - 9 = 15$

Solution: To get r alone, add 9 to each side.

$r - 9 = 15$ **Check:** $r - 9 = 15$
$r - 9 + 9 = 15 + 9$ $24 - 9$
$r = 24$ 15

Self-Check *Solve and check.*

1. $x - 5 = 18$ **2.** $y - 7 = 11$ **3.** $t - 4 = 1$

NOTE: In solving an equation, you write equivalent equations. **Equivalent equations** have the same solution(s). Thus, in the solution of Example 1, these equations are equivalent.

$r - 9 = 15$
$r - 9 + 9 = 15 + 9$ ◀ **Equivalent equations.**
$r = 24$

The three equations have the same solution, **24.**

Solving Equations

EXAMPLE 2 Solve and check.

a. $19.7 = t - 8.2$

b. $\frac{1}{2} = r - \frac{3}{4}$

Solutions:

a. $19.7 = t - 8.2$ ◀ Add 8.2 to each side

$19.7 + 8.2 = t - 8.2 + 8.2$

$27.9 = t$

b. $\frac{1}{2} = r - \frac{3}{4}$ ◀ Add $\frac{3}{4}$ to each side.

$\frac{1}{2} + \frac{3}{4} = r - \frac{3}{4} + \frac{3}{4}$

$\frac{2}{4} + \frac{3}{4} = r$

$\frac{5}{4} = r$

The checks are left for you.

Self-Check Solve and check.

4. $x - 1.1 = 4.8$
5. $\frac{5}{6} = r - \frac{1}{6}$
6. $y - 1\frac{1}{8} = 3\frac{3}{8}$

CLASSROOM EXERCISES

Evaluate each expression. (Table)

1. $a - 5 + 5$
2. $t - 10 + 10$
3. $r - 12 + 12$
4. $s - 37 + 37$
5. $x - 2.8 + 2.8$
6. $n - 0.8 + 0.8$
7. $a - \frac{1}{4} + \frac{1}{4}$
8. $p - 6\frac{1}{2} + 6\frac{1}{2}$
9. $3\frac{1}{4} - x + x$

Write the equivalent equation formed as the first step in solving each equation. (Step 1, Examples 1–3)

10. $x - 7 = 19$
11. $a - 12 = 5$
12. $23 = s - 27$
13. $47 = t - 83$
14. $q - 2.7 = 3.9$
15. $12.9 = b - 5.6$
16. $w - \frac{3}{4} = 2\frac{1}{2}$
17. $4\frac{1}{4} = n - 1\frac{7}{8}$
18. $q - \frac{3}{4} = 6\frac{1}{2}$

WRITTEN EXERCISES

Goal: To use addition to solve equations

For Exercises 1–51, solve and check each equation. Show all steps. (Example 1)

1. $x - 32 = 26$
2. $r - 23 = 36$
3. $t - 48 = 53$
4. $p - 59 = 47$
5. $75 = n - 128$
6. $87 = b - 136$
7. $185 = a - 207$
8. $95 = p - 235$
9. $m - 169 = 169$
10. $378 = k - 378$
11. $y - 4021 = 86$
12. $x - 17 = 8013$

(Example 2)

13. $s - 4.2 = 5.9$
14. $q - 7.9 = 6.7$
15. $0.86 = w - 0.27$
16. $0.54 = n - 0.89$
17. $19.7 = g - 28.3$
18. $28.5 = h - 42.7$
19. $r - 9.08 = 15.29$
20. $x - 12.37 = 9.09$
21. $c - 10.8 = 9.7$
22. $d - 90.6 = 7.9$
23. $6.08 = p - 104.1$
24. $113.2 = t - 8.4$
25. $b - 2\frac{1}{2} = 5\frac{1}{2}$
26. $a - 3\frac{1}{4} = 2\frac{1}{2}$
27. $3\frac{3}{4} = n - 10\frac{3}{4}$
28. $5\frac{1}{8} = q - 12\frac{3}{8}$
29. $r - \frac{7}{16} = 1\frac{3}{16}$
30. $t - \frac{11}{12} = \frac{7}{12}$
31. $2\frac{1}{3} = m - 3\frac{1}{6}$
32. $c - 7\frac{1}{4} = 3\frac{1}{8}$
33. $11\frac{1}{7} = d - 2\frac{6}{7}$
34. $s - \frac{4}{9} = \frac{1}{3}$
35. $\frac{5}{8} = q - 3\frac{1}{4}$
36. $g - \frac{8}{11} = 3\frac{7}{22}$

MIXED PRACTICE

37. $49 = r - 78$
38. $g - 24.9 = 13.9$
39. $k - 263 = 105$
40. $57 = t - 96$
41. $4\frac{5}{6} = n - 1\frac{5}{12}$
42. $5\frac{5}{8} = p - 2\frac{3}{4}$
43. $t - \frac{3}{4} = 2\frac{15}{16}$
44. $m - 135.8 = 94.7$
45. $2.97 = y - 0.08$
46. $s - \frac{4}{5} = \frac{13}{15}$
47. $b - 276.5 = 57.9$
48. $1356 = f - 867$

MORE CHALLENGING EXERCISES

The set or collection of numbers that a variable represents is called the **replacement set.**

EXAMPLE: Solve the equation $n + 9 = n$ when the replacement set is {whole numbers}.

Solution: Since 0 is the only whole number for which $n + 9 = 9$, the solution is **0**.

Find the solution(s). The replacement set is {2, 4, 6, 8, 10}.

49. $s + 2 = 10$
50. $t + 2 = 2 + t$
51. $r + r = 2r$
52. $p = 2$
53. $q - 5 = 3$
54. $21 - a = 15$

REVIEW CAPSULE FOR SECTION 7-3

Solve each equation. (Pages 190–192)

1. $x + 4 = 18$
2. $p + 3.5 = 16.2$
3. $36 = x + 21$
4. $n + 24 = 86$
5. $m + 0.4 = 3.8$
6. $t + \frac{4}{5} = 2\frac{1}{2}$
7. $s + 2\frac{1}{2} = 4\frac{2}{3}$
8. $25.4 = p + 15.5$

The answers are on page 212.

Solving Equations **195**

PROBLEM SOLVING AND APPLICATIONS

7-3 Using Equations: Addition/Subtraction

The following word rule and formula can be used to find net pay.

Word Rule: Net Pay + Deductions = Gross Earnings
Formula: $n + d = g$

Net pay, or **take-home pay,** is the amount left when all deductions (such as taxes) have been subtracted from the amount actually earned. The amount actually earned is called **gross earnings.**

EXAMPLE 1 Brian's weekly gross earnings are $385. His total deductions amount to $58.60. Find the net pay.

Solution:
$$n + d = g$$
$d = 58.60$
$g = 385$
$$n + 58.60 = 385$$
$$n + 58.60 - 58.60 = 385 - 58.60$$
$$n = 326.40$$
Brian's net pay is **$326.40.**

Check: Does $326.40 + 58.60 = $385?   Yes ✔

Self-Check   1. Find the net pay: Gross earnings: $410;   Deductions: $73.80

The word rule and formula below can be used to find a plane's air speed in miles per hour (mi/h).

Word Rule: Air Speed − Head Wind Speed = Ground Speed
Formula: $a - h = g$

Air speed is the speed of the plane in still air. **Ground speed** is the speed of the plane relative to the ground. A **head wind** is a wind blowing in the opposite direction in which the plane is heading.

EXAMPLE 2 A plane is flying into a head wind of 30 mi/h. The ground speed of the plane is 615 mi/h. Find the air speed.

Solution:
$$a - h = g$$
$h = 30$ mi/h;
$g = 615$ mi/h
$$a - 30 = 615$$
$$a - 30 + 30 = 615 + 30$$
$$a = 645$$
The air speed is **645 mi/h.**

Check: Does $645 - 30 = 615?$ Yes ✔

196   Chapter 7

Self-Check   2. Find the air speed: Head Wind: 25 mi/h; Ground speed: 630 mi/h

CLASSROOM EXERCISES

Use the formula $n + d = g$ to find the net pay.   (Example 1)

1. Gross earnings: $285
   Deductions: $52
2. Gross earnings: $152
   Deductions: $27
3. Gross earnings: $600
   Deductions: $137.89

Use the formula $a - h = g$ to find the air speed.   (Example 2)

4. g: 430 mi/h
   h: 30 mi/h
5. g: 350 mi/h
   h: 25 mi/h
6. g: 512 mi/h
   h: 15.4 mi/h

WRITTEN EXERCISES

Goal: To use equations to solve word problems

Use the formula $n + d = g$ to find the net pay.   (Example 1)

	Gross Earnings	Total Deductions	Net Pay
1.	$560.00	$121.00	?
3.	$600.00	$108.32	?
5.	$392.50	$ 55.27	?

	Gross Earnings	Total Deductions	Net Pay
2.	$ 80.00	$ 9.18	?
4.	$370.00	$49.00	?
6.	$295.82	$54.48	?

7. Last week, Scott's gross earnings were $386.15. The total deductions were $94.38. Find the net pay.

8. Lisa's gross earnings each week are $460.85. The total deductions are $125.09. Find the net pay.

Use the formula $a - h = g$ to find the air speed.   (Example 2)

	Ground Speed	Head Wind	Air Speed
9.	320 mi/h	17 mi/h	?
11.	865 mi/h	10.4 mi/h	?
13.	356.8 mi/h	20 mi/h	?

	Ground Speed	Head Wind	Air Speed
10.	650 mi/h	24 mi/h	?
12.	212 mi/h	8.7 mi/h	?
14.	1042 mi/h	15 mi/h	?

15. A plane is flying into a head wind of 23 miles per hour. The ground speed is 452 mi/h. Find the air speed.

16. A plane is flying into a head wind of 42.6 mi/h. The ground speed is 1030.4 mi/h. Find the air speed.

Solving Equations

MORE CHALLENGING EXERCISES

Solve each problem.

17. A **tail wind** is a wind blowing in the same direction in which a plane is heading. Write a formula to find the ground speed, using the air speed (a) and the tail wind speed (t).

18. A plane is flying with a tail wind of 30 miles per hour. The ground speed is 425 miles per hour. Write an addition equation to find the air speed (see Exercise 17). Then solve the equation.

REVIEW: SECTIONS 7-1 – 7-3

Solve and check each equation. Show all steps. (Section 7-1)

1. $x + 12 = 27$
2. $43 = n + 21$
3. $b + 10.2 = 18.8$
4. $0.61 + y = 0.94$
5. $r + \frac{5}{6} = 1\frac{1}{6}$
6. $1\frac{1}{2} + c = 4\frac{1}{8}$
7. $n + 6.7 = 11$
8. $p + \frac{2}{3} = \frac{5}{6}$
9. $7\frac{1}{2} = t + 3\frac{1}{5}$

Solve and check each equation. Show all steps. (Section 7-2)

10. $y - 28 = 10$
11. $h - 65 = 65$
12. $7.2 = s - 6.4$
13. $x - 21.9 = 86.1$
14. $4\frac{1}{2} = t - 3\frac{1}{2}$
15. $m - 1\frac{3}{4} = \frac{3}{8}$
16. $v - 8.25 = 1.75$
17. $6\frac{1}{3} = w - 5\frac{3}{4}$
18. $d - \frac{3}{5} = 5\frac{1}{4}$

For Exercises 19–20, use the formula $n + d = g$. (Section 7-3)

19. Issac's gross earnings each week are $325.86. The total deductions are $68.17. Find his net pay.

20. Sonya's gross earnings are $160 each week. The total deductions are $11.20. What amount does Sonya take home each week?

For Exercises 21–22, use the formula $a - h = g$. (Section 7-3)

21. Airline Flight 101 is flying into a head wind of 94.7 miles per hour. The ground speed is 475.2 miles per hour. Find the air speed of Flight 101.

22. Airline Flight 102 has a ground speed of 525.5 miles per hour. The plane is flying into a head wind of 72 miles per hour. What is the air speed of Flight 102?

CONSUMER APPLICATION
Wallpaper and Estimation

To find the number of rolls of wallpaper needed to paper a room, use these rules.

RULES
1. One roll of wallpaper covers 3 square meters of wall area.
2. Subtract one roll of paper for every door and one roll of paper for every two windows.

This formula can be used to estimate the amount of wallpaper needed.

$R = \dfrac{Ph}{3} - W$

R = number of rolls of paper
P = perimeter of room
h = height of room
W = door and window allowance

EXAMPLE A rectangular room is 3.4 meters long, 2.6 meters wide, and 2.9 meters high. The room has 1 door and 2 windows. At $6.45 per roll, how much will it cost to paper the four walls?

SOLUTION

1 Find the amount of paper needed.
$P = 2(3.4 + 2.6); h = 2.9; W = 2$

$R = \dfrac{Ph}{3} - W$

$R = \dfrac{(3.4 + 2.6)(2)(2.9)}{3} - 2$ ← Use a calculator.

3 . 4 [+] 2 . 6 [=] [×] 2 [×]
2 . 9 [÷] 3 [−] 2 [=]

[9.6]

$R = $ **10 rolls** ← Round up to the next whole roll.

2 Find the total cost: $6.45(10) = **$64.50**

Exercises

Complete the table to find the cost of papering each room.

	Length	Width	Height	Doors	Windows	Number of Rolls	Cost Per Roll	Total Cost
1.	4.1 m	3.2 m	2.7 m	2	2	?	$5.79	?
2.	3.6 m	2.4 m	2.7 m	1	2	?	$6.50	?
3.	4.4 m	3.4 m	2.8 m	2	4	?	$7.99	?
4.	5.3 m	3.7 m	2.6 m	1	1	?	$9.43	?

REVIEW CAPSULE FOR SECTION 7-4

Evaluate each expression. (Pages 64–66)

1. $8 \cdot 5 \div 5$
2. $6 \cdot 3 \div 3$
3. $9 \cdot 6 \div 6$
4. $10 \cdot 2 \div 2$
5. $0.4 \times 4 \div 4$
6. $3.7 \times 8 \div 8$
7. $1.6 \times 2 \div 2$
8. $19.5 \times 7 \div 7$
9. $6\frac{1}{2} \cdot 3\frac{1}{3} \div 3\frac{1}{3}$
10. $8 \cdot 2\frac{1}{2} \div 2\frac{1}{2}$
11. $\frac{1}{3} \cdot \frac{2}{5} \div \frac{2}{5}$
12. $4\frac{3}{8} \cdot 2\frac{2}{3} \div 2\frac{2}{3}$

The answers are on page 212.

7-4 Solving Equations by Division

The following table shows that division "undoes" multiplication.

① Choose a number.	② Multiply by any number.	③ Divide by the same number.	④ Result: The original number
7	$7 \cdot 3$	$7 \cdot 3 \div 3$	7
25	$25 \cdot 4$	$25 \cdot 4 \div 4$	25
n	$8n$	$\frac{8n}{8}$, or $\frac{8}{8} \cdot n$	n

Thus, division is the **opposite,** or **inverse operation,** of multiplication. To solve an equation such as

$$12x = 180,$$

divide **each side** of the equation by 12.

> **PROCEDURE**
> To solve the equation $ax = c$, $a \neq 0$, divide **each side** of the equation by a.

EXAMPLE 1 Solve and check: $12x = 180$

Solution: To get x alone, divide each side by 12.

$12x = 180$
$\frac{12x}{12} = \frac{180}{12}$ $\frac{12}{12} = 1$
$x = 15$

Check: $12x = 180$
$12 \cdot 15$
180

Self-Check *Solve and check.*

1. $8x = 112$
2. $3t = 258$
3. $3n = 48$
4. $4y = 128$

200   Chapter 7

NOTE: In Example 1, these equations are equivalent because they have the same solution, 15.

$$12x = 180, \quad \frac{12x}{12} = \frac{180}{12}, \quad \text{and} \quad x = 15$$

EXAMPLE 2   Solve and check.

a. $21 = 0.7w$ 　　　　　　　　　 b. $16w = 200$

Solutions: a. $21 = 0.7w$ ◀ Divide each side by 0.7.　　b. $16w = 200$ ◀ Divide each side by 16.

$$\frac{21}{0.7} = \frac{0.7w}{0.7} \qquad\qquad \frac{16w}{16} = \frac{200}{16}$$

$$30 = w \qquad\qquad\qquad w = \frac{25}{2}$$

$$w = 12\frac{1}{2}, \text{ or } 12.5$$

The checks are left for you.

Self-Check   *Solve and check.*

5. $147 = 2.1p$ 　　**6.** $5w = 79$ 　　**7.** $1.5q = 3$ 　　**8.** $100 = 8p$

CLASSROOM EXERCISES

Evaluate each expression.   (Step 1, Examples 1–2)

1. $\frac{10x}{10}$ 　　**2.** $\frac{27y}{27}$ 　　**3.** $\frac{0.9n}{0.9}$ 　　**4.** $\frac{128k}{128}$

Write the equivalent equation formed as the first step in solving each equation.   (Step 1, Examples 1–2)

5. $4n = 32$ 　　**6.** $9x = 72$ 　　**7.** $44 = 11y$ 　　**8.** $0.7w = 21$

9. $2.6a = 33.8$ 　　**10.** $63.6 = 53t$ 　　**11.** $96 = 19.2x$ 　　**12.** $108b = 1296$

Determine whether the value of x shown is the solution for the given equation. Answer Yes or No.   (Check, Example 1)

13. $4x = 8$ 　　　　**14.** $12 = 36x$ 　　**15.** $8x = 30$ 　　　**16.** $12.4 = 0.2x$
　　$x = \frac{1}{2}$ 　　　　　　$x = \frac{1}{3}$ 　　　　　　$x = 3\frac{3}{4}$ 　　　　　$x = 6.2$

17. $14x = 2$ 　　　**18.** $9x = 55$ 　　　**19.** $13x = 10$ 　　　**20.** $15x = 45$
　　$x = \frac{1}{7}$ 　　　　　　$x = 6\frac{1}{9}$ 　　　　　$x = 1\frac{3}{10}$ 　　　　$x = \frac{1}{3}$

Solving Equations

WRITTEN EXERCISES

Goal: To use division to solve equations

For Exercises 1–44, solve and check. Show all steps. (Example 1)

1. $6x = 54$
2. $9x = 63$
3. $60 = 12n$
4. $2p = 120$
5. $256 = 8y$
6. $7a = 161$
7. $80 = 16q$
8. $18t = 72$

(Example 2)

9. $8.4 = 0.6r$
10. $22.5 = 0.9n$
11. $5.2k = 213.2$
12. $2.8m = 64.4$
13. $1.9x = 38$
14. $2.7y = 162$
15. $9.6 = 0.08t$
16. $7.2 = 0.03p$
17. $88 = 2.2w$
18. $0.04z = 4.8$
19. $92 = 10c$
20. $108 = 6e$
21. $5x = 84$
22. $15y = 126$
23. $85 = 8x$
24. $111 = 4r$
25. $32x = 24$
26. $45y = 15$
27. $416 = 13s$
28. $504 = 14t$

MIXED PRACTICE

29. $168 = 14x$
30. $8.6t = 215$
31. $2.4p = 128.4$
32. $15s = 80$
33. $375 = 3n$
34. $2.9k = 391.5$
35. $12t = 150$
36. $18x = 8$
37. $1.47 = 0.06a$
38. $16x = 34$
39. $25y = 10$
40. $17y = 493$
41. $7b = 154$
42. $4.3d = 34.4$
43. $138 = 9u$
44. $48v = 30$

MORE CHALLENGING EXERCISES

Solve and check each equation. Show all steps.

45. $4x = 2\frac{1}{2}$
46. $8p = 5\frac{1}{3}$
47. $\frac{3}{4}m = 9$
48. $1\frac{2}{3}n = 10$
49. $\frac{2}{3}y = \frac{3}{5}$
50. $\frac{7}{8}t = 2\frac{1}{2}$
51. $3\frac{2}{3}s = 8\frac{1}{4}$
52. $6\frac{2}{3}w = 6\frac{1}{4}$
53. $1\frac{3}{4}r = 49$
54. $2x = 4\frac{4}{5}$
55. $2\frac{1}{5}q = 3\frac{3}{10}$
56. $\frac{4}{5}v = 5\frac{3}{5}$

REVIEW CAPSULE FOR SECTION 7-5

Evaluate each expression.

(Pages 18–21, 38–41, 44–46, 129–131, 148–151)

1. $16 \div 4 \cdot 4$
2. $84 \div 12 \cdot 12$
3. $\frac{10}{8}(8)$
4. $\frac{240}{3}(3)$
5. $21 \div 4.2 \times 4.2$
6. $\frac{31.6}{1.6}(1.6)$
7. $\frac{136}{6.8}(6.8)$
8. $22.2 \div 3.7 \times 3.7$

The answers are on page 212.

7-5 Solving Equations by Multiplication

The following table shows that multiplication "undoes" division.

① Choose a number.	② Divide by any number (not zero).	③ Multiply by the same number.	④ Result: The original number
24	$24 \div 6$, or $\frac{24}{6}$	$\frac{24}{6} \cdot 6$	24
63	$\frac{63}{5}$	$\frac{63}{5} \cdot 5$	63
n	$\frac{n}{4}$	$\frac{n}{4} \cdot 4$	n

Thus, multiplication is the **inverse operation** of division. To solve an equation such as

$$\frac{x}{5} = 24,$$

multiply each side by 5.

PROCEDURE

To solve the equation $\frac{x}{b} = c$, $b \neq 0$, multiply **each side** by b.

EXAMPLE 1 Solve and check: $\frac{x}{5} = 24$

Solution: To get x alone, multiply each side by 5.

$$\frac{x}{5} = 24$$
$$\frac{x}{5} \cdot 5 = 24 \cdot 5$$
$$x = 120$$

Check: $\frac{x}{5} = 24$
$$\frac{120}{5}$$
$$24$$

Self-Check Solve and check.

1. $\frac{v}{3} = 9$ 2. $\frac{z}{8} = 1$ 3. $\frac{q}{16} = 21$ 4. $\frac{w}{9} = 12$

NOTE: In Example 1, these equations are equivalent because they have the <u>same</u> solution, **120**.

$$\frac{x}{5} = 24, \quad \frac{x}{5} \cdot 5 = 24 \cdot 5, \quad \text{and} \quad x = 120$$

Solving Equations   203

EXAMPLE 2 Solve and check: **a.** $10 = \dfrac{d}{12.5}$ **b.** $\dfrac{n}{6} = 12\dfrac{1}{4}$

Solutions: **a.** $10 = \dfrac{d}{12.5}$ ◀ Multiply each side by 12.5.

$10(12.5) = \dfrac{d}{12.5}(12.5)$

$125 = d$

Check: $10 = \dfrac{d}{12.5}$

$\dfrac{125}{12.5}$

10

b. $\dfrac{n}{6} = 12\dfrac{1}{4}$ ◀ Multiply each side by 6.

$\dfrac{n}{6} \cdot 6 = 12\dfrac{1}{4} \cdot 6$

$\dfrac{n}{\cancel{6}} \cdot \cancel{6} = \dfrac{49}{\cancel{4}} \cdot \cancel{6}^{3}$

$n = \dfrac{147}{2}$, or $73\dfrac{1}{2}$

Check: $\dfrac{n}{6} = 12\dfrac{1}{4}$

$\dfrac{\frac{147}{2}}{6}$

$\dfrac{147}{12}$

$12\dfrac{1}{4}$

Self-Check *Solve and check.*

5. $\dfrac{t}{6.3} = 20$ **6.** $2\dfrac{1}{4} = \dfrac{c}{8}$ **7.** $\dfrac{m}{3} = 3\dfrac{1}{3}$ **8.** $\dfrac{r}{1.5} = 8.2$

CLASSROOM EXERCISES

Evaluate each expression. (Table)

1. $q \div 3 \cdot 3$ **2.** $\dfrac{z}{2} \cdot 2$ **3.** $a \div 19 \cdot 19$ **4.** $\dfrac{s}{89} \cdot 89$

5. $\dfrac{r}{9.3}(9.3)$ **6.** $n \div 8 \cdot 8$ **7.** $\dfrac{p}{11} \cdot 11$ **8.** $\left(\dfrac{y}{4.6}\right)4.6$

Write the equivalent equation formed as the first step in solving each equation. (Step 1, Examples 1–2)

9. $\dfrac{n}{5} = 6$ **10.** $\dfrac{x}{20} = 9$ **11.** $\dfrac{y}{10} = 3$ **12.** $\dfrac{m}{8} = 15$

13. $\dfrac{s}{5} = 1.1$ **14.** $4.7 = \dfrac{t}{7}$ **15.** $2\dfrac{1}{2} = \dfrac{y}{4}$ **16.** $\dfrac{p}{9} = 3\dfrac{1}{3}$

17. $\dfrac{b}{17} = 5$ **18.** $6.2 = \dfrac{c}{3}$ **19.** $3\dfrac{3}{4} = \dfrac{d}{8}$ **20.** $\dfrac{e}{10} = 5\dfrac{2}{5}$

21. $\dfrac{f}{3} = 4\dfrac{5}{6}$ **22.** $7.8 = \dfrac{h}{11}$ **23.** $\dfrac{r}{7} = 12$ **24.** $\dfrac{x}{8} = 6\dfrac{1}{2}$

WRITTEN EXERCISES

Goal: To use multiplication to solve equations

Solve and check each equation. Show all steps. (Example 1)

1. $\frac{t}{8} = 17$
2. $\frac{r}{12} = 9$
3. $12 = \frac{x}{23}$
4. $15 = \frac{x}{34}$
5. $\frac{w}{7} = 123$
6. $\frac{k}{9} = 208$
7. $346 = \frac{p}{13}$
8. $529 = \frac{a}{16}$
9. $\frac{x}{2} = 5$
10. $4 = \frac{y}{3}$
11. $\frac{n}{8} = 2$
12. $5 = \frac{r}{6}$

(Example 2)

13. $\frac{a}{1.8} = 70$
14. $\frac{n}{2.3} = 9$
15. $83 = \frac{s}{0.09}$
16. $56 = \frac{t}{0.07}$
17. $\frac{w}{3.7} = 0.8$
18. $\frac{p}{5.2} = 0.4$
19. $23.5 = \frac{x}{2.4}$
20. $40.4 = \frac{y}{2.5}$
21. $\frac{a}{2.5} = 4$
22. $\frac{k}{1.3} = 2.3$
23. $9 = \frac{m}{1.1}$
24. $2 = \frac{w}{3.1}$
25. $\frac{x}{8} = 3\frac{1}{2}$
26. $\frac{k}{18} = 6\frac{1}{4}$
27. $4\frac{3}{8} = \frac{n}{16}$
28. $3\frac{5}{6} = \frac{m}{18}$
29. $\frac{p}{10} = \frac{2}{5}$
30. $\frac{r}{30} = \frac{5}{6}$
31. $12\frac{1}{2} = \frac{y}{13}$
32. $1\frac{1}{4} = \frac{a}{5}$

MIXED PRACTICE

33. $2.8 = \frac{r}{12}$
34. $\frac{p}{16} = 3\frac{1}{4}$
35. $\frac{n}{26} = 35$
36. $3.6 = \frac{t}{15}$
37. $\frac{q}{12} = 5\frac{1}{6}$
38. $\frac{w}{42} = 93$
39. $102 = \frac{a}{47}$
40. $1\frac{3}{5} = \frac{b}{5}$
41. $\frac{m}{9.6} = 0.5$
42. $\frac{r}{7.8} = 9.2$
43. $2\frac{2}{3} = \frac{d}{9}$
44. $\frac{p}{11} = 122$

MORE CHALLENGING EXERCISES

Solve and check each equation. Show all steps.

45. $\frac{x}{\frac{3}{4}} = 12$
46. $\frac{x}{\frac{5}{6}} = 3$
47. $\frac{x}{\frac{2}{5}} = 8$
48. $\frac{x}{1\frac{1}{4}} = \frac{3}{10}$
49. $\frac{x}{2\frac{1}{3}} = \frac{2}{7}$
50. $\frac{x}{\frac{1}{6}} = 1\frac{1}{2}$
51. $\frac{x}{1\frac{1}{3}} = 2\frac{3}{4}$
52. $\frac{x}{3\frac{1}{9}} = 4\frac{1}{5}$

REVIEW CAPSULE FOR SECTION 7-6

Solve and check each equation. (Pages 200–202)

1. $2x = 56$ **2.** $7y = 4.2$ **3.** $3p = 5$ **4.** $12n = 1440$
5. $4.5q = 15$ **6.** $17r = 52$ **7.** $36 = 1.8t$ **8.** $12.4 = 31s$

Evaluate each expression. (Pages 22–24)

9. $x \div y$ when $x = 12$ and $y = 0.5$
10. pq when $p = 4.5$ and $q = 0.6$
11. mr when $m = 7\frac{1}{3}$ and $r = \frac{1}{2}$
12. $\frac{s}{t}$ when $s = 3\frac{1}{3}$ and $t = 4$

The answers are on page 212.

PROBLEM SOLVING AND APPLICATIONS

7-6 Using Equations: Multiplication/Division

In Chapter 1, you used this formula to find distance.

$$d = rt$$

d = distance
r = rate
t = time

You can use the same formula to find rate or time.

EXAMPLE 1 A diving bell moves up and down in the water at the rate of 1.4 meters per second. How long would it take for the diving bell to descend 655.2 meters?

Solution:
1. Formula: $d = rt$
2. Known value: $d = 655.2$ m; $r = 1.4$ m/sec
3. $655.2 = 1.4t$ *Divide each side by 1.4.*
   $\frac{655.2}{1.4} = \frac{1.4t}{1.4}$
   $468 = t$ It will take **468** seconds.

Check: Does $655.2 = 1.4(468)$?   Yes ✓

Self-Check
1. How long will it take the diving bell in Example 1 to rise 357 meters?

The following word rule and formula can be used to find a baseball player's batting average.

Word Rule: Batting Average = Number of Hits ÷ Number of Times at Bat

Formula: $A = \dfrac{h}{t}$

A = Batting Average
h = number of hits
t = times at bat

When you know a player's batting average and the number of times at bat, you can use the formula to find the number of hits.

EXAMPLE 2 A player's batting average is 0.400 for 30 times at bat. Find the number of hits.

Solution:
[1] Formula: $A = \dfrac{h}{t}$

[2] Known values: $A = .400$; $t = 30$

[3] $0.400 = \dfrac{h}{30}$ Multiply each side by 30.

$0.400(30) = \dfrac{h}{30}(30)$

$12 = h$ The player had **12** hits.

Check: Does $0.400 = \dfrac{12}{30}$?   Yes ✔

Self-Check

2. Diana Lopez has a batting average of 0.250 for 104 times at bat. Find the number of hits.

CLASSROOM EXERCISES

Use the formula $d = rt$ to find the time t. (Example 1)

1. Distance: 400 miles
   Rate: 20 miles per hour

2. Distance: 750 meters
   Rate: 25 meters per minute

Use the formula $d = rt$ to find the rate, r. (Example 1)

3. Distance: 48 kilometers
   Time: 2 hours

4. Distance: 500 miles
   Time: 10 hours

Use the formula $A = \dfrac{h}{t}$ to find the number of hits, h. (Example 2)

5. Times at Bat: 100 times
   Batting Average: 0.280

6. Times at Bat: 50 times
   Batting Average: 0.180

Solving Equations

WRITTEN EXERCISES

Goal: To use equations to solve word problems

For Exercises 1–10, use the formula $d = rt$ to find the rate or the time as indicated. (Example 1)

	Rate	Time	Distance
1.	50 mph	? h	300 mi
3.	? m/sec	65 sec	340.6 m
5.	2 ft/sec	? sec	$16\frac{2}{5}$ ft

	Rate	Time	Distance
2.	? km/h	5 h	56.0 km
4.	8 ft/sec	? sec	$42\frac{2}{3}$ ft
6.	? km/h	16 h	1280 km

7. The Concorde SST flies 3468 miles from London, England to New York. The flight takes 3.2 hours. What is the average speed of the Concorde?

8. A peregrine falcon in a vertical fall takes about 16 seconds to dive 5000 feet. What is the falcon's average speed?

9. The distance from St. Louis to Dallas is about 1050 kilometers. How many hours will it take to drive this distance at an average speed of 70 kilometers per hour?

10. A roller skater set a world's speed record by skating 1320 feet at a rate of $37\frac{5}{7}$ feet per second. How long did it take the skater to complete the distance?

For Exercises 11–20, use the formula $A = \frac{h}{t}$ to find the number of hits. (Example 2)

	Times at Bat	Batting Average	Number of Hits
11.	250	0.200	?
13.	100	0.300	?
15.	150	0.260	?
17.	550	0.320	?

	Times at Bat	Batting Average	Number of Hits
12.	250	0.380	?
14.	340	0.350	?
16.	200	0.285	?
18.	605	0.238	?

19. Gilda Sanchez has a batting average of .250. She has been at bat 104 times. Find her number of hits.

20. Nick Demarco has been at bat 124 times. He has a batting average of .250. Find his number of hits.

MORE CHALLENGING EXERCISES

21. In a triathalon, Sue Lin must swim $1\frac{1}{2}$ miles, bicycle 25 miles, and run 6 miles. She hopes to average 2 miles per hour swimming, 20 miles per hour bicycling, and 10 miles per hour running. How long should it take her to complete the triathalon?

22. Southern Skies Flight 224 flies north from Miami to Boston and back. The distance each way is about 2400 miles. The plane maintains a constant airspeed of 570 miles per hour in a 70 mile-per-hour wind blowing from north to south. How much longer does it take to fly from Miami to Boston than from Boston to Miami?

23. During the softball season, Norma had 57 hits and a batting average of 0.285. How many times was she at bat during the season?

24. The earth travels 934,240,000 kilometers around the sun in 8767 hours. How fast (nearest whole number) does the earth travel?

REVIEW: SECTIONS 7-4 – 7-6

Solve and check each equation. Show all steps. (Section 7-4)

1. $8x = 56$
2. $14b = 84$
3. $1.2m = 7.2$
4. $0.6p = 9$
5. $7q = 30$
6. $8s = 3$
7. $156 = 6t$
8. $4.4 = 0.4y$

Solve and check each equation. Show all steps. (Section 7-5)

9. $\frac{x}{9} = 13$
10. $\frac{y}{4} = 25$
11. $\frac{p}{15} = 1.6$
12. $\frac{t}{3.5} = 4.82$
13. $\frac{s}{8} = 1\frac{1}{3}$
14. $\frac{m}{7} = 5\frac{2}{3}$
15. $104 = \frac{n}{16}$
16. $2.6 = \frac{r}{0.04}$

Solve each problem. (Section 7-6)

17. The deepest point in the Pacific Ocean is 11,023 meters. Sound waves take 7.55 seconds to reach the bottom at this point. Use the formula $d = rt$ to find the speed of sound in water.

18. Michael has a batting average of 0.425 for 40 times at bat. Use

$$A = \frac{h}{t}$$

to find the number of hits.

Solving Equations 209

CHAPTER REVIEW

PART 1: VOCABULARY

For Exercises 1–10, choose from the box at the right the word(s) that best correspond to each description.

1. Amount of pay left after deductions __?__ (p. 196)
2. The speed of a plane in still air __?__ (p. 196)
3. An equation such as $n + 4 = 8$ is an __?__ equation. (p. 190)
4. The number of hits divided by the number of times at bat __?__ (p. 207)
5. To solve a multiplication equation, you __?__ both sides by the same number. (p. 200)
6. Amount of pay actually earned __?__ (p. 196)
7. A mathematical sentence containing the symbol "=" __?__ (p. 190)
8. A wind blowing in the opposite direction in which a plane is heading __?__ (p. 196)
9. Equations having the same solution or solutions __?__ (p. 193)
10. To solve a division equation, you __?__ both sides by the same number. (p. 203)

> multiply
> equation
> gross earnings
> addition
> head wind
> equivalent equations
> batting average
> net pay
> air speed
> divide

PART 2: SKILLS

Solve and check each equation. Show all steps. (Section 7-1)

11. $x + 8 = 42$
12. $r + 28 = 73$
13. $32.1 = y + 19.3$
14. $t + 4.39 = 6.2$
15. $p + 3\frac{1}{2} = 14\frac{1}{4}$
16. $15\frac{1}{4} = n + 7\frac{3}{5}$

Solve and check each equation. Show all steps. (Section 7-2)

17. $a - 9 = 13$
18. $w - 109 = 58$
19. $k - 7.8 = 28.6$
20. $47.8 = b - 23.4$
21. $6\frac{1}{8} = m - 5\frac{1}{4}$
22. $t - 6\frac{3}{4} = 13\frac{1}{3}$

For Exercises 23–26, use the formula $n + d = g$ to find the net pay, n. (Section 7-3)

	Gross Earnings	Total Deductions	Net Pay
23.	$280.00	$56.00	__?__
25.	$312.30	$72.85	__?__

	Gross Earnings	Total Deductions	Net Pay
24.	$185.00	$37.15	__?__
26.	$524.00	$108.09	__?__

210 Chapter 7

For Exercises 27–30, use the formula $a - h = g$ to find the airspeed, a. (Section 7-3)

	Ground-speed	Head Wind	Airspeed
27.	425 mi/h	15 mi/h	?
29.	256.4 mi/h	21 mi/h	?

	Ground-speed	Head Wind	Airspeed
28.	708 mi/h	8.4 mi/h	?
30.	384.2 mi/h	31.6 mi/h	?

Solve and check each equation. Show all steps. (Section 7-4)

31. $126 = 14p$ **32.** $0.8a = 344$ **33.** $8t = 70$ **34.** $105 = 9c$

Solve and check each equation. Show all steps. (Section 7-5)

35. $\dfrac{d}{14} = 13$ **36.** $\dfrac{n}{9.7} = 4.8$ **37.** $\dfrac{b}{20} = 6\dfrac{1}{3}$ **38.** $27 = \dfrac{r}{1.3}$

For Exercises 39–44, complete as indicated. (Section 7-6)

	Rate	Time	Distance
39.	6 ft/sec	?	360 ft
40.	?	18 sec	82.8 m
41.	?	6.4 hr	332.8 km

	Times at Bat	Batting Average	Number of Hits
42.	72	0.250	?
43.	240	0.425	?
44.	586	0.314	?

PART 3: APPLICATIONS

45. Albert Chin's gross earnings last week were $382. The total deductions were $82.55. Use the formula $n + d = g$ to find the net pay. (Section 7-4)

46. A plane is flying into a head wind of 27.2 miles per hour. The ground speed is 456 miles per hour. Use the formula $a - h = g$ to find the air speed. (Section 7-4)

47. The longest recorded distance for a woman in a 24 hour walking race is $122\frac{1}{2}$ miles. Use the formula $d = rt$ to find the rate. Round the answer to the nearest tenth of a mile per hour. (Section 7-6)

48. The highest lifetime batting average in professional baseball is 0.367. The record is held by Ty Cobb who was at bat 11,420 times. Use the formula $A = \dfrac{h}{t}$ to find the number of hits. Round the answer to the nearest whole number. (Section 7-6)

CHAPTER TEST

Solve and check each equation. Show all steps.

1. $m + 28 = 53$
2. $t + 11.5 = 36.4$
3. $7.1 = a + 4.21$
4. $e + 15\frac{2}{3} = 18\frac{1}{4}$
5. $52 = w - 25$
6. $q - 10.73 = 19.5$
7. $c - \frac{3}{5} = 1\frac{1}{5}$
8. $1\frac{5}{8} = h - 3\frac{2}{3}$
9. $12x = 132$
10. $8.4 = 5.6p$
11. $42 = 4x$
12. $15a = 130$
13. $\frac{b}{16} = 257$
14. $\frac{f}{3.8} = 1.7$
15. $40.6 = \frac{y}{5.4}$
16. $\frac{m}{18} = 5\frac{2}{9}$

Solve each problem.

17. Monica's gross earnings each week are $458.16. The total deductions are $106.05. Use the formula $n + d = g$ to find the net pay.

18. A jet helicopter is flying into a headwind of 52.7 miles per hour. The ground speed is 214.8 miles per hour. Use the formula $a - h = g$ to find the air speed.

19. In 1984, the Daytona 500 winner, Cale Yarborough, completed the 500 miles in about 3.2 hours. Use the formula $d = rt$ to find his average speed.

20. Maria Hernandez has a batting average of 0.300. She has been at bat 90 times. Use the formula $A = \frac{h}{t}$ to find the number of hits.

ANSWERS TO REVIEW CAPSULES

Page 192 1. 7 2. 6 3. $9\frac{1}{2}$ 4. 0.9 5. q 6. c 7. f 8. 1800
Page 195 1. 14 2. 12.7 3. 15 4. 62 5. 3.4 6. $1\frac{7}{10}$ 7. $2\frac{1}{6}$ 8. 9.9
Page 200 1. 8 2. 6 3. 9 4. 10 5. 0.4 6. 3.7 7. 1.6 8. 19.5 9. $6\frac{1}{2}$ 10. 8 11. $\frac{1}{3}$ 12. $4\frac{3}{8}$
Page 202 1. 16 2. 84 3. 10 4. 240 5. 21 6. 31.6 7. 136 8. 22.2
Page 206 1. 28 2. 0.6 3. $1\frac{2}{3}$ 4. 120 5. $3\frac{1}{3}$ 6. $3\frac{1}{17}$ 7. 20 8. 0.4 9. 24 10. 2.7 11. $3\frac{2}{3}$ 12. $\frac{5}{6}$

ENRICHMENT

Repeating Decimals

You learned in Chapter 6 that every fraction can be represented by an infinite repeating decimal or by a terminating decimal.

$\frac{2}{3} = 0.666\ldots$, or $0.\overline{6}$ ◄ **The bar indicates that "6" repeats** $\frac{3}{4} = 0.75$ ◄ **Terminating decimal**

This Example shows how to write an infinite repeating decimal in the form $\frac{a}{b}$.

EXAMPLE Write each infinite repeating decimal in the form $\frac{a}{b}$.

    a. $0.\overline{1}$ **b.** $9.\overline{27}$

Solutions: **a.** Let $n = 0.\overline{1}$ ◄ **Multiply each side by 10^1, or 10.**

$10n = 1.\overline{1}$ ◄ **Subtract $n = 0.\overline{1}$ from this equation.**

$\underline{n = 0.\overline{1}}$

$9n = 1.0$

$\frac{9n}{9} = \frac{1}{9}$

$n = \frac{1}{9}$ Thus, $0.\overline{1} = \frac{1}{9}$.

  b. Let $n = 9.\overline{27}$ ◄ **Multiply each side by 10^2, or 100.**

$100n = 927.\overline{27}$ ◄ **Subtract $n = 9.\overline{27}$ from this equation.**

$\underline{n = 9.\overline{27}}$

$99n = 918$

$\frac{99n}{99} = \frac{918}{99}$

$n = \frac{918}{99}$, or $\frac{102}{11}$ Thus, $9.\overline{27} = \frac{102}{11}$.

EXERCISES

Write each infinite repeating decimal in the form $\frac{a}{b}$.

1. $0.\overline{3}$
2. $0.\overline{5}$
3. $0.\overline{18}$
4. $0.\overline{81}$
5. $0.\overline{72}$
6. $0.\overline{4}$
7. $1.\overline{6}$
8. $2.\overline{36}$
9. $3.\overline{45}$
10. $4.\overline{7}$
11. $0.\overline{918}$
12. $0.\overline{369}$
13. $6.\overline{25}$
14. $3.\overline{16}$
15. $7.\overline{32}$

Solving Equations

ADDITIONAL PRACTICE

SKILLS

Solve and check. Show all steps. (Pages 190–192).

1. $x + 108 = 413$
2. $t + 45 = 198$
3. $1.47 = x + 0.35$
4. $7.05 = v + 3.49$
5. $v + 1\frac{7}{8} = 3\frac{1}{4}$
6. $x + 2\frac{3}{4} = 9\frac{7}{16}$

(Pages 193–195)

7. $x - 38 = 51$
8. $t - 84 = 58$
9. $v - 12.3 = 17.9$
10. $m - 117.8 = 95.6$
11. $1\frac{13}{16} = t - \frac{3}{4}$
12. $4\frac{3}{8} = P - 4\frac{3}{4}$

Complete the tables.
(Pages 196–198)

	Gross Earnings	Total Deductions	Net Pay
13.	$640.00	$125.00	?
14.	$847.95	$258.13	?

Complete the table.
(Pages 206–209)

	Rate	Time	Distance
15.	? km/h	5 h	375 km
16.	8.8 ft/sec	?	21.12 ft

Solve and check. (Pages 200–202)

17. $13x = 208$
18. $3w = 231$
19. $455 = 7t$
20. $234 = 18v$
21. $0.9w = 11.07$
22. $1.3y = 18.2$
23. $15.2 = 10v$
24. $2.94 = 0.06a$

(Pages 203–205)

25. $\frac{n}{14} = 32$
26. $\frac{p}{13} = 120$
27. $2.3 = \frac{m}{12}$
28. $1.4 = \frac{x}{15}$
29. $4.2 = \frac{r}{6.4}$
30. $2.1 = \frac{t}{12}$
31. $\frac{p}{15} = 3\frac{2}{3}$
32. $\frac{c}{12} = 2\frac{1}{4}$

APPLICATIONS

33. Alfredo has a batting average of 0.325. He has been at bat 600 times. How many hits does Alfredo have? (Pages 206–209)

34. A plane is flying into a head wind of 27 miles per hour. The ground speed of the plane is 475 miles per hour. Find the air speed. (Pages 196–198)

35. Joan's gross earnings each week are $350.75. The total deductions are $87.83. Find the net pay. (Pages 196–198)

36. The distance from New York to San Francisco is about 5002 kilometers. How many hours will it take a plane to fly this distance at an average speed of 820 kilometers per hour? (Pages 206–209)

CHAPTER

8 Ratio, Proportion, and Percent

FEATURES
Calculator Application: *Simple Interest*

Computer Application: *Problem Solving*

Enrichment: *Compound Interest*

SECTIONS
8-1 Ratio
8-2 Proportions
8-3 Problem Solving and Applications: Using Proportions
8-4 Meaning of Percent
8-5 Percents and Decimals
8-6 Finding a Percent of a Number
8-7 Problem Solving and Applications: Formulas: Simple Interest
8-8 Problem Solving and Applications: Formulas: Discount

8-1 Ratio

In 1956, the cost for the first ounce of a first-class letter was 3¢. In 1985, the cost was 22¢. You can use this ratio to compare these costs.

$$\frac{3}{22} \begin{array}{l}\longleftarrow \text{Cost in 1956} \\ \longleftarrow \text{Cost in 1985}\end{array}$$

A **ratio** uses division to compare two numbers.

3 to 22,   3 : 22,   or $\frac{3}{22}$

22 to 3,   22 : 3,   or $\frac{22}{3}$

3 : 22 or $\frac{3}{22}$ is read "3 is to 22."

NOTE: The ratios 3 : 22 and 22 : 3 do <u>not</u> have the same meaning. They are not equal.

Just as with fractions, ratios can be written in lowest terms.

EXAMPLE 1 In 1968, the cost of the first ounce of a first-class letter was 6¢. In 1974, the cost was 10¢. Write a ratio in lowest terms to compare the costs.

Solution: $\frac{\text{Cost in 1968}}{\text{Cost in 1974}} = \frac{6}{10} = \frac{3}{5}$ ◀ **Lowest terms**

Self-Check *Write a fraction in lowest terms for each ratio.*

1. 5 to 15   **2.** 12 to 9   **3.** 3 : 21   **4.** 25 : 5

The ratios $\frac{6}{10}$ and $\frac{3}{5}$ are equivalent.

Equivalent ratios are equal.

PROCEDURE

To determine whether two ratios are equal:

1 Find the cross-products.

2 Compare the cross-products. **Equivalent ratios have equal cross-products.**

---

216   Chapter 8

EXAMPLE 2 Determine whether each pair of ratios is equivalent.

  a. $\frac{3}{20}$ and $\frac{9}{40}$ **b.** $\frac{7}{12}$ and $\frac{21}{36}$

The loops show the cross-products.

Solutions:

a. $\frac{3}{20} \stackrel{?}{=} \frac{9}{40}$

$3 \cdot 40 \stackrel{?}{=} 20 \cdot 9$

$120 \stackrel{?}{=} 180$ **No**

The ratios are **not equivalent.**

b. $\frac{7}{12} \stackrel{?}{=} \frac{21}{36}$

$7 \cdot 36 \stackrel{?}{=} 12 \cdot 21$

$252 \stackrel{?}{=} 252$ **Yes**

The ratios are **equivalent.**

Self-Check *Determine whether each pair of ratios is equivalent.*

5. $\frac{2}{3}$ and $\frac{3}{4}$ **6.** $\frac{1}{2}$ and $\frac{3}{6}$ **7.** $\frac{2}{25}$ and $\frac{9}{58}$ **8.** $\frac{4}{5}$ and $\frac{7}{9}$

Variables can be used in one, or both, terms of a ratio. However, the variables cannot have a value that will make the second term equal zero.

> In the ratio $\frac{a}{b}$, $b \neq 0$, a is the first term and b is the second term.

EXAMPLE 3 Write each ratio in lowest terms.

  a. $\frac{9q}{12q}$, $q \neq 0$ **b.** $\frac{3}{12d}$, $d \neq 0$

Solutions:

a. $\frac{9q}{12q} = \frac{\cancel{9}^{3}}{\cancel{12}_{4}} \cdot \frac{\cancel{q}^{1}}{\cancel{q}_{1}}$

$= \frac{3 \cdot 1}{4 \cdot 1}$

$= \frac{3}{4}$

b. $\frac{3}{12d} = \frac{\cancel{3}^{1}}{\cancel{12}_{4}} \cdot \frac{1}{d}$

$= \frac{1 \cdot 1}{4 \cdot d}$

$= \frac{1}{4d}$

Self-Check *Write each ratio in lowest terms. No variable equals 0.*

9. $\frac{6y}{15y}$ **10.** $\frac{25t}{35t}$ **11.** $\frac{4}{16n}$ **12.** $\frac{9}{21p}$

CLASSROOM EXERCISES

Write each ratio in lowest terms. (Example 1)

1. $\frac{9}{12}$
2. $\frac{50}{35}$
3. 20 to 4
4. 36 to 15
5. 8 : 14
6. 300 : 100

Determine whether each pair of ratios is equivalent. Answer Yes or No. (Example 2)

7. $\frac{1}{2}$ and $\frac{6}{12}$
8. $\frac{3}{4}$ and $\frac{15}{22}$
9. $\frac{2}{7}$ and $\frac{21}{63}$
10. $\frac{2}{3}$ and $\frac{14}{21}$

Write each ratio in lowest terms. No variable equals 0. (Example 3)

11. $\frac{3x}{5x}$
12. $\frac{16t}{8t}$
13. $\frac{42p}{12}$
14. $\frac{8}{14c}$
15. $\frac{24d}{16d}$
16. $\frac{8m}{2m}$
17. $\frac{7}{14q}$
18. $\frac{4}{36t}$
19. $\frac{16a}{48a}$
20. $\frac{77d}{11d}$
21. $\frac{s}{s}$
22. $\frac{r}{r}$

WRITTEN EXERCISES

Goals: To write ratios in lowest terms
To determine whether pairs of ratios are equivalent

Write each ratio in lowest terms. (Example 1)

1. 14 to 42
2. 16 to 25
3. 30 : 80
4. 5 : 25
5. 10 to 100
6. 36 to 20
7. $\frac{7}{21}$
8. $\frac{4}{20}$
9. $\frac{15}{60}$
10. $\frac{16}{64}$
11. $\frac{17}{51}$
12. $\frac{6}{54}$

Write the cross-product to determine whether each pair of ratios is equivalent. Answer Yes or No. (Example 2)

13. $\frac{4}{1}$ and $\frac{28}{7}$
14. $\frac{12}{1}$ and $\frac{132}{13}$
15. $\frac{3}{16}$ and $\frac{9}{48}$
16. $\frac{3}{2}$ and $\frac{27}{18}$
17. $\frac{7}{10}$ and $\frac{54}{80}$
18. $\frac{7}{12}$ and $\frac{36}{60}$
19. $\frac{6}{5}$ and $\frac{17}{12}$
20. $\frac{1}{3}$ and $\frac{16}{51}$

Write each ratio in lowest terms. No variable equals 0. (Example 3)

21. $\frac{10a}{12a}$
22. $\frac{21y}{14y}$
23. $\frac{15r}{6}$
24. $\frac{8}{20k}$
25. $\frac{40d}{28d}$
26. $\frac{12}{30p}$
27. $\frac{15t}{35t}$
28. $\frac{48c}{18}$
29. $\frac{5a}{12a}$
30. $\frac{20c}{5}$
31. $\frac{15a}{8}$
32. $\frac{18x}{60x}$
33. $\frac{9r}{9r}$
34. $\frac{3d}{3d}$
35. $\frac{8t}{6t}$
36. $\frac{14a}{49a}$
37. $\frac{5r}{10}$
38. $\frac{60y}{15}$
39. $\frac{21r}{21}$
40. $\frac{47r}{47}$
41. $\frac{t}{9t}$
42. $\frac{q}{7q}$
43. $\frac{40p}{15p}$
44. $\frac{25x}{75x}$

REVIEW CAPSULE FOR SECTION 8-2

Solve each equation. (Pages 200–202)

1. $5x = 75$
2. $13n = 130$
3. $546 = 21p$
4. $90 = 8w$
5. $4p = 50$
6. $4k = 45$
7. $105 = 8t$
8. $80 = 25q$

The answers are on page 249.

8-2 Proportions

A **proportion** is an equation that shows equivalent ratios.

$$\frac{r}{7} = \frac{15}{21}$$ ◀ Proportion $$\frac{3}{4} = \frac{n}{8}$$ ◀ Proportion

A ratio has two terms. A proportion has four terms.

First term ⟶ $\dfrac{3}{4} = \dfrac{15}{20}$ ⟵ Third term
Second term ⟶   ⟵ fourth term

Solving a proportion is similar to solving a multiplication equation.

Property of Proportions

If $\dfrac{a}{b} = \dfrac{c}{d}$, then $ad = cb$, $b \neq 0$, $d \neq 0$.

If $ad = cb$, then $\dfrac{a}{b} = \dfrac{c}{d}$, $b \neq 0$, $d \neq 0$.

PROCEDURE

To solve a proportion:

1 Write the cross-products.
2 Solve the resulting equation for the variable.

EXAMPLE 1 Solve and check: $\dfrac{t}{30} = \dfrac{5}{6}$

Solution: $\dfrac{t}{30} \bowtie \dfrac{5}{6}$ Check: $\dfrac{25}{30} \stackrel{?}{=} \dfrac{5}{6}$

1   $t \cdot 6 = 30 \cdot 5$ $25 \cdot 6 \mid 30 \cdot 5$
    $6t = 150$ **150** | **150**

2   $\dfrac{6t}{6} = \dfrac{150}{6}$

    $t = 25$

Ratio, Proportion, and Percent   **219**

Self-Check *Solve and check.*

1. $\dfrac{r}{8} = \dfrac{3}{24}$
2. $\dfrac{p}{7} = \dfrac{6}{42}$
3. $\dfrac{w}{20} = \dfrac{65}{100}$
4. $\dfrac{p}{13} = \dfrac{1}{5}$

A variable can appear in any term of the proportion.

EXAMPLE 2 Solve and check: $\dfrac{10}{3} = \dfrac{x}{2}$

Solution:
$\dfrac{10}{3} = \dfrac{x}{2}$

[1] $10 \cdot 2 = 3 \cdot x$

$20 = 3x$

[2] $\dfrac{20}{3} = \dfrac{3x}{3}$

$6\dfrac{2}{3} = x$

Check: $\dfrac{10}{3} = \dfrac{6\frac{2}{3}}{2}$

$10 \cdot 2 \;\bigg|\; 3 \cdot 6\dfrac{2}{3}$

$20 \;\bigg|\; 3 \cdot \dfrac{20}{3}$

$20 \;\bigg|\; 20$

Self-Check *Solve and check.*

5. $\dfrac{3}{10} = \dfrac{w}{8}$
6. $\dfrac{14}{35} = \dfrac{q}{10}$
7. $\dfrac{5}{p} = \dfrac{15}{24}$
8. $\dfrac{10}{t} = \dfrac{5}{8}$

CLASSROOM EXERCISES

Use cross-products to write an equation. Do not solve the equation. (Examples 1 and 2, Step 1)

1. $\dfrac{n}{8} = \dfrac{10}{16}$
2. $\dfrac{p}{3} = \dfrac{6}{18}$
3. $\dfrac{15}{x} = \dfrac{7}{3}$
4. $\dfrac{8}{6} = \dfrac{y}{5}$
5. $\dfrac{3}{4} = \dfrac{m}{8}$
6. $\dfrac{3}{2} = \dfrac{9}{r}$

Solve and check each proportion. (Examples 1 and 2)

7. $\dfrac{x}{6} = \dfrac{1}{2}$
8. $\dfrac{n}{7} = \dfrac{20}{35}$
9. $\dfrac{5}{p} = \dfrac{15}{24}$
10. $\dfrac{9}{s} = \dfrac{3}{7}$
11. $\dfrac{6}{27} = \dfrac{k}{9}$
12. $\dfrac{6}{7} = \dfrac{m}{56}$

WRITTEN EXERCISES

Goal: To solve proportions

Solve and check each proportion. (Examples 1 and 2)

1. $\dfrac{x}{30} = \dfrac{3}{5}$
2. $\dfrac{y}{24} = \dfrac{5}{6}$
3. $\dfrac{n}{8} = \dfrac{3}{24}$
4. $\dfrac{p}{9} = \dfrac{24}{54}$
5. $\dfrac{m}{3} = \dfrac{10}{15}$
6. $\dfrac{r}{84} = \dfrac{3}{14}$

7. $\dfrac{6}{x} = \dfrac{18}{21}$ 8. $\dfrac{10}{t} = \dfrac{20}{30}$ 9. $\dfrac{12}{s} = \dfrac{4}{9}$ 10. $\dfrac{24}{k} = \dfrac{6}{5}$ 11. $\dfrac{12}{a} = \dfrac{9}{12}$ 12. $\dfrac{n}{3} = \dfrac{10}{15}$

13. $\dfrac{9}{7} = \dfrac{p}{42}$ 14. $\dfrac{2}{3} = \dfrac{w}{21}$ 15. $\dfrac{35}{90} = \dfrac{b}{18}$ 16. $\dfrac{25}{75} = \dfrac{c}{15}$ 17. $\dfrac{2}{5} = \dfrac{m}{45}$ 18. $\dfrac{8}{70} = \dfrac{d}{35}$

19. $\dfrac{3}{12} = \dfrac{15}{n}$ 20. $\dfrac{3}{5} = \dfrac{21}{x}$ 21. $\dfrac{15}{33} = \dfrac{5}{y}$ 22. $\dfrac{30}{70} = \dfrac{6}{q}$ 23. $\dfrac{7}{9} = \dfrac{28}{m}$ 24. $\dfrac{30}{50} = \dfrac{3}{t}$

25. $\dfrac{c}{10} = \dfrac{8}{25}$ 26. $\dfrac{7}{y} = \dfrac{8}{15}$ 27. $\dfrac{9}{16} = \dfrac{x}{8}$ 28. $\dfrac{4}{5} = \dfrac{10}{p}$ 29. $\dfrac{3}{4} = \dfrac{k}{15}$ 30. $\dfrac{7}{5} = \dfrac{6}{n}$

APPLICATIONS: USING PROPORTIONS

Solve each problem.

31. The scale on a map compares distance on the map to actual distance. The map of California on the right uses 1 inch to represent 145 miles. How many inches represent 290 miles?

32. On the map of California, how many miles do 3 inches represent?

33. On a map of California, 3 centimeters represent 100 kilometers. How many kilometers do 12 centimeters represent?

34. On a map of California, 2.5 centimeters represent 40 kilometers. How many centimeters represent 200 kilometers?

35. On a map, $\dfrac{1}{2}$ inch represents 200 miles. How many inches would represent 500 miles?

REVIEW CAPSULE FOR SECTION 8-3

Write each ratio in lowest terms. (Pages 216–218)

1. 12 to 30   2. 6 to 58   3. 20 : 18   4. 21 : 35   5. $\dfrac{7}{91}$ 6. $\dfrac{42}{15}$

Solve each equation. (Pages 200–202)

7. $6x = 24$ 8. $12p = 108$ 9. $350 = 14m$ 10. $73 = 8n$

The answers are on page 249.

Ratio, Proportion, and Percent

PROBLEM SOLVING AND APPLICATIONS

8-3 Using Proportions

Proportions are useful in solving many kinds of problems.

PROCEDURE

To use proportions to solve word problems:

1. Choose a variable to represent the unknown.
2. Write two ratios. Compare the quantities <u>in the same order</u> in both ratios.
3. Use the ratios to write a proportion.
4. Solve the proportion.

EXAMPLE 1 To obtain a certain shade of green paint, Anna mixed yellow and blue paint in the ratio 4 : 3. She plans to use 12 quarts of yellow paint. How much blue paint will be needed?

Solution: 1 Let x = the number of quarts of blue paint.

2
	One Ratio	Ratio with the Unknown
	$\dfrac{4}{3}$ ← Yellow paint ← Blue paint	$\dfrac{12}{x}$ ← Quarts of yellow paint ← Quarts of blue paint

3 $\dfrac{4}{3} = \dfrac{12}{x}$ Use the table to write a proportion.

4 $4 \cdot x = 3 \cdot 12$
   $4x = 36$
   $x = 9$ Thus, **9 quarts** of blue paint will be needed.

Self-Check 1. In Example 1, suppose that Anna plans to use 16 quarts of blue paint. How much yellow paint will she need?

EXAMPLE 2 In checking the heartbeat of a patient, a nurse counted 19 beats in 15 seconds. Find the patient's heartbeat for 60 seconds.

Solution: ▢1 Let $h=$ the heartbeat for 60 seconds.

▢2
	One Ratio	Ratio with the Unknown
	$\frac{19}{15}$ ← Number of beats ← Number of seconds	$\frac{h}{60}$ ← Number of beats ← Number of seconds

▢3 $\quad \dfrac{19}{15} = \dfrac{h}{60}$

▢4 $\quad 19 \cdot 60 = 15 \cdot h$

$\qquad 1140 = 15h$

$\qquad \dfrac{1140}{15} = \dfrac{15h}{15}$

$\qquad 76 = h$

There will be **76 beats** in 60 seconds.

Self-Check 2. In Example 2, find the patient's heartbeat for 75 seconds.

CLASSROOM EXERCISES

Complete each table. (Examples 1 and 2)

1. 2 pounds of bananas at 96¢
5 pounds at x cents

One Ratio	Ratio with the Unknown
$\frac{2}{96}$ ← pounds ← cents	?

2. 20 pounds of dog food at $8
35 pounds at x dollars

One Ratio	Ratio with the Unknown
$\frac{20}{8}$ ← pounds ← dollars	?

3. 64 feet of rope weigh 20 pounds
90 feet weigh x pounds

One Ratio	Ratio with the Unknown
$\frac{64}{20}$ ← feet ← pounds	?

4. 2 meters of cable weigh 5 kg.
12 meters weigh x kg

One Ratio	Ratio with the Unknown
$\frac{2}{5}$ ← meters ← kilograms	?

Ratio, Proportion, and Percent

WRITTEN EXERCISES

Goal: To use proportions to solve word problems

1. On a map, the distance from San Francisco to Pasadena is 9 centimeters. On the map, 2 centimeters represent 130 kilometers. Find the actual distance between the cities.

2. On a blue print for a house, 2 inches represent 5 feet. If the actual length of the family room is 32 feet, what is its length on the blueprint?

3. A painter mixed yellow paint with blue paint in the ratio of 2:3 to obtain a shade of green. How many gallons of blue paint are needed to mix with 12 gallons of yellow paint to obtain the same shade of green?

4. A certain shade of paint can be obtained by mixing yellow with red in the ratio of 1:2. How many gallons of red paint are needed to mix with 5 gallons of yellow to obtain the same shade of paint?

5. The Santini family drove 280 miles on a tank of gasoline. The tank held about 18 gallons. How many miles could they drive on 5 gallons?

6. In checking a patient's heartbeat, a nurse counted 28 beats in 20 seconds. Find the patient's heartbeat for 60 seconds.

7. A photograph has a length of 7 inches and a width of 5 inches. The photograph is enlarged so that the width is 7.5 inches. What is the length of the enlarged photograph?

8. A photograph measuring 200 millimeters wide and 250 millimeters long must be reduced to a width of 85 millimeters. What will be its reduced length?

9. Alan is building a radio controlled model airplane. One inch on the model represents 12 inches on the actual plane. If the actual plane is 360 inches long, what is the length of the model airplane?

10. Virginia is building a model of an old sailing ship. The scale of the model is in the ratio of 1:15. If the length of the actual ship is 22.5 meters, what is the length of the model?

11. A 20-acre field yields 610 bushels of soybeans. About how many bushels would a 25-acre field yield under similar conditions?

12. A farmer had a yield of 228 bushels of corn for every 3 acres. What yield could he expect for 11 acres under similar conditions?

13. There are about 60 calories in 4 ounces of apple juice. How many calories are there in 20 ounces of apple juice?

14. Jogging for 15 minutes uses about 105 calories. How many minutes of jogging would be needed to use 525 calories?

REVIEW: SECTIONS 8-1—8-3

Write a fraction in lowest terms for each ratio. (Section 8-1)

1. 12 to 15 **2.** 42 to 28 **3.** 16 : 10 **4.** 5 : 25

5. $\dfrac{18}{30}$ **6.** $\dfrac{65}{25}$ **7.** $\dfrac{18t}{75t}$, $t \neq 0$ **8.** $\dfrac{16k}{24}$

Determine whether each pair of ratios is equivalent. Answer Yes or No. (Section 8-1)

9. $\dfrac{2}{3}$ and $\dfrac{16}{24}$ **10.** $\dfrac{6}{13}$ and $\dfrac{18}{39}$ **11.** $\dfrac{4}{5}$ and $\dfrac{8}{20}$ **12.** $\dfrac{35}{59}$ and $\dfrac{5}{7}$

Solve each proportion. (Section 8-2)

13. $\dfrac{4}{12} = \dfrac{5}{d}$ **14.** $\dfrac{15}{x} = \dfrac{40}{16}$ **15.** $\dfrac{9}{12} = \dfrac{b}{8}$ **16.** $\dfrac{a}{21} = \dfrac{4}{5}$

Solve by using a proportion. (Section 8-3)

17. An 8 ounce glass of whole milk contains about 160 calories. How many calories are there in 24 ounces of whole milk?

18. Bicycling for 10 minutes uses about 54 calories of energy. How many minutes of bicycling would be needed to use 270 calories?

REVIEW CAPSULE FOR SECTION 8-4

Solve each proportion. (Pages 219-221)

1. $\dfrac{2}{5} = \dfrac{x}{100}$ **2.** $\dfrac{3}{10} = \dfrac{p}{100}$ **3.** $\dfrac{4}{25} = \dfrac{a}{100}$ **4.** $\dfrac{23}{50} = \dfrac{b}{100}$

5. $\dfrac{3}{4} = \dfrac{c}{100}$ **6.** $\dfrac{9}{20} = \dfrac{t}{100}$ **7.** $\dfrac{23}{10} = \dfrac{r}{100}$ **8.** $\dfrac{12}{5} = \dfrac{y}{100}$

9. $\dfrac{x}{100} = \dfrac{3}{8}$ **10.** $\dfrac{a}{100} = \dfrac{7}{12}$ **11.** $\dfrac{b}{100} = \dfrac{7}{8}$ **12.** $\dfrac{d}{100} = \dfrac{1}{3}$

The answers are on page 249.

COMPUTER APPLICATIONS Problem Solving

In baseball, a player's **batting average** is defined as a ratio of the **number of hits** to the **number of times-at-bat**. This ratio is then expressed as a decimal.

PROBLEM Write a program which accepts a baseball player's times-at-bat and the number of hits. Then have the program print the player's batting average.

Solution: There are four logical phases for writing a program to solve a problem.

Phase 1 Identify the input variables and the output variables.

B = number of times-at-bat ⟵ INPUT variable
H = number of hits ⟵ INPUT variable
A = batting average ⟵ OUTPUT variable

Phase 2 Write a formula for calculating the value of the output variable.

Batting Average (A) = Number of Hits (H) ÷ Times at Bat (B)

Phase 3 Write a program.

```
10 PRINT "ENTER TIMES-AT-BAT."
20 INPUT B
30 PRINT "ENTER NUMBER OF HITS."
40 INPUT H
50 LET A = H / B
60 PRINT "BATTING AVERAGE = "; A
70 END
```

Phase 4 Check the program by using test values. In the following output, $B = 50$ and $H = 15$.

```
ENTER TIMES-AT-BAT.
? 50
ENTER NUMBER OF HITS.
? 15
BATTING AVERAGE = .3
```
⟵ Final zeros after the decimal point are not usually printed.

Exercises

Write a program for each problem.

1. Given the number of hours worked and the pay per hour, compute the worker's pay.

2. Given the number of miles a car has traveled and the number of gallons of gasoline used, compute the average miles-per-gallon.

8-4 Meaning of Percent

The ratio 77 out of 100 or $\frac{77}{100}$ can be read as 77 percent. It is written as 77%.

> **A percent is a ratio in which the second term is 100.**
>
> $\frac{3}{100} = 3\%$
>
> $\frac{y}{100} = y\%$

If the second term of a ratio is not 100, you can use a proportion to write a percent for the ratio.

PROCEDURE

To write a percent for a ratio whose denominator is not 100:

1. Choose a variable for the first term of the percent ratio. Using 100 as the second term, write the percent ratio.
2. Write a proportion.
3. Solve the proportion.

EXAMPLE 1 Write a percent for $\frac{17}{20}$.

Solution:

1. Let p = the first term of the percent ratio.

   Percent ratio: $\frac{p}{100}$ ◀ The second term is 100.

2. Solve for p.

   $\frac{p}{100} = \frac{17}{20}$

3. $20p = (100)(17)$

   $20p = 1700$

   $\frac{20p}{20} = \frac{1700}{20}$

   $p = 85$ Thus, $\frac{17}{20} = \frac{85}{100} = 85\%$

Self-Check *Write a percent for each ratio.*

1. $\frac{1}{4}$ 2. $\frac{3}{5}$ 3. $\frac{7}{10}$ 4. $\frac{0}{3}$ 5. $\frac{9}{20}$ 6. $\frac{41}{50}$

EXAMPLE 2 *Write a percent for $\frac{1}{6}$.*

Solution: ☐1 Let m = the first term of the percent ratio.

Percent ratio: $\frac{m}{100}$

☐2 $\frac{m}{100} = \frac{1}{6}$

☐3 $6m = 100$

$m = 16\frac{4}{6} = 16\frac{2}{3}$

Thus, $\frac{1}{6} = 16\frac{2}{3}\%$.

Self-Check *Write a percent for each ratio.*

7. $\frac{1}{3}$ **8.** $\frac{5}{6}$ **9.** $\frac{3}{7}$ **10.** $\frac{2}{9}$ **11.** $\frac{5}{12}$ **12.** $\frac{9}{16}$

If the first and second terms of a ratio are the same, the ratio *equals* 100%.

$$\frac{50}{50} = 100\% \qquad \frac{7.3}{7.3} = 100\% \qquad \frac{2\frac{1}{2}}{2\frac{1}{2}} = 100\%$$

If the first term of a ratio is greater than the second term, the ratio is greater than 100%.

EXAMPLE 3 *Write a percent for $\frac{11}{8}$.*

Solution: ☐1 Let q = the first term of the percent ratio.

Percent ratio: $\frac{q}{100}$

☐2 $\frac{q}{100} = \frac{11}{8}$

☐3 $8q = 1100$

$q = 137\frac{4}{8} = 137\frac{1}{2}$

Thus, $\frac{11}{8} = \frac{137\frac{1}{2}}{100} = 137\frac{1}{2}\%$.

Self-Check *Write a percent for each ratio.*

13. $\frac{100}{100}$ **14.** $\frac{1.5}{1.5}$ **15.** $\frac{7}{2}$ **16.** $\frac{25}{4}$ **17.** $\frac{13}{6}$ **18.** $\frac{39}{12}$

228 *Chapter 8*

CLASSROOM EXERCISES

Write a percent for each ratio. (Definition)

1. $\dfrac{9}{100}$
2. $\dfrac{13}{100}$
3. $\dfrac{7.6}{100}$
4. $\dfrac{113.6}{100}$
5. $\dfrac{16\frac{2}{3}}{100}$
6. $\dfrac{\frac{1}{2}}{100}$

Write the proportion you would use in writing a percent for each ratio. Use n for the variable. (Examples 1–3, Steps 1 and 2)

7. $\dfrac{3}{4}$
8. $\dfrac{3}{10}$
9. $\dfrac{7}{20}$
10. $\dfrac{2}{5}$
11. $\dfrac{1}{2}$
12. $\dfrac{3}{8}$
13. $\dfrac{2}{3}$
14. $\dfrac{5}{12}$
15. $\dfrac{4}{15}$
16. $\dfrac{4}{9}$
17. $\dfrac{1}{6}$
18. $\dfrac{3}{7}$
19. $\dfrac{16}{5}$
20. $\dfrac{21}{10}$
21. $\dfrac{20}{7}$
22. $\dfrac{11}{7}$
23. $\dfrac{20}{19}$
24. $\dfrac{24}{11}$

WRITTEN EXERCISES

Goal: To write a percent for a ratio

For Exercises 1–54, write a percent for each ratio. (Example 1)

1. $\dfrac{1}{4}$
2. $\dfrac{3}{10}$
3. $\dfrac{1}{5}$
4. $\dfrac{1}{20}$
5. $\dfrac{2}{25}$
6. $\dfrac{3}{50}$
7. $\dfrac{4}{5}$
8. $\dfrac{29}{50}$
9. $\dfrac{4}{25}$
10. $\dfrac{11}{20}$
11. $\dfrac{17}{20}$
12. $\dfrac{37}{50}$

(Example 2)

13. $\dfrac{5}{8}$
14. $\dfrac{1}{12}$
15. $\dfrac{1}{40}$
16. $\dfrac{3}{16}$
17. $\dfrac{5}{7}$
18. $\dfrac{8}{9}$
19. $\dfrac{1}{50}$
20. $\dfrac{9}{11}$
21. $\dfrac{7}{11}$
22. $\dfrac{5}{9}$
23. $\dfrac{13}{40}$
24. $\dfrac{9}{16}$

(Example 3)

25. $\dfrac{12}{10}$
26. $\dfrac{14}{10}$
27. $\dfrac{30}{20}$
28. $\dfrac{15}{8}$
29. $\dfrac{21}{16}$
30. $\dfrac{17}{6}$
31. $\dfrac{51}{50}$
32. $\dfrac{7}{4}$
33. $\dfrac{11}{5}$
34. $\dfrac{76}{25}$
35. $\dfrac{17}{5}$
36. $\dfrac{167}{20}$

MIXED PRACTICE

37. $\dfrac{4}{10}$
38. $\dfrac{18}{5}$
39. $\dfrac{7}{9}$
40. $\dfrac{1}{8}$
41. $\dfrac{22}{15}$
42. $\dfrac{9}{2}$
43. $\dfrac{9}{50}$
44. $\dfrac{15}{4}$
45. $\dfrac{7}{11}$
46. $\dfrac{3}{40}$
47. $\dfrac{1}{12}$
48. $\dfrac{16}{40}$
49. $\dfrac{61}{10}$
50. $\dfrac{8}{9}$
51. $\dfrac{119}{10}$
52. $\dfrac{11}{12}$
53. $\dfrac{7}{10}$
54. $\dfrac{25}{4}$

Ratio, Proportion, and Percent

APPLICATIONS: FROM RATIOS TO PERCENTS

Write a percent for each ratio.

55. In 1970, $\frac{2}{5}$ of the population of the United States was under age 21.

56. By 1990, $\frac{8}{25}$ of the population of the United States will be under age 21.

57. By 1990, $\frac{22}{25}$ of the population of the United States will be under age 65.

58. By 1990, $\frac{19}{50}$ of the population of the United States will be 40 years old or older.

59. Refer to your answer to Exercise 56 to determine what percent of the population will be 21 years old or older by 1990.

60. Refer to your answer to Exercise 57 to determine what percent of the population will be 65 years old or older by 1990.

61. Refer to your answer to Exercise 58 to determine what percent of the population will be under 40 years old by 1990.

REVIEW CAPSULE FOR SECTION 8-5

Multiply each number by 100. (Pages 38–41)

1. 0.35	**2.** 0.71	**3.** 0.08	**4.** 0.05	**5.** 0.875	**6.** 0.125
7. 1.45	**8.** 1.97	**9.** 0.005	**10.** 0.001	**11.** 0.305	**12.** 0.109

Divide each number by 100. (Pages 44–46, 173–175)

13. 18	**14.** 26	**15.** 4	**16.** 7	**17.** 19.5	**18.** 50.9
19. 119	**20.** 256	**21.** 103	**22.** 405	**23.** 1.5	**24.** 47.6

The answers are on page 249.

8-5 Percents and Decimals

A manufacturer found that, on the average, 2 out of every 100 TV's produced by a plant were defective. That is, 0.02 of the total number produced were defective. Recall that

$$\frac{2}{100} = 0.02 = 2\%.$$

Thus, **2% of the TV's produced** by the plant were defective.

To write a percent for a decimal, first write the decimal as a fraction with 100 as the denominator.

TABLE 1

Decimal	Fraction	Percent
0.16	$\frac{16}{100}$	16%
0.9	$\frac{90}{100}$	90%
0.005	$\frac{0.5}{100}$	0.5%
3.8	$\frac{380}{100}$	380%

Table 1 suggests this procedure.

PROCEDURE

To write a percent for a decimal:

1 Move the decimal point two places to the right. Annex zeros as needed.

2 Write the percent symbol.

EXAMPLE 1 Write a percent for each decimal.

    a. 0.35   **b.** 0.3   **c.** 9   **d.** 0.006

Solutions:

	a.	b.	c.	d.
1	0.35 = 0.35	0.3 = 0.30	9 = 9.00	0.006 = 0.006
2	= 35%	= 30%	= 900%	= 0.6%

Ratio, Proportion, and Percent

Self-Check *Write a percent for each decimal.*

1. 0.08 **2.** 0.63 **3.** 0.501 **4.** 9.4 **5.** 3.58

To write a decimal for a percent, first write the percent ratio.

TABLE 2

Percent	Percent Ratio	Decimal
22%	$\frac{22}{100}$	0.22
8%	$\frac{8}{100}$	0.08
0.6%	$\frac{0.6}{100}$	0.006
100%	$\frac{100}{100}$	1.00
225%	$\frac{225}{100}$	2.25

Table 2 suggests this procedure.

PROCEDURE

To write a decimal for a percent:

1. Move the decimal point two places to the left. Insert zeros as needed.
2. Drop the percent symbol.

EXAMPLE 2 Write a decimal for each percent.

    a. 17% **b.** 2% **c.** 0.1% **d.** 150%

Solutions:

	a.	b.	c.	d.
1	17% = 17%	2% = 02%	0.1% = 00.1%	150% = 150%
2	= 0.17	= 0.02	= 0.001	= 1.50

Self-Check *Write a decimal for each per cent.*

6. 15% **7.** 6% **8.** 21.5% **9.** 0.4% **10.** 113%

CLASSROOM EXERCISES

Write a fraction with 100 as the denominator for each decimal. (Table 1)

1. 0.41
2. 0.8
3. 0.97
4. 0.08
5. 7.9
6. 0.223

Write a decimal for each ratio. (Table 2)

7. $\frac{21}{100}$
8. $\frac{1}{100}$
9. $\frac{0.7}{100}$
10. $\frac{110}{100}$
11. $\frac{6.3}{100}$
12. $\frac{15.76}{100}$
13. $\frac{200}{100}$
14. $\frac{900}{100}$
15. $\frac{1.5}{100}$
16. $\frac{8.9}{100}$
17. $\frac{0.05}{100}$
18. $\frac{0.01}{100}$

WRITTEN EXERCISES

Goals: To write a percent for a decimal
To write a decimal for a percent

Write a percent for each decimal. (Example 1)

1. 0.4
2. 3.6
3. 9.4
4. 0.15
5. 0.203
6. 4.57
7. 1.5
8. 0.9
9. 0.26
10. 5.1
11. 0.1
12. 2.04
13. 8.6
14. 0.005
15. 0.04
16. 0.03
17. 3.62
18. 0.8

Write a decimal for each percent. (Example 2)

19. 18%
20. 91%
21. 66.1%
22. 17.8%
23. 3%
24. 10%
25. 101%
26. 507%
27. 0.5%
28. 0.1%
29. 20%
30. 80%
31. 0.25%
32. 0.66%
33. 400%
34. 900%
35. 4.5%
36. 9.8%

MIXED PRACTICE

Complete the tables below. Write all fractions in lowest terms. (Tables 1 and 2, Examples 1 and 2)

	Percent	Decimal	Fraction
37.	?	0.14	?
39.	33%	?	?
41.	?	?	$\frac{17}{20}$
43.	60%	?	?
45.	?	?	$\frac{9}{10}$

	Percent	Decimal	Fraction
38.	?	0.05	?
40.	118%	?	?
42.	?	?	$\frac{16}{25}$
44.	?	4.08	?
46.	51%	?	?

Ratio, Proportion, and Percent

APPLICATIONS: FROM PERCENT TO DECIMALS

This circle graph below shows the percent (by weight) which each element forms of the earth's surface.

Write a decimal for each percent.

47. oxygen
48. silicon
49. iron
50. magnesium
51. potassium
52. calcium
53. sodium
54. aluminum
55. oxygen and iron
56. silicon and oxygen
57. iron and aluminum
58. calcium and sodium

Oxygen 47%
Silicon 28%
Aluminum 8.1%
Iron 5%
Calcium 3.6%
Sodium 2.5%
Potassium 2.5%
Magnesium 2.1%
All others 1.2%

REVIEW CAPSULE FOR SECTION 8-6

Evaluate each expression. (Pages 18–21)

1. $(1 \div 4) \times 100$
2. $(2 \div 5) \times 100$
3. $(1 \div 2) \times 100$
4. $(3 \div 10) \times 100$
5. $(3 \div 5) \times 100$
6. $(7 \div 20) \times 100$
7. $(8 \div 25) \times 100$
8. $(3 \div 4) \times 100$

Multiply. (Pages 38–41)

9. $800(0.15)$
10. $912(0.38)$
11. $461(0.03)$
12. $53(0.21)$
13. $2644(0.45)$
14. $8678(0.15)$
15. $6990(0.12)$
16. $9003(0.48)$

(Pages 129–131)

17. $920 \cdot \frac{7}{8}$
18. $618 \cdot \frac{2}{3}$
19. $804 \cdot \frac{5}{6}$
20. $275 \cdot \frac{4}{15}$
21. $680 \cdot \frac{3}{50}$
22. $400 \cdot \frac{19}{20}$
23. $760 \cdot \frac{11}{25}$
24. $405 \cdot \frac{3}{5}$
25. $927 \cdot \frac{2}{3}$
26. $3072 \cdot \frac{1}{6}$
27. $3072 \cdot \frac{5}{8}$
28. $3128 \cdot \frac{1}{8}$

The answers are on page 249.

8-6 Finding a Percent of a Number

You can use an equation to find a percent of a number.

The **variable** in the equation represents the unknown (what you are trying to find).

> Inflation Rises 1.5%
> 11.37% YIELD ON NEW BONDS!
> JUNIORS FALL SALE 30%–40% off

PROCEDURE

To find a percent of a number:

1. Write an equation for the problem.
2. Solve the equation.

EXAMPLE 1 What number is 5% of 120?

Solution:
1. $n = 0.05 \cdot 120$
2. $n = 6$

Replace "is" with "=".
Replace "of" with "·" (times).

Thus, **6** is 5% of 120.

Self-Check

1. What number is 1% of 30?
2. What number is 12% of 20?

Sometimes it is easier to write a fraction for a percent. This table can help you. You should memorize the equivalent fractions and percents in the table.

TABLE

Equivalent Fractions and Percents

$\frac{1}{4} = 25\%$	$\frac{1}{2} = 50\%$	$\frac{3}{4} = 75\%$	
$\frac{1}{5} = 20\%$	$\frac{2}{5} = 40\%$	$\frac{3}{5} = 60\%$	$\frac{4}{5} = 80\%$
$\frac{1}{6} = 16\frac{2}{3}\%$	$\frac{1}{3} = 33\frac{1}{3}\%$	$\frac{2}{3} = 66\frac{2}{3}\%$	$\frac{5}{6} = 83\frac{1}{3}\%$
$\frac{1}{8} = 12\frac{1}{2}\%$	$\frac{3}{8} = 37\frac{1}{2}\%$	$\frac{5}{8} = 62\frac{1}{2}\%$	$\frac{7}{8} = 87\frac{1}{2}\%$

Ratio, Proportion, and Percent

EXAMPLE 2 What number is $87\frac{1}{2}\%$ of 1600?

Solution:
1. $n = \frac{7}{8} \cdot 1600$ $\quad \frac{7}{\cancel{8}_1} \cdot \cancel{1600}^{200} = \frac{7 \cdot 200}{1}$
2. $n = 7 \cdot 200$
$n = 1400$ **1400** is $87\frac{1}{2}\%$ of 1600.

Self-Check

3. What number is 25% of 8?
4. What number is $66\frac{2}{3}\%$ of 90?

Follow a similar procedure to solve word problems involving percent.

PROCEDURE
To solve word problems involving percent:
1. Identify what is given and what you are to find.
2. Use the given information to restate the problem.
3. Write an equation for the problem.
4. Solve the equation.

Example 3 shows how to find commission. **Commission** is earnings based on a percent of sales.

EXAMPLE 3 In addition to her regular salary at Central Auto Sales, Louise Wong receives a commission of 6% on total sales. Her total sales last week amounted to $32,000. How much commission did she receive?

Solution:
1. Given: Rate of commission = 6%   Total sales = $32,000
   Find: Amount of commission, or 6% of $32,000
2. Restate the problem.   What number is 6% of 32,000?
3. Write an equation. $n = 0.06 \cdot 32,000$
4. Solve the equation. $n = 1920$

Louise's commission was **$1920.00.**

Self-Check

5. A real estate agent receives a 6% commission of sales. What is the agent's commission on the sale of a house for $65,000?

236   Chapter 8

CLASSROOM EXERCISES

Write an equation for each percent problem.
(Examples 1 and 2, step 1)

1. n is 7% of 121.
2. q is 1% of 30.
3. r is 15% of 144.
4. t is 30% of 612.
5. w is 60% of 75.
6. a is 105% of 60.
7. $r = 33\frac{1}{3}\%$ of 96.
8. $p = 2\frac{1}{2}\%$ of 16.
9. c is 75% of 4.

Write an equation for each word problem. Do NOT solve.
(Example 3, steps 1, 2, and 3)

10. What is the commission on sales of $4,600 if the rate of commission is 1%?

11. A baseball team won 75% of the 30 games played. How many games were won?

12. A state has a 4% sales tax. What is the sales tax on a suit that sells for $118.50?

WRITTEN EXERCISES

Goals: To find a percent of a number

To apply the skill of finding a percent of a number to solving word problems

For Exercises 1–6, write an equation for each problem.
(Examples 1 and 2, step 1)

1. What is 35% of 400?
2. What is 9% of 800?
3. What is 83% of 100?
4. What is 16% of 900?
5. Find $37\frac{1}{2}\%$ of 16.
6. Find $66\frac{2}{3}\%$ of 60.

Solve. (Example 1)

7. What is 10% of 516?
8. What is 30% of 505?
9. What is 45% of 120?
10. What is 18% of 150?
11. 4% of 7.8 is n. Find n.
12. 3% of 24 is w. Find w.
13. Find 12% of 900.
14. Find 1% of 69.

Ratio, Proportion, and Percent **237**

(Example 2)

15. What is 25% of 80?
16. What is 50% of 88?
17. What is 75% of 64?
18. What is 10% of 38?
19. 60% of 155 is q. Find q.
20. What is $16\frac{2}{3}$% of 12?
21. What is $66\frac{2}{3}$% of 12?
22. 30% of 102 is r. Find r.
23. What is $33\frac{1}{3}$% of 102?
24. What is $87\frac{1}{2}$% of 64?
25. Find $83\frac{1}{3}$% of 144.
26. Find $62\frac{1}{2}$% of 144.

APPLICATIONS: FINDING A PERCENT OF A NUMBER

Solve each problem. (Example 3)

27. An auctioneer sold goods amounting to $125,000. She received a commission of 6% on the total sales. How much was her commission?

28. A real estate agent received a 5% on the sales of a house. The house sold for $112,000. What was the agent's commission?

29. After a meal in a restaurant, Eileen gave the server a tip which was 15% of the total bill. The total bill amounted to $20.00. How much was the tip?

30. Of 120 students in a high school band, 20% play woodwind instruments. How many play woodwind instruments?

31. A basketball team played 25 games this year. The team won 60% of the games played. How many games did the team win?

32. The **depreciation** (loss in value) of a car was 14% of its original value of $10,200. Find the amount of depreciation.

33. A family budgets 28% of its take home pay for food. The family's monthly take home pay is $912. How much is their monthly food budget?

34. An automobile supply store found that 2% of the batteries it buys are defective. The store recently ordered 250 batteries. How many would the store expect to be defective?

MORE CHALLENGING EXERCISES

Solve each problem.

35. There are 15 girls on the tennis team at Jansen Senior High. Of these, 20% are seniors. How many are not seniors?

36. Ceanne Brophy bought a suit for $89.50. She also paid a sales tax which was 4% of the purchase price. Find the total cost of the suit.

37. A worker who earns $5.00 per hour receives a 9% raise. What is the new hourly wage?

38. Luis Hernandez earns a base salary of $225 per week. He also earns a 3% commission on all sales. What is his total salary for a week in which his total sales amount to $4500?

REVIEW CAPSULE FOR SECTION 8-7

Write a decimal for each percent. (Pages 231–234)

1. 6%   2. 9%   3. 10%   4. 60%   5. 12%   6. 14%

Write as a fraction in lowest terms. (Pages 124–125)

7. $\frac{6}{12}$ 8. $\frac{18}{12}$ 9. $\frac{38}{12}$ 10. $\frac{9}{12}$ 11. $\frac{46}{12}$ 12. $\frac{10}{12}$

Multiply. (Pages 38–41, 129–131)

13. $900 \times 0.04 \times \frac{2}{3}$ 14. $2500 \times 0.06 \times \frac{3}{4}$ 15. $700 \times 0.13 \times \frac{1}{2}$

16. $(4500)(0.085)\frac{20}{3}$ 17. $(9400)(0.12)\frac{9}{4}$ 18. $(6000)(0.10)\frac{5}{2}$

The answers are on page 249.

PROBLEM SOLVING AND APPLICATIONS

8-7 Formulas: Simple Interest

An important use of percent is in computing interest.

> **Interest** is an amount of money charged for the use of borrowed money. The amount of money borrowed is called the **principal**.
>
> **Interest** is also paid on money deposited in certain bank accounts.

Interest is usually expressed as a percent of the principal. This percent is called the **interest rate**.

Use this formula to compute simple interest, i.

$$i = prt$$

p = principal
r = rate (expressed as a decimal or a fraction)
t = time (expressed in years)

PROCEDURE

To compute simple interest:

1. Write the formula.
2. Identify known values.
3. Substitute and solve for the unknown value.

EXAMPLE 1 Find the interest on $400 for one year at a yearly rate of 13%.

Solution:
1. $i = prt$
2. $p = 400$; $r = 13\% = 0.13$; $t = 1$
3. $i = 400\,(0.13)(1)$
   $i = 52$

The interest on $400 for one year at a yearly rate of 13% is $52.00.

Self-Check 1. Find the interest on $1000 borrowed for 2 years at a rate of 12%.

240 Chapter 8

When using the interest formula, you must express the time given in months in terms of years.

Expressed in Months	Expressed in Years
6	$\frac{6}{12}$, or $\frac{1}{2}$
11	$\frac{11}{12}$
40	$\frac{40}{12}$, or $\frac{10}{3}$

EXAMPLE 2 The Velez family opened a savings account with $2400. The yearly interest rate is 8.5%. How much will the account earn in 4 months?

Solution:
1. $i = prt$
2. $p = 2400$; $r = 8.5\% = 0.085$; $t = \frac{4}{12} = \frac{1}{3}$
3. $i = (2400)(0.085)\left(\frac{1}{3}\right)$
   $i = (204)\left(\frac{1}{3}\right)$
   $i = 68$ The account will earn **$68** in 4 months.

Self-Check 2. Find the interest on $1000 at $5\frac{1}{2}\%$ for 6 months.

EXAMPLE 3 Carla borrowed $45,000 from her credit union to open a custom furniture shop. The yearly interest rate for the loan is $9\frac{1}{2}\%$ and the loan is to be paid back in 40 months. Find the total interest.

Credit Union
9½%
INTEREST

Solution:
1. $i = prt$
2. $p = 45{,}000$; $r = 9\frac{1}{2}\% = 0.09\frac{1}{2} = 0.095$; $t = \frac{40}{12} = \frac{10}{3}$ years
3. $i = (45{,}000)(0.095)\frac{10}{3}$
   $i = (4275)\frac{10}{3}$
   $i = 14{,}250$ Carla will pay **$14,250** in interest.

Self-Check 3. Find the interest on $2500 at $12\frac{1}{2}\%$ for 30 months.

Ratio, Proportion, and Percent **241**

CLASSROOM EXERCISES

Replace the variables in $i = prt$ with the known values.
(Examples 1, 2, and 3, step 3)

1. Interest on $950 at 8% for one year
2. Interest on $10,000 for one year at 12%
3. Interest on $1956 at a yearly rate of 5% for 6 months
4. Interest on $864 at a yearly interest rate of 9% for 3 months
5. Interest on $2575 borrowed for 2 years at a yearly rate of $12\frac{1}{2}\%$
6. Interest on $900 for 18 months at a yearly interest rate of 13%

For Exercises 7–18, find the simple interest. (Example 1)

7. $100 borrowed for one year at 5%
8. $100 borrowed for one year at 6%
9. $200 borrowed for one year at 8%
10. $100 borrowed for one year at 9%

(Example 2)

11. $100 borrowed for 3 months at 12%
12. $200 borrowed for 9 months at 10%
13. $100 borrowed for 8 months at 12%
14. $800 borrowed for 3 years at 11%

(Example 3)

15. $500 borrowed for 2 years at 10%
16. $300 borrowed for 10 months at 13%
17. $400 borrowed for 6 months at $9\frac{1}{2}\%$
18. $1000 borrowed for 30 months at $12\frac{1}{2}\%$

WRITTEN EXERCISES

Goals: To use the formula $i = prt$ to compute simple interest
To apply computing simple interest to solving word problems

For Exercises 1–18, find the simple interest. (Example 1)

1. $500 borrowed for one year at 8%
2. $600 borrowed for one year at 10%
3. $100 borrowed for one year at 12%
4. $200 borrowed for one year at 13%
5. $100 borrowed for one year at 6%
6. $300 borrowed for one year at 9%

(Example 2)

7. $500 borrowed for 6 months at 9%
8. $900 borrowed for 3 months at 8%
9. $1200 borrowed for 3 months at 10%
10. $5000 borrowed for 8 months at 6%
11. $1800 borrowed for 4 months at 12.5%
12. $720 borrowed for 2 months at 13.5%

(Example 3)

13. $600 borrowed for 2 years at 12%
14. $1700 borrowed for 30 months at 13%
15. $7500 borrowed for 3 years at $12\frac{1}{2}$%
16. $9000 borrowed for 3 years at $11\frac{1}{2}$%
17. $15,000 borrowed for 4 months at $14\frac{1}{2}$%
18. $3200 borrowed for 3 months at $13\frac{1}{2}$%

Solve each problem.   (Examples 1–3)

19. George and Clara Lightfoot borrow $20,000 to start a business. The yearly interest rate is $11\frac{1}{2}$%. How much interest is due on the loan in 3 months?
20. Miriam Taylor borrowed $950 to operate a stand at the state fair. The yearly interest rate is 18%. How much is the interest for one month?
21. The Nishifue family obtained a loan of $6500 to buy a car. The yearly interest rate is 12%. Find the interest paid on the loan at the end of 4 years.
22. A drama club has $3500 in a savings account. The account pays a yearly interest rate of 6.5%. How much interest will the account earn in 2 years?

SIMPLE INTEREST

This Example uses a calculator to compute simple interest.

EXAMPLE   Find the simple interest on a loan of $1950 borrowed for 10 months at 13%.

Solution   Remember to write a fraction for 10 months.

10 months = $\frac{10}{12} = \frac{5}{6}$

1 9 5 0 [×] . 1 3 [×] 5 [÷] 6 [=]   211.25

EXERCISES

Use a calculator to check Classroom Exercises 1–18 on page 242.

REVIEW CAPSULE FOR SECTION 8-8

Write a decimal for each percent. (Pages 231–234)

1. 3%   2. 2%   3. 9%   4. 10%   5. 15%   6. $4\frac{1}{2}\%$

Multiply. (Pages 38–41)

7. (55.80)(0.33)   8. (151.80)(0.40)   9. (41.90)(0.10)

Subtract. (Pages 35–37)

10. 18.91 − 2.48   11. 10.05 − 1.96   12. 93.80 − 11.72

The answers are on page 249.

PROBLEM SOLVING AND APPLICATIONS

8-8 Formulas: Discount

Phil Maquire works as a salesperson in Central Department Store. As an employee, he receives a 20% discount on all items he purchases.

A **discount** is the amount that an article of merchandise is reduced in price. Discounts are often expressed as a percent of the list price. The **list price** is the regular price of an article.

Use this formula to find the amount of discount.

$$d = rp$$

d = amount of discount
r = rate of discount
p = list price

EXAMPLE 1 Phil decided to purchase a bicycle he saw at Central Department Store. The list price was $117.50 and his employee discount was 20%. How much was the discount?

Solution: $d = rp$ $p = 117.50;$
$r = 20\% = 0.20$

$d = (0.20)(117.50)$

$d = \$23.50$

The amount of discount was **$23.50**.

244 Chapter 8

Self-Check 1. An item with a list price of $50 is marked for a 25% discount. Find the amount of discount.

The price of an item after the discount is deducted is the **net price** or **sale price.**

Use this formula to find the sale price.

$$s = p - d$$

p = list price
d = discount
s = sale price

To find the sale price when the list price and rate of discount are given, you must first answer the question.

What is the amount of discount?

This is the **hidden question** in the problem. Then you use subtraction to find the sale price.

EXAMPLE 2 A calculator with a list price of $26.50 is on sale at a 40% discount. Find the sale price.

Solution: ⓵ Find the amount of discount.

$$d = rp$$

p = $26.50;
r = 40% = 0.40

$$d = (26.50)(0.40)$$

$$d = 10.60$$

The amount of discount is $10.60.

⓶ Find the sale price.

$$s = p - d$$

p = $26.50;
d = $10.60

$$s = 26.50 - 10.60$$

$$s = 15.90$$

The sale price is **$15.90.**

$26.50
ON SALE
40% DISCOUNT

Self-Check *Find the sale price.*
2. List price: $52.90; discount: 20%
3. List price: $17.32; discount: 25%
4. List price: $149.00; discount: 33%

Ratio, Proportion, and Percent   **245**

CLASSROOM EXERCISES

Find the amount of discount on each item. (Example 1)

1. A baseball glove listed at $24 with a discount rate of 25%
2. A hockey stick listed at $18.50 with a discount rate of 10%
3. A $78 dress on sale at 35% discount
4. A washer listed at $198 with a $33\frac{1}{3}\%$ discount

Find the sale price of each item. (Example 2)

5. A shirt listed at $19.75 with a 20% discount
6. A book listed at $15.60 with a 15% discount
7. Running shoes listed at $35.90 with a 30% discount
8. An automobile listed at $8995 with a 10% discount

WRITTEN EXERCISES

Goals: To find the amount of discount

To apply the skill of finding the amount of discount and the technique of the "hidden question" to solving word problems

Find the amount of discount on each item. (Example 1)

1. SALE! 15% off REGULARLY $380
2. 40% off regularly $38.95
3. CLOSE-OUT warm-up suits ORIGINALLY $32.50 NOW 60% OFF
4. 14 KARAT GOLD CHAINS 30% off REGULARLY $78
5. 3-Day Sale 20% off REGULAR PRICE $264.00
6. JEANS SALE 30% OFF REGULAR PRICE $45.90

246 Chapter 8

Find the sale price of each item. Refer to the advertisements in Exercise 1-6. (Example 2)

7. Typewriter   **8.** Golf Clubs   **9.** Warm-up Suit

For Exercises 10–15, find the sale price. (Example 2)

10. A theater ticket purchased for $25 is sold at a 20% discount.

11. A sailboat with a list price of $3200 is offered on sale at a 33% discount.

12. Senior citizens receive a special discount of 15% off at Wilson's Pharmacy. The regular price of a certain prescription is $25.80.

13. A motorbike listed at $576 is on sale at a $12\frac{1}{2}\%$ discount.

REVIEW: SECTIONS 8-4 – 8-8

Write a percent for each ratio. (Section 8-4)

1. $\frac{1}{2}$ **2.** $\frac{3}{20}$ **3.** $\frac{27}{50}$ **4.** $\frac{21}{25}$ **5.** $\frac{49}{40}$ **6.** $\frac{31}{5}$

Write a percent for each decimal. (Section 8-5)

7. 0.9   **8.** 3.8   **9.** 0.17   **10.** 1.04   **11.** 0.1   **12.** 0.01

Write a decimal for each percent. (Section 8-5)

13. 12%   **14.** 2%   **15.** 85%   **16.** 70%   **17.** 9.5%   **18.** 15.3%

Solve. (Section 8-6)

19. What number is 15% of 44?

20. What is 2% of 64?

21. A real estate agent receives a 5% commission on sales. Find the commission on $82,600.

Solve. (Section 8-7)

22. $p = \$550$; $r = 6\%$, $t = 2$ years

23. $p = \$8000$; $r = 12\%$, $t = 6$ months

24. Clare borrowed $2000 at an interest rate of 8% for 9 months. How much interest will she pay?

(Section 8-8)

25. Compute the discount on a radio if the rate of discount is 22% and the list price is $65.50.

26. Jim Barnes bought slacks on sale at 24% off. The list price was $27. How much did Jim pay for the slacks?

Ratio, Proportion, and Percent

CHAPTER REVIEW

PART 1: VOCABULARY

For Exercises 1–10, choose from the box at the right the word(s) that best corresponds to each description.

1. Earnings based on a percent of total sales __?__ (p. 236)
2. Uses division to compare two numbers __?__ (p. 216)
3. The amount that an article of merchandise is reduced in price __?__ (p. 244)
4. A ratio in which the second term is 100 __?__ (p. 227)
5. An equation that shows equivalent ratios __?__ (p. 219)
6. An amount charged for the use of borrowed money __?__ (240)
7. In Exercise 6, the amount of money borrowed __?__ (p. 240)
8. The regular price of an article of merchandise __?__ (p. 244)
9. The price of an item after the discount is subtracted __?__ (245)
10. The amount paid for borrowing money expressed as a percent of the principal __?__ (p. 240)

> ratio
> proportion
> percent
> commission
> interest
> interest rate
> principal
> discount
> list price
> net price

PART 2: SKILLS

Write each ratio in lowest terms. No variable equals zero. (Section 8–1)

11. 12 to 20
12. 6 to 8
13. 15 to 10
14. 9:21
15. 24:14
16. 7:28
17. $\dfrac{15}{30}$
18. $\dfrac{21}{35}$
19. $\dfrac{18}{42}$
20. $\dfrac{6}{15p}$
21. $\dfrac{30a}{50a}$
22. $\dfrac{36c}{24c}$

Determine whether each pair of ratios is equivalent. Answer Yes or No. (Section 8–1)

23. $\dfrac{1}{2}$ and $\dfrac{4}{6}$
24. $\dfrac{2}{3}$ and $\dfrac{16}{24}$
25. $\dfrac{8}{20}$ and $\dfrac{4}{5}$
26. $\dfrac{6}{14}$ and $\dfrac{2}{7}$
27. $\dfrac{5}{12}$ and $\dfrac{15}{36}$
28. $\dfrac{4}{5}$ and $\dfrac{23}{30}$
29. $\dfrac{24}{9}$ and $\dfrac{8}{3}$
30. $\dfrac{49}{35}$ and $\dfrac{7}{5}$

Solve and check each proportion. (Section 8–2)

31. $\dfrac{n}{14} = \dfrac{2}{7}$
32. $\dfrac{x}{12} = \dfrac{5}{6}$
33. $\dfrac{6}{8} = \dfrac{a}{4}$
34. $\dfrac{10}{15} = \dfrac{p}{18}$
35. $\dfrac{21}{28} = \dfrac{12}{c}$
36. $\dfrac{4}{8} = \dfrac{5}{y}$
37. $\dfrac{8}{n} = \dfrac{10}{15}$
38. $\dfrac{24}{p} = \dfrac{16}{18}$
39. $\dfrac{3}{5} = \dfrac{m}{8}$
40. $\dfrac{9}{10} = \dfrac{4}{b}$
41. $\dfrac{d}{15} = \dfrac{4}{9}$
42. $\dfrac{7}{g} = \dfrac{12}{5}$

Complete. (Sections 8-4, 8-5)

	Ratio	Decimal	Percent
43.	?	0.05	?
45.	?	?	2.5%

	Ratio	Decimal	Percent
44.	$\frac{1}{100}$	?	?
46.	?	1.05	?

Complete. (Section 8-6)

47. What number is 15% of 300?
48. What is 1% of 82?
49. What number is 9% of 812?

	Amount of Sales	Rate of Commission	Commission
50.	$80,000	7%	?
51.	$92,000	2%	?

Find the interest. (Section 8-7)

	Principal	Rate	Time
52.	$2000	9%	3 years
53.	$5000	6%	4 months

Find the discount. (Section 8-8)

	List Price	Rate of Discount
54.	$55.00	20%
55.	$29.80	30%

PART 3: APPLICATIONS

56. A car travels 225 kilometers on 15 liters of fuel. At this rate, how many liters of fuel are needed to travel 375 kilometers? (Section 8–3)

57. Juan obtained a loan of $5400. The yearly interest rate is 12%. Find the interest paid on the loan at the end of 4 months. (Section 8–7)

58. A real estate agent receives a 6% commission on sales. How much commission does the agent receive on an acre of land that sells for $2400. (Section 8–6)

59. Senior citizens receive a 20% discount on bus fares during off-peak hours. If a one-way fare is 75¢, find how much a senior citizen pays during off-peak hours. (Section 8–8)

ANSWERS TO REVIEW CAPSULES

Page 219   **1.** 15   **2.** 10   **3.** 26   **4.** $11\frac{1}{4}$ **5.** $12\frac{1}{2}$ **6.** $11\frac{1}{4}$ **7.** $13\frac{1}{8}$ **8.** $3\frac{1}{5}$

Page 221   **1.** $\frac{2}{5}$ **2.** $\frac{3}{29}$ **3.** $\frac{10}{9}$ **4.** $\frac{3}{5}$ **5.** $\frac{1}{13}$ **6.** $\frac{14}{5}$ **7.** 4   **8.** 9   **9.** 25   **10.** $9\frac{1}{8}$

Page 225   **1.** 40   **2.** 30   **3.** 16   **4.** 46   **5.** 75   **6.** 45   **7.** 230   **8.** 240   **9.** $37\frac{1}{2}$ **10.** $58\frac{1}{3}$ **11.** $87\frac{1}{2}$ **12.** $33\frac{1}{3}$

Page 230   **1.** 35   **2.** 71   **3.** 8   **4.** 5   **5.** 87.5   **6.** 12.5   **7.** 145   **8.** 197   **9.** 0.5   **10.** 0.1   **11.** 30.5   **12.** 10.9   **13.** 0.18   **14.** 0.26   **15.** 0.04   **16.** 0.07   **17.** 0.195   **18.** 0.509   **19.** 1.19   **20.** 2.56   **21.** 1.03   **22.** 4.05   **23.** 0.015   **24.** 0.476

Page 234   **1.** 25   **2.** 40   **3.** 50   **4.** 30   **5.** 60   **6.** 35   **7.** 32   **8.** 75   **9.** 120   **10.** 346.56   **11.** 13.83   **12.** 11.13   **13.** 1189.8   **14.** 1301.7   **15.** 838.8   **16.** 4321.44   **17.** 805   **18.** 412   **19.** 670   **20.** $73\frac{1}{3}$ **21.** $40\frac{4}{5}$ **22.** 380   **23.** $334\frac{2}{5}$ **24.** 243   **25.** 618   **26.** 512   **27.** 1920   **28.** 391

Page 239   **1.** 0.06   **2.** 0.09   **3.** 0.10   **4.** 0.60   **5.** 0.12   **6.** 0.14   **7.** $\frac{1}{2}$ **8.** $\frac{3}{2}$ **9.** $\frac{19}{6}$ **10.** $\frac{3}{4}$ **11.** $\frac{23}{6}$ **12.** $\frac{5}{6}$ **13.** 24   **14.** 112.5   **15.** 45.5   **16.** 2550   **17.** 2538   **18.** 1500

Page 244   **1.** 0.03   **2.** 0.02   **3.** 0.09   **4.** 0.10   **5.** 0.15   **6.** 0.045   **7.** 18.414   **8.** 60.72   **9.** 4.19   **10.** 16.43   **11.** 8.09   **12.** 82.08

Ratio, Proportion, and Percent

CHAPTER TEST

Write the cross-product to determine whether each pair of ratios is equivalent. Answer Yes or No.

1. $\frac{2}{3}$ and $\frac{14}{21}$
2. $\frac{2}{7}$ and $\frac{21}{63}$
3. $\frac{8}{3}$ and $\frac{18}{48}$
4. $\frac{39}{26}$ and $\frac{3}{2}$

Solve and check each proportion.

5. $\frac{5}{8} = \frac{15}{x}$
6. $\frac{9}{21} = \frac{b}{7}$
7. $\frac{y}{48} = \frac{7}{8}$
8. $\frac{16}{c} = \frac{2}{9}$

Write a percent for each of the following.

9. $\frac{3}{5}$
10. $\frac{11}{5}$
11. 0.7
12. 0.07

Write a decimal for each percent.

13. 18%
14. 3%
15. 8.5%
16. 140%

17. What number is 28% of 75?
18. What number is 36% of 25?
19. What is 3% of 27?
20. What is 90% of 82?
21. On a blueprint for a house, 2 inches represent 6 feet. If the actual length of the living room is 24 feet, what is its length on the blueprint?
22. Elizabeth borrowed $1200 at a yearly interest rate of 12%. How much interest will she owe at the end of 4 months?
23. A tent is on sale at a 30% discount. The list price is $425. How much is the discount?
24. A stereo system with a list price of $620.50 is on sale at a 20% discount. Find the sale price.
25. After a meal in a restaurant, David gave the server a tip which was 15% of the total bill. The total bill amounted to $25.00. How much was the tip?

ENRICHMENT

Compound Interest

Compound interest is computed on the principal plus the interest previously earned. The compound interest formula shown below is used to calculate the compounded amount (principal plus interest).

$$A = P(1 + r)^n$$

A: compounded amount
P: principal
r: yearly interest rate
n: number of years

EXAMPLE Four years ago, Erica Caulo put $400 into a savings account. The account earns 9% interest compounded yearly. Find the compounded amount.

Solution: $A = P(1 + r)^n$ P = $400; r = 9% = 0.09; n = 4

$A = 400(1 + 0.09)^4$

$A = 400(1.09)^4$ 1 + 0.09 = 1.09

$A = 400 \times 1.09 \times 1.09 \times 1.09 \times 1.09$ Use a calculator.

4 0 0 [×] 1 . 0 9 [=][=][=][=]   564.63264

The compounded amount after 4 years is **$564.63** (rounded <u>down</u> to the nearest cent).

EXERCISES

Use the compound interest formula to find the compounded amount. The interest is compounded yearly. Round down to the nearest cent.

	Principal	Rate	Years
1.	$900	7%	2
3.	$750	8%	3
5.	$1000	10%	4

	Principal	Rate	Years
2.	$8000	9%	2
4.	$4000	7%	3
6.	$12,000	12%	4

7. Five years ago, Ed Fong put $3000 in a savings account at 7% interest compounded yearly. What is the compounded amount?

ADDITIONAL PRACTICE

SKILLS *Write each ratio in lowest terms.* (Pages 216–218)

1. $\frac{8}{40}$ 2. 21:35   3. 12 to 27   4. $\frac{9}{18}$ 5. 84 to 21   6. 70:90

Determine whether each pair of ratios is equivalent. Answer Yes or No. (Pages 216–218)

7. $\frac{2}{3}$ and $\frac{11}{16}$ 8. $\frac{5}{6}$ and $\frac{60}{72}$ 9. $\frac{5}{7}$ and $\frac{30}{42}$ 10. $\frac{3}{7}$ and $\frac{8}{19}$

Write each ratio in lowest terms. No variable equals zero. (Pages 216–218)

11. $\frac{10x}{25x}$ 12. $\frac{8x}{10}$ 13. $\frac{40t}{8}$ 14. $\frac{15y}{45y}$ 15. $\frac{12}{50y}$ 16. $\frac{7}{56t}$

Solve and check each proportion. (Pages 219–221)

17. $\frac{7}{y} = \frac{3}{5}$ 18. $\frac{11}{c} = \frac{4}{9}$ 19. $\frac{x}{99} = \frac{4}{11}$ 20. $\frac{x}{91} = \frac{7}{13}$ 21. $\frac{4}{7} = \frac{5}{y}$ 22. $\frac{132}{y} = \frac{11}{5}$

Write a percent for each ratio. (Pages 227–230)

23. $\frac{4}{11}$ 24. $\frac{7}{12}$ 25. $\frac{11}{4}$ 26. $\frac{5}{9}$ 27. $\frac{7}{50}$ 28. $\frac{23}{20}$

Complete the tables. Write all fractions in lowest terms. (Pages 231–234)

	Percent	Decimal	Fraction
29.	?	0.25	?
31.	42%	?	?
33.	?	?	$\frac{17}{20}$

	Percent	Decimal	Fraction
30.	?	0.15	?
32.	?	?	$\frac{3}{5}$
34.	12%	?	?

Solve. (Pages 235–239)

35. What is 37% of 143?

36. What is $66\frac{2}{3}$% of 480?

APPLICATIONS *Find the simple interest.* (Pages 240–243)

37. $400 borrowed for one year at 6%

38. $800 borrowed for 7 years at $12\frac{1}{2}$%

39. After running a 100 meter race, a runner had a heartbeat of 22 beats in 10 seconds. Find the heartbeat for one minute. (Pages 222–225)

40. Jeans are on sale at a 25% discount. The regular price of the jeans is $28.60. Find the sale price of the jeans. (Pages 244–247)

CHAPTER 9
Percent and Applications

SECTIONS
- **9-1** Finding What Percent One Number Is of Another
- **9-2** Problem Solving and Applications: Percent of Increase and Decrease
- **9-3** Finding a Number Given a Percent
- **9-4** Using Proportions to Solve Percent Problems
- **9-5** Bar Graphs
- **9-6** Circle Graphs
- **9-7** Problem Solving and Applications: Using Estimation

FEATURES
Calculator Application: *Finding Discount*
Consumer Application: *Nutrition*
Enrichment: *Sequences*
Common Errors

9-1 Finding What Percent One Number Is of Another

Uranium is a radioactive element producing enormous amounts of energy. One pellet of uranium (worth $7 in a recent year) has the energy of one ton of coal (worth $28 in a recent year).

You can use an equation to find what percent 7 is of 28.

PROCEDURE

To find what percent one number is of another:

1. Write an equation for the problem.
2. Solve the equation for the variable.
3. Write a percent for the value of the variable.

EXAMPLE 1 What percent of 28 is 7?

Solution:

1. Write an equation.

   What percent of 28 is 7?
   $$p \cdot 28 = 7$$
   $$28p = 7$$

2. Solve the equation.
   $$\frac{28p}{28} = \frac{7}{28}$$
   $$p = \frac{1}{4}$$

3. Write a percent for $\frac{1}{4}$.
   $$\frac{1}{4} = \frac{25}{100} = 25\%$$

Thus, 7 is **25%** of 28.

Self-Check

1. What percent of 85 is 17?
2. What percent of 90 is 36?

Some word problems involve finding a percent. Follow the procedure for solving word problems involving percent given on page 236. Remember to write a percent for the value of the variable.

254 Chapter 9

EXAMPLE 2 In a recent year, the United States produced 9 million barrels of oil. That same year, the United States imported 6 million barrels of oil. What percent of the number of barrels imported is the number of barrels produced?

Solution:
1. Barrels produced: 9 million   Barrels imported: 6 million
   Find: What percent of the number of barrels imported is the number of barrels produced?

2. Restate the problem.   What percent of 6 is 9?

3. Write an equation. $p \cdot 6 = 9$
   $6p = 9$

4. Solve the equation.
   $\frac{6p}{6} = \frac{9}{6}$
   $p = \frac{9}{6} = \frac{3}{2}$

5. Write a percent for $\frac{3}{2}$. $p = 1.50 = 150\%$

The number of barrels produced is **150%** of the number of barrels imported.

Self-Check 3. Karen borrowed $300 from Fred. He charged her $24 interest. What percent of the amount borrowed is the interest?

CLASSROOM EXERCISES

Write an equation for each percent problem. (Example 1, step 1)

1. What percent of 36 is 20?
2. What percent of 50 is 5?
3. What percent of 108 is 81?
4. What percent of 32 is 8?
5. 120 is what percent of 12?
6. 12 is what percent of 8?

Write an equation for each word problem. Do not solve. (Example 2, steps 1, 2, and 3)

7. Mike spent $4000 on his car last year. Of this amount, $200 was spent on tune-ups and oil change. What percent was spent on tune-ups and oil changes?

8. In May, the depth of water in a reservoir was 30 feet. In August, the depth had fallen 12 feet. What percent of the depth in May was the amount of decrease?

Percent and Applications **255**

WRITTEN EXERCISES

Goal: To determine what percent one number is of another

Write an equation for each percent problem. (Example 1, step 1)

1. What percent of 20 is 15?
2. What percent of 12 is 18?
3. What percent of 15 is 9?
4. What percent of 4 is 1?
5. 56 is what percent of 64?
6. 64 is what percent of 56?

Solve. (Example 1)

7. What percent of 40 is 15?
8. What percent of 100 is 250?
9. What percent of 30 is 75?
10. What percent of 75 is 60?
11. 16 is what % of 48?
12. 36 is what percent of 72?
13. What percent of 70 is 14?
14. What percent of 96 is 72?
15. What percent of 80 is 24?
16. What percent of 50 is 5?
17. 25 is what percent of 30?
18. 3 is what percent of 4?
19. 6 is what percent of 9?
20. 4 is what percent of 12?

Solve each problem. (Example 2)

21. In a survey, 45 out of 75 persons chose Daisy vacuum cleaners. What percent was this?
22. The fee for selling a $40,000 house is $4800. What percent of the sales price is the fee?
23. A concert by the Supersounds had an audience of 24,000 people. The next night, a concert by the same group drew a crowd of 30,000 people. Find what percent the smaller crowd was of the larger.
24. A salesperson received commission of $324 during one month. The sales for the month were $7200. Find what percent commission the salesperson receives.
25. The population of a city of 2,500,000 dropped by 400,000 in ten years. Find what percent of its original population the city lost.
26. The discount on a table listed at $95 is $19. What is the rate of discount?

REVIEW CAPSULE FOR SECTION 9-2

Write a percent for each fraction. (Pages 227–230)

1. $\frac{1}{4}$
2. $\frac{1}{5}$
3. $\frac{7}{10}$
4. $\frac{1}{3}$
5. $\frac{7}{8}$
6. $\frac{5}{6}$

Round each decimal to the nearest whole number. (Pages 47–49)

7. 6.05
8. 11.9
9. 7.28
10. 25.1
11. 18.55
12. 9.05

The answers are on page 286.

PROBLEM SOLVING AND APPLICATIONS

9-2 Percent of Increase and Decrease

You can save on the cost of energy during the warmer months of the year by keeping the thermostat of your air conditioner set at about 80°F.

The figure at the right shows the approximate percent of increase in energy costs when the thermostat of the air conditioner is set at 79°F, 78°F, 77°F, 76°F, and 75°F.

Temperature	Increase
79°	10% more
78°	19% more
77°	29% more
76°	40% more
75°	52% more

The **percent of increase** is the ratio of the amount of increase to the original amount.

PROCEDURE

To find the percent of increase:

1. Find the amount of increase.
2. Write this ratio in lowest terms: $\dfrac{\text{Amount of Increase}}{\text{Original Amount}}$
3. Write a percent for the ratio.

EXAMPLE 1 The Bassett family paid $120 in cooling costs for the month of June. During August, they lowered the thermostat setting by 1°. Cooling costs for August amounted to $132. Find the percent of increase.

Solution:
1. Amount of increase: $132 − 120 = $12

2. $\dfrac{\text{Amount of increase}}{\text{Original amount}}$: $\dfrac{12}{120}$, or $\dfrac{1}{10}$ ◀ Lowest terms

3. Percent of increase: Let $n =$ the first term.

$$\dfrac{n}{100} = \dfrac{1}{10}$$

$$10n = 100$$

$$n = 10 \quad \text{The increase is 10%.}$$

Self-Check 1. The Wilson's paid $140 in cooling costs last month. Their bill increased to $196 this month. Find the percent of increase.

Percent and Applications 257

You can save on heating costs during the cooler months of the year by lowering the thermostat setting to about 65°F.

The **percent of decrease** is the ratio of the amount of decrease to the original amount.

> **PROCEDURE**
>
> **To find the percent of decrease:**
>
> 1. Find the amount of decrease.
> 2. Write this ratio in lowest terms: $\dfrac{\text{Amount of Decrease}}{\text{Original Amount}}$
> 3. Write a percent for the ratio.

EXAMPLE 2 The Gadbois family paid $90 in heating costs last month. This month's bill was reduced by lowering the thermostat setting 3°. Heating costs for this month were $82. Find the percent of decrease (nearest whole percent).

Solution:
1. Amount of decrease: $90 − 82 = $8
2. $\dfrac{\text{Amount of decrease}}{\text{Original amount}}$: $\dfrac{8}{90}$, or $\dfrac{4}{45}$ ◀ **Lowest terms**
3. Percent of decrease: Let p = the first term.

$$\dfrac{p}{100} = \dfrac{4}{45}$$
$$45p = 400$$
$$p = 8.8$$ ◀ **Decrease: 9%**

Self-Check 2. The Lopez family paid $123 in heating costs during December. Their bill for January was $105. Find the percent of decrease (nearest whole percent).

CLASSROOM EXERCISES

Complete. (Example 1)

	Original Cost	Present Cost	Amount of Increase	$\dfrac{\text{Amount of Increase}}{\text{Original Amount}}$	Percent of Increase
1.	$100	$119	?	?	?
2.	$150	$210	?	?	?
3.	$ 80	$ 96	?	?	?
4.	$120	$156	?	?	?

(Example 2)

	Original Cost	Present Cost	Amount of Decrease	Amount of Decrease / Original Amount	Percent of Decrease
5.	$80	$68	?	?	?
6.	$100	$88	?	?	?
7.	$150	$141	?	?	?

WRITTEN EXERCISES

Goals: To find the percent of increase and decrease

To apply this skill to solving word problems

For Exercises 1–8, find the percent of increase. Round each answer to the nearest whole percent. (Example 1)

	Original Cost	Present Cost	Percent of Increase
1.	$110	$121	?
3.	$135	$150	?
5.	$140	$160	?

	Original Cost	Present Cost	Percent of Increase
2.	$125	$175	?
4.	$90	$137	?
6.	$180	$216	?

7. The Donaldson family paid $150 in cooling costs last month. This month, their bill was $179.

8. The Haas family paid $121 in cooling costs for the month of June. Cooling costs for the month of July amounted to $169.

For Exercises 9–16, find the percent of decrease. Round each answer to the nearest whole percent. (Example 2)

	Original Cost	Present Cost	Percent of Decrease
9.	$100	$91	?
11.	$80	$75	?
13.	$106	$87	?

	Original Cost	Present Cost	Percent of Decrease
10.	$140	$119	?
12.	$120	$106	?
14.	$200	$164	?

15. The Emanuel family paid $150 in heating costs one month. The next month, their heating costs amounted to $118.

16. The Caron family paid $89 in heating costs during January. Their bill for February was $78.

Percent and Applications

MIXED PRACTICE

Solve. Round each answer to the nearest whole percent.

17. The Stafford Company paid $330 in heating costs one month. The next month, after lowering the thermostat setting, their bill decreased to $290. Find the percent of decrease.

18. Beverly Kindell paid $104 in cooling costs for the month of July. For the month of August, her bill increased to $135. What was the percent of increase?

19. Don Tran paid $95 for cooling this month. Last month his thermostat was set at a higher temperature and his bill was $85. Find the percent of increase.

MORE CHALLENGING EXERCISES

The rate schedule at the right shows the cost for electricity in a certain area based on the number of kilowatt-hours used. Use this rate schedule and the information in the table at the right below to solve each problem.

Rate Schedule	
First 10 kwh or less	$2.79
Next 90 kwh	0.13 per kwh
Next 600 kwh	0.11 per kwh

Andros Family's Heating Costs		
Month	Kilowatt-hours Used	Cost
December	620	?
January	580	?
February	420	?

20. Find the Andros family's heating costs for the month of December.

21. Find the percent of decrease between the heating costs for January and the heating costs for February. Round the answer to the nearest whole percent.

22. The table shows the number of kilowatt-hours used this December. Last December, the Andros family used 570 kilowatt-hours. Using the same rate schedule, find the percent of increase between the heating costs for last December and the heating costs for this December. Round the answer to the nearest whole percent.

23. During March, the Andros family used their new fireplace. They spent $25 for wood and used 104 kilowatt-hours of electricity for the month. Find the percent of decrease between the heating cost for February and the total heating cost for March. Round the answer to the nearest whole percent.

REVIEW CAPSULE FOR SECTION 9-3

Write a decimal for each percent. (Pages 231–234)

1. 8% **2.** 1% **3.** 10% **4.** 80% **5.** 21% **6.** 19%

Divide. (Pages 44–46)

7. 26 ÷ 0.20 **8.** 84 ÷ 0.56 **9.** 26 ÷ 0.05 **10.** 45 ÷ 1.2

Solve each equation. (Pages 200–202)

11. $0.5r = 90$ **12.** $0.7n = 140$ **13.** $0.75t = 18$ **14.** $1.15q = 34.5$

The answers are on page 286.

9-3 Finding a Number Given a Percent

It is estimated that by 1991, Texas will have 2,600,000 more residents than it had in 1981. This number, 2,600,000, is 20% of the 1981 population. To find the population in 1981, follow this procedure.

Texas

PROCEDURE

To find a number when a percent of it is known:

1 Write an equation.
2 Solve the equation.

EXAMPLE 1   2,600,000 is 20% of what number?

Solution:
1 $2{,}600{,}000 = 20\% \cdot n$
   $2{,}600{,}000 = 0.20n$ Solve for *n*.
2 $\dfrac{2{,}600{,}000}{0.20} = \dfrac{0.20n}{0.20}$
   $13{,}000{,}000 = n$ Thus, 2,600,000 is 20% of **13,000,000**.

Self-Check **1.** 3 is 8% of what number?   **2.** 45 is 12% of what number?

The **wholesale price** is the price that a store pays for an item it will sell to the public. The public pays the **retail price** for these items.

Percent and Applications **261**

EXAMPLE 2 The wholesale price of a watch is $84.00. This is 58% of the retail price. Find the retail price. Round your answer to the nearest dollar.

Solution:

1. Wholesale price: $84.00; $84.00 is 58% of the retail price.
   Find: Retail price

2. Restate the problem.   84 is 58% of what number?

3. Write an equation. $84 = 58\% \cdot n$, or
                              $84 = 0.58n$

4. Solve the equation. $\dfrac{84}{0.58} = \dfrac{0.58n}{0.58}$

   $144.827 = n$ ◄ Round to the nearest dollar.

The retail price is about **$145.00**.

Self-Check 3. The wholesale price of a pair of running shoes is $32.80. This is 62% of the retail price. What is the retail price? Round your answer to the nearest dollar.

CLASSROOM EXERCISES

Write an equation for each problem. Do not solve the equation.
(Example 1, step 1)

1. 50 is 20% of n.
2. 4 is 80% of q.
3. 9 is 15% of t.
4. 10 is 15% of a.
5. 26 is 15% of p.
6. 45 is 120% of r.

Write an equation for each word problem. Do not solve.
(Example 2, steps 1, 2, and 3)

7. Kozo won 18 chess games. This was 45% of the games he played. How many games did he play?

8. Natalie received a raise of $1800. This is 10% of her present yearly salary. What is her present salary?

9. Cora Whitefeather spent 6 vacation days at the beach. This was 30% of her total number of vacation days. Find the number of vacation days.

10. A baseball player got a hit 32% of the times at bat. The player's record showed 24 hits. Find the number of times at bat.

11. Solve the equations you wrote for Exercises 7-10.

WRITTEN EXERCISES

Goal: To find a number when a percent of it is known.

Solve. (Example 1)

1. 20% of r is 15.
2. 30% of x is 21.
3. 42 is 21% of a.
4. 19 is 50% of n.
5. 42 is 25% of what number?
6. 6 is 15% of what number?
7. 9 is 2% of what number?
8. 93 is 10% of what number?
9. 16 is 80% of what number?
10. 27% of what number is 81?
11. 20% of what number is 19.5?
12. 16.8 is 15% of what number?
13. 54 is 45% of what number?
14. 204 is 85% of what number?
15. 65% of what number is 195?
16. 52% of what number is 234?

Solve each problem. (Example 2)

17. The wholesale price of a camera is $60. This is 55% of the retail price. Find the retail price. Round your answer to the nearest dollar.

18. A typewriter has a wholesale price of $355. This is 48% of the retail price. Find the retail price. Round your answer to the nearest dollar.

19. A carpenter spends 18% of his monthly take-home pay for rent. If he spends $350 a month for rent, what is his monthly take-home pay (to the nearest dollar)?

20. In the 1984 presidential elections, 53% of the population of voting age voted. About 93,000,000 people voted. Find the population of voting age that year. Round the answer to the nearest ten million.

MORE CHALLENGING EXERCISES

Solve Exercises 21–24. Write a fraction for each percent.

21. 7 is $16\frac{2}{3}$% of what number?
22. 50 is $33\frac{1}{3}$% of what number?
23. 80 is $12\frac{1}{2}$% of what number?
24. 12 is $66\frac{2}{3}$% of what number?

25. It takes $4\frac{1}{5}$ minutes to play the longest song on a record. This is 20% of the total time it takes to play the entire record. What is the total playing time of the record?

26. By 1991, it is projected that the population of Vermont will be 60,000 more than it was in 1981. The amount of increase is $12\frac{1}{2}$%. Find the 1981 population.

Percent and Applications

REVIEW CAPSULE FOR SECTION 9-4

Multiply. (Pages 38–41)

1. $39 \cdot 100$ **2.** $68 \cdot 100$ **3.** $10.8 \cdot 100$ **4.** $0.09 \cdot 100$ **5.** $45.5 \cdot 100$

Find each answer. (Pages 18–21)

6. $35 \times 84 \div 100$ **7.** $27 \times 4 \div 36$ **8.** $15 \times 32 \div 16$ **9.** $3 \times 28 \div 21$

Solve each proportion. (Pages 219-221)

10. $\dfrac{x}{30} = \dfrac{3}{5}$ **11.** $\dfrac{p}{10} = \dfrac{27}{15}$ **12.** $\dfrac{1}{3} = \dfrac{p}{15}$ **13.** $\dfrac{1}{4} = \dfrac{q}{28}$

The answers are on page 286.

9-4 Using Proportions to Solve Percent Problems

The three basic types of percent problems can also be solved by using proportions. To help you write the proportion, use this statement.

100% of a number is the number.

When writing a proportion, you will also find it helpful to write each percent problem in the same general form, $x\%$ of $a = b$.

Percent Problem	General Form: $x\%$ of $a = b$
1. A number is 5% of 120.	1. 5% of 120 is some number.
2. What percent of 28 is 42?	2. Some % of 28 is 42.
3. 45.5 is 91% of what number?	3. 91% of some number is 45.5.

To solve percent problems by the proportion method, follow this procedure.

PROCEDURE

To solve percent problems by using the proportion method:

[1] Write the problem in general form.
[2] Use a variable to represent the unknown.
[3] Write a statement using 100%.
[4] Use the statements in steps 2 and 3 to write a proportion.
[5] Solve the proportion.

EXAMPLE 1 A number is 5% of 120. Find the number.

Solution:

1	Write the problem in general form.	5% of 120 is some number.
2	Use a variable to represent the unknown.	5% of 120 is n.
3	Write a statement using 100%.	100% of 120 is 120.
4	Use the statements in steps 2 and 3 to write a proportion.	$\dfrac{5}{100} = \dfrac{n}{120}$ ◄ n corresponds to 5; 120 corresponds to 100.
5	Solve the proportion.	$(5)(120) = 100n$
		$600 = 100n$
		$6 = n$

Thus, 5% of 120 is **6**.

Self-Check 1. Use the proportion method to find 8% of 120.

Example 2 shows how to use the proportion method to solve the other two types of percent problems.

EXAMPLE 2 **a.** What percent of 28 is 42?   **b.** 45.5 is 91% of what number?

Solutions:

	a.	**b.**
1	Some % of 28 is 42.	91% of some number is 45.5.
2	p % of 28 is 42.	91% of t is 45.5.
3	100 % of 28 is 28.	100% of t is t.
4	$\dfrac{p}{100} = \dfrac{42}{28}$	$\dfrac{91}{100} = \dfrac{45.5}{t}$
5	$28p = (100)(42)$	$91t = (100)(45.5)$
	$28p = 4200$	$91t = 4550$
	$p = 150$	$t = 50$
	42 is **150**% of 28.	45.5 is 91% of **50**.

Self-Check *Use the proportion method to solve.*

2. What percent of 14 is 49?   3. 16 is 80% of what number?

Percent and Applications   **265**

CLASSROOM EXERCISES

Write each percent statement in general form. Use the variable n to represent the unknown. (Examples 1 and 2, Steps 1 and 2)

1. What number is 8% of 21?
2. 4 is 20% of what number?
3. 17 is what percent of 85?
4. What percent of 60 is 20?
5. 4 is 15% of what number?
6. Find 30% of 612.

For Exercises 7–14, first write a statement using 100%. Then write a proportion. Do NOT *solve.* (Examples 1 and 2, Steps 3 and 4)

7. 300% of k is 86.
8. w% of 4 is 1.
9. 128% of t is 45.
10. 5% of q is 1.55.
11. 3% of 126 is h.
12. 8% of 90 is g.
13. 15% of a is 10.
14. z% of 156 is 23.

WRITTEN EXERCISES

Goal: To solve percent problems by using proportions

For Exercise 1–32, use the proportion method to solve. (Example 1)

1. 25% of 28 is x.
2. A number is 10% of 261.
3. 12% of 90 is a certain number.
4. p is 8% of 75.
5. 21% of 500 = r.
6. Find 23% of 500.
7. What number is 1% of 512?
8. 13% of 651 is some number.

(Example 2)

9. r% of 92 is 69.
10. 30% of b is 21.
11. 12% of t is 3.
12. 30 is z% of 48.
13. 20 is 80% of b.
14. 14 is 40% of p.
15. 7 is 16% of what number?
16. What percent of 65 is 39?
17. What percent of 52 is 78?
18. 91.84 is 82% of what number?

MIXED PRACTICE

19. 42 is 84% of what number?
20. What number is 28% of 30?
21. What percent of 36 is 9?
22. A number is 2% of 72.
23. 28 is 5% of what number?
24. What percent of 102 is 56.1?
25. 102 is 30% of what number?
26. Find 42% of 510.

APPLICATIONS: SOLVING PERCENT PROBLEMS

Solve each problem.

27. In three hours, there were 80 landings at an air terminal. 68 of the planes landing were jets. What percent of planes landing were jets?

28. At 7 years of a age, a girl is about 74% of adult height. If a 7 year old girl is 50 inches tall, how tall can she expect to be as an adult? Round the answer to the nearest inch.

29. A certain cereal contains 12% protein. How many ounces of protein are there in a 26-ounce package of the cereal?

30. One season, a softball player got a hit 30% of her times at bat. She had 30 hits in all. How many times was she at bat?

31. The Blouin family spends 15% of total income on food each month. Last month, they spent $312 on food. What was their total income?

32. An appliance salesperson receives a 10% commission on sales. What is the commission on sales of $1500?

33. An architect charged a fee of $2250 for designing a house. The cost of the house was $90,000. What percent of the cost was the fee?

34. Lisa earns a 5% commission on sales. Last week, her commission was $35.50. What were her sales?

REVIEW: SECTIONS 9-1—9-4

Solve. (Section 9-1)

1. What percent of 30 is 27?
2. What percent of 75 is 90?
3. 25 is what percent of 60?
4. 42 is what percent of 30?
5. The cruising range of a car is 432 miles for city driving and 675 miles for highway driving. What percent of the highway driving range is the city driving range?
6. Warren spelled 30 words correctly on a 40-word spelling test. What percent of the words on the test did Warren correctly spell?

Percent and Applications

Find the percent of increase or decrease as indicated. Round each answer to the nearest whole percent. (Section 9-2)

7. Original Cost: $98
   Present Cost: $92
   Percent of Decrease: __?__

8. Original Cost: $85
   Present Cost: $129
   Percent of increase: __?__

Solve. Round each answer to the nearest whole percent. (Section 9-2)

9. The Ramirez family paid $115 last month for cooling. This month their cooling bill was $150. Find the percent of increase.

10. The Tobin family's heating cost for January was $92. After lowering the thermostat setting, their bill for February was $84. Find the percent of decrease.

Solve. (Section 9-3)

11. 18 is 20% of what number?
12. 18 is 60% of what number?
13. 60 is 75% of what number?
14. 4 is 2% of what number?
15. 25% of what number is 15?
16. 40% of what number is 25?

Solve. (Section 9-4)

17. A salesperson received a commission of $450 during one month. The sales for the month were $7500. Find what percent commission the salesperson receives.

18. A family budgets 25% of their take-home pay for food. The family's monthly take-home pay is $1200. How much is their monthly food budget?

REVIEW CAPSULE FOR SECTION 9-5

Complete. Write a percent for each fraction. (Pages 227-230)

1. $\frac{3}{4} =$ __?__
2. $\frac{2}{5} =$ __?__
3. $\frac{5}{6} =$ __?__
4. $\frac{1}{8} =$ __?__
5. $\frac{2}{3} =$ __?__

(Pages 227-230)

6. $\frac{85}{170} =$ __?__
7. $\frac{900}{3000} =$ __?__
8. $\frac{150}{400} =$ __?__
9. $\frac{355}{568} =$ __?__
10. $\frac{1800}{8000} =$ __?__

The answers are on page 286.

CONSUMER APPLICATION
Nutrition

The **RDA** (recommended daily allowance) indicates the total amount of vitamins, minerals, and protein that should be included in a daily diet. The table below shows the adult **RDA** for certain vitamins and minerals.

Item	Recommended Daily Allowance
Vitamin A	5000 International Units
Vitamin C	45 milligrams
Vitamin D	400 International Units
Calcium	800 milligrams
Phosphorus	800 milligrams

The four basic food groups are shown: protein products (upper left), carbohydrates (upper right), dairy products (lower left) and fruits and vegetables (lower right).

Exercises

Solve each problem. Refer to the table above for the adult RDA.

1. Find the amount of Vitamin A in 100 grams of milk when 100 grams of milk contain 3% of the RDA for Vitamin A.

2. Find the amount of calcium in 100 grams of milk when 100 grams of milk contain 15% of the total RDA for calcium.

3. One large apple contains 20% of the total RDA for Vitamin C. Find the amount of Vitamin C in an apple.

4. One tablespoon of butter contains 1¼% of the total RDA for Vitamin D. Find the amount of Vitamin D in one tablespoon of butter.

5. One large banana contains 16 milligrams of calcium. What per cent of the total RDA for calcium is this?

6. One cup of cooked green beans contains 0.9 milligrams of iron. This is 5% of the total RDA for iron. Find the total RDA.

7. One slice of enriched white bread contains 0.5 milligrams or 2.5% of the total RDA for niacin. Find the total RDA for niacin.

8. One cup of shredded carrots contains 12,000 International Units of Vitamin A. What per cent of the total RDA for Vitamin A is this?

9-5 Bar Graphs

A **bar graph** has two axes, a horizontal axis and a vertical axis. On a bar graph, the length of each bar represents a number. The **scale** on one of the axes tells you how to find the number.

A graph is one way to present information or **data**. This **bar graph** shows the number of cars sold during the month of July by four automobile manufacturers.

JULY AUTO SALES

(Vertical Axis — Number of Autos: 0 to 1200; Horizontal Axis — Auto Companies: W, X, Y, Z)

EXAMPLE 1 Use the bar graph above to answer each question.

Question	Solutions	
a. What was the total number of cars sold in July?	a. $600 + 700 + 1200 + 500 = \mathbf{3000}$	From the graph
b. What percent of the cars was sold by company W?	b. $\frac{600}{3000} = \frac{1}{5} = \mathbf{20\%}$	
c. What percent of the cars was sold by company Z?	c. $\frac{500}{3000} = \frac{1}{6} = \mathbf{16\frac{2}{3}\%}$	See the table on page 235
d. By what percent did the sales of company W exceed the sales of company Z?	d. $20\% - 16\frac{2}{3}\% = \mathbf{3\frac{1}{3}\%}$	

Self-Check

1. Use the bar graph above to find what percent of the cars was sold by company Y?

Since the graph shown above has vertical bars, it is called a **vertical bar graph**. Bar graphs may also be **horizontal**.

PROCEDURE

To construct a bar graph:

1. Draw the vertical and horizontal axes. Mark off a convenient scale at equal intervals on each axis. Label the axes.
2. Draw vertical or horizontal bars to represent the data. (The bars should be the same width and distance apart.)

EXAMPLE 2 Use the given data to construct a horizontal bar graph.

BUSINESS SCHOOL GRADUATES

Fields of Study	Number
Managerial	600
Technical	450
Business	250
Medical	400

Self-Check

2. Use the data in the table below to construct a horizontal bar graph.

Average July Temperatures

City	Phoenix	Chicago	Boston	Los Angeles	Cheyenne
Average Temperature (Degrees Celsius)	34°	24°	22°	22°	20°

CLASSROOM EXERCISES

For Exercises 1–4, refer to the bar graph below. (Example 1)

1. What was the total number of cats sold?

2. What percent of the cats was sold in November?

3. By what percent did the September sales exceed the October sales?

4. The percent of profit each month was in the same ratio as the percent of sales. The total profit for the four months was $3600. What was the amount of profit for the month of September?

Percent and Applications

WRITTEN EXERCISES

Goals: To read bar graphs

To construct bar graphs using data from tables

The horizontal bar graph at the right below shows the number of metric tons of wheat produced by 6 countries in a recent year.

Refer to this graph for Exercises 1-4. (Example 1)

1. What was the total number of metric tons of wheat produced by the six countries?

2. What percent of the wheat was produced by the United States?

3. What percent of the wheat was produced by Canada?

4. By what percent did the production of wheat by Canada exceed the production of wheat by Turkey?

Six Wheat Producing Countries in a Recent Year

(Bar graph showing: Canada ~23, China ~43, France ~16, India ~30, Turkey ~17, United States ~58; x-axis: Thousands of Metric Tons, 0 to 60)

This bar graph shows predicted energy sources for the United States in 1985.

Refer to this graph for Exercises 5-8.

5. What is the total number of predicted energy source units for the United States in 1990?

6. What percent of our predicted energy will come from coal?

7. What percent of our predicted energy source will come from coal and gas?

8. By what percent does the predicted energy source of gas exceed the predicted energy source of hydropower?

Predicted Energy Sources for the United States in 1990

(Bar graph showing: Coal ~23, Gas ~21, Hydro-Power ~4, Nuclear ~7, Oil ~28; x-axis: Quadrillion Units of Energy, 0 to 30; y-axis: Energy Sources)

272   Chapter 9

For Exercises 9–12. use the given data to construct a bar graph. (Example 2)

9. Book Sales

Month	Books Sold
January	150
February	200
March	300
April	400
May	175
June	500

10. Land Use in a Town

Use	Per Cent
Roads	25%
Railroad	3%
Public Use	12%
Farm	8%
Housing	52%

11. Approximate Heights in Meters of Six Waterfalls

Waterfall	Height
Yosemite Lower Falls	100
Yosemite Upper Falls	435
Nevada Falls	180
Great Falls	255
Ribbon Falls	490
Silver Strand Falls	360

12. Approximate Yearly Amount of Food Consumed by the Average Person

Food	Pounds
Beef	75
Poultry	100
Fruit	130
Vegetables	150
Grains	130
Dairy	285

REVIEW CAPSULE FOR SECTION 9-6

For Exercises 1–10, write a decimal for each percent. (Pages 231–234)

1. 40%
2. 65%
3. 8%
4. 10%
5. $17\frac{1}{2}\%$
6. 7%
7. 81%
8. $12\frac{3}{4}\%$
9. 92%
10. $9\frac{1}{2}\%$

For Exercise 11–16:

    a. *Write an equation for the problem.*
    b. *Solve the equation.* (Pages 235–239)

11. What is 10% of 516?
12. Find 45% of 200.
13. What is $66\frac{2}{3}\%$ of 90?
14. Find 60% of 75.
15. What is $37\frac{1}{2}\%$ of 400?
16. Find $83\frac{1}{3}\%$ of 144.

The answers are on page 286.

9-6 Circle Graphs

Circle graphs are often used to show data given as percents. The entire circle represents 100%.

The circle graph below shows what percent of each type of vehicle was serviced at a station during a certain day.

Vehicles Serviced in a Day

- Full-size Cars 25%
- Compact Cars 20%
- Pickups 5%
- Motorcycles 10%
- Vans 18%
- Trucks 22%

EXAMPLE 1 The circle graph above represents a total of 500 vehicles. Find how many of these vehicles were compact cars.

Solution:
1. Percent of compact cars: 20%
2. Write an equation. What is 20% of 500?
   $n = 0.20 \, (500)$
3. Solve the equation. $n = 100$

Thus, **100** of the **vehicles** serviced were compact cars.

Self-Check

1. Refer to the circle graph above to find how many of the 500 vehicles serviced were pickups.

To construct a circle graph, you use a protractor.

PROCEDURE

To construct a circle graph:

1. Write a decimal for each percent.
2. Multiply the decimal by 360°. Round the answer to the nearest degree.
3. Use a protractor to construct the graph.

To construct a circle graph, you use this fact from geometry.

There are 360° in a circle.

EXAMPLE 2 This table shows the percent of each of four types of tissues in the human body. Construct a circle graph to show the data.

Body Tissue

Type	Percent
Muscle	47%
Supporting	33%
Blood	7%
Surface	13%

Solution:

[1] Write a decimal for each percent.

$47\% = 0.47$ $\qquad 33\% = 0.33$
$13\% = 0.13$ $\qquad 7\% = 0.07$

[2] Multiply each decimal by 360°. Round to the nearest degree.

$0.47 \times 360° = 169.2$, or 169° $\qquad 0.33 \times 360° = 118.8$, or 119°
$0.13 \times 360° = 46.8$, or 47° $\qquad 0.07 \times 360° = 25.2$, or 25°

[3] Construct the graph.

Draw a radius. Place the protractor on the radius. Draw the angle for the muscle tissue.

Place the protractor on the "new" radius. Draw the angle for supporting tissue.

Place the protractor on the "new" radius. Draw the angle for surface tissue.

The remaining angle represents the blood tissue.

Self-Check 2. Construct a circle graph for the following data. Round each angle measure to the nearest degree.

Popular Vacation Cities

City	New York	Los Angeles	San Francisco	Washington, D.C.	Orlando
Percent of Total	37%	19%	17%	14%	13%

Percent and Applications **275**

CLASSROOM EXERCISES

The circle graph at the right represents the total sales of gasoline and other items at a service station for one day. The total sales were $900.

Refer to the graph to complete Exercises 1-5.
(Example 1)

Sales for One Day

- Unleaded 42%
- Premium 18%
- Oil 5%
- Regular 20%
- Miscellaneous 15%

	Item	Percent of Total Sales	Related Equation	Dollar Amount Sold
1.	Unleaded Gas	42%	$n = 0.42(900)$	?
2.	Premium Gas	18%	$n = 0.18(900)$	?
3.	Regular Gas	20%	?	?
4.	Oil	?	?	?
5.	Miscellaneous	?	?	?

Find the decimal equivalent of the percent and the number of degrees needed to represent each percent on a circle graph. Round each answer to the nearest degree. (Example 2, Steps 1 and 2)

	Percent	Decimal	Degrees
6.	30%	0.3	?
8.	42%	?	?
10.	8%	?	?

	Percent	Decimal	Degrees
7.	25%	0.25	?
9.	18%	?	?
11.	21%	?	?

12. Construct a circle graph for the following data. (Example 2)

Kinds of Farms in the United States

Farm	Grain	Cattle	Produce	Poultry	Other
Percent	34%	41%	12%	3%	10%

WRITTEN EXERCISES

Goals: To read and interpret circle graphs

To construct circle graphs

The circle graph at the right shows how a school club raised $3000 for a special project. Refer to this graph for Exercises 1-4. (Example 1)

Funds Raised by Math Club
(Car Wash 10%, Concert 20%, Fruit Sale 40%, Newspaper Sales 30%)

1. How much did the club earn selling newspapers?
2. How much did the club earn from the concert?
3. How much did the club earn washing cars and selling newspapers?
4. How much greater was the income from the fruit sale than the income from the car wash?

The circle graph at the right shows the percent of each type of book found in a public library. The collection consists of 4000 books. Refer to this graph for Exercises 5-8. (Example 1)

Books in a Library
(Fiction 45%, History 25%, Science 10%, Reference 8%, Other 12%)

5. How many are history books?
6. How many are reference books?
7. How many are either science or fiction books?
8. What is the difference between the number of fiction and the number of history books?

Use the given data to construct a circle graph. Round each angle measure to the nearest degree. (Example 2)

9.

City Budget

Item	Schools	Public Works	Health and Safety	Other
Percent	35%	15%	40%	10%

Percent and Applications **277**

10. United States Energy Sources

Source	Percent
Oil	45%
Natural Gas	25%
Coal	21%
Nuclear	4%
Water	5%

11. Causes of Forest Fires

Cause	Percent
Lightning	34%
Children	17%
Campers	11%
Auto Passengers	10%
Other	28%

12. Uses of Paper

Use	Percent
Packaging	48%
Writing Paper	29%
Tissues	7%
Other	16%

13. Garbage Recycled

Year	Percent
1976	22%
1977	23%
1980	25%
1985	30%

REVIEW CAPSULE FOR SECTION 9-7

Round to the nearest whole number. (Pages 47-49)

1. 58.75
2. 16.50
3. 11.21
4. 32.95
5. 70.55
6. 0.085
7. 4.38
8. 5.499
9. 6.3
10. 19.6
11. 0.927
12. 69.5

Round to the nearest ten. (Pages 47–49)

13. 58.75
14. 16.50
15. 11.21
16. 32.95
17. 70.55
18. 83.05
19. 39.5
20. 46.81
21. 50.05
22. 86.91
23. 70.62
24. 78.62

Multiply. (Pages 129–131)

25. $\frac{1}{10} \times 900$
26. $\frac{1}{8} \times 512$
27. $\frac{1}{6} \times 426$
28. $\frac{1}{12} \times 1560$
29. $\frac{2}{5} \times 75$
30. $\frac{2}{3} \times 960$
31. $\frac{1}{4} \times 188$
32. $\frac{5}{8} \times 200$
33. $\frac{3}{10} \times 900$
34. $\frac{5}{6} \times 900$

Write a fraction for each percent. (Pages 235-239)

35. 75%
36. 40%
37. $12\frac{1}{2}\%$
38. $66\frac{2}{3}\%$
39. $16\frac{2}{3}\%$

The answers are on page 286.

PROBLEM SOLVING AND APPLICATIONS

9-7 Using Estimation

To estimate a percent of a number, you can sometimes round the number only.

EXAMPLE 1 A car radio with a list price of $78.69 is on sale at a 25% discount. Estimate the amount of discount.

Solution:
1. $25\% = \frac{1}{4}$

2. Round to a <u>convenient</u> price; that is, to a price that is close to $78.69 <u>and</u> is easy to multiply by $\frac{1}{4}$.

3. Estimated discount: $\frac{1}{4} \cdot 80 = 20$

◄ The amount of discount is about $20.

Self-Check 1. A sport coat with a list price of $62.25 is on sale at a 20% discount. Estimate the amount of discount.

Sometimes you can round to a convenient percent.

EXAMPLE 2 The Athlete's Shoe Shop has running shoes on sale at a 38% discount. Estimate the amount of discount on the running shoes if the list price is $45.

Solution:
1. Round to a <u>convenient</u> percent; that is, to a percent close to 38% and easy to multiply by 45.

   38% is about 40%. $\quad 40\% = \frac{2}{5}$

2. Estimated discount: $\frac{2}{\cancel{5}} \cdot \cancel{45}^{9} = 18$

◄ The amount of discount is about $18.

Self-Check 2. A canoe with a list price of $200 is on sale at a 18% discount. Estimate the amount of discount on the canoe.

It is sometimes useful to estimate what percent a number is of another.

Percent and Applications **279**

EXAMPLE 3 The total cost of repairs on Mario's car was $58. The cost of parts alone was $15. Estimate what percent of the total cost was the cost of parts.

Solution:

1 Restate the problem.   What percent of 58 is 15?

2 Think: 15 is about $\frac{1}{4}$ of 58.   ◀ 58 is about 60 and 15 is $\frac{1}{4}$ of 60.

3 Therefore, 15 is about **25%** of 58.

Self-Check 3. Of the 150 students in a school band, 29 play woodwind instruments. Estimate what percent play woodwind instruments.

CLASSROOM EXERCISES

Choose the best estimate. Choose a, b, or c.   (Example 1)

1. 25% of $58.56 **a.** 25% of $50 **b.** 25% of $55 **c.** 25% of $60
2. 10% of $29.95 **a.** 10% of $20 **b.** 10% of $25 **c.** 10% of $30
3. 50% of $91.30 **a.** 50% of $80 **b.** 50% of $90 **c.** 50% of $100
4. 20% of $34.75 **a.** 20% of $35 **b.** 20% of $30 **c.** 20% of $40
5. 30% of $10.06 **a.** 30% of $20 **b.** 30% of $15 **c.** 30% of $10

(Example 2)

6. 21% of $150 **a.** 30% of $150 **b.** 25% of $150 **c.** 20% of $150
7. 9.5% of $700 **a.** 5% of $700 **b.** 10% of $700 **c.** 20% of $700
8. 26% of $400 **a.** 20% of $400 **b.** 25% of $400 **c.** 30% of $400
9. 33% of $90 **a.** $33\frac{1}{3}$% of $90 **b.** 30% of $90 **c.** 40% of $90
10. 12% of $24 **a.** 20% of $24 **b.** $16\frac{2}{3}$% of $24 **c.** $12\frac{1}{2}$% of $24

(Example 3)

11. What percent of 73 is 12? **a.** 10% **b.** $16\frac{2}{3}$% **c.** 25%
12. What percent of 41 is 8? **a.** 10% **b.** 15% **c.** 20%
13. What percent of 14 is 3? **a.** 10% **b.** 20% **c.** 30%
14. What percent of 68 is 23? **a.** 50% **b.** 40% **c.** $33\frac{1}{3}$%
15. What percent of 95 is 46? **a.** 40% **b.** 50% **c.** 60%
16. 58 is what percent of 79? **a.** $66\frac{2}{3}$% **b.** 75% **c.** 80%
17. 39 is what percent of 60? **a.** 40% **b.** 60% **c.** 50%

WRITTEN EXERCISES

Goal: To use estimation to find the percent of a number

To use estimation to find what percent a number is of another

Choose the best estimate. Choose a, b, or c. (Example 1)

1. 50% of $121.90   a. $60   b. $65   c. $70
2. 20% of $98.50   a. $15   b. $20   c. $25
3. 25% of $20.98   a. $5   b. $6   c. $7
4. 75% of $101.17   a. $65   b. $85   c. $75
5. $33\frac{1}{3}$% of $134.90   a. $33   b. $50   c. $45

(Example 2)

6. 19% of $80   a. $8   b. $16   c. $20
7. 8.5% of $40   a. $4   b. $2   c. $8
8. 41% of $1600   a. $800   b. $480   c. $640
9. 49% of $500   a. $200   b. $250   c. $300
10. 24.2% of $60   a. $15   b. $18   c. $12

Estimate each percent. (Example 3)

11. What percent of 83 is 20?
12. What percent of 50 is 9?
13. What percent of 73 is 15?
14. What percent of 88 is 9?
15. What percent of 130 is 12?
16. What percent of 60 is 5.5?

Choose the best estimate for each discount. Choose a, b, or c.

17. **SALE!** 20% off **LAWN CHAIR** REGULARLY—$19.50
   a. $4   b. $15   c. $20

18. **Sale!** 40% off **Tennis Racquet** REGULARLY $49.95
   a. $18   b. $24   c. $20

19. **SALE!** Snow Tires 9% Off REGULARLY $39.95
   a. $4   b. $5   c. $6

Percent and Applications

Solve each problem.

20. A quarterback completed 10 of 28 passes in a football game. Estimate what percent of the passes he completed.

21. A reporter spends $300 a month for rent. Her monthly salary is $1235. Estimate what percent of her monthly salary she spends on rent.

22. A rain coat with a list price of $91.20 is on sale at a 20% discount. Estimate the amount of discount.

23. A tape deck with a list price of $47.88 is on sale at a $33\frac{1}{3}$% discount. Estimate the amount of discount.

24. In a class of 32 students, 5 students were absent. Estimate what percent of the class was absent.

25. A tennis player made 32 of 61 first serves. Estimate what percent of first serves were made.

FINDING DISCOUNT

Before using a calculator to find the amount of discount:

1. Estimate the answer.
2. Use the calculator to find the exact answer.
3. Compare the exact answer with the estimate to see whether the exact answer is reasonable.

EXAMPLE: Regular Price: $300.00   Rate of Discount: 9.8%   Discount: __?__

Solution: Estimated Discount: $\frac{1}{10} \times \$300 = \$30$ ◀ 9.8% is about 10%.

Exact Discount: 300 × 9.8%

3 0 0 [×] 9 . 8 [%] [=]   →   29.4

Since the estimate is $30, **$29.40** is a reasonable answer.

EXERCISES

First estimate the discount. Then find the exact discount.

	Regular Price	Rate of Discount	Estimated Discount	Exact Discount
1.	$800.00	26%	?	?
2.	$401.50	20%	?	?
3.	$ 87.90	25%	?	?
4.	$996.85	30%	?	?
5.	$ 25.00	8.5%	?	?

REVIEW: SECTIONS 9-5—9-7

1. Use the data below to construct a horizontal bar graph. (Section 9-5)

Renell High School Enrollment

Year	1976	1977	1978	1979	1980
Number of Students	1030	1020	1095	1090	1075

The circle graph at the right shows how a library club raised $2000. Refer to this graph for Exercises 2–4. (Section 9-6)

Funds Raised by Library Club

- Car Washes 15%
- Collecting Newspapers 40%
- Selling Calendars 5%
- Bake Sale 10%
- Magazine Sales 30%

2. How much did the library club earn by selling calendars?

3. How much did the library club earn by washing cars?

4. By how much did the amount received by collecting newspapers exceed the amount made by selling magazines?

5. Use the data below to construct a circle graph. (Section 9-6)

Population of a City by Age

Age	65 and over	45–64	20–44	19 and under
Percent	11%	20%	35%	34%

Choose the best estimate. Choose a, b, or c. (Section 9-7)

6. 20% of $61.01   a. $12   b. $10   c. $14
7. 40% of $148.71   a. $80   b. $70   c. $60
8. 58% of $91.20   a. $48   b. $54   c. $60
9. 19% of $50   a. $1   b. $5   c. $10

Estimate what percent a number is of another. (Section 9-7)

10. What percent of 75 is 24?

11. What percent of 49 is 8?

12. The regular price of a television set is $510.00. The set is marked at 25% off. Estimate the amount of discount. (Section 9-7)

13. A warm-up suit with a list price of $61.50 is on sale at a discount of 40%. Estimate the amount of discount. (Section 9-7)

Percent and Applications

CHAPTER REVIEW

PART 1: VOCABULARY

For Exercises 1–5, choose from the box at the right the word(s) that best corresponds to each description.

1. The ratio of the amount of decrease to the original amount __?__ (p. 258)
2. The amount that the public pays for an item __?__ (p. 261)
3. A graph is a method of presenting __?__ (p. 270)
4. The amount that a store pays for an item it sells to the public __?__ (p. 261)
5. The ratio of the amount of increase to the original amount __?__ (p. 256)

> data
> retail price
> percent of increase
> percent of decrease
> wholesale price

PART 2: SKILLS

Solve. (Section 9–1)

6. What percent of 16 is 10?
7. What percent of 10 is 6?
8. What percent of 16 is 24?
9. What percent of 35 is 30?
10. What percent of 30 is 12?
11. What percent of 72 is 9?

Find the percent of increase or decrease. Identify each answer as a percent of increase or decrease. Round each answer to the nearest whole percent. (Section 9–2)

	Original Cost	Present Cost
12.	$100	$120
14.	$130	$104
16.	$250	$322
18.	$174	$156

	Original Cost	Present Cost
13.	$75	$89
15.	$112	$109
17.	$97	$74
19.	$150	$212

Solve. (Section 9–3)

20. 14 is 40% of what number?
21. 42 is 35% of what number?
22. 17 is 85% of what number?
23. 9 is 15% of what number?
24. 80% of what number is 56?
25. 36% of what number is 18?

(Section 9-4)

26. 35 is what percent of 50?
27. What number is 60% of 315?
28. 75 is 30% of what number?
29. 20 is what percent of 200?
30. What number is 11% of 100?
31. 64 is 25% of what number?

For Exercises 32-34, refer to the graph below.
(Section 9-5)

32. What was the total number of yearbooks sold?
33. What percent was sold by the eighth grade?
34. By what percent did the seventh grade yearbook sales exceed the eighth grade yearbook sales?

Yearbook Sales

(Section 9-6)

35. Use this data to construct a circle graph.

Area of World's Oceans

Ocean	Pacific	Atlantic	Indian	Arctic	Other
Per Cent	46%	23%	20%	4%	7%

Choose the best estimate. Choose a, b, or c. (Section 9-7)

36. 50% of $23.85   a. $10   b. $12   c. $15
37. 25% of $198.50   a. $50   b. $40   c. $30
38. 21% of $50   a. $8   b. $10   c. $12
39. 32% of $3300   a. $700   b. $900   c. $1100

Estimate each percent. (Section 9-7)

40. What percent of 98 is 24?
41. What percent of 149 is 50?
42. What percent of 62 is 30?
43. What percent of 80 is 7?

PART 3: APPLICATIONS

44. The Pelletier family budgets $480 per month for food. The family's monthly income is $1920. What percent of the monthly income is the amount budgeted for food? (Section 9-1)

45. The Haber Family paid $109 for cooling in June. After lowering the thermostat setting, their bill increased to $119.90 in July. Find the percent of increase. (Section 9-2)

Percent and Applications

46. The Las–Tech Corporation's heating costs last month were $526. After lowering the thermostat setting, the heating costs this month are $468.14. Find the percent of decrease. (Section 9–2)

47. The wholesale price of a gold watch is $150. This is 60% of the retail price. What is the retail price? (Sections 9–3 and 9–4)

48. A stereo with a list price of $795 is on sale at a 25% discount. Estimate the amount of discount. (Section 9–7)

ANSWERS TO REVIEW CAPSULES

Page 256   **1.** 25%   **2.** 20%   **3.** 70%   **4.** $33\frac{1}{3}$%   **5.** $87\frac{1}{2}$%   **6.** $83\frac{1}{3}$%   **7.** 6   **8.** 12   **9.** 7   **10.** 25   **11.** 19   **12.** 9

Page 261   **1.** 0.08   **2.** 0.01   **3.** 0.10   **4.** 0.80   **5.** 0.21   **6.** 0.19   **7.** 130   **8.** 150   **9.** 520   **10.** 37.5   **11.** 180   **12.** 200   **13.** 24   **14.** 30

Page 264   **1.** 3900   **2.** 6800   **3.** 1080   **4.** 9   **5.** 4550   **6.** 29.4   **7.** 3   **8.** 30   **9.** 4   **10.** 18   **11.** 18   **12.** 5   **13.** 7

Page 268   **1.** 75%   **2.** 40%   **3.** $83\frac{1}{3}$%   **4.** $12\frac{1}{2}$%   **5.** $66\frac{2}{3}$%   **6.** 50%   **7.** 30%   **8.** $37\frac{1}{2}$%   **9.** $62\frac{1}{2}$%   **10.** $22\frac{1}{2}$%

Page 273   **1.** 0.40   **2.** 0.65   **3.** 0.08   **4.** 0.10   **5.** 0.175   **6.** 0.07   **7.** 0.81   **8.** 0.1275   **9.** 0.92   **10.** 0.095   **11.** $n = 0.10 \cdot 516$; 51.6   **12.** $n = 0.45 \cdot 200$; 90   **13.** $n = \frac{2}{3} \cdot 90$; 60   **14.** $n = 0.60 \cdot 75$; 45   **15.** $n = \frac{3}{8} \cdot 400$; 150   **16.** $n = \frac{5}{6} \cdot 144$; 120

Page 278   **1.** 59   **2.** 17   **3.** 11   **4.** 33   **5.** 71   **6.** 0   **7.** 4   **8.** 5   **9.** 6   **10.** 20   **11.** 1   **12.** 70   **13.** 60   **14.** 20   **15.** 10   **16.** 30   **17.** 70   **18.** 80   **19.** 40   **20.** 50   **21.** 50   **22.** 90   **23.** 70   **24.** 80   **25.** 90   **26.** 64   **27.** 71   **28.** 130   **29.** 30   **30.** 640   **31.** 47   **32.** 125   **33.** 270   **34.** 750   **35.** $\frac{3}{4}$ **36.** $\frac{2}{5}$ **37.** $\frac{1}{8}$ **38.** $\frac{2}{3}$ **39.** $\frac{1}{6}$

CHAPTER TEST

Solve.

1. What number is 32% of 25?
2. What is 45% of 780?
3. 16 is 80% of what number?
4. 27 is what percent of 90?
5. 7 is 50% of what number?
6. What percent of 84 is 14?
7. 64 is 25% of what number?
8. 35 is what percent of 50?

Complete. Round each answer to the nearest whole percent.

9. Original Cost: $150
   Present Cost: $179
   Percent of Increase: __?__

10. Original Cost: $104
    Present Cost: $94
    Percent of Decrease: __?__

11. In last week's football game, Philip threw 20 passes. He completed 6 of them. What percent did he complete?

12. Theresa bought a new car at a 12% discount. The regular price was $9600. How much money did she save on the purchase?

13. Luis earned $30 on Monday. This was 20% of his weekly earnings. What does he earn per week?

14. The table at the right shows the mileage from San Francisco to five cities in the United States. Use the data from the table to construct a vertical bar graph.

15. Company A has an advertising budget of $8000. Refer to the data below to construct a circle graph.

Mileage from San Francisco

City	Miles
To Los Angeles	250
To Denver	1300
To Dallas	1800
To Chicago	2200
To New York City	3000

Advertising Budget

Item	Newspaper	TV	Radio	Other
Percent	15%	55%	20%	10%

Percent and Applications

ENRICHMENT

Sequences

Suppose a new sports car is worth $25,600. This table shows how the value of the car depreciates (lessens) for each of four years.

New	After One Year	After Two Years	After Three Years	After Four Years	After Five Years
$25,600	$19,200	$14,400	$10,800	$8100	?

The pattern of numbers shown in the table is a **sequence**. Each number in the sequence is three-fourths of the preceding number. Thus, to find the value of the car after five years, multiply the value after four years by $\frac{3}{4}$.

$$\overset{2025}{\cancel{8100}} \times \frac{3}{\underset{1}{\cancel{4}}} = 6075$$

◀ Value after 5 years: $6075

EXAMPLES   Find the next three numbers in each sequence.

  a. 1, 4, 7, 10, 13, · · ·   b. 1, 3, 9, 27, · · ·

Solutions: a. Think: Each number is <u>3 more than</u> the preceding number.

  1, 4, 7, 10, 13, **16, 19, 22,** · · ·

  b. Think: Each number is <u>3 times</u> the preceding number.

  1, 3, 9, 27, **81, 243, 729,** · · ·

EXERCISES

Write the next three numbers in each sequence.

1. 1, 3, 5, 7, · · ·   **2.** 5, 10, 15, 20, · · ·   **3.** 2, 8, 32, 128, · · ·

4. 12, 60, 300, 1500, · · ·   **5.** 256, 128, 64, 32, · · ·   **6.** 2187, 729, 243, 81, · · ·

Each number in this **Fibonacci sequence** is the sum of the two preceding numbers.

  1, 1, 2, 3, 5, 8, 13, · · ·

Write the next three numbers in each sequence.

7. 2, 3, 5, 8, 13, · · ·   **8.** 4, 4, 8, 12, 20, · · ·   **9.** 4, 9, 13, 22, · · ·

ADDITIONAL PRACTICE

SKILLS

Solve. (Pages 254–256)

1. What percent of 40 is 20?

2. What percent of 72 is 12?

Complete the table. Round each answer to the nearest whole percent. (Pages 257–260)

	Original Cost	Present Cost	Percent of Increase
3.	$40	$60	?
5.	$57	$76	?

	Original Cost	Present Cost	Percent of Decrease
4.	$12	$ 6	?
6.	$72	$60	?

Solve. (Pages 261–263)

7. 12 is 25% of x.

8. 6 is 12% of n.

9. 108.8 is 16% of y.

Use the given data to construct a bar graph. (Pages 270–273)

10. Car Sales

Month	Number Sold
July	5285
August	4950
September	4800
October	6350

Use the given data to construct a circle graph. (Pages 274–278)

11. Boutique Sales

Item	Percent
Suits	40%
Dresses	35%
Skirts	15%
Shoes	10%

Choose the best estimate. Choose a, b, or c. (Pages 279–281)

12. 24% of 160   **a.** 40   **b.** 20   **c.** 60

13. What percent of 41 is 8?   **a.** 50%   **b.** 20%   **c.** 15%

14. 49% of 281   **a.** 150   **b.** 130   **c.** 140

APPLICATIONS

15. Ted scored 180 points during last year's basketball season. This is 15% of the total number of points his team scored. How many points did Ted's team score? (Pages 261–263)

16. René was earning $3.80 per hour as a cook in a restaurant. After working for one year, her wage was increased to $4.75 per hour. Find the percent of increase. (Pages 257–260)

COMMON ERRORS

Each of these problems contains a common error.

a Find the correct answer.

b Find the error.

Solve each equation.

1. $x + 7 = 18$
   $x + 7 - 7 = 18 + 7$
   $x = 25$

2. $y - 5 = 20$
   $y - 5 + 5 = 20$
   $y = 20$

3. $3x = 6$
   $\dfrac{3x}{6} = \dfrac{6}{6}$
   $x = 1$

Solve each problem.

4. Write a proportion.
   2 apples for 68¢
   6 apples for x cents
   $\dfrac{2}{68} = \dfrac{x}{6}$

5. $\dfrac{t}{20} = \dfrac{5}{6}$; $t = $ ___?___
   $\dfrac{t}{20} = \dfrac{5}{6}$
   $5t = 120$
   $t = 24$

6. Write a percent for $\dfrac{2}{3}$.
   $\dfrac{2}{3} = \dfrac{100}{n}$
   $2n = 300$
   $n = 150$ ◀ $\dfrac{2}{3} = 150\%$

7. What number is 5% of 120?
   $n = 0.5(120)$
   $n = 60$

8. What percent of 16 is 48?
   $p \cdot 48 = 16$
   $48p = 16$
   $p = \dfrac{1}{3}$, or $33\dfrac{1}{3}\%$

9. 36 is 12% of what number?
   $36 = 12 \cdot p$
   $36 = 12p$
   $3 = p$

10. Original cost: $80
    Present cost: $60
    Percent of decrease: ___?___
    $\dfrac{n}{100} = \dfrac{60}{80}$
    $80n = 6000$
    $n = 75$ ◀ Percent of Decrease: 75%

11. Original cost: $160
    Present cost: $200
    Percent of increase: ___?___
    $\dfrac{p}{100} = \dfrac{40}{200}$
    $200p = 4000$
    $p = 20$ ◀ Percent of Increase: 20%

CUMULATIVE REVIEW: CHAPTERS 7–9

Choose the correct answer. Choose a, b, c, or d.

1. Solve for c: $13.7 = c + 8.9$
   a. 21.6 b. 22.6 c. 5.2 d. 4.8

2. Solve for y: $y - 24 = 33$
   a. 9 b. 57 c. 65 d. 14

3. Last month, Frank's gross earnings were $1158.45. The total deductions were $283.14. Use the formula $n + d = g$ to find the net pay.
   a. $1341.59 b. $975.31 c. $875.31 d. $935.31

4. A plane is flying into a head wind of 35 kilometers per hour. The ground speed is 477 kilometers per hour.
   Use the formula $a - h = g$ to find the air speed.
   a. 442 b. 502 c. 412 d. 512

5. Solve for t: $11t = 132$
   a. 121 b. 127 c. 12 d. 143

6. Solve for n: $\frac{n}{2} = 1\frac{2}{3}$
   a. $\frac{5}{12}$ b. $3\frac{1}{3}$ c. $\frac{1}{3}$ d. $3\frac{2}{3}$

7. A plane flies the 3000 miles from New York to San Francisco in $3\frac{3}{4}$ hours. Find the average speed.
   a. 495 b. 925 c. 750 d. 800

8. Write this fraction in lowest terms: $\frac{16x}{50x}$
   a. $\frac{8x}{25}$ b. $\frac{8}{25}$ c. $\frac{4x}{25}$ d. $\frac{4}{25}$

9. Solve for m: $\frac{9}{27} = \frac{m}{6}$
   a. 3 b. 2 c. 4 d. 5

10. On a map of California, 1 centimeter equals 40 kilometers. How many kilometers do 4.2 centimeters represent?
    a. 16.8 b. 1.68 c. 168 d. 1680

11. The Evans family drove 368 miles on 16 gallons of gas. How many miles could they drive on 7 gallons?
    a. 181 b. 161 c. 171 d. 191

12. Write a percent for $\frac{9}{8}$.
  a. $112\frac{1}{2}\%$ b. 108%   c. 109%   d. $110\frac{1}{2}\%$

13. Write a decimal for 63%.
  a. 0.063   b. 0.63   c. 6.3   d. 630

14. On a math test, Rene answered 75% of the 80 questions asked. How many questions did Rene answer?
  a. 65   b. 55   c. 70   d. 60

15. The Garbey family invested $1400 at a yearly interest rate of 10.5%. How much interest will the account earn in 6 months?
  a. $73.50   b. $147.00   c. $95.75   d. $355.00

16. A camera sells at a regular price of $60. It is on sale at a 20% discount. Find the sale price.
  a. $48.80   b. $48   c. $45   d. $58.80

17. What percent of 20 is 4?
  a. 25%   b. 15%   c. 40%   d. 20%

18. 27 is 30% of what number?
  a. 8.1   b. 81   c. 900   d. 90

19. Thirty of forty club members attended a meeting. What percent of the members attended the meeting?
  a. 75%   b. 65%
  c. 70%   d. 80%

20. Use the bar graph at the right to estimate the percent of increase between the books sold in January and the books sold in April.
  a. 40%   b. 45%
  c. 3.9%   d. 4.5%

21. In the circle graph at the right, how many degrees represent the section "Ages 12–14"?
  a. 108°   b. 30°
  c. 100°   d. 106°

22. Choose the best estimate: 24.7% of 1019.5
  a. 250   b. 260   c. 350   d. 200

CHAPTER 10
Rational Numbers: Addition/Subtraction

SECTIONS
10-1 Integers
10-2 Rational Numbers
10-3 Addition on the Number Line
10-4 Adding Rational Numbers
10-5 Subtracting Rational Numbers
10-6 Properties of Addition
10-7 Adding Polynomials

FEATURES
Calculator Application: *Adding Rational Numbers*
Career Application: *Travel Agent*
Consumer Application: *Driving Safety*
Enrichment: *Clock Arithmetic*

10-1 Integers

Positive and negative numbers are used to represent temperatures above and below zero. Positive and negative numbers can also be used to describe other situations.

EXAMPLE 1 Write a positive or negative number to represent each word description.

Word Description	Number
a. 54 meters below sea level	a. ⁻54
b. 3 minutes before rocket lift-off	b. ⁻3
c. A gain of 25 yards	c. +25, or 25

Self-Check Write a positive or negative number to represent each description.

1. A profit of $200
2. A loss in altitude of 120 meters

Numbers such as ⁻54, ⁻3, 25, and 0 are called <u>integers</u>.

> The set of **integers** is made up of the positive integers, zero, and the negative integers.
>
> **Integers:** $\{\cdots ⁻3, ⁻2, ⁻1, 0, 1, 2, 3, \cdots\}$

◀ The 3 dots mean the numbers continue indefinitely.

To compare integers, it is often helpful to graph them. Numbers become larger as you move <u>from left to right</u> on the number line.

> **PROCEDURE**
>
> **To graph on a number line:**
>
> ① Draw and label a number line.
> ② Draw a "dot" at the point that corresponds to the integer being graphed.

EXAMPLE 2 a. Graph: 5; ⁻2; 0; 3; ⁻7
b. Write the integers in order from least to greatest.

Solutions: a. [number line from ⁻8 to 7 with dots at ⁻7, ⁻2, 0, 3, 5]

◀ Each dot is the graph of an integer.

b. Write the integers in order from left to right: ⁻7, ⁻2, 0, 3, 5

294 Chapter 10

Self-Check *Graph the integers.*
Then write the integers in order from least to greatest.

3. ⁻6; ⁻7; 4; ⁻5; 3   **4.** 10; 8; ⁻9; ⁻3; 0

Statements that use > or < are **inequalities**.

"**>**" means **"is greater than."**   "**<**" means **"is less than."**

EXAMPLE 3 Replace each ___?___ with > or <. Refer to the number line in Example 2.

Problem	Think	Solution
a. ⁻6 _?_ 0	⁻6 is to the left of 0.	⁻6 < 0
b. 5 _?_ ⁻9	5 is to the right of ⁻9.	5 > ⁻9
c. ⁻2 _?_ ⁻8	⁻2 is to the right of ⁻8.	⁻2 > ⁻8

Self-Check *Replace each ___?___ with > or <. Refer to the number line in Example 2.*

5. 0 _?_ ⁻4   **6.** 7 _?_ ⁻8   **7.** ⁻10 _?_ ⁻2   **8.** ⁻5 _?_ 1

NOTE: The inequality symbol always points toward the smaller number.

CLASSROOM EXERCISES

Write a positive or negative number to represent each description. (Example 1)

1. A loss of 8 yards
2. A gain of 15 yards
3. A temperature of 20°F below zero
4. A drop of 6 degrees in temperature

For each of Exercises 5–10:
a. *Graph the integers on a number line.*
b. *Write the integers in order from least to greatest.* (Example 2)

5. ⁻3, 2, 7, 1, 3   **6.** 4, 8, ⁻7, 3, 0   **7.** 2, 0, 11, ⁻8, ⁻1
8. 2, ⁻5, 3, 0, ⁻2   **9.** 6, 12, ⁻12, 0, 13   **10.** 11, 7, ⁻9, 8, ⁻14

Replace each ___?___ with > or <. (Example 3)

11. ⁻4 _?_ 8   **12.** 0 _?_ ⁻1   **13.** 11 _?_ ⁻1   **14.** ⁻1 _?_ 1

WRITTEN EXERCISES

Goal: To graph and compare integers

Write a positive or negative number to represent each description. (Example 1)

1. A rise of 10° in temperature
2. A pay raise of 65¢ per hour
3. A loss of 2000 feet in altitude
4. 156 meters above sea level
5. Two years ago
6. Nine years from now

For each of Exercises 7–12:
   a. *Graph the integers on a number line.*
   b. *Write the integers in order from least to greatest.* (Example 2)

7. 0; 3; ⁻3; 6
8. ⁻6; 2; 4; ⁻4
9. 0; 3; ⁻2; 1
10. 3; 1; ⁻2; 2
11. ⁻4; 8; 1; 3; ⁻2
12. 5; ⁻2; 1; ⁻1; 0

For Exercises 13-28, replace each __?__ with > or <. (Example 3)

13. 0 __?__ 4
14. 0 __?__ ⁻4
15. 3 __?__ ⁻3
16. ⁻4 __?__ 2
17. ⁻6 __?__ ⁻5
18. 4 __?__ 0
19. ⁻7 __?__ 7
20. 7 __?__ ⁻7
21. ⁻1 __?__ ⁻2
22. 1 __?__ ⁻2
23. ⁻1 __?__ 2
24. 0 __?__ ⁻8

APPLICATIONS: USING INTEGERS

25. The lowest temperature ever recorded in Florida was ⁻2°F. The lowest temperature ever recorded in Georgia was ⁻17°F.

    Complete: ⁻2° __?__ ⁻17°

26. The lowest temperature ever recorded in Alaska was ⁻80°F. The lowest temperature ever recorded in Hawaii was 12°F.

    Complete: ⁻80° __?__ 12°

27. The lowest temperature ever recorded in Vermont was ⁻50°F. The lowest temperature ever recorded in Maryland was ⁻40°F.

    Complete: ⁻50° __?__ ⁻40°

28. The lowest temperature ever recorded in Illinois was ⁻35°F. The lowest temperature ever recorded in Iowa was ⁻47°F.

    Complete: ⁻35° __?__ ⁻47°

REVIEW CAPSULE FOR SECTION 10-2

Write a fraction for each mixed numeral. (Pages 135-137)

1. $2\frac{1}{5}$
2. $5\frac{2}{3}$
3. $8\frac{7}{9}$
4. $3\frac{1}{6}$
5. $10\frac{1}{2}$
6. $1\frac{11}{12}$
7. $6\frac{1}{4}$
8. $1\frac{2}{5}$
9. $3\frac{1}{10}$
10. $6\frac{3}{10}$
11. $4\frac{1}{7}$
12. $3\frac{2}{9}$

The answers are on page 320.

10-2 Rational Numbers

Numbers such as these are **positive rational numbers**.

$$\frac{2}{3} \quad 1 \quad \frac{5}{4} \quad 2\frac{2}{5} \quad 5.6 \quad 27$$

Numbers such as these are **negative rational numbers**.

$$\frac{-1}{4} \quad -2 \quad \frac{-8}{5} \quad -4\frac{2}{3} \quad -3.4 \quad -32$$

Any <u>rational number</u> can be written as a fraction in the form $\frac{a}{b}$, $b \neq 0$.

EXAMPLE 1 Write a fraction for each rational number.

a. 6 b. $-3\frac{1}{3}$ c. 0 d. -4.2

Solutions: a. $6 = \frac{6}{1}$ b. $-3\frac{1}{3} = \frac{-10}{3}$ c. $0 = \frac{0}{1}$ d. $-4.2 = -4\frac{2}{10} = \frac{-42}{10}$

Self-Check *Write a fraction for each rational number.*

1. 4
2. $-4\frac{1}{2}$
3. 3.8
4. -5
5. $2\frac{1}{4}$

NOTE: In Example 1c, any number except 0 could have been written in the denominator. Thus,

$$0 = \frac{0}{4} = \frac{0}{10} = \frac{0}{21} = \frac{0}{100}, \text{ and so on.}$$

The rational numbers 4 and -4 are the same distance from 0.

```
|←— 4 units —→|←— 4 units —→|
-5  -4  -3  -2  -1  0  1  2  3  4  5
```

Therefore, 4 and -4 are <u>opposites</u>. The number 0 is its own opposite.

> Two numbers that are the same distance from 0 and are in opposite directions from 0 are **opposites**.

Rational Numbers: Addition/Subtraction

EXAMPLE 2 Write the opposite of each number.
    a. 7   b. $^-2\frac{1}{2}$ c. 3.5   d. $\frac{^-3}{4}$

Solutions: a. $^-7$ b. $2\frac{1}{2}$ c. $^-3.5$ d. $\frac{3}{4}$

Self-Check   *Write the opposite of each number.*

6. $^-15$ 7. $\frac{^-8}{3}$ 8. 0   9. $^-2.7$ 10. $\frac{4}{5}$

A dash, −, is used to represent "the opposite of." For example,

$$^-4 = -4$$

Read: "Negative 4 equals the opposite of 4."

> Since $^-4 = -4$, we will use the symbol for "the opposite of" to indicate a negative number. Thus, -4 can be read "negative 4."

CLASSROOM EXERCISES

Write a fraction for each rational number. (Example 1)

1. 4   2. 12   3. -10 4. -9 5. 3.2   6. 4.5
7. $-2\frac{1}{5}$ 8. $-7\frac{1}{6}$ 9. $2\frac{1}{4}$ 10. $1\frac{1}{3}$ 11. -1.8 12. -0.4

Write the opposite of each number. (Example 2)

13. 12   14. 9   15. -6 16. -15 17. 3.2   18. 6.8
19. -1.5 20. -12.4 21. $\frac{1}{4}$ 22. $2\frac{1}{2}$ 23. $-1\frac{4}{5}$ 24. $-\frac{3}{8}$

WRITTEN EXERCISES

Goals: To write a fraction for a rational number
          To write the opposite of a rational number

Write a fraction for each rational number. (Example 1)

1. 17   2. 8   3. -7 4. -5 5. 3.8   6. 6.2
7. $-3\frac{1}{3}$ 8. $-2\frac{4}{5}$ 9. $3\frac{1}{4}$ 10. $3\frac{1}{10}$ 11. $5\frac{2}{5}$ 12. $-1\frac{1}{6}$
13. 0   14. -1.7 15. -9.3 16. -1 17. 7.5   18. -0.8

Write the opposite of each number. (Example 2)

19. 12 **20.** 15 **21.** −20 **22.** −3 **23.** $2\frac{1}{3}$ **24.** $6\frac{2}{5}$
25. $\frac{3}{4}$ **26.** $-\frac{9}{4}$ **27.** $\frac{2}{5}$ **28.** $\frac{5}{6}$ **29.** $-8\frac{1}{2}$ **30.** $-9\frac{2}{3}$
31. 4.8 **32.** 7.25 **33.** −8.9 **34.** −12.6 **35.** 0 **36.** 0.3

APPLICATIONS: USING RATIONAL NUMBERS

Write a rational number to represent each word description.

37. A growth of $2\frac{1}{2}$ inches
38. A loss of $8.60
39. 32.6 meters below sea level
40. A gain of $3\frac{1}{2}$ pounds
41. An increase of $4.35
42. 7.6° below 0
43. 132.5 meters above sea level
44. A fall of $6\frac{3}{4}$ feet
45. A loss of 5.5 kilograms
46. A deduction of $4.75
47. 12.5° above 0
48. An increase of 3.6 centimeters

MORE CHALLENGING EXERCISES

Arrange the numbers in order from least to greatest.

49. $1; \frac{1}{4}; -2\frac{1}{5}; 3\frac{1}{3}; -5\frac{1}{5}; -\frac{1}{4}$
50. $-8; 0; 4\frac{1}{3}; -7\frac{1}{2}; -\frac{2}{5}; 2$
51. $2; -\frac{3}{4}; 1; -\frac{2}{3}; \frac{4}{5}; -1\frac{1}{2}$
52. $0; -\frac{2}{3}; 1\frac{1}{2}; -2\frac{3}{4}; \frac{2}{3}; -3\frac{1}{5}$
53. $2.25; -1.83; -1.875; 1.5; 0; -0.687$
54. $-0.875; 1; -2; 0.937; -1; 0.88$
55. $-1.125; -3; -2.5; -1; -0.25; -2.875$
56. $0.2; 0; -0.25; 1.5; -2.7; -5$

REVIEW CAPSULE FOR SECTION 10-3

For Exercises 1–4:
a. *Graph the integers on a number line.*
b. *Write the integers in order from least to greatest.*
(Pages 294–296)

1. $3; 0; 1; -1$ **2.** $-5; 1; 2; -3$ **3.** $0; 4; -2; 6$ **4.** $-4; 5; -6; 7$

Replace each _?_ *with* > *or* <. (Pages 294–296)

5. 0 _?_ −4 **6.** −3 _?_ −5 **7.** −4 _?_ 4 **8.** 9 _?_ −13
9. −6 _?_ 5 **10.** 0 _?_ 7 **11.** 3 _?_ −5 **12.** −8 _?_ −10

The answers are on page 320.

10-3 Addition on the Number Line

You can use a number line to find the sum of two rational numbers.

> **PROCEDURE**
>
> **To add on the number line:**
> 1. Graph the first number.
> 2. Draw an arrow to represent the second number.
>   a. Move to the right if the second number is positive.
>   b. Move to the left if the second number is negative.
> 3. Read the answer at the tip of the arrow.

EXAMPLE 1 Use a number line to add $-1 + (-4)$.

Solution:
1. Graph the first number, -1.
2. From this point, draw an arrow 4 units to the left.
3. Read the answer at the tip of the arrow.

$$-1 + (-4) = -5$$

Self-Check *Use a number line to add.*

1. $-1 + (-3)$ 2. $-3 + (-2)$ 3. $-4 + (-5)$ 4. $-8 + (-6)$

Examples 2 and 3 show how to add a positive and a negative integer on the number line.

EXAMPLE 2 Use a number line to add $-4 + 9$.

Solution:
1. Graph the first number, -4.
2. From this point, draw an arrow 9 units to the right.
3. Read the answer at the tip of the arrow.

$$-4 + 9 = 5$$

Self-Check *Use a number line to add.*

5. $-2 + 3$ 6. $-5 + 2$ 7. $-7 + 7$ 8. $-12 + 5$

EXAMPLE 3 Use a number line to add $5 + (-8)$.

Solution: ① and ②

③ $5 + (-8) = -3$

Self-Check *Use a number line to add.*

9. $2 + (-5)$ 10. $6 + (-6)$ 11. $8 + (-5)$ 12. $9 + (-4)$

CLASSROOM EXERCISES

For Exercises 1–24, use a number line to add. (Example 1)

1. $-2 + (-4)$
2. $-7 + (-4)$
3. $-5 + (-5)$
4. $-2 + (-7)$
5. $-6 + (-4)$
6. $-8 + (-3)$
7. $-5 + (-2)$
8. $-3 + (-4)$

(Example 2)

9. $-7 + 10$
10. $-3 + 8$
11. $-1 + 4$
12. $-7 + 5$
13. $-6 + 2$
14. $-4 + 1$
15. $-2 + 5$
16. $-8 + 4$

(Example 3)

17. $2 + (-7)$
18. $5 + (-6)$
19. $1 + (-4)$
20. $8 + (-10)$
21. $3 + (-2)$
22. $6 + (-5)$
23. $7 + (-2)$
24. $4 + (-4)$

WRITTEN EXERCISES

Goal: To use a number line to add integers

For Exercises 1–48, use a number line to add. (Example 1)

1. $-5 + (-4)$
2. $-6 + (-5)$
3. $-1 + (-6)$
4. $-4 + (-1)$
5. $-3 + (-7)$
6. $-8 + (-4)$
7. $-2 + (-2)$
8. $-3 + (-3)$
9. $-7 + (-8)$
10. $-1 + (-5)$
11. $-3 + (-2)$
12. $-4 + (-6)$

(Example 2)

13. $-4 + 7$
14. $-3 + 6$
15. $-5 + 10$
16. $-6 + 14$
17. $-2 + 8$
18. $-1 + 5$
19. $-3 + 7$
20. $-6 + 8$
21. $-8 + 4$
22. $-5 + 1$
23. $-4 + 4$
24. $-7 + 5$

Rational Numbers: Addition/Subtraction

(Example 3)

25. 3 + (−5)
26. 4 + (−7)
27. 5 + (−3)
28. 6 + (−2)
29. 1 + (−6)
30. 3 + (−7)
31. 7 + (−2)
32. 4 + (−1)
33. 2 + (−8)
34. 6 + (−7)
35. 9 + (−2)
36. 10 + (−6)

MIXED PRACTICE

37. −7 + 4
38. −3 + (−4)
39. 6 + (−9)
40. −1 + 7
41. −2 + (−3)
42. 10 + (−5)
43. 9 + (−3)
44. −8 + 8
45. −1 + (−5)
46. −8 + 2
47. 6 + (−3)
48. −4 + (−5)

APPLICATIONS: USING ADDITION OF INTEGERS

Use a number line to solve each problem.

49. Ian scored 10 points by answering his first quiz-bowl question correctly. He missed the second question and lost 15 points. What was his score after two questions?
50. Andrea lost 8 pounds last month. This month she lost 6 pounds. How much did she lose in all?
51. A football team lost 12 yards on one play and gained 8 yards on the next play. Find the total gain or loss for the two plays.

REVIEW: SECTIONS 10-1 – 10-3

Graph the integers on a number line. Then write the integers in order from least to greatest. (Section 10-1)

1. 5; −3; 0; 2; −1
2. −8; 9; 1; 7; −4
3. 4; 0; −2; 1; −4

Write a fraction for each rational number. (Section 10-2)

4. 15
5. −8
6. −5.8
7. $2\frac{1}{8}$
8. −2.2
9. 0

For Exercises 10–18, use a number line to solve. (Section 10-3)

10. −2 + (−1)
11. −5 + (−8)
12. −3 + 11
13. −10 + 2
14. 4 + (−15)
15. 9 + (−7)
16. 10 + (−10)
17. −25 + 25

18. The temperature was −5°C and rose 7 degrees. Find the new temperature.

CONSUMER APPLICATION: Driving Safety

The ability of a motorist to stop quickly in an emergency depends greatly on the vehicle's speed, brakes, and the condition of the road.

This formula can be used to estimate the stopping distance of a vehicle.

$$s = 1.1r + \frac{r^2}{30F}$$

- s = stopping distance (ft)
- r = rate in miles per hour
- F = driving surface factor

The table at the right shows several values for the driving surface factor.

Type of Surface	Driving Surface Factor Dry Road	Wet Road
Asphalt	0.85	0.65
Concrete	0.90	0.60
Gravel	0.65	0.65
Packed Snow	0.45	0.45

EXAMPLE Carlos Dorman was driving 36 miles per hour on a dry concrete road when he saw a tree branch lying in the road. Estimate the stopping distance to the nearest foot.

SOLUTION $s = 1.1r + \frac{r^2}{30F}$ $r = 36$
$F = 0.9$ (from the table)

$s = 1.1(36) + \frac{(36)^2}{30(0.9)}$ $(36)^2$ **means** $36 \cdot 36$.

Using a calculator, first compute $1.1r$. Store the value in memory.

1 . 1 [×] 3 6 [=] [M+]   M 39.6

Compute $\frac{r^2}{30F}$. Add the value stored in memory.

36 [×] 36 [÷] 30 [÷] .9 [=] [+] [MR] [=]   87.6

The stopping distance is about **88 feet**.

Exercises

Complete the table. Round each answer to the nearest foot.

Rate	Road Condition	Stopping Distance		Rate	Road Condition	Stopping Distance
1. 39 mi/hr	wet asphalt	?		**2.** 27 mi/hr	packed snow	?
3. 45 mi/hr	wet concrete	?		**4.** 54 mi/hr	dry concrete	?
5. 35 mi/hr	dry gravel	?		**6.** 39 mi/hr	dry asphalt	?

REVIEW CAPSULE FOR SECTION 10-4

Replace the __?__ with > or <. (Pages 294-296)

1. 14 __?__ 8
2. 16 __?__ 2
3. 0 __?__ 5
4. 24 __?__ 25
5. 37 __?__ 29
6. 101 __?__ 110
7. 3.5 __?__ 3.45
8. 9.06 __?__ 9.6

Write the opposite of each number. (Pages 297-299)

9. 6
10. 15
11. −8
12. 3
13. 0
14. −7

The answers are on page 320.

10-4 Adding Rational Numbers

You use **absolute value notation** when you want to indicate distance, but not direction, from 0. The symbol for absolute value is | |.

> The **absolute value** of a number is the distance of the number from 0.

Some examples are given in the table below.

TABLE

Absolute Value	Meaning	Solution
$|5|$	Distance of 5 from 0	$|5| = 5$
$|-6.7|$	Distance of −6.7 from 0	$|-6.7| = 6.7$
$|0|$	Distance of 0 from 0	$|0| = 0$
$\left|-\frac{1}{2}\right|$	Distance of $-\frac{1}{2}$ from 0	$\left|-\frac{1}{2}\right| = \frac{1}{2}$

You can use absolute value to add two negative numbers.

> **PROCEDURE**
>
> To add two negative numbers:
> 1. Add the absolute values.
> 2. Write the opposite of the result.

304 Chapter 10

EXAMPLE 1 Add: $-6 + (-9)$

Solution:
1. Add the absolute values. $\quad |-6| + |-9| = 6 + 9$
$$= 15$$
2. Write the opposite of the result. $\longrightarrow \quad -15$

Thus, $-6 + (-9) = \mathbf{-15}$

Self-Check *Add:* **1.** $-12 + (-6)$ **2.** $-7 + (-15)$ **3.** $-4 + (-9)$

Follow this procedure to add a positive and a negative number.

PROCEDURE

To add a positive and a negative number:
1. Subtract the absolute values.
2. If the positive number has the greater absolute value, write the difference.

  If the negative number has the greater absolute value, write the opposite of the difference.

EXAMPLE 2 Add: $8 + (-3)$

Solution:
1. Subtract the absolute values. $\quad |8| - |-3| = 8 - 3$
2. Since $|8| > |-3|$, write the difference. $\qquad\qquad = 5$

$8 + (-3) = \mathbf{5}$ ◀ The positive number has the greater absolute value.

Self-Check *Add:* **4.** $12 + (-6)$ **5.** $-13 + 21$ **6.** $-8 + 15$

EXAMPLE 3 Add: $-20 + 8$

Solution:
1. Subtract the absolute values. $\quad |-20| - |8| = 20 - 8$
2. Since $|-20| > |8|$, write the opposite of $\qquad\qquad = 12$
the difference. $\longrightarrow \quad -12$

$-20 + 8 = \mathbf{-12}$ ◀ The negative number has the greater absolute value.

Self-Check *Add:* **7.** $8 + (-9)$ **8.** $-18 + 12$ **9.** $-16 + 10$

Rational Numbers: Addition/Subtraction **305**

CLASSROOM EXERCISES

Evaluate. (Table)

1. $|8|$
2. $|-8|$
3. $|-6|$
4. $|10|$
5. $|-15|$
6. $|24|$

For Exercises 7–30, add the given numbers. (Example 1)

7. $-4 + (-14)$
8. $-6 + (-3)$
9. $-7 + (-15)$
10. $-9 + (-4)$
11. $16 + 13$
12. $7 + 18$
13. $-8 + (-14)$
14. $-5 + (-15)$

(Example 2)

15. $6 + (-4)$
16. $7 + (-1)$
17. $-4 + 15$
18. $-5 + 21$
19. $20 + (-16)$
20. $-10 + 21$
21. $30 + (-18)$
22. $-7 + 24$

(Example 3)

23. $6 + (-9)$
24. $-16 + 7$
25. $-22 + 14$
26. $12 + (-20)$
27. $9 + (-13)$
28. $-31 + 11$
29. $-18 + 3$
30. $7 + (-19)$

WRITTEN EXERCISES

Goals: To write the absolute value of a number
To add two rational numbers
To apply the skill of adding rational numbers to solving word problems

Evaluate. (Table)

1. $|7|$
2. $|18|$
3. $|-5|$
4. $|-21|$
5. $|15|$
6. $|36|$
7. $|-8|$
8. $|-53|$
9. $|32|$
10. $|-16|$
11. $|0|$
12. $|1|$
13. $|1.3|$
14. $|-3.4|$
15. $|-6.7|$
16. $|0.9|$
17. $|\frac{1}{3}|$
18. $|-\frac{3}{5}|$

For Exercises 19–34, add the given numbers. (Example 1)

19. $-11 + (-5)$
20. $-9 + (-3)$
21. $6 + 22$
22. $15 + 14$
23. $-22 + (-14)$
24. $-8 + (-10)$
25. $-15 + (-3)$
26. $-32 + (-21)$
27. $-1.3 + (-4.2)$
28. $-1.6 + (-1.2)$
29. $-3.4 + (-1.8)$
30. $-6.3 + (-0.4)$
31. $-\frac{1}{3} + \left(-\frac{1}{3}\right)$
32. $-\frac{2}{7} + \left(-\frac{3}{7}\right)$
33. $-\frac{1}{2} + \left(-\frac{1}{4}\right)$
34. $-\frac{1}{3} + \left(-\frac{1}{5}\right)$

For Exercises 35–86, add the given numbers. (Example 2)

35. $21 + (-17)$
36. $8 + (-7)$
37. $-4 + 12$
38. $-1 + 21$
39. $-16 + 25$
40. $13 + (-4)$
41. $8 + (-3)$
42. $-12 + 7$
43. $2.5 + (-1.8)$
44. $-1.7 + 3.2$
45. $8.7 + (-4.8)$
46. $-0.8 + 3.5$
47. $-\frac{1}{5} + \frac{2}{5}$
48. $\frac{3}{4} + \left(-\frac{1}{4}\right)$
49. $\frac{1}{2} + \left(-\frac{1}{3}\right)$
50. $-\frac{1}{8} + \frac{1}{2}$

(Example 3)

51. $-12 + 7$
52. $-21 + 14$
53. $8 + (-20)$
54. $4 + (-19)$
55. $-27 + 6$
56. $7 + (-8)$
57. $22 + (-30)$
58. $-15 + 10$
59. $1.3 + (-3.2)$
60. $-2.5 + 1.2$
61. $1.1 + (-2.2)$
62. $-4.6 + 0.8$
63. $-\frac{3}{8} + \frac{1}{8}$
64. $\frac{1}{3} + \left(-\frac{2}{3}\right)$
65. $\frac{1}{2} + \left(-\frac{5}{8}\right)$
66. $-\frac{4}{5} + \frac{1}{4}$

MIXED PRACTICE

67. $-16 + 21$
68. $-17 + 30$
69. $-24 + (-10)$
70. $-2 + (-17)$
71. $-13 + 8$
72. $-18 + 7$
73. $27 + (-12)$
74. $16 + 32$
75. $-4 + (-21)$
76. $-14 + 12$
77. $30 + (-24)$
78. $-9 + (-22)$
79. $-8.4 + 2.7$
80. $-7.3 + (-6.2)$
81. $3.5 + 0.4$
82. $-4.6 + 1.4$
83. $-\frac{1}{5} + \left(-\frac{2}{5}\right)$
84. $\frac{3}{8} + \left(-\frac{5}{8}\right)$
85. $-\frac{1}{4} + \frac{1}{2}$
86. $\frac{3}{10} + \left(-\frac{3}{4}\right)$

APPLICATIONS: USING ADDITION OF RATIONAL NUMBERS

87. A submarine at a depth of 40 meters below the ocean's surface fired a missile which rose 225 meters. How far above the ocean's surface did the missile rise?
88. A plane was flying at an altitude of 4000 feet. Then it dropped 320 feet. What was the plane's new altitude?
89. The temperature was $-2°C$ and dropped 6 degrees. Find the new temperature.

REVIEW CAPSULE FOR SECTION 10-5

Write the opposite of each number. (Pages 297–299)

1. 12
2. 4
3. -28
4. -15
5. 3.7
6. $-1\frac{1}{3}$

The answers are on page 320.

10-5 Subtracting Rational Numbers

Notice that the answers to the two problems below are the same.

Subtraction	Related Addition
$7 - 4 = 3$	$7 + (-4) = 3$

Subtracting an integer is the same as adding its opposite.

> **PROCEDURE**
>
> To write an addition problem for a subtraction problem, <u>add</u> the <u>opposite</u> of the number you are subtracting.

EXAMPLE 1 Write an addition problem for each subtraction problem.

Subtraction Problem	Think	Addition Problem
a. $9 - 4$	The opposite of 4 is -4.	$9 + (-4)$
b. $5 - (-2)$	The opposite of -2 is 2.	$5 + 2$
c. $-6 - 3$	The opposite of 3 is -3.	$-6 + (-3)$
d. $-7 - (-1)$	The opposite of -1 is 1.	$-7 + 1$

Self-Check *Write an addition problem for each subtraction problem.*

1. $4 - 9$ 2. $-10 - 18$ 3. $-1 - (-12)$ 4. $17 - (-21)$

> **PROCEDURE**
>
> To subtract two rational numbers:
> 1. Write the related addition problem.
> 2. Follow the rules for addition.

EXAMPLE 2 Subtract: $-5 - 6$

Solution:
1. Write the related addition problem. $\quad -5 - 6 = -5 + (-6)$
2. Add. $\quad = -11$

Self-Check *Subtract.*

5. $-2 - 4$ 6. $-47 - 38$ 7. $-0.5 - 2.1$ 8. $-4.5 - 1.4$

Example 3 shows how to subtract a negative number.

EXAMPLE 3 Subtract: **a.** $8 - (-9)$ **b.** $-6 - (-4)$
Solutions: ☐1 **a.** $8 - (-9) = 8 + 9$ **b.** $-6 - (-4) = -6 + 4$
☐2 $= 17$ $= -2$

Self-Check *Subtract.*
9. $12 - (-17)$ **10.** $24 - (-39)$ **11.** $-10 - (-25)$ **12.** $-9 - (-16)$

CLASSROOM EXERCISES

Complete. (Example 1)
1. $14 - 5 = 14 + \underline{\ ?\ }$ **2.** $2 - 9 = 2 + \underline{\ ?\ }$ **3.** $5 - (-14) = 5 + \underline{\ ?\ }$
4. $-6 - 8 = -6 + \underline{\ ?\ }$ **5.** $0 - 17 = 0 + \underline{\ ?\ }$ **6.** $-1 - (-10) = (-1) + \underline{\ ?\ }$

For Exercises 7–22, subtract. (Example 2)
7. $5 - 8$ **8.** $3 - 13$ **9.** $-5 - 17$ **10.** $-6 - 10$
11. $6 - 9$ **12.** $8 - 17$ **13.** $-4 - 21$ **14.** $-12 - 30$

(Example 3)
15. $4 - (-5)$ **16.** $12 - (-4)$ **17.** $-6 - (-3)$ **18.** $-4 - (-10)$
19. $10 - (-11)$ **20.** $15 - (-32)$ **21.** $-13 - (-19)$ **22.** $-21 - (-43)$

WRITTEN EXERCISES

Goal: To subtract two rational numbers

Write an addition problem for each subtraction problem.
(Example 1)
1. $11 - 3$ **2.** $14 - 2$ **3.** $14 - (-6)$ **4.** $12 - (-9)$
5. $-6 - 1$ **6.** $-9 - 7$ **7.** $-8 - (-5)$ **8.** $-12 - (-1)$
9. $-7 - 7$ **10.** $-8 - 8$ **11.** $4.7 - 8.6$ **12.** $-1\frac{1}{2} - \left(-2\frac{1}{2}\right)$

For Exercises 13–52, subtract. (Example 2)
13. $11 - 15$ **14.** $17 - 22$ **15.** $-19 - 22$ **16.** $-16 - 24$
17. $0 - 24$ **18.** $18 - 43$ **19.** $-37 - 84$ **20.** $-10 - 26$
21. $2.7 - 3.6$ **22.** $1.8 - 5.1$ **23.** $4\frac{3}{4} - 6\frac{1}{4}$ **24.** $3\frac{1}{2} - 5\frac{1}{4}$

Rational Numbers: Addition/Subtraction

(Example 3)

25. $-3 - (-7)$
26. $-8 - (-15)$
27. $6 - (-2)$
28. $25 - (-17)$
29. $18 - (-7)$
30. $24 - (-29)$
31. $9 - (-48)$
32. $5 - (-76)$
33. $9.5 - (-13.9)$
34. $-8.9 - (-17.3)$
35. $3\frac{3}{4} - \left(-1\frac{3}{4}\right)$
36. $\frac{1}{2} - \left(-\frac{5}{6}\right)$

MIXED PRACTICE

37. $-127 - (-127)$
38. $-4 - (-8)$
39. $0 - (-7)$
40. $-128 - 43$
41. $48 - 236$
42. $39 - 124$
43. $-254 - 97$
44. $-64 - (-64)$
45. $-24 - (-36)$
46. $27 - (-15)$
47. $-36 - 45$
48. $15 - 82$
49. $4.8 - (-6.7)$
50. $-5.4 - 2.9$
51. $2\frac{3}{8} - 5\frac{1}{4}$
52. $-1\frac{1}{6} - 2\frac{1}{2}$

APPLICATIONS: USING SUBTRACTION OF RATIONAL NUMBERS

53. The greatest recorded temperature change in one day in North America took place in Montana. The change was from 7°C to −49°C. Find the difference in temperatures. (Hint: $7 - (-49) = $?)

54. The actual temperature is 32°F. The wind chill temperature is −9°F. How much colder than the actual temperature is the wind chill temperature? (Hint: $32 - (-9) = $?)

55. The highest point in California is 14,494 feet above sea level (+). The lowest point is 282 feet below sea level (−). Find the difference in altitude between the highest and lowest point.

REVIEW CAPSULE FOR SECTION 10-6

Evaluate each expression. (Pages 18–21)

1. $2 + 17 - 9$
2. $16 + 32 + 25$
3. $32 + 57 - 65 + 10$
4. $23 - 16 + 10 + 52$
5. $81 + 125 + 51$
6. $13 + 42 + 180 - 57$

Add. (Pages 300–302)

7. $-5 + (-12)$
8. $16 + (-20)$
9. $-32 + 55$
10. $-48 + 19$
11. $25 + (-17)$
12. $-19 + (-13)$
13. $18 + (-18)$
14. $-54 + 54$

The answers are on page 320.

CAREER APPLICATIONS
Travel Agent

Travel agents must be able to determine the difference in time between any two cities in the world. The longitude of the cities can be used to find this difference.

The figure at the right shows the longitude of New York City to be 74°W. This means that New York City is 74° west of the prime meridian. A point east of the prime meridian has east longitude.

This table shows the relation between the longitude of a point and its time zone.

One hour of time = 15° in longitude

A **zone description (ZD)** of +5 (west longitude) means that the time is 5 hours earlier than the time at Zone 0. A ZD of −3 (east longitude) means that the time is 3 hours later than the time at Zone 0. To find the ZD of a point, first divide its longitude by 15°. Then round to the nearest whole number.

Zone Description Table

	$37\frac{1}{2}°$W	$22\frac{1}{2}°$W	$7\frac{1}{2}°$W	$7\frac{1}{2}°$E	$22\frac{1}{2}°$E	$37\frac{1}{2}°$E
	Zone +2	Zone +1	Zone 0	Zone −1	Zone −2	
	30°W	15°W	0°	15°E	30°E	

EXAMPLE At point A with longitude 0°, the time is 10:45 A.M. Find the time at point B with longitude 74°W.

SOLUTION Find the ZD of point B.

$$\frac{74}{15} = 4.9\overline{3} \qquad ZD = 5$$

▶ West longitude is positive. East longitude is negative.

The time at point B is 5 hours <u>earlier</u> than the time at point A.

$$10:45 - 5 = \textbf{5:45 A.M.}$$

Exercises

Find the zone description (ZD) of each city. Then find its time when it is 6:05 P.M. in London (longitude 0°).

1. Chicago (87.6°W)
2. Seattle (122.3°W)
3. Houston (95.4°W)
4. Moscow (37.7°E)
5. New Delhi (77.2°E)
6. Rome (12.5°E)

10-6 Properties of Addition

The properties of addition also apply to addition with rational numbers.

Addition Property of Zero
For any rational number a,
$a + 0 = a$ and $0 + a = a$

$4 + 0 = 4$
$0 + (-2.7) = -2.7$

Addition Property of Opposites
(Additive Inverse Property)
For any rational number a,
$a + (-a) = 0$ and $-a + a = 0$

$35 + (-35) = 0$
$-2\frac{1}{4} + 2\frac{1}{4} = 0$

Commutative Property of Addition
For any rational numbers a and b,
$a + b = b + a$

$17 + (-21) = -21 + 17$
$-\frac{1}{2} + \frac{2}{5} = \frac{2}{5} + (-\frac{1}{2})$

Associative Property of Addition
For any rational numbers a, b, and c
$(a + b) + c = a + (b + c)$

$(5 + 3) + (-2) = 5 + (3 + (-2))$
$(1.2 + 5.8) + 4 = 1.2 + (5.8 + 4)$

The Commutative and Associative Properties can be used when adding three or more numbers.

EXAMPLE 1 Add: $(-3) + 8 + (-6)$

Solution:

Method 1
$-3 + 8 + (-6)$
$5 + (-6)$
-1

◀ Add from left to right.

Method 2
$-3 + 8 + (-6)$
$-3 + (-6) + 8$
$-9 + 8$
-1

◀ First add the negative numbers. Then add 8 to the sum.

Self-Check *Add.*

1. $3 + (-6) + (-7)$
2. $9 + (-4) + 7$

Example 2 shows how the Commutative and Associative Properties can be used to add four rational numbers.

EXAMPLE 2 Add: $5 + (-1) + 8 + (-3)$

Solution:

Method 1
$$5 + (-1) \mid 8 \mid (-3)$$
$$\quad\quad 4 \quad\quad + \quad 5$$
$$\quad\quad\quad\quad 9$$

Method 2
$$5 + (-1) + 8 + (-3)$$
$$5 + 8 + (-1) + (-3)$$
$$\quad 13 \quad + \quad (-4)$$
$$\quad\quad\quad 9$$

◀ Group the positive and negative numbers. Then add.

Self-Check *Add.*

3. $-4 + 9 + 6 + (-5)$ **4.** $-3 + 4 + (-5) + 8$

CLASSROOM EXERCISES

For Exercises 1–6, complete. (Example 1)

1. $12 + (-3) + 2 = 12 + \underline{\ ?\ } = \underline{\ ?\ }$ **2.** $9 + (-6) + 3 = \underline{\ ?\ } + (-6) = \underline{\ ?\ }$
3. $-6 + 4 + (-8) = -14 + \underline{\ ?\ } = \underline{\ ?\ }$ **4.** $-8 + 3 + (-1) = \underline{\ ?\ } + 3 = \underline{\ ?\ }$
5. $4 + (-1) + (-2) = 3 + \underline{\ ?\ } = \underline{\ ?\ }$ **6.** $3 + (-13) + 5 = \underline{\ ?\ } + (-13) = \underline{\ ?\ }$

Add. (Example 2)

7. $1 + (-3) + (-2) + 4$ **8.** $-15 + 4 + 7 + (-6)$ **9.** $-10 + 9 + 4 + (-6)$
10. $-1 + 1 + (-1) + 9$ **11.** $-12 + 7 + 13 + (-8)$ **12.** $-6 + (-12) + 9 + 3$

WRITTEN EXERCISES

Goal: To add more than two rational numbers

For Exercises 1–36, add. (Example 1)

1. $-8 + 6 + (-3)$ **2.** $-12 + 5 + (-2)$ **3.** $7 + (-13) + 8$
4. $5 + (-12) + 13$ **5.** $-6 + (-9) + (-14)$ **6.** $(-3) + (-9) + (-15)$
7. $-10 + 18 + (-32)$ **8.** $-9 + 21 + (-21)$ **9.** $16 + (-8) + 8$
10. $-31 + (-19) + (-7)$ **11.** $-51 + 27 + (-40)$ **12.** $16 + (-29) + 54$

(Example 2)

13. $-5 + (-2) + (-7) + 4$ **14.** $-6 + 2 + (-11) + 8$ **15.** $3 + (-9) + 16 + 3$
16. $42 + (-6) + 6 + (-9)$ **17.** $-8 + (-3) + 9 + 7$ **18.** $9 + (-7) + 13 + (-6)$

19. $-9 + 12 + 4 + (-7)$ **20.** $-8 + 8 + (-12) + (-4)$ **21.** $-7 + 13 + (-16) + 4$
22. $-7 + 14 + 6 + (-5)$ **23.** $13 + (-3) + (-9) + 7$ **24.** $-10 + 2 + (-6) + 4$

MIXED PRACTICE

25. $-14 + 3 + (-5) + 7$ **26.** $-14 + 3 + (-5) + 12$ **27.** $3 + (-6) + 4 + (-10)$
28. $-12 + 14 + (-6)$ **29.** $5 + 9 + (-5) + 17$ **30.** $-35 + 25 + (-15)$
31. $-8 + 9 + (-5)$ **32.** $-40 + 15 + (-25)$ **33.** $-8 + 3 + (-11) + 4$
34. $-5 + 28 + (-37)$ **35.** $2 + (-5) + 6 + (-1)$ **36.** $5 + (-12) + 4$

MORE CHALLENGING EXERCISES

Add.

37. $0.3 + (-5.8) + 4.2 + (-9.4)$ **38.** $-6.8 + 3.02 + (-9.46) + 14.5 + (-7.23)$
39. $-\frac{1}{4} + \frac{3}{8} + \frac{1}{4} + (-\frac{5}{8})$ **40.** $-\frac{3}{8} + 1\frac{1}{4} + (-1) + (-\frac{1}{2})$
41. $0.8 + (-1.5) + 2.3 + 1.2 + (-3.5)$ **42.** $-0.3 + 0.47 + 0.28 + (-0.31)$
43. $\frac{1}{3} + (-\frac{1}{2}) + 1\frac{2}{3} + (-\frac{1}{4})$ **44.** $-\frac{3}{8} + (-1\frac{1}{2}) + (-\frac{1}{6}) + 1\frac{1}{8}$

ADDING RATIONAL NUMBERS

A calculator with a "+/−" or "sign change" key (sometimes indicated as "SC") allows you to operate with negative numbers. This key enables you to add rational numbers from left to right without grouping.

EXAMPLE Add: $-6.9 + (-2.3) + 4.01 + (-7.2)$

Solution Press the sign change key for any negative number <u>after</u> you enter the value of the number.

6 . 9 [+/−] [+] 2 . 3 [+/−] [+]

4 . 0 1 [+] 7 . 2 [+/−] [=]   `-12.39`

EXERCISES Add.

1. $-17 + 52 + (-94) + (-38)$ **2.** $-52 + (-16) + 31 + (-64)$
3. $-6.0 + 3.23 + (-5.2) + 7.3$ **4.** $-0.2 + 0.41 + 0.37 + (-0.25)$
5. $-6.8 + 3.02 + (-9.46) + (-7.23) + 14.5 + 6.9 + (-7.92)$
6. $-104 + (-602) + (-523) + 344 + 681 + (-712) + 942$

REVIEW CAPSULE FOR SECTION 10-7

For Exercises 1–8, combine like terms. (Pages 76–80)

1. $4a + 3a$
2. $5m - 2m$
3. $8x^2 - x^2$
4. $10b^2 + 4b^2$
5. $2a + 4a + 5$
6. $3x + 5x + 9$
7. $3t^2 + 12 + 2t^2$
8. $5n^2 + 6 + 7n^2$

Add. (Pages 304–307)

9. $-5 + 12$
10. $-4 + (-3)$
11. $9 + (-6)$
12. $10 + (-1)$
13. $1 + (-5)$
14. $-7 + (-3)$
15. $-8 + 10$
16. $7 + (-11)$

The answers are on page 320.

10-7 Adding Polynomials

Each of the following is an example of a **monomial**.

a. $5y^3$ b. $-q$ c. $\frac{1}{3}r$ d. -9.6 e. g^2

A **binomial** is the sum or difference of two monomials.

a. $t + 5$ b. $2g - 1$ c. $5x^2 - 4$ d. $a + b$

A **trinomial** is the sum or difference of three monomials.

a. $x^2 + 5x - 9$ b. $2t - t^2 - 8$ c. $5p^3 - 1 + p^2$

Monomials and binomials belong to a larger class of expressions called polynomials.

> A **polynomial** is a monomial or the sum or difference of two or more monomials.
>
> $-5v$
> $8t - 9$
> $x^2 - 2x + 3$

Adding polynomials is similar to combining like terms.

EXAMPLE 1   Add: $(-3x^2 + 2x) + (5x^2 - 3x)$

Solution:
1. Write without parentheses and combine like terms.

   $-3x^2 + 2x + 5x^2 - 3x$

   $-3x^2 + 5x^2 + 2x - 3x$

2. Arrange terms in order with the greatest exponent first.

   $2x^2 - x$

Self-Check   1. $(3x + 7) + (x - 4) =$ __?__   2. $(-2p^2 + 5p) + (-3p^2 - 8p) =$ __?__

Rational Numbers: Addition/Subtraction **315**

It is sometimes helpful to arrange polynomials in vertical order.

EXAMPLE 2 Add: $(2x^2 - 5x + 6) + (-2x^2 + 3x - 10)$

Solution: Write like terms in the same column.

$$\begin{array}{r} 2x^2 - 5x + 6 \\ -2x^2 + 3x - 10 \\ \hline -2x - 4 \end{array}$$ ◀ Add the columns.

Self-Check **3.** $(3y - 5) + (-6y + 3)$ **4.** $(4n^2 + 6n + 3) + (n^2 - 3n + 4)$

The **degree of a polynomial** is the highest degree of its monomials.

Polynomial	Degree
$3x^2 - 5x + 9$	2
$5y + 7$	1
$a^3 + 3a + 8$	3

◀ Read the degree from the exponent of the monomial.

CLASSROOM EXERCISES

Classify each of the following as a monomial, binomial, or trinomial. (Definition)

1. $-5h$ **2.** $3x^2 + 5x - 4$ **3.** $3p - 4$ **4.** $7t^2 - 5t + 21$

Write the degree of each polynomial.

5. $t^2 - 5$ **6.** $-9p^3$ **7.** $q^3 + q^2$ **8.** $-x + x^4 - 21$

WRITTEN EXERCISES

Goal: To add polynomials

Add. (Examples 1 and 2)

1. $(3p - 2) + (4p + 5)$
2. $(-5t^2 + 1) + (4t^2 - 3)$
3. $(2c^2 - c) + (4c^2 - 5c)$
4. $(3x + 7) + (-5x - 2)$
5. $(4d - 3) + (3d + 1)$
6. $(-2y^2 - 7) + (5y + 10)$
7. $(4n^2 - 3n + 1) + (n^2 - 3n + 4)$
8. $(-5y^2 + 4y - 7) + (-2y^2 - 3y + 6)$
9. $(x^2 - 3x + 1) + (2x^2 + x - 3)$
10. $(4g^2 - g + 5) + (-g^2 + 6g - 3)$
11. $(5r^2 - 2r + 3) + (-r^2 + 6r - 5)$
12. $(3p^2 - 2p + 10) + (2p^2 + 4p - 6)$

MORE CHALLENGING EXERCISES

Subtract.

EXAMPLE: $5t - (-6t)$

Solution: $5t - (-6t) = 5t + 6t$
$= 11t$

Add the opposite of the number you are subtracting.

13. $8x - (-3x)$
14. $7y - (-5y)$
15. $-4p^2 - 5p^2$
16. $-3b^2 - 7b^2$
17. $-9y - (-3y)$
18. $-2n^2 - (-7n^2)$
19. $6t^2 - 15t^2$
20. $4s - 10s$
21. $3x^2 - (-5x^2)$
22. $-4ab - 6ab$
23. $-6y - (-4y)$
24. $b - 8b$

REVIEW: SECTIONS 10-4 – 10-7

Add. (Section 10-4)

1. $-6 + 12$
2. $-14 + (-7)$
3. $-18 + 13$
4. $-11 + 22$
5. $-3 + 14$
6. $16 + 34$
7. $16 + (-30)$
8. $-31 + (-8)$

Subtract. (Section 10-5)

9. $25 - 18$
10. $23 - 47$
11. $6 - (-18)$
12. $-3 - 14$
13. $18 - (-21)$
14. $-4 - 12$
15. $-5 - (-11)$
16. $-15 - (-3)$

Solve.

17. On a certain day, the temperature at noon was 8°F. By midnight it had dropped 17°. What was the temperature at midnight? (Section 10-4)

18. The average daytime temperature on Mars is about −30°C. On Earth it is about 20°C. About how much warmer is the average daytime temperature on Earth? (Section 10-5)

Add. (Section 10-6)

19. $-10 + 4 + (-12)$
20. $-5 + (-12) + (-9)$
21. $12 + (-10) + (-32) + 16$
22. $-4 + 15 + 16 + (-24)$
23. $-6 + 18 + (-34) + 25$
24. $-12 + (-25) + 40$

Add. (Section 10-7)

25. $(x + 3) + (-2x + 5)$
26. $(-2y^2 + 4y - 5) + (-4y^2 - 5y + 2)$
27. $(12t - 3) + (8t + 4)$
28. $(2p^2 - p) + (p^2 + 5p)$
29. $(y^2 + 2y - 3) + (4y^2 - 6y + 5)$
30. $(3n - 2n^2 - 4) + (5n^2 + n - 3)$

Rational Numbers: Addition/Subtraction

CHAPTER REVIEW

PART 1: VOCABULARY

For Exercises 1–11, choose from the box at the right below, the word(s) or numerical expression(s) that best corresponds to each description.

1. Commutative Property of Addition __?__ (p. 294)
2. Distance of a number from zero __?__ (304)
3. Numbers that are the same distance from zero <u>and</u> are in opposite directions from zero __?__ (p. 297)
4. Addition Property of Zero __?__ (p. 312)
5. A monomial or the sum or difference of two or more monomials __?__ (p. 315)
6. Associative Property of Addition __?__ (312)
7. Integers less than zero __?__ (p. 294)
8. Any number that can be written as a fraction __?__ (p. 297)
9. Addition Property of Opposites __?__ (p. 312)
10. Set of positive integers, zero, and negative integers __?__ (p. 294)
11. Integers to the right of zero __?__ (p. 294)

> integers
> opposites
> $5 + 0 = 5$
> $-2 + (5 + 3) = (-2 + 5) + 3$
> $3 + (-3) = 0$
> $4 + (-7) = -7 + 4$
> absolute value
> positive integers
> rational number
> negative integers
> polynomial

PART 2: SKILLS

Graph the integers on a number line. (Section 10-1)

12. $0; -1; 3; -2$
13. $4; 2; -5; 1$
14. $-3; -4; 6; 0$
15. $5; -1; 0; -3$

Replace each __?__ with < or >. (Section 10-1)

16. 4 __?__ -4
17. 6 __?__ -2
18. -4 __?__ 3
19. 0 __?__ 4
20. 0 __?__ -1
21. -3 __?__ 3
22. -7 __?__ 6
23. -2 __?__ -4

Write a fraction for each rational number. (Section 10-2)

24. 3
25. -5
26. $2\frac{1}{2}$
27. $-3\frac{1}{3}$
28. 4.5
29. -2.3

Write the opposite of each number. (Section 10-2)

30. 6
31. -6
32. 0
33. 8
34. 12
35. -7

Add. (Sections 10–3 and 10–4)

36. $9 + (-2)$ **37.** $-4 + 12$ **38.** $16 + (-20)$ **39.** $-10 + 5$
40. $-4 + (-8)$ **41.** $-13 + (-9)$ **42.** $9 + 17$ **43.** $-8 + 24$
44. $-15 + 7$ **45.** $-11 + (-22)$ **46.** $-1 + 6$ **47.** $-4 + 4$

Subtract. (Section 10–5)

48. $6 - 8$ **49.** $-3 - 14$ **50.** $-9 - (-12)$ **51.** $13 - 8$
52. $15 - (-9)$ **53.** $3 - 15$ **54.** $-4 - (-3)$ **55.** $8 - (-8)$
56. $19 - 32$ **57.** $-7 - (-20)$ **58.** $8 - (-5)$ **59.** $-3 - 14$

Add. (Section 10–6)

60. $3 + (-8) + 12$ **61.** $-9 + (-4) + 6$
62. $-12 + 15 + (-3)$ **63.** $-7 + (-11) + 8 + (-13)$
64. $12 + 20 + (-18) + 30$ **65.** $12 + 16 + 22 + (-14)$

(Section 10–7)

66. $(4r - 7) + (-r + 3)$ **67.** $(-6x^2 + 5x) + (2x^2 - 8x)$
68. $(-2y^2 - 5y + 7) + (5y^2 + 11y + 3)$ **69.** $(8b^2 - 5b + 2) + (-2b^2 + 6b - 3)$

PART 3: APPLICATIONS

Write a rational number to represent each word description.
(Sections 10–1 and 10–2)

70. 4° below zero

71. $32\frac{1}{2}$ feet below sea level

72. A gain of 2.5 kilograms

73. An increase of $245

Solve each problem.

74. The lowest temperature ever recorded in Los Angeles, California, was 28°F. The lowest temperature ever recorded in Atlanta, Georgia, was −3°F. Replace the __?__ with < or >: 28° __?__ −3°. (Section 10–1)

75. A football team lost 8 yards on one play and gained 10 yards on the next play. Use a number line to find the total yardage on the two plays. (Section 10–3)

76. A submarine dove to a level of 320 feet below the ocean's surface. It later descended 150 feet. What was the new depth? (Section 10–4)

77. The lowest temperature ever recorded in Louisiana was −16°F. The highest temperature ever recorded was 114°F. Find the difference. (Section 10–5)

Rational Numbers: Addition/Subtraction

CHAPTER TEST

Replace each __?__ with < or >.

1. 5 __?__ −2
2. 2 __?__ −2
3. 0 __?__ −6
4. −9 __?__ −3

Add.

5. 21 + (−14)
6. −16 + 9
7. −43 + (−29)
8. −48 + 60
9. 9 + (−18)
10. −28 + 35
11. −11 + (−17)
12. 27 + (−56)

Subtract.

13. −4 − 21
14. −14 − (−32)
15. 9 − 22
16. 5 − (−12)
17. 20 − 35
18. 14 − (−8)
19. −8 − 12
20. −43 − (−18)

For Exercises 21–23, add.

21. −5 + 4 + (−12)
22. 8 + (−3) + 2 + (−7)
23. 12 + (−9) + 18

24. A football team lost 4 yards on one play and gained 7 yards on the next play. Find the total yardage for the two plays.

25. The actual temperature is 23°F. The wind chill temperature is −11°F. How much colder than the actual temperature is the wind chill temperature?

ANSWERS TO REVIEW CAPSULES

Page 297 1. $\frac{11}{5}$ 2. $\frac{17}{3}$ 3. $\frac{79}{9}$ 4. $\frac{19}{6}$ 5. $\frac{21}{2}$ 6. $\frac{23}{12}$ 7. $\frac{25}{4}$ 8. $\frac{7}{5}$ 9. $\frac{31}{10}$ 10. $\frac{63}{10}$ 11. $\frac{29}{7}$ 12. $\frac{29}{9}$

Page 299 1. The points on the number line from left to right in the following order: −1; 0; 1; 3 2. The points on the number line from left to right in the following order: −5; −3; 1; 2 3. The points on the number line from left to right in the following order: −2; 0; 4; 6 4. The points on the number line from left to right in the following order: −6; −4; 5; 7 5. > 6. > 7. < 8. > 9. < 10. < 11. > 12. >

Page 304 1. > 2. > 3. < 4. < 5. > 6. < 7. > 8. < 9. −6 10. −15 11. 8 12. −3 13. 0 14. 7

Page 307 1. −12 2. −4 3. 28 4. 15 5. −3.7 6. $1\frac{1}{3}$

Page 310 1. 10 2. 73 3. 34 4. 69 5. 257 6. 178 7. −17 8. −4 9. 23 10. −29 11. 8 12. −32 13. 0 14. 0

Page 315 1. $7a$ 2. $3m$ 3. $7x^2$ 4. $14b^2$ 5. $6a + 5$ 6. $8x + 9$ 7. $5t^2 + 12$ 8. $12n^2 + 6$ 9. 7 10. −7 11. 3 12. 9 13. −4 14. −10 15. 2 16. −4

320 Chapter 10

ENRICHMENT

Clock Arithmetic

Think of a 7-hour clock such as that shown at the right. Addition on this clock is the same as on a 12-hour clock. That is,

$4 + 5 = 2$ ◀ **Start at 4. Move 5 units in a clockwise direction.**

You learned in algebra that the sum of opposites is 0.

In Algebra: $7 + (-7) = 0$ $(-1) + 1 = 0$ $0 + 0 = 0$

On this 7-hour clock, the sum of opposites is also 0.

$1 + 6 = 0$ $2 + 5 = 0$ $3 + 4 = 0$ $0 + 0 = 0$ ◀ **Opposites: 1 and 6; 2 and 5; 3 and 4; 0 and 0**

NOTE: On the 7-hour clock, opposites are <u>not directly opposite</u> each other.

To subtract in algebra, you add the opposite of the number you are subtracting. To subtract on a 7-hour clock, you follow a similar procedure.

EXAMPLES Evaluate on the 7-hour clock: **a.** $6 + 6$ **b.** $1 - 3$

Solutions: **a.** Start at 6. Move 6 units in a clockwise direction. $6 + 6 = 5$

b. $1 - 3 = 1 + 4$ ◀ **Add 4, the opposite of 3.**
$= 5$

EXERCISES

Evaluate. Use a 7-hour clock.

1. $5 + 3$ **2.** $5 + 5$ **3.** $4 + 2$ **4.** $6 + 1$ **5.** $2 + 0$ **6.** $6 + 5$
7. $1 - 5$ **8.** $3 - 6$ **9.** $3 - 4$ **10.** $2 - 2$ **11.** $0 - 1$ **12.** $6 - 4$

Multiplication on a 7-hour clock can be defined as "repeated addition." That is, $4 \times 3 = 4 + 4 + 4 = 1 + 4 = 5$. Multiply.

13. 4×2 **14.** 3×6 **15.** 3×5 **16.** 4×5 **17.** 6×6 **18.** 3×3

Rational Numbers: Addition/Subtraction

ADDITIONAL PRACTICE

SKILLS

For Exercises 1–3:
  a. Graph the integers on a number line.
  b. Write the integers in order from least to greatest.
     (Pages 294–296)

1. 0; −4; −6; 2   **2.** −5; 1; 3; −3   **3.** 4; 5; 3; 2; −1

Replace each __?__ with > or <. (Pages 294–296)

4. 0 __?__ −6   **5.** 3 __?__ −1   **6.** −10 __?__ −2   **7.** −13 __?__ −1

Write a fraction for each rational number. (Pages 297–299)

8. −4   **9.** 0.6   **10.** 7.2   **11.** −3.5   **12.** $-2\frac{1}{2}$ **13.** $3\frac{1}{3}$

Write the opposite of each number. (Pages 297–299)

14. −6.1   **15.** 4   **16.** 0   **17.** −7   **18.** $3\frac{1}{3}$ **19.** −2.5

Use a number line to add. (Pages 300–302)

20. 5 + 3   **21.** −8 + 4   **22.** 4 + (−8)   **23.** −4 + (−2)

Evaluate. (Pages 304–306)

24. |6| **25.** |−3| **26.** |−17| **27.** |32| **28.** |1| **29.** |0|

Perform the indicated operations. (Pages 304–314)

30. −27 + 31   **31.** −54 + 63   **32.** −12 + (−36)   **33.** −22 + (−56)
34. 18.4 + (−7.9)   **35.** 19.2 + (−27.1)   **36.** −3 − (−4)   **37.** −4 − (−6)
38. −2 − 7   **39.** −5 − 9   **40.** 7 − (−4)   **41.** 5 − (−4)
42. $-\frac{2}{3} + \frac{1}{6}$ **43.** $-\frac{1}{4} + \frac{3}{8}$ **44.** $\frac{1}{2} - \left(-\frac{3}{4}\right)$ **45.** $\frac{1}{3} - \left(-\frac{1}{6}\right)$
46. (−5) + (−3) + 6 + (−1)   **47.** 8 + (−3) + 4 + (−9)
48. −6 + 17 + (−19) + 7   **49.** (−1) + (−6) + (−3) + 9

APPLICATIONS

50. The high temperature for the day was 8°C and the low temperature was −6°C. How much warmer was the high temperature than the low temperature? (Pages 308–310)

51. A football team lost 7 yards on first down, gained 13 yards on second down, and lost 4 yards on third down. Find the total yardage for the three downs. (Pages 312–314)

322   Chapter 10

CHAPTER 11
Equations: Addition/Subtraction

SECTIONS
- 11-1 Equations: Addition/Subtraction
- 11-2 Like Terms/More Than One Operation
- 11-3 Variable on Both Sides
- 11-4 Problem Solving and Applications: Words to Symbols: Addition and Subtraction
- 11-5 Inequalities on the Number Line
- 11-6 Using Addition and Subtraction to Solve Inequalities

FEATURES
Calculator Application: *Checking Equations*

Calculator Application: *Checking Inequalities*

Consumer Application: *Banking*

Computer Application: *Adding Fractions*

Enrichment: *Compound Inequalities*

11-1 Equations: Addition/Subtraction

You used the following properties when you solved equations in Chapter 7.

> **Addition Property for Equations**
>
> Adding the same number to each side of an equation forms an equivalent equation.
>
> $p - 8 = -10$
> $p - 8 + 8 = -10 + 8$
> $p = -2$

EXAMPLE 1 Solve and check: $x - 6 = -3$

Solution: To get x alone, add 6 to each side.

$x - 6 = -3$
$x - 6 + 6 = -3 + 6$
$x = 3$

Check: $x - 6 = -3$
$3 - 6$
$3 + (-6)$
-3

Self-Check Solve and check.

1. $t - 4 = -8$
2. $a - 13 = 8$
3. $-15 = b - 12$

> **Subtraction Property for Equations**
>
> Subtracting the same number from each side of an equation forms an equivalent equation.
>
> $t + 4 = -15$
> $t + 4 - 4 = -15 - 4$
> $t = -15 + (-4)$
> $t = -19$

EXAMPLE 2 Solve and check: $y + 5 = -8$

Solution:
$y + 5 = -8$
$y + 5 - 5 = -8 - 5$ ◄ $-8 - 5$ means $-8 + (-5)$
$y = -8 + (-5)$
$y = -13$

Check: $y + 5 = -8$
$-13 + 5$
-8

Self-Check Solve and check.

4. $p + 6 = -9$
5. $m + 8 = 4$
6. $-12 = r + 5$

CLASSROOM EXERCISES

Write the equivalent equations formed as the first step in solving each equation. (Examples 1 and 2)

1. $y + 18 = 7$
2. $x + 24 = 20$
3. $a - 5 = 22$
4. $q - 18 = -30$
5. $-15 = 23 + b$
6. $-28 = y + 39$

For Exercises 7–18 solve and check. (Example 1)

7. $a - 4 = -5$
8. $r - 8 = -20$
9. $q - 9 = 20$
10. $b - 4 = 10$
11. $-8 = y - 6$
12. $20 = m - 14$

(Example 2)

13. $p + 4 = -8$
14. $c + 5 = -19$
15. $n + 1 = 4$
16. $x + 3 = 8$
17. $0 = d + 7$
18. $-6 = m + 22$

WRITTEN EXERCISES

Goal: To solve equations by using the Addition and Subtraction Properties for Equations

For Exercises 1–39, solve and check each equation. (Example 1)

1. $x - 10 = 16$
2. $y - 42 = 42$
3. $a - 6 = -4$
4. $c - 10 = -15$
5. $d - 3 = -16$
6. $b - 8 = -11$
7. $x - 14 = 25$
8. $-15 = d - 24$
9. $13 = k - 18$
10. $p - 12 = -15$
11. $-6.2 = q - 4.4$
12. $t - 5.5 = 9.8$

(Example 2)

13. $h + 2 = 8$
14. $k + 8 = 2$
15. $n + 5 = -4$
16. $b + 15 = -32$
17. $t + 9 = -12$
18. $w + 8 = 3$
19. $p + 29 = 16$
20. $t + 8 = -19$
21. $-23 = m + 12$
22. $27 = q + 35$
23. $b + 13.7 = 19.6$
24. $5.1 = c + 9.4$

MIXED PRACTICE

25. $a + 48 = 23$
26. $p + 38 = -5$
27. $x - 47 = -29$
28. $-10 = g + 4$
29. $7 = t - 15$
30. $12 = y + 14$
31. $12 + a = -90$
32. $t - 15 = -50$
33. $t - 16 = -16$
34. $-19.5 = 8.7 + c$
35. $n + 22.9 = 15.4$
36. $a - 11.9 = -27.4$
37. $33.4 = k + 22.9$
38. $y - 5\frac{3}{4} = -3\frac{1}{4}$
39. $m + 2\frac{3}{5} = 1\frac{1}{5}$

Equations: Addition/Subtraction

REVIEW CAPSULE FOR SECTION 11-2

Combine like terms. (Pages 76–78)   The answers are on page 346.

1. $3x + 2x$
2. $7p + 6p$
3. $8q - 6q$
4. $4t - 2t$
5. $8x + x$
6. $9b + b$
7. $2m - m$
8. $5q - 4q$

Solve each equation. (Pages 200–202)

9. $8t = 352$
10. $15 = 60z$
11. $1.2q = 1.32$
12. $16 = 0.04r$

11-2 Like Terms/More Than One Operation

When solving an equation, combine like terms on each side of the equation before applying the Addition and Subtraction Properties for Equations. Then divide each side by the same number (if necessary) as you did in Chapter 7 (see pages 200-202).

EXAMPLE 1 Solve and check. $5x + 3x = -8 + 24$

Solution:
[1] Combine like terms. $8x = 16$
[2] Divide each side by 8. $\dfrac{8x}{8} = \dfrac{16}{8}$, $x = 2$

Check:
$$5x + 3x = -8 + 24$$
$$5(2) + 3(2) = 16$$
$$10 + 6$$
$$16$$

Self-Check *Solve and check.*

1. $2x + 4x = -5 + 29$
2. $-x + 4x = 8 + 7$

EXAMPLE 2 Solve and check: $2x - 4 + 3x = -3 + 14$

Solution:
[1] Combine like terms. $2x - 4 + 3x = -3 + 14$
$5x - 4 = 11$
[2] To get $5x$ alone, add $5x - 4 + 4 = 11 + 4$
4 to each side. $5x = 15$
[3] Divide each side by 5. $\dfrac{5x}{5} = \dfrac{15}{5}$, $x = 3$

The check is left for you

Self-Check *Solve and check.*

3. $x - 3 + 2x = -2 + 17$ **4.** $5x + 6 - x = -7 + 25$

CLASSROOM EXERCISES

*Combine like terms on both sides of each equation. Do **not** solve the equation.* (Example 1, step 1)

1. $6x + 2x = -2 + 34$
2. $y + 2y = -7 + 10$
3. $-c + 2c = 25 - 6$
4. $4n - n = -4 + 13$

(Example 2, step 1)

5. $3x - 5 + 4x = -3 + 19$
6. $p - 2 + 3p = -4 + 18$
7. $-a + 4 + 3a = -5 + 11$
8. $4g + 3 - g = -6 + 36$

Solve and check. (Example 1)

9. $6x + 3x = -4 + 58$
10. $y + 2y = -12 + 21$
11. $5h - h = 34 - 2$
12. $8k - k = 54 - 5$

(Example 2)

13. $4x - 7 + 2x = -2 + 37$
14. $q - 5 + 4q = -11 + 26$
15. $-b + 3 + 2b = -1 + 8$
16. $7a + 8 - a = -9 + 35$

WRITTEN EXERCISES

Goal: To solve equations containing like terms

Solve and check. (Example 1)

1. $4a + 5a = 37 + 44$
2. $8c + 9c = -12 + 80$
3. $8e + 12e = -15 + 95$
4. $2x + 11x = 144 + 25$
5. $5p + 6p = 110 - 11$
6. $2y + y = -5 + 32$
7. $11m - m = -4 + 34$
8. $9v - v = -25 + 89$
9. $-n + 8n = -7 + 168$
10. $-q + 2q = -9 + 36$
11. $-b + 3b = -17 + 21$
12. $5.2g + 6.3g = -2 + 71$
13. $2\frac{3}{4}f - \frac{1}{4}f = -6 + 16$
14. $4\frac{1}{2}d - 1\frac{1}{2}d = 107 - 17$

Equations: Addition/Subtraction

Solve and check. (Example 2)

15. $3x - 8 + x = -3 + 19$
16. $w - 5 + 4w = -10 + 15$
17. $2h + 8 + 3h = -9 + 27$
18. $k + 6 + 6k = -13 + 68$
19. $4z - 12 - z = -6 + 9$
20. $13y - 9 - y = -10 + 85$
21. $7m - 7 + 4m = -15 + 74$
22. $3n - 14 + 5n = -23 + 105$
23. $-p + 13 + 3p = -7 + 58$
24. $-c + 9 + 10c = -17 + 125$
25. $-k - 21 + 6k = -11 + 155$
26. $12a - 52 + 13a = -12 + 135$
27. $3\frac{1}{2}g - 8 + 1\frac{1}{2}g = -3\frac{3}{5} + 23\frac{3}{5}$
28. $2\frac{1}{3}x - 5 + 1\frac{2}{3}x = -3\frac{1}{4} + 7\frac{3}{4}$

MIXED PRACTICE

29. $-q + 3q = -8 + 15$
30. $3y - 8 + y = -7 + 50$
31. $3n - 14 + 5n = -23 + 113$
32. $4x + 2x = -5 + 43$
33. $4\frac{1}{2}a + 5\frac{1}{2}a = -6 + 71$
34. $\frac{3}{4}k - 8 + 6\frac{1}{4}k = -12 + 60$
35. $3x + 5x = 17 + 31$
36. $t + 3t = 18 + 6$
37. $11\frac{1}{8}m - 5\frac{1}{8} - 3\frac{1}{8}m = -5\frac{1}{8} + 64$
38. $-b + 3b = -17\frac{1}{2} + 21$
39. $8.5v - 2.1v = -25 + 89$
40. $12c - 52.5 + 13c = -11.5 + 134$
41. $4m - 7 + 5m = 29 + 9$
42. $5z - 3 + 2z = 11 + 7$
43. $6.5x - 9.5 - 1.3x = 32 + 10.5$
44. $-2.3y + 7.2 + 6.8y = 25.2$
45. $-13.1 + 3.2n + 2.8n = 3.2 + 7.7$
46. $-11.4 - 3.7v + 6.7v = 3.1 - 5.5$
47. $y + 3 + 3y = 14 + 11$
48. $5w - 11 - w = 26 + 8$

REVIEW CAPSULE FOR SECTION 11-3

Combine like terms. (Pages 79–80)

1. $4d + 2d + 6$
2. $3p - 2p + 8$
3. $4x - 6 - 2x$
4. $9 + 6x - 2x$
5. $4 + 6 + 2r + 3r$
6. $7 + 2b - 5 + 3b$
7. $2 + 3 + 5y - y$
8. $8c + 4 - 3c + 9$
9. $-x + 3 + 4x - 5$
10. $2x + 3 - 5x + 8$
11. $n - 7 - 14n + 2$
12. $3y + 8 + 9y - 5$

Solve. (Pages 324–325)

13. $x + 3 = 12$
14. $y - 4 = 12$
15. $a - 4 = -6$
16. $b + 7 = -9$
17. $m + 6 = 3$
18. $n + 15 = 4$
19. $p - 5 = -1$
20. $q - 9 = -3$
21. $8 + x = 12$
22. $7 + 2y = 19$
23. $-5 + 3g = -20$
24. $-12 + 2t = -18$
25. $3\frac{1}{2} + 2d = 7\frac{1}{2}$
26. $5\frac{1}{5} + 3x = 11\frac{1}{5}$
27. $n - 4\frac{2}{3} = -5\frac{2}{3}$

The answers are on page 346.

11-3 Variable on Both Sides

Some equations have a variable on both sides. Use the Addition and Subtraction Properties for Equations to get the variable <u>alone on one side</u>.

EXAMPLE 1 Solve and check: $y = 14 - y$

Solution:

⬚1 To get rid of $-y$ on the right side, add y to both sides.

$y = 14 - y$
$y + y = 14 - y + y$
$2y = 14$

⬚2 To get y alone, divide both sides by 2.

$\dfrac{2y}{2} = \dfrac{14}{2}$
$y = 7$

Check: $y = 14 - y$
$7 \mid 14 - 7$
$7 \mid 7$

Self-Check *Solve and check.*

1. $n = 10 - n$ **2.** $2m = 9 - m$ **3.** $4b = 60 - b$

Combine like terms on each side of an equation before applying the Addition and Subtraction Properties for Equations.

EXAMPLE 2 Solve and check: $2x + 9 = 38 + x$

Solution:

⬚1 To get rid of x on the right side, subtract x from both sides.

$2x + 9 = 38 + x$
$2x - x + 9 = 38 + x - x$
$x + 9 = 38$

⬚2 To get x alone, subtract 9 from both sides.

$x + 9 - 9 = 38 - 9$
$x = 29$

Check: $2x + 9 = 38 + x$
$2(29) + 9 \mid 38 + 29$
$58 + 9 \mid 38 + 29$
$67 \mid 67$

Self-Check *Solve and check:* **4.** $3e + 7 = 11 + 2e$ **5.** $6 + 2k = k + 10$

Equations: Addition/Subtraction **329**

CLASSROOM EXERCISES

Complete. (Example 1)

1. $v = 6 - v$
    $v + \underline{\ ?\ } = 6 - v + \underline{\ ?\ }$
    $\underline{\ ?\ } = 6$
    $v = \underline{\ ?\ }$

2. $3a = 20 - a$
    $3a + \underline{\ ?\ } = 20 - a + \underline{\ ?\ }$
    $\underline{\ ?\ } = 20$
    $a = \underline{\ ?\ }$

3. $5d = 72 - d$
    $5d + \underline{\ ?\ } = 72 - d + \underline{\ ?\ }$
    $\underline{\ ?\ } = 72$
    $d = \underline{\ ?\ }$

(Example 2)

4. $4g + 8 = 11 + 3g$
    $4g - \underline{\ ?\ } + 8 = 11 + 3g - \underline{\ ?\ }$
    $\underline{\ ?\ } + 8 = 11$
    $g = \underline{\ ?\ }$

5. $6h + 5 = 12 + 5h$
    $6h - \underline{\ ?\ } + 5 = 12 + 5h - \underline{\ ?\ }$
    $\underline{\ ?\ } + 5 = \underline{\ ?\ }$
    $h = \underline{\ ?\ }$

WRITTEN EXERCISES

Goal: To solve an equation with the variable on both sides

For Exercises 1–32, solve and check. (Example 1)

1. $a = 20 - a$
2. $b = 28 - b$
3. $2c = 21 - c$
4. $3d = 36 - d$
5. $4e = 80 - e$
6. $5f = 54 - f$
7. $g = 15 - g$
8. $h = 23 - h$
9. $7k = 28 - k$
10. $2m = 29 - m$
11. $3.6n = 100 + 1.6n$
12. $6.3p = 81 + 3.3p$

(Example 2)

13. $5q + 5 = 12 + 4q$
14. $3r + 3 = 5 + 2r$
15. $2s - 3 = 7 + s$
16. $7t - 7 = 5 + 6t$
17. $4u - 12 = -4 + 3u$
18. $10v - 15 = -10 + 9v$
19. $6.5w + 10 = 17 + 5.5w$
20. $8.3x - 21 = -12 + 7.3x$
21. $3y - 5\frac{1}{2} = 2\frac{1}{2} + 2y$
22. $15z + 3\frac{1}{8} = 7\frac{1}{8} + 14z$
23. $5\frac{1}{2}a + 14\frac{1}{4} = 19\frac{1}{4} + 4\frac{1}{2}a$
24. $\frac{13}{3}b - \frac{11}{3} = \frac{4}{3} + \frac{10}{3}b$

MIXED PRACTICE

25. $15c = 48 - c$
26. $12e + 10 = 16 + 11e$
27. $3f + 6 = 8 + 2f$
28. $11d = 132 - d$
29. $4.7g = 90 + 1.7g$
30. $3.8h = 28 + 1.8h$
31. $13k - 4\frac{1}{3} = 3\frac{2}{3} + 11k$
32. $6m - 5\frac{1}{4} = 2\frac{3}{4} + 2m$

CHECKING EQUATIONS

You can use a calculator with a $+/-$ key to check solutions of equations. First, use the solution to evaluate the left side of the equation. Store this value in memory. Then use the solution to evaluate the right side of the equation. Compare with the value in memory.

EXAMPLE Check the solution $x = -16$ for the equation $8 - 2x = 24 - x$.

Left side: 8 [−] 2 [×] 1 6 [+/−] [=] [M+] → 40.

Right side: 2 4 [−] 1 6 [+/−] [=] → 40.

EXERCISES Check the given solution to each equation.

1. $-14m - 35 = 46 - 13m$; $m = -3$
2. $-12v + 40 = -13v + 112$; $v = 72$
3. $4.7x - 1.1 = 5.7x - 2.5$; $x = 1.5$
4. $19t - 80 = -16t - 1025$; $t = 27$

REVIEW: SECTIONS 11-1 – 11-3

Solve and check. (Section 11-1)

1. $n + 5 = 12$
2. $p - 7 = 3$
3. $q + 9 = -15$
4. $r - 3 = -7$
5. $s - 6 = -4$
6. $t + 10 = 3$
7. $a - 12 = -15$
8. $d - 9 = -12$

Solve and check. (Section 11-2)

9. $3p + 8 + 2p = 5 + 8$
10. $8a - 14 + 7a = -4 + 65$
11. $22 - 5 = -7 - y + 5y$
12. $-c + 14 + 7c = -8 + 40$

Solve and check. (Section 11-3)

13. $p = 6 - p$
14. $2a = 27 - a$
15. $5v + 9 = 17 + 4v$
16. $3r - 15 = -10 + 2r$

REVIEW CAPSULE FOR SECTION 11-4

Solve and check. (Pages 324-325)

1. $n + 6 = 18$
2. $t + 9 = 15$
3. $b + 14 = 27$
4. $r + 21 = 55$
5. $m - 9 = 3$
6. $n - 4 = 6$
7. $a - 17 = 8$
8. $c - 32 = 15$

The answers are on page 346.

Equations: Addition/Subtraction

CONSUMER APPLICATION Banking

When you borrow money from a bank, the bank charges you interest for the use of the money.

You can estimate the yearly percentage rate of interest by using this formula.

$$r = \frac{2Mi}{B(n+1)}$$

r: yearly interest rate
i: interest charge
B: amount of loan
M: number of payment periods per year
n: number of payments

The formula can only be used when the same number of payments are made each year.

EXAMPLE A $5000 loan was repaid in 36 monthly payments. The interest charge was $800. Estimate to the nearest tenth the yearly percentage rate of interest.

SOLUTION $r = \dfrac{2Mi}{B(n+1)}$ ← $M = 12$; $i = \$800$; $B = \$5000$; $n = 36$

$r = \dfrac{2(12)(800)}{5000(36+1)}$ ← Use a calculator.

2 × 12 × 800 ÷ 5000 ÷ 37 =   0.1037837

The annual percentage rate of interest is about **10.4%**.

Exercises

Complete the table. Round each answer to the nearest tenth.

	Loan Amount	Interest Charge	Payment Periods per year	Number of Payments	Yearly Percentage Rate of Interest
1.	$700	$30	12	6	?
2.	$900	$55	12	12	?
3.	$1100	$150	12	18	?
4.	$2500	$410	12	30	?
5.	$3400	$800	12	36	?

COMPUTER APPLICATIONS Adding Fractions

You can find the sum of two fractions by using the following program.

Program:
```
100 INPUT A, B, C, D
110 IF A*B*C*D <> 0 THEN 140
120 PRINT "NO ZERO VALUES.  TRY AGAIN."
130 GOTO 100
140 LET N = A*D + B*C
150 LET D1 = B*D
160 IF N = 0 THEN 290
170 LET N1 = ABS(N)
180 LET D2 = ABS(D1)
190 IF N1 > D2 THEN 220
200 LET S = N1
210 GOTO 230
220 LET S = D2
230 FOR Y = S TO 1 STEP -1
240 IF N1/Y <> INT(N1/Y) THEN 260
250 IF D2/Y = INT(D2/Y) THEN 270
260 NEXT Y
270 LET N = N/Y
280 LET D1 = D1/Y
290 PRINT "(";A;"/";B;") + (";C;"/";D;") =";
300 IF ABS(D1) = 1 THEN 360
310 IF N * D1 < 0 THEN 340
320 PRINT ABS(N);"/";ABS(D1)
330 GOTO 370
340 PRINT "-";ABS(N);"/";ABS(D1)
350 GOTO 370
360 PRINT N * D1
370 PRINT
380 END
```

This program is based on the following.

In Algebra

$$\frac{a}{b} + \frac{c}{d} = \frac{ad}{bd} + \frac{bc}{bd} = \frac{ad + bc}{bd}$$

In Arithmetic

$$\frac{2}{3} + \frac{5}{7} = \frac{2 \cdot 7}{3 \cdot 7} + \frac{5 \cdot 3}{3 \cdot 7} = \frac{(2 \cdot 7) + (5 \cdot 3)}{3 \cdot 7}$$

Exercises Use the program to find each sum.

1. $\frac{1}{2} + \frac{1}{5}$
2. $\frac{2}{3} + \frac{7}{8}$
3. $\frac{4}{9} + \left(-\frac{1}{3}\right)$
4. $\frac{5}{16} + \frac{4}{5}$
5. $\left(-\frac{5}{6}\right) + \frac{1}{3}$
6. $\frac{3}{8} + \left(-\frac{1}{10}\right)$
7. $\left(-\frac{9}{5}\right) + \left(-\frac{3}{5}\right)$
8. $\left(-\frac{7}{2}\right) + \left(-\frac{12}{15}\right)$

PROBLEM SOLVING AND APPLICATIONS

11-4 Words to Symbols: Addition and Subtraction

Jenna's basketball coach told her that she had scored 12 more points this season than last season. Jenna thought:

Points last season: p

12 more points than last season: $p + 12$

The following table shows some word expressions with the corresponding algebraic expressions.

TABLE

Operation	Word Expression	Algebraic Expression
Addition	Twelve **more than** the number of points, p, scored	$p + 12$
	The **sum** of the distance, d, and 20	$d + 20$
	Carol's score, s, **plus** 8	$s + 8$
	The number of problems, p, **increased by** 10	$p + 10$
	The number of hours, h, **added to** 40	$40 + h$
	The **total** of 16 and $-r$	$16 + (-r)$
Subtraction	The **difference** between Juan's age, a, and 18	$a - 18$
	The number of days, d, **minus** 6	$d - 6$
	Thirty **decreased by** the number of years, y	$30 - y$
	Fifteen **less than** the number of meters, m	$m - 15$
	t **decreased** by -4	$t - (-4)$

Jenna knows that she scored 240 points this season. That is 12 points more than she scored last season. She can use an equation to find how many points she scored last season.

> **PROCEDURE**
>
> **To use an equation to solve word problems:**
> 1. Choose a variable to represent the unknown.
> 2. Write an equation for the problem.
> 3. Solve the equation.
> 4. Check the solution with the statements in the problem. Answer the question.

EXAMPLE 1 Twelve more than the number of points Jenna scored last season is 240. Find how many points she scored last season.

Solution:
1. Choose a variable.   Let p = the number of points.
2. Write an equation.   12 more than the number of points is 240.

$$p + 12 = 240$$

3. Solve the equation. $p + 12 - 12 = 240 - 12$
$$p = \mathbf{288}$$

4. Check: Does $228 + 12 = 240$?   Yes ✔

Jenna scored **288 points.**

Self-Check

1. The original cost of a radio increased by $15 is $68. Find the original cost of the radio.

EXAMPLE 2 Juan's weekly salary decreased by $24 in deductions is $198. What is his weekly salary?

Solution:
1. Let x = Juan's weekly salary.
2. The weekly salary decreased by $24 is $198.

$$x - 24 = 198$$

3. $x - 24 + 24 = 198 + 24$
$$x = 222$$

4. Check: Does $222 - 24 = 198$?   Yes ✔

Juan's weekly salary is **$222.**

Self-Check

2. Twelve pounds less than Laura's weight is 92 pounds. Find Laura's weight.

Equations: Addition/Subtraction

CLASSROOM EXERCISES

For Exercises 1–10, choose from the box at the right the algebraic expression that represents each word expression. More than one answer may be possible in some cases. (Table)

1. y minus 6
2. 6 plus y
3. y increased by 6
4. the total of 6 and y
5. y more than 6
6. 6 less than y
7. the sum of 6 and y
8. y plus 6
9. y decreased by 6
10. 6 decreased by y

| $y + 6$ | $6 + y$ |
| $y - 6$ | $6 - y$ |

Complete. (Examples 1 and 2, step 2)

Word Sentence	Equation
11. Twelve dollars more than the cost, c, is $27	__?__ = 27
12. The number of club members, m, increased by 5 is 24	__?__ = 24
13. Sixty workdays decreased by the number of vacation days, d, is 24	__?__
14. The number of desks, n, increased by 40 is 98	__?__
15. The number of kilometers, k, on the odometer minus 100 is 425	__?__
16. Nine is the difference between Tom's score, s, and 45	__?__

WRITTEN EXERCISES

Goals: To write an algebraic equation for a word sentence
To apply the skill of writing algebraic equations to solving word problems

Write an algebraic expression for each word expression. (Table)

1. The difference of 25 and n
2. 16 more than p
3. The total of x and 5
4. y decreased by 17
5. Evelyn's age, e, minus 6
6. Eleven added to n
7. 32 plus b
8. 45 decreased by w
9. The sum 85 and c
10. t minus 25
11. 14.5 less than d
12. a increased by 0.6
13. 25 years more than y
14. r minus 2.6
15. 42 decreased by $(-p)$
16. $(-t)$ added to 5

For Exercises 17–24, write an equation for each word sentence. Use t for the variable in each exercise. Do not solve the equation.
(Examples 1 and 2, step 2)

17. Nineteen more than a certain number is 42.
18. Twelve dollars more than the cost of a jacket is $57.
19. The number of tickets sold increased by 32 is 78.
20. The total of Rick's age and 15 is 31.
21. The difference of 24 and a certain number is 15.
22. Fourteen less than the number of sandwiches is 264.
23. The number of apples picked minus 12 is 52.
24. The length of a room decreased by 3 meters is 15 meters.

For Exercises 25–36, solve each problem. (Example 1)

25. Thirty-nine kilometers more than the distance traveled by the Blair family is 721 kilometers. How far did the Blair family travel?
26. Last Tuesday's average temperature plus 7° is 88°. What was last Tuesday's average temperature?
27. The amount of the Ohira family's electric bill increased by $8.50 is $45.25. Find the amount.
28. The number of passengers increased by a flight crew of 7 is 192. Find the number of passengers.

(Example 2)

29. The original cost of a pair of skis decreased by a discount of $30 is $135. Find the original cost of the skis.
30. The number of words Raymond types per minute decreased by 13 is 75. How many words does Raymond type per minute?
31. Sixty minutes minus the number of minutes for commercials is 52. How many minutes are used for commercials?
32. Thirteen dollars less than the original cost of a hair dryer is $18. Find the original cost of the hair dryer.

MIXED PRACTICE

33. The amount of snow plus 7 centimeters of rain is 16 centimeters. Find the amount of snow.
34. The number of library books minus 43 science books is 325. Find the number of library books.
35. The number of students in a karate class decreased by 15 is 19. Find the number of students.
36. Thirty-one more than the number of marathon runners is 124. Find the number of marathon runners.

Equations: Addition/Subtraction **337**

REVIEW CAPSULE FOR SECTION 11-5

Complete. Replace each __?__ with < or >. (Pages 294–296)

1. 4 __?__ 6
2. 4 __?__ −6
3. −3 __?__ −5
4. 6 __?__ 2
5. 0 __?__ 5
6. −1 __?__ −9
7. 0 __?__ −3
8. 15 __?__ −10
9. −9 __?__ −8
10. 7 __?__ 70

The answers are on page 346.

11-5 Inequalities on the Number Line

The fuel gauge on a plane shows that it has enough fuel for almost 4 hours of flying time. Thus, the flying time of the plane can be expressed by the <u>inequality</u> $t < 4$, where t is the number of hours.

The symbols < or > are used to write inequalities.

"<" means "is less than."   ">" means "is greater than."

You can graph inequalities on a number line.

EXAMPLE 1 Graph each inequality: **a.** $x > -6$ **b.** $x < 4$

Solutions: a.

The open circle at −6 means −6 is *not* on the graph.

The heavy line and arrow pointing to the right indicate that the graph includes all numbers greater than −6. Thus, **all numbers greater than −6 are** solutions of the inequality $x > -6$.

b.

The heavy line and arrow pointing to the left indicate that the graph includes all numbers less than 4. Thus, **all numbers less than 4 are** solutions of the inequality $x < 4$.

Self-Check Graph each inequality: **1.** $y > 2$ **2.** $y < -1$ **3.** $x < 0$

The symbol "≤" means "is less than or equal to."

The symbol "≥" means "is greater than or equal to."

EXAMPLE 2 Graph each inequality.

  a. $n \leq -3$ **b.** $n \geq 1$

Solutions: **a.** [number line with solid dot at -3, shaded left, from -8 to 6]

The large dot at −3 means −3 *is* a solution.

b. [number line with solid dot at 1, shaded right, from -4 to 10]

Self-Check *Graph each inequality:* **4.** $n \leq 3$ **5.** $n \geq -5$ **6.** $y \geq 0$

CLASSROOM EXERCISES

Which of these are solutions for $d > -3$? Write Yes or No.
(Example 1)

| 1. -4 | 2. 3 | 3. 5 | 4. -3 | 5. 0 | 6. -8 |
| 7. -1.5 | 8. $-\frac{2}{3}$ | 9. -3.01 | 10. $-10\frac{1}{4}$ | 11. 6.17 | 12. $\frac{1}{2}$ |

Which of these are solutions for $x \leq 3$? Write Yes or No.
(Example 2)

| 13. 2 | 14. 4 | 15. 8.7 | 16. 3 | 17. -1.9 | 18. 0 |
| 19. -3 | 20. $2\frac{1}{2}$ | 21. $3\frac{1}{9}$ | 22. 6.1 | 23. $\frac{4}{3}$ | 24. -1 |

WRITTEN EXERCISES

Goal: To graph inequalities on a number line

For Exercises 1–35, graph each inequality. (Example 1)

| 1. $a > 2$ | 2. $n < 3$ | 3. $s < -1$ | 4. $y > -4$ | 5. $m < 4$ |
| 6. $e > 5$ | 7. $x > 0$ | 8. $p < -6$ | 9. $r > 1$ | 10. $t < 6$ |

(Example 2)

| 11. $y \geq 1$ | 12. $l \leq 6$ | 13. $c \geq -5$ | 14. $x \geq 0$ | 15. $z \leq 0$ |
| 16. $s \geq 3$ | 17. $t \leq -2$ | 18. $m \leq 5$ | 19. $n \geq -2$ | 20. $p \leq 1$ |

MIXED PRACTICE

21. $x > 3$	22. $y \leq 2$	23. $a \geq -3$	24. $c < -4$	25. $s \leq -1$
26. $t > -1$	27. $b \geq 5$	28. $d < 7$	29. $e > -5$	30. $g \leq 7$
31. $h < -3$	32. $j \geq -7$	33. $k \leq 4$	34. $n > 6$	35. $m \geq -4$

REVIEW CAPSULE FOR SECTION 11-6

Solve and check. (Pages 324–325)

1. $x - 5 = 7$
2. $u - 2 = 9$
3. $16 = y - 8$
4. $235 = r - 36$
5. $b - 64 = -7$
6. $t - 15 = -4$
7. $x + 4 = 8$
8. $a + 12 = 17$
9. $w + 5 = 4$
10. $n + 11 = 3$
11. $d + 6 = 7$
12. $c + 15 = -1$
13. $y - 3 = -5$
14. $d + 3 = -3$
15. $y + 2 = -4$
16. $t - 5 = -7$

The answers are on page 346.

11-6 Using Addition and Subtraction to Solve Inequalities

The Addition Property for solving inequalities is similar to the Addition Property for solving equations.

> **Addition Property for Inequalities**
>
> Adding the same number to each side of an inequality forms an equivalent inequality.
>
> $k - 5 > 7$
> $k - 5 + 5 > 7 + 5$
> $k > 12$

EXAMPLE 1 Solve and graph: $x - 3 > 2$

Solution:
1. To get x alone, add 3 to each side.
$x - 3 + 3 > 2 + 3$
$x > 5$
2. Graph the inequality.

The solution is **all numbers greater than 5.**

Self-Check Solve and graph: **1.** $p - 3 > 4$ **2.** $b - 7 < -2$ **3.** $0 > t - 1$

There is also a Subtraction Property for Inequalities.

> **Subtraction Property for Inequalities**
>
> Subtracting the same number from each side of an inequality forms an equivalent inequality.
>
> $q + 6 < 8$
> $q + 6 - 6 < 8 - 6$
> $q < 2$

Chapter 11

EXAMPLE 2 Solve and graph: $\quad c + 2 \leq 6$

Solution:
1. To get c alone, subtract 2 from each side.
$$c + 2 - 2 \leq 6 - 2$$
$$c \leq 4$$

2. Graph the inequality.

The solution is **all numbers less than 4.**

Self-Check *Solve and graph.*

4. $x + 7 < 10$ **5.** $c + 5 > -1$ **6.** $a + 7 > 2$

CLASSROOM EXERCISES

Write an equivalent inequality as in Step 1 of Example 1. Do <u>not</u> solve the inequality. (Step 1, Examples 1 and 2)

1. $a - 5 > 3$
2. $t - 7 > -5$
3. $w - 4 < 8$
4. $p - 11 < -9$
5. $y - 2 > 6$
6. $h - 4 < -7$
7. $m + 12 \leq 3$
8. $x + 14 \leq -2$
9. $t + 15 \geq -5$
10. $b + 3 \geq -4$
11. $c + 5 < 12$
12. $d + 4 \geq -8$

Solve and graph. (Example 1)

13. $g - 5 > 7$
14. $t - 8 \geq 11$
15. $w - 1 \leq -4$
16. $a - 4 < -9$
17. $-3 > b - 8$
18. $6 < c - 12$

(Example 2)

19. $p + 3 < 9$
20. $x + 5 \leq -4$
21. $y + 8 \geq 12$
22. $m + 7 \geq -15$
23. $14 < b + 4$
24. $-15 > t + 6$

WRITTEN EXERCISES

Goal: To solve and graph inequalities by using the Addition and Subtraction Properties

For Exercises 1–9, solve and graph. (Example 1)

1. $t - 3 > 2$
2. $r - 7 > -13$
3. $p - 5 < -3$
4. $x - 2 < -4$
5. $x - 12 > 4$
6. $-8 > q - 2$
7. $10 < c - 7$
8. $x - 9 \leq -9$
9. $b - 3 \leq -9$

Equations: Addition/Subtraction 341

For Exercises 10–48, solve and graph. (Example 2)

10. $x + 3 < 8$
11. $z + 5 < 6$
12. $y + 5 > 7$
13. $n + 6 < 10$
14. $r + 4 < -2$
15. $k + 5 > 3$
16. $t + 7 > 3$
17. $w + 14 < 8$
18. $m + 24 \geq 32$
19. $79 \geq 68 + d$
20. $15 \leq 18 + p$
21. $18 + b \geq 3$

MIXED PRACTICE

22. $t + 4 < 1$
23. $n + 3 < 4$
24. $y - 5 > 7$
25. $x - 15 \geq -22$
26. $r + 16 < 9$
27. $a - 8 > 8$
28. $x - 3 > -5$
29. $t - 15 \leq 35$
30. $m + 1 \leq -3$
31. $x + 7 \geq -2$
32. $4 > t - 6$
33. $19 < y + 20$
34. $2n + 5 \leq 19$
35. $5x - 1 < 19$
36. $23 \geq 2x + 3$

MORE CHALLENGING EXERCISES

37. $5n - 13 - 4n < -17$
38. $-8t + 14 + 9t > 25$
39. $3r - 6 - 2r < -4$
40. $-23 - 2y + 3y > -15$
41. $12w - 15 - 11w < 28$
42. $-18b - 42 + 19b > -56$
43. $4p - 3 \leq 5p + 6$
44. $-4y - 2 \geq 8 - 3y$
45. $3x + 5 > 2x - 2$
46. $4 + 6d < 5d - 3$
47. $-3c - 1 \leq 8 - 2c$
48. $6p + 2p + 4 \geq 7p - 12$

CHECKING INEQUALITIES

You can use a calculator to check whether a given number is a solution of a given inequality.

EXAMPLE Is -117 a solution of $t + 49 < -78$?

Solution Substitute -117 in the left side of the inequality.

1 1 7 [+/−] [+] 4 9 [=]   −68.

Since $-68 > -78$, -117 is **not** a solution.

EXERCISES Check whether the given number is a solution for the inequality.

1. $-47 > x - 94$; 46
2. $-133 < y - 51$; -81
3. $s - 163 < -85$; 79
4. $72 < x + 37$; 100
5. $y + 71 > -84$; -155
6. $m - 13.4 < -47.7$; -61.1

REVIEW: SECTIONS 11-4 – 11-6

Write an algebraic expression for each word expression. (Section 11-4)

1. Some number, n, increased by 5
2. The number of days, d, minus 7
3. Five pounds less than the weight, w
4. The number of books, b, plus 31
5. The price of a coat, c, decreased by $5
6. The number, y, reduced by 27
7. The quantity, x, increased by 14
8. A number, n, increased by 3

Write an equation for each word sentence. (Section 11-4)

9. Eight dollars more than the cost, c, is $33.
10. The sum of 12 and the number of points, p, is 56.
11. Robert's age, a, decreased by 3 is 13.
12. The distance, d, minus 10 meters is 36 meters.
13. The number, x, increased by 11 is 34.
14. The difference of y and 5 is 19.
15. Twenty reduced by n is 13.
16. The price of a suit, s, decreased by $28 is $103.

Solve. (Section 11-4)

17. The amount of Gina's gas bill minus $13 is $52. Find the amount of her gas bill.
18. Acme Appliances sold washers for $103. The regular price was reduced $38 for the sale. Find the regular price.
19. James grew $2\frac{1}{2}$ inches since his birthday last year. He is now 51 inches tall. What was his height a year ago?
20. Twenty-two kilometers more than the distance traveled on Tuesday is 630 kilometers. Find the distance traveled on Tuesday.

For Exercises 21–24, graph each inequality. (Section 11-5)

21. $x > -8$
22. $m < 5$
23. $p \leq -3$
24. $n \geq 6$

For Exercises 25–28, solve and graph. (Section 11-6)

25. $y - 5 < 4$
26. $c - 8 > -6$
27. $t + 1 > 0$
28. $5 \leq k + 1$

Equations: Addition/Subtraction

CHAPTER REVIEW

PART 1: VOCABULARY

For Exercises 1–5, choose from the box at the right the word(s) that best correspond to each description.

1. Statements that use $<, >, \leq,$ and $\geq$ are called __?__ (p. 338)
2. The words difference, minus, less than, and decreased by suggest the operation of __?__. (p. 334)
3. "Eight points more than 45" is an example of a __?__ expression. (p. 334)
4. The words sum, more than, plus, and increased by suggest the operation of __?__. (p. 334)
5. "$c - 6$" is an example of an __?__ expression. (p. 334)

addition
inequalities
algebraic
word
subtraction

PART 2: SKILLS

Solve and check. (Section 11-1)

6. $t - 8 = 2$
7. $p - 13 = -9$
8. $n - 47 = -28$
9. $a + 29 = -15$
10. $-18 = y + 36$
11. $4 = b + 8$
12. $x + 25 = 13$
13. $y - 9 = -12$

Solve and check. (Section 11-2)

14. $13m - 14 + 12m = -13 + 149$
15. $-5 - 18 = -c - 95 + 9c$
16. $3d - 8 - d = 18 - 12$
17. $-p + 6p - 3 = -4 + 31$
18. $19 - 33 = 4q - 91 + 3q$
19. $29 - x + 11x = -7 + 66$

Solve and check. (Section 11-3)

20. $11s = 27 + 2s$
21. $6d = 28 - d$
22. $17x = 90 - x$
23. $y = 18 - y$
24. $15 + 10m = 24 + 9m$
25. $39 + 36p = 44 + 35p$
26. $8 + 5\frac{1}{3}y = 17 + 2\frac{1}{3}y$
27. $9 + 4\frac{1}{2}t = 10 + 3\frac{1}{2}t$

Write an algebraic expression for each word expression. (Section 11-4)

28. The difference of 4 and p
29. The sum of 35 apples and t
30. y increased by 32
31. The cost, c, decreased by $6
32. Twelve feet less than w
33. The distance, d, plus 5

Write an equation for each word sentence. Do not solve.
(Section 11-4)

34. Three dollars more than the original cost, c, is $19.
35. The number of runners, r, at the starting line decreased by 15 is 42.
36. The difference between $10 and the price, p, of a ticket is $4.
37. The width, w, plus 16 centimeters is 84 centimeters.

Graph each inequality. (Section 11-5)

38. $n > -2$
39. $r < 5$
40. $t < -3$
41. $y > -4$
42. $p \leq 2$
43. $t \leq -4$
44. $a \geq -5$
45. $q \geq 0$

Solve and graph. (Section 11-6)

46. $x + 2 < 5$
47. $-15 > t - 19$
48. $-37 + m > -41$
49. $y + 4 \geq 1$
50. $n - 18 \leq -23$
51. $-24 > t - 19$
52. $-19 + r > -25$
53. $-27 < k - 24$

PART 3: APPLICATIONS

Solve each problem. (Section 11-4)

54. Eighty-five calories plus the number of calories in a fruit salad totals 164. How many calories are there in the fruit salad?
55. Twelve dollars more than the cost of a pair of jeans is $28. Find the cost of the jeans.
56. A house painter's weekly salary minus $72 in deductions is $208. What is the painter's weekly salary?
57. Twenty dollars less than the amount of the Lopez family's gas bill is $68. Find the amount of the gas bill.

CHAPTER TEST

Solve and check.

1. $x - 12 = 15$
2. $d - 9 = -1$
3. $t + 10 = -5$
4. $n + 1.8 = -2.9$
5. $2x + 5x = 5 + 9$
6. $x + 2 + 3x = 10$
7. $3x - 9 = 3 - x$
8. $7 + 3k = 19 - k$
9. $5\frac{3}{4}w = 1\frac{3}{4}w + 16$

Write an algebraic expression for each word expression.

10. Nineteen weeks increased by x weeks
11. Forty days decreased by d days.
12. Fifteen less than v
13. Eleven more than n apples

Solve.

14. The amount of Ted's grocery bill minus a refund of $5 is $32. Find the amount of his bill.

15. James bought a sweater and a tie for $37. The sweater cost $12 more than the tie. Find the cost of the sweater.

16. The number of people in an elevator increased by 6 is 14. Find the number of people in the elevator.

17. When $105 is subtracted from the amount in Jill's checking account, the result is $48.20. How much is in the account?

Solve and graph.

18. $b \geq -5$
19. $z - 13 > -5$
20. $p + 8 < 3$

ANSWERS TO REVIEW CAPSULES

Page 326 1. $5x$ 2. $13p$ 3. $2q$ 4. $2t$ 5. $9x$ 6. $10b$ 7. m 8. q 9. 44 10. $\frac{1}{4}$ 11. 1.1 12. 400

Page 328 1. $6d + 6$ 2. $p + 8$ 3. $2x - 6$ 4. $9 + 4x$ 5. $10 + 5r$ 6. $2 + 5b$ 7. $5 + 4y$ 8. $5c + 13$ 9. $3x - 2$ 10. $-3x + 11$ 11. $-13n - 5$ 12. $12y + 3$ 13. 9 14. 16 15. -2 16. -16 17. -3 18. -11 19. 4 20. 6 21. 4 22. 6 23. -5 24. -3 25. 2 26. 2 27. -1

Page 331 1. 12 2. 6 3. 13 4. 34 5. 12 6. 10 7. 25 8. 47

Page 338 1. < 2. > 3. > 4. > 5. < 6. > 7. > 8. > 9. < 10. <

Page 340 1. 12 2. 11 3. 24 4. 271 5. 57 6. 11 7. 4 8. 5 9. -1 10. -8 11. 1 12. -16 13. -2 14. -6 15. -6 16. -2

346 Chapter 11

ENRICHMENT

Compound Inequalities

In a high school wrestling weight class, wrestlers must weigh more than 167 pounds but no more than 185 pounds. If w represents the weight of a wrestler, you can write an inequality for this guideline.

$$w > 167 \text{ and } w \leq 185$$

A sentence containing two inequalities with the connecting words <u>and</u> or <u>or</u> is a **compound inequality.**

EXAMPLES Graph each compound inequality.

a. $w > 167$ <u>and</u> $w \leq 185$ **b.** $x \leq -2$ <u>or</u> $x > 1$

Solutions: **a.** When the connective <u>and</u> is used, the solutions are those numbers that make <u>both</u> inequalities true.

$w > 167$ and $w \leq 185$

b. When the connective <u>or</u> is used, the solutions are those numbers that make either inequality true.

$x \leq -2$ <u>or</u> $x > 1$

EXERCISES

Graph the solution for each compound inequality.

1. $x > 3$ <u>and</u> $x < 4$
2. $x \geq -3$ <u>and</u> $x < 1$
3. $x \geq -2$ <u>and</u> $x < 3$
4. $x > 1$ <u>or</u> $x < -1$
5. $x > 0$ <u>or</u> $x \leq -3$
6. $x > 2$ <u>or</u> $x \leq 0$

The weight of each player on a high school football team is greater than 150 pounds and no greater than 220 pounds.

7. Represent this weight as a compound inequality.
8. Graph the solution of the compound inequality.

Equations: Addition/Subtraction

ADDITIONAL PRACTICE

SKILLS

Solve and check. (Pages 324-328)

1. $x + 24 = 27$
2. $p + 41 = -17$
3. $c + 12 = -81$
4. $t + 13 = -13$
5. $a - 13 = -54$
6. $t - 42 = -6$
7. $a + 11.5 = -33.1$
8. $n + 22.7 = 14.6$
9. $x + 27.8 = -12.9$
10. $3a + 2a = 9 + 11$
11. $5c + 4c = 18 + 9$
12. $12x - 11x = 14 - 22$
13. $17b - 16b = 8 - 21$
14. $10y - 2y = -25 + 89$
15. $12m - 2m = -4 + 34$
16. $6x - 12 - 5x = -12 + 18$
17. $14r - 9 - 2r = -10 + 85$
18. $4n - 8 + 6n = -23 + 105$

(Pages 329-330)

19. $c = 21 - 2c$
20. $x = 36 - 3x$
21. $4.6n = 100 + 2.6n$
22. $7.3p = 80 + 3.3p$
23. $6q + 7 = 13 + 4q$
24. $4y + 3 = 5 + 3y$
25. $6x - 6 = 13 + 5x$
26. $8t - 9 = 5 + 6t$
27. $9.3a - 21 = 12 + 8.3a$
28. $4.3w - 9 = 21 + 1.3w$

Graph each inequality on a number line. (Pages 338-339)

29. $x > 5$
30. $m < 7$
31. $c \geq -4$
32. $p \leq -2$

Solve and graph. (Pages 340-342)

33. $n + 6 < 9$
34. $m + 3 < -6$
35. $a - 7 > 7$
36. $y - 5 < 2$

APPLICATIONS

Write an equation for each word sentence. (Pages 334-337)

37. The sum of 3 and n is 84.
38. 16 more than y is 43.
39. The difference of c and 24 is 36.
40. 16.1 less than x is 24.3.
41. The original cost of a television decreased by $55 is $326. Find the original cost. (Pages 334-337)
42. Thirteen pounds more than Carl's weight is 171 pounds. Find Carl's weight. (Pages 334-337)

CHAPTER 12
Rational Numbers: Multiplication/Division

SECTIONS
- **12-1** Multiplication: Unlike Signs
- **12-2** Multiplication: Like Signs
- **12-3** Properties of Multiplication
- **12-4** Dividing Rational Numbers
- **12-5** Factoring

FEATURES
Calculator Application:
Multiplying Rational Numbers

Calculator Application:
Distributive Property

Consumer Application:
Conserving Energy

Enrichment:
Multiplying Binomials

12-1 Multiplication: Unlike Signs

Study the patterns in each column.

Column 1

2 × 2 = 4
2 × 1 = 2
2 × 0 = 0
2 × (−1) = −2
2 × (−2) = −4
2 × (−3) = −6

The products are decreasing by 2.

Column 2

2 × 3 = 6
1 × 3 = 3
0 × 3 = 0
(−1) × 3 = −3
(−2) × 3 = −6
(−3) × 3 = −9

The products are decreasing by 3.

These patterns suggest that the product of two numbers having unlike signs is negative.

PROCEDURE

To multiply two numbers with <u>unlike</u> signs:

1. Multiply as with whole numbers.
2. Write the opposite of the result.

EXAMPLE 1 Multiply: (12)(−8)

Solution:
1. Multiply.   12 · 8 = 96
2. Write the opposite.   −96

Thus, (12)(−8) = **−96**

Self-Check *Multiply:* **1.** (14)(−5)   **2.** (9)(−25)   **3.** (6)(−16)

Example 2 shows the product of a negative number and a positive number.

EXAMPLE 2 Multiply: (−8)(11)

Solution:
1. Multiply.   8 · 11 = 88
2. Write the opposite.   −88

Thus, (−8)(11) = **−88**

Self-Check *Multiply:* **4.** (−4)(9)   **5.** −12(6)   **6.** (−18)(3)

The following properties that apply to whole numbers are also true for rational numbers.

Multiplication Property of Zero

For any rational number a,
$a \cdot 0 = 0$ and $0 \cdot a = 0$

$0 \cdot (-3) = 0$

$\left(-1\frac{3}{4}\right) \cdot 0 = 0$

Multiplication Property of One

For any rational number a,
$a \cdot 1 = a$ and $1 \cdot a = a$

$(-10)(1) = -10$

$(1)(-2.51) = -2.51$

CLASSROOM EXERCISES

For Exercises 1–4, complete the pattern. (Columns 1 and 2)

1. $8 \times 1 = 8$
   $8 \times 0 = 0$
   $8 \times (-1) = -8$
   $8 \times (-2) = \underline{}$

2. $2 \times 3 = 6$
   $1 \times 3 = 3$
   $0 \times 3 = 0$
   $-1 \times 3 = \underline{}$

3. $0 \times 5 = 0$
   $-1 \times 5 = -5$
   $-2 \times 5 = \underline{}$
   $-3 \times 5 = \underline{}$

4. $16 \times 1 = 16$
   $16 \times 0 = 0$
   $16 \times (-1) = \underline{}$
   $16 \times (-2) = \underline{}$

For Exercises 5–24, multiply the given numbers. (Example 1)

5. $0(-10)$
6. $1(-15)$
7. $5(-13)$
8. $7(-18)$
9. $11(-21)$
10. $19(-32)$
11. $27(-128)$
12. $349(-2)$

(Example 2)

13. $-1(6)$
14. $-3(2)$
15. $-4(5)$
16. $-7(8)$
17. $-5(0)$
18. $-15(10)$
19. $(-32)(12)$
20. $(-20)(8)$
21. $(-25)(17)$
22. $(-16)(13)$
23. $(-41)(27)$
24. $(-15)(52)$

WRITTEN EXERCISES

Goal: To multiply rational numbers having unlike signs

For Exercises 1–36, multiply the given numbers. (Example 1)

1. $6(-5)$
2. $9(-3)$
3. $15(-12)$
4. $11(-17)$
5. $19(-35)$
6. $34(-21)$
7. $42(-50)$
8. $52(-48)$
9. $2.1(-1.5)$
10. $0.7(-3.2)$
11. $\frac{1}{2}(-6)$
12. $\left(1\frac{1}{4}\right)\left(-1\frac{2}{5}\right)$

Rational Numbers: Multiplication/Division

(Example 2)

13. $-8(0)$ **14.** $-5(1)$ **15.** $-9(9)$ **16.** $-7(6)$
17. $-12(13)$ **18.** $-18(20)$ **19.** $-35(24)$ **20.** $-42(60)$
21. $-8.2(4)$ **22.** $-0.13(1.5)$ **23.** $\left(-\frac{1}{3}\right)\left(\frac{1}{4}\right)$ **24.** $-\frac{3}{4}\left(\frac{6}{7}\right)$

MIXED PRACTICE

25. $-75(0)$ **26.** $-81(1)$ **27.** $36(-56)$ **28.** $47(-63)$
29. $224(-28)$ **30.** $156(-18)$ **31.** $-31(142)$ **32.** $-258(37)$
33. $12.4(-7)$ **34.** $(-0.4)(3.6)$ **35.** $\left(1\frac{1}{5}\right)\left(-\frac{2}{3}\right)$ **36.** $\left(-\frac{1}{8}\right)\left(2\frac{4}{5}\right)$

APPLICATIONS: USING MULTIPLICATION WITH RATIONAL NUMBERS

Use positive and negative numbers to represent each situation by a multiplication problem. Then find the product.

37. Three penalties of 5 yards each

38. Five times a debt of $100

39. Six times a depth of 15 meters below sea level

40. Twice a loss of 200 feet in altitude

41. Four times a temperature drop of 6 degrees

42. Three times a loss of 4 yards in rushing

43. Two debts of $78 each

44. Twice a drop in temperature of 3 degrees

MORE CHALLENGING EXERCISES

Evaluate each expression.

45. $4(-29) + 8(-3) + (-1)(5)$ **46.** $2(-1) + 7(-2) + 3(-4)$
47. $9(-2) + 3(-22) + (6)(-3)$ **48.** $(-7)(10) + (-5)(3) + (-2)(5)$
49. $12(-21 + 19) + (-3)$ **50.** $8 + (-3) \cdot (-6 + 12)$

Multiply the monomials.

51. $-10(5y)$ **52** $4(-12x)$ **53.** $(-2a)(4b)$ **54.** $(6p)(-3q)$
55. $(-8a)(7a)$ **56.** $(2x)(-3x)$ **57.** $(-3m^2)(2m)$ **58.** $(2rt)(-3t^2)$
59. $2v^2(-v^3)$ **60.** $(-3c^3)(4c^4)$ **61.** $(5n)(-6n^2)$ **62.** $(-e^2)(3e^2)$

REVIEW CAPSULE FOR SECTION 12-2

Evaluate each expression. (Pages 22-24)

1. $a + a$ when $a = -3$
2. $b + b$ when $b = -6$
3. $x + x + x$ when $x = -5$
4. $m + m + m$ when $m = -1$
5. $p + p + p + p$ when $p = 12$
6. $t + t + t + t$ when $t = -10$
7. rs when $r = 4$ and $s = 12$
8. xy when $x = 13$ and $y = 11$

The answers are on page 370.

12-2 Multiplication: Like Signs

Study the pattern in each column.

Column 1

$2 \times (-4) = -8$
$1 \times (-4) = -4$
$0 \times (-4) = 0$
$-1 \times (-4) = 4$
$-2 \times (-4) = 8$
$-3 \times (-4) = 12$

The products are increasing by 4.

Column 2

$-6 \times 2 = -12$
$-6 \times 1 = -6$
$-6 \times 0 = 0$
$-6 \times (-1) = 6$
$-6 \times (-2) = 12$
$-6 \times (-3) = 18$

The products are increasing by 6.

These patterns suggest that the product of two rational numbers having like signs is positive.

PROCEDURE

To multiply two rational numbers with <u>like</u> signs:

1. Multiply as with whole numbers.
2. Write the result.

EXAMPLE Multiply: $(-24)(-3)$

Solution:
1. Multiply. $24 \cdot 3 = 72$
2. Result. $(-24)(-3) = \mathbf{72}$

Self-Check *Multiply.*

1. $-2(-7)$
2. $-10(-13)$
3. $(-12)(-25)$
4. $(-7)(-11)$

Rational Numbers: Multiplication/Division **353**

CLASSROOM EXERCISES

For Exercises 1–4, complete the pattern. (Columns 1 and 2)

1. $2 \times (-6) = -12$
   $1 \times (-6) = -6$
   $0 \times (-6) = 0$
   $-1 \times (-6) = 6$
   $-2 \times (-6) = \underline{\ ?\ }$

2. $-8 \times 2 = -16$
   $-8 \times 1 = -8$
   $-8 \times 0 = 0$
   $-8 \times (-1) = 8$
   $-8 \times (-2) = \underline{\ ?\ }$

3. $-9 \times 2 = -18$
   $-9 \times 1 = -9$
   $-9 \times 0 = 0$
   $-9 \times (-1) = \underline{\ ?\ }$
   $-9 \times (-2) = \underline{\ ?\ }$

4. $2 \times (-7) = -14$
   $1 \times (-7) = -7$
   $0 \times (-7) = 0$
   $-1 \times (-7) = \underline{\ ?\ }$
   $-2 \times (-7) = \underline{\ ?\ }$

WRITTEN EXERCISES

Goal: To multiply two negative rational numbers

Multiply. (Example)

1. $(-7)(-5)$
2. $(-4)(-8)$
3. $(-1)(-9)$
4. $(-3)(-6)$
5. $(-21)(-35)$
6. $(-32)(-28)$
7. $(-45)(-18)$
8. $(-14)(-55)$
9. $(-5)(-125)$
10. $(-221)(-9)$
11. $(-8)(-374)$
12. $(-458)(-7)$
13. $(-14)(-107)$
14. $(-186)(-25)$
15. $(-36)(-217)$
16. $(-412)(-52)$
17. $(-3)(-2.2)$
18. $(-1.4)(-2)$
19. $(-2.07)(-4)$
20. $(-0.9)(-5)$
21. $(-1.5)(-3.8)$
22. $(-0.6)(-0.8)$
23. $(-5.07)(-4.2)$
24. $(-2.7)(-21.51)$
25. $\left(-\frac{2}{3}\right)\left(-\frac{3}{4}\right)$
26. $\left(-\frac{3}{8}\right)\left(-\frac{4}{5}\right)$
27. $\left(-4\frac{1}{4}\right)\left(-2\frac{2}{5}\right)$
28. $\left(-3\frac{3}{10}\right)\left(-\frac{5}{11}\right)$
29. $\left(-\frac{1}{2}\right)\left(-\frac{1}{4}\right)$
30. $\left(-\frac{2}{5}\right)\left(-\frac{3}{4}\right)$
31. $\left(-1\frac{1}{4}\right)\left(-3\frac{1}{3}\right)$
32. $\left(-2\frac{2}{5}\right)\left(-1\frac{3}{8}\right)$

MORE CHALLENGING EXERCISES

Evaluate each expression.

33. $(-2)^2 + (-3)^2$
34. $(-1.1)^2 + (-0.2)^2$
35. $(-9 + 6)(-12)$
36. $(-1)^2 + 12(-13 + 10)$
37. $-8(-8 + 2) + (-2)^2$
38. $(-21 + 7)(-3) + (-4)^2$
39. $(-1)^2 \cdot (-1) \cdot (-1)$
40. $(-5)(-5)(-5)(-1)(-2)$

Multiply the monomials.

41. $(-6)(-3a)$
42. $(-2b)(-9)$
43. $(-3x)(-5x)$
44. $(-10c)(-4c)$
45. $(-5ab)(-2ab)$
46. $(-6p^2)(-8p)$
47. $(-3xy)(-4x^2)$
48. $(-2pt)(-7pt)$

MULTIPLYING RATIONAL NUMBERS

A calculator with a [+/−] or [sc] key enables you to multiply two or more rational numbers.

EXAMPLE 1 Multiply: (−928)(57)(−1239)

Solution 9 2 8 [+/−] [×] 5 7

[×] 1 2 3 9 [+/−] [=]   **65538144.**

NOTE: To multiply without the "sign change" key, observe that the answer is:
a. <u>positive</u>, when the number of multiplied negative numbers is <u>even</u>,
b. <u>negative</u>, when the number of multiplied negative numbers is <u>odd</u>.

EXAMPLE 2 Multiply: (−3.6)(−4)(−6.5)   Do not use the "sign change" key.

Solution 3 . 6 [×] 4 [×] 6 . 5 [=]   **93.6**

Since <u>three</u> negative numbers are multiplied, the answer is **−93.6.**

EXERCISES Multiply

1. (−128)(209)
2. (436)(−157)
3. (−89.6)(−56.3)
4. (−37.59)(−92.47)
5. (−0.33829)(0.48)
6. (−0.5286)(−0.8025)
7. (−48)(−29)(37)
8. (−65)(44)(−23)
9. (−28.6)(−42.3)(−38.7)

REVIEW CAPSULE FOR SECTION 12-3

Evaluate each expression.   (Pages 350-352)

1. ab when $a = 16$ and $b = -3$
2. xy when $x = -4$ and $y = 12$
3. cd when $c = -15$ and $d = 20$
4. rt when $r = 14$ and $t = -11$
5. pq when $p = 7$ and $q = 12$
6. fg when $f = 24$ and $g = -8$

Multiply.   (Pages 67-69)

7. $10(5y)$
8. $4(12t)$
9. $(9r)9$
10. $(4k)7$
11. $(8a)(7a)$
12. $(2x)(x)$
13. $(3z)(6z)$
14. $(5p)(3p)$

The answers are on page 370.

Rational Numbers: Multiplication/Division   **355**

12-3 Properties of Multiplication

The following properties that apply to whole numbers are also true for rational numbers.

Commutative Property of Multiplication

For any rational numbers a and b,
$$ab = ba. \qquad (-7)(4) = 4(-7)$$

Associative Property of Multiplication

For any rational numbers a and b,
$$(ab)c = a(bc). \qquad (-3 \cdot 5)6 = -3(5 \cdot 6)$$

Distributive Property of Multiplication over Addition

For any rational numbers a, b, and c,
1. $a(b + c) = ab + ac$, and $\qquad 2(-4 + 8) = 2(-4) + 2(8)$
2. $(b + c)a = ba + ca.$ $\qquad (x + 0.5)3 = x(3) + 0.5(3)$

Distributive Property of Multiplication over Subtraction

For any rational numbers a, b, and c,
1. $a(b - c) = ab - ac$, and $\qquad 3(4 - 1) = 3(4) - 3(1)$
2. $(b - c)a = ba - ca.$ $\qquad (x - 5)2 = x(2) - 5(2)$

The Distributive Property can be used to write products as sums or differences.

EXAMPLE 1 Multiply: $4(2x + 3)$

Solution:
1. Use the Distributive Property. $\quad 4(2x + 3) = 4(2x) + 4(3)$
2. Multiply. Write in simplest form. $\qquad\qquad = 8x + 12 \quad$ ◄ Sum

Self-Check Multiply: **1.** $2(5n + 4)$ **2.** $3(-8y + 6)$ **3.** $(p + 7)(4)$

EXAMPLE 2 Multiply: $-2(-6x + 5)$

Solution:
1. $-2(-6x + 5) = (-2)(-6x) + (-2)(5)$
2. $\qquad\qquad\qquad = 12x + (-10)$
$\qquad\qquad\qquad = 12x - 10$

Self-Check *Multiply:* **4.** $(4p + 7)(-3)$ **5.** $(x + 12)(-4)$ **6.** $-2(4x + 1)$

This example shows multiplication by a term containing a variable.

EXAMPLE 3 Multiply: $(4y - 1)3y$

Solution:
① $(4y - 1)3y = (4y)(3y) - (1)(3y)$
② $\quad\quad\quad\quad\quad = 12y^2 - 3y$ ◀ **Difference**

Self-Check *Multiply:* **7.** $(4x - 2)6x$ **8.** $(5t - 2)(3t)$ **9.** $-2r(3r + 1)$

CLASSROOM EXERCISES

For Exercises 1-12, complete each problem. (Example 1)

1. $2(x + 5) = 2(x) + 2(5)$
$\quad\quad\quad\quad = \underline{\ ?\ } + 10$

2. $(3g + 2)5 = \underline{\ ?\ } + \underline{\ ?\ }$
$\quad\quad\quad\quad = 15g + 10$

(Example 2)

3. $-2(y + 4) = -2(y) + (-2)(4)$
$\quad\quad\quad\quad = -2y + \underline{\ ?\ }$
$\quad\quad\quad\quad = -2y - 8$

4. $-6(p + 5) = \underline{\ ?\ } + (-6)(5)$
$\quad\quad\quad\quad = -6p + (-30)$
$\quad\quad\quad\quad = -6p - 30$

5. $(2x + 6)(-3) = 2x(-3) + \underline{\ ?\ }$
$\quad\quad\quad\quad = \underline{\ ?\ } + (-18)$
$\quad\quad\quad\quad = -6x - 18$

6. $(-3c + 2)(-4) = \underline{\ ?\ } + 2(-4)$
$\quad\quad\quad\quad = 12c + \underline{\ ?\ }$
$\quad\quad\quad\quad = 12c - 8$

(Example 3)

7. $2n(n - 5) = (2n)(n) - (2n)(5)$
$\quad\quad\quad\quad = \underline{\ ?\ } - 10n$

8. $(5b - 6)4b = (5b)(4b) - \underline{\ ?\ }$
$\quad\quad\quad\quad = 20b^2 - 24b$

WRITTEN EXERCISES

Goal: To use the Distributive Property to express a product as a sum or difference

For Exercises 1-39, multiply. Show all work. (Example 1)

1. $5(3s + 10)$ **2.** $4(m + 9)$ **3.** $2(4x + 7)$
4. $3(5p + 6)$ **5.** $(2n + 3)7$ **6.** $(y + 4)8$
7. $6(4q + 1)$ **8.** $8(-a + 9)$ **9.** $9(-3c + 2)$

Rational Numbers: Multiplication/Division **357**

(Example 2)

10. $-1(t+4)$
11. $-2(r+7)$
12. $-4(x+2)$
13. $-3(-2p+1)$
14. $-2(-3y+5)$
15. $(-4c+3)(-1)$
16. $(-2m+6)(-4)$
17. $-5(z+(-2))$
18. $-6(2d+(-5))$

(Example 3)

19. $(a-1)a$
20. $(r-3)r$
21. $(q-2)q$
22. $2x(x-3)$
23. $3y(y-4)$
24. $(n-6)(5n)$
25. $(p-2)(6p)$
26. $2b(2b-1)$
27. $4x(3x-7)$

MIXED PRACTICE

28. $-7(p+1)$
29. $4(2x+5)$
30. $y(2y-3)$
31. $2(t+9)$
32. $2m(m-1)$
33. $(p+5)(-3)$
34. $(a-6)(3a)$
35. $-4(3n+1)$
36. $8(2c+6)$
37. $10(4b+7)$
38. $(-1)(-2q+7)$
39. $4d(4d-2)$

MORE CHALLENGING EXERCISES

Multiply, then combine like terms.

40. $2(x+10)+4x$
41. $3(2n+4)-2n$
42. $-4(3a-4)+-5a$
43. $-4b+3(b-2)$
44. $6c+2(3c-5)$
45. $-9p-4(2p+-3)+3p$
46. $a(3a-1)+2a$
47. $m(m+d)+2md$
48. $-5z-z(4-3z)+6z$

REVIEW: SECTIONS 12-1—12-3

For Exercises 1-16, multiply. (Section 12-1)

1. $8(-5)$
2. $10(-6)$
3. $(-15)9$
4. $(-17)(21)$
5. $-42(16)$
6. $35(-24)$
7. $(-126)(25)$
8. $(57)(-206)$

(Section 12-2)

9. $(-8)(-4)$
10. $(-5)(-3)$
11. $(-9)(-7)$
12. $(-16)(-32)$
13. $(-25)(-40)$
14. $(-38)(-118)$
15. $(-12)(-308)$
16. $(-429)(-67)$

Multiply. Show all work. (Section 12-3)

17. $2(5a+3)$
18. $(b+8)6$
19. $-4(x+2)$
20. $(3t+1)(-5)$
21. $y(y-5)$
22. $(2r-1)4r$
23. $(-5c)(4c+3)$
24. $6v(7-3v)$

CONSUMER APPLICATION
Conserving Energy

In warm weather, outside temperatures are usually higher than inside temperatures. The following formula is used to estimate the rate at which heat passes from the outside to the inside of a building through the walls and windows. This **heat transfer** is measured in British Thermal Units (BTU's) per hour.

Heat transfer $= A \cdot U(i - o)$

A = surface area in ft^2
U = heat transfer factor
i = inside temperature in °F
o = outside temperature in °F

The variable U in the formula represents a number that varies depending on the type of surface through which the heat passes.

Surface	Value of U
Concrete, 6 inches thick	0.58
Glass, single pane	1.13
Brick, 8 inches thick	0.41
Wood, 2 inches thick	0.43

EXAMPLE Estimate the rate of heat transfer through a 200-ft^2 wall of 8-inch brick. The inside temperature is 74°F and the outside temperature is 88°F.

SOLUTION

Heat transfer $= A \cdot U(i - o)$ ← $A = 200$; $U = 0.41$; $i = 74$; $o = 88$

Heat transfer $= (200)(0.41)(74 - 88)$

Heat transfer $= 82(-14)$

Heat transfer $= -1148$ ← The negative number means that heat is gained.

Thus, heat is being gained at a rate of **1148 BTU's per hour.**

Exercises

Find the rate of heat transfer.

1. Surface: glass
   $A = 50$ ft^2; $i = 78°$F; $o = 90°$F

2. Surface: concrete
   $A = 300$ ft^2; $i = 75°$F; $o = 95°$F

3. Surface: wood
   $A = 400$ ft^2; $i = 72°$F; $o = 85°$F

4. Surface: brick
   $A = 100$ ft^2; $i = 70°$F; $o = 84°$F

REVIEW CAPSULE FOR SECTION 12-4

Evaluate each expression. (Pages 124-125, 148-151)

1. $m \div p$ when $m = 81$ and $p = 9$
2. $\dfrac{c}{d}$ when $c = 186$ and $d = 3$
3. $\dfrac{375}{t}$ when $t = 15$
4. $a \div b$ when $a = 6$ and $b = \dfrac{2}{3}$
5. $\dfrac{p}{q}$ when $p = \dfrac{3}{4}$ and $q = 9$
6. $\dfrac{t}{r}$ when $t = \dfrac{1}{3}$ and $r = 6$

The answers are on page 370.

12-4 Dividing Rational Numbers

A multiplication problem has two related division problems.

TABLE 1

Multiplication	Related Division
$4 \cdot 8 = 32$	$32 \div 4 = 8$ and $32 \div 8 = 4$

You can use this fact to find the pattern for dividing a positive number by a negative number.

TABLE 2

Multiplication	Related Division
$(-3)(-8) = 24$	$24 \div (-3) = -8$ and $24 \div (-8) = -3$
$(-5)(-7) = 35$	$35 \div (-5) = -7$ and $35 \div (-7) = -5$

Conclusion: The quotient of a positive number and a negative number is negative.

Now study these patterns.

TABLE 3

Multiplication	Related Division
$-5(3) = -15$	$-15 \div (-5) = 3$ and $-15 \div 3 = -5$
$-9(6) = -54$	$-54 \div (-9) = 6$ and $-54 \div 6 = -9$

Conclusion: The quotient of two negative numbers is positive.

Conclusion: The quotient of a negative number and a positive number is negative.

You can use these patterns to divide integers.

> **PROCEDURE**
>
> **To divide two rational numbers:**
>
> 1. Divide.
> 2. a. When the two numbers have like signs, the quotient is positive.
>   b. When the two numbers have unlike signs, the quotient is negative.

EXAMPLE 1 Divide: **a.** $-36 \div (-4)$ **b.** $-27 \div 9$ **c.** $12 \div (-2)$

Solutions: a.   b.   c.

1. $36 \div 4 = 9$ 1. $27 \div 9 = 3$ 1. $12 \div 2 = 6$
2a. $-36 \div (-4) = 9$ 2b. $-27 \div 9 = -3$ 2b. $12 \div (-2) = -6$

Self-Check Divide: **1.** $-18 \div (-2)$ **2.** $-28 \div 4$ **3.** $\frac{-36}{6}$ **4.** $\frac{24}{-3}$

Recall that two numbers are reciprocals if their product is 1.

$$\tfrac{2}{1} \cdot \tfrac{1}{2} = 1 \qquad \left(-\tfrac{3}{8}\right)\left(-\tfrac{8}{3}\right) = 1 \qquad \left(-\tfrac{5}{4}\right)\left(-\tfrac{4}{5}\right) = 1$$

Reciprocals: 2 and $\tfrac{1}{2}$ $-\tfrac{3}{8}$ and $-\tfrac{8}{3}$ $-\tfrac{5}{4}$ and $-\tfrac{4}{5}$

Dividing by a number is the same as multiplying by its reciprocal.

EXAMPLE 2 Divide: **a.** $-30 \div \tfrac{1}{6}$ **b.** $-16 \div \left(-\tfrac{4}{3}\right)$

Solutions:
a. $-30 \div \tfrac{1}{6} = -30 \cdot 6$
$\phantom{-30 \div \tfrac{1}{6}} = -180$

b. $-16 \div \left(-\tfrac{4}{3}\right) = -\tfrac{\overset{4}{\cancel{16}}}{1} \cdot \left(-\tfrac{3}{\underset{1}{\cancel{4}}}\right)$
$\phantom{-16 \div \left(-\tfrac{4}{3}\right)} = \left(-\tfrac{4}{1}\right)\left(-\tfrac{3}{1}\right)$
$\phantom{-16 \div \left(-\tfrac{4}{3}\right)} = \tfrac{12}{1}$, or **12**

Self-Check Divide: **5.** $-4 \div \tfrac{1}{3}$ **6.** $\tfrac{4}{5} \div (-8)$ **7.** $\tfrac{7}{8} \div \left(-\tfrac{1}{4}\right)$

In Algebra	In Arithmetic
$a \div b = a \cdot \tfrac{1}{b} \, (b \neq 0)$	$8 \div (-3) = 8 \cdot \left(-\tfrac{1}{3}\right)$

Rational Numbers: Multiplication/Division

CLASSROOM EXERCISES

Complete. (Tables 1–3)

1. $6 \cdot 8 = 48$
   $48 \div 6 = \underline{\ ?\ }$
   $48 \div 8 = \underline{\ ?\ }$

2. $-4(-6) = 24$
   $24 \div (-4) = \underline{\ ?\ }$
   $24 \div (-6) = \underline{\ ?\ }$

3. $-8(7) = -56$
   $-56 \div (-8) = \underline{\ ?\ }$
   $-56 \div 7 = \underline{\ ?\ }$

4. $5(-4) = -20$
   $-20 \div 5 = \underline{\ ?\ }$
   $-20 \div (-4) = \underline{\ ?\ }$

Divide. (Example 1)

5. $(-54) \div (-9)$
6. $-24 \div (-8)$
7. $63 \div (-7)$
8. $81 \div (-9)$
9. $\frac{-100}{10}$
10. $\frac{42}{-6}$
11. $\frac{-99}{-9}$
12. $\frac{-20}{2}$
13. $30 \div (-5)$
14. $-72 \div (-6)$
15. $-44 \div 11$
16. $19 \div (-1)$

(Example 2)

17. $-3 \div \frac{1}{4}$
18. $-5 \div \frac{1}{3}$
19. $-6 \div \left(-\frac{3}{5}\right)$
20. $-8 \div \left(-\frac{4}{5}\right)$
21. $-\frac{1}{4} \div \frac{5}{4}$
22. $\frac{8}{9} \div \left(-\frac{4}{3}\right)$
23. $-\frac{12}{5} \div \left(-\frac{8}{5}\right)$
24. $-\frac{6}{7} \div \left(-\frac{2}{3}\right)$

WRITTEN EXERCISES

Goal: To divide two rational numbers

Divide. (Example 1)

1. $-36 \div (-4)$
2. $-30 \div (-6)$
3. $-63 \div (-9)$
4. $-35 \div (-5)$
5. $21 \div (-7)$
6. $81 \div (-9)$
7. $49 \div (-7)$
8. $36 \div (-9)$
9. $-39 \div 13$
10. $-45 \div 5$
11. $-72 \div 4$
12. $-14 \div 2$
13. $\frac{-15}{-3}$
14. $\frac{-24}{-8}$
15. $\frac{-75}{5}$
16. $\frac{-48}{12}$
17. $\frac{64}{-8}$
18. $\frac{100}{-20}$
19. $\frac{-138}{6}$
20. $\frac{-98}{-14}$
21. $200 \div (-40)$
22. $-98 \div (-7)$
23. $133 \div (-19)$
24. $-200 \div (-8)$

(Example 2)

25. $-16 \div \frac{1}{3}$
26. $-25 \div \frac{5}{6}$
27. $-\frac{2}{3} \div (-18)$
28. $-\frac{5}{8} \div (-12)$
29. $\frac{4}{3} \div \left(-\frac{2}{9}\right)$
30. $\frac{1}{2} \div \left(-\frac{3}{4}\right)$
31. $-\frac{3}{8} \div \frac{1}{4}$
32. $-\frac{9}{10} \div \frac{2}{5}$
33. $-\frac{11}{12} \div \left(-\frac{3}{4}\right)$
34. $-\frac{7}{10} \div \left(-\frac{3}{4}\right)$
35. $4 \div \left(-\frac{2}{5}\right)$
36. $9 \div \left(-\frac{3}{7}\right)$

MORE CHALLENGING EXERCISES

EXAMPLE: $\dfrac{w}{-3} \div \dfrac{2w}{9} = \dfrac{w}{-3} \cdot \dfrac{9}{2w}$ ◀ The reciprocal of $\dfrac{9}{2w}$ is $\dfrac{2w}{9}$.

$$= \dfrac{\cancel{w}^1}{\cancel{-3}_{-1}} \cdot \dfrac{\cancel{9}^3}{2\cancel{w}_2}$$

$$= \dfrac{1 \cdot 3}{-1 \cdot 2} = \dfrac{3}{-2}, \text{ or } -1\tfrac{1}{2}$$

37. $\dfrac{a}{-4} \div 2$
38. $\dfrac{2a}{5} \div a$
39. $\dfrac{3e}{-7} \div \dfrac{-9}{14}$
40. $\dfrac{4}{9} \div \dfrac{-22}{27g}$

41. $\dfrac{b}{-6} \div (-3)$
42. $\dfrac{5v}{-7} \div v$
43. $\dfrac{6h}{-5} \div \dfrac{12}{25}$
44. $\dfrac{-8}{13} \div \dfrac{-4}{39k}$

45. $\dfrac{7}{x} \div (-7)$
46. $\dfrac{z}{-3} \div \dfrac{1}{3}$
47. $\dfrac{-1}{9} \div \dfrac{1}{-9}$
48. $\dfrac{11}{-3m} \div \dfrac{22}{3}$

49. $\dfrac{y}{4} \div \dfrac{y}{c}$
50. $(-d) \div \dfrac{1}{d}$
51. $(-10) \div \dfrac{f}{10}$
52. $(-6) \div \dfrac{18}{-5s}$

The <u>Closure Property for Addition</u> states that the sum of two rational numbers is a rational number. The <u>Closure Property for Multiplication</u> states that the product of two rational numbers is a rational number.

53. Does the Closure Property hold for subtraction of whole numbers; that is, is the difference of two whole numbers always a whole number? Give a reason for your answer.

54. Does the Closure Property hold for division of natural numbers; that is, is the quotient of two natural numbers always a natural number? Give a reason for your answer.

55. Does the Closure Property hold for division of rational numbers?

56. Does the Closure Property hold for multiplication of integers?

REVIEW CAPSULE FOR SECTION 12-5

Multiply. (Pages 356-358) The answers are on page 370.

1. $2(a + 4)$
2. $3(b + 1)$
3. $6(x - 3)$
4. $5(y - 5)$
5. $4(a - b)$
6. $7(x + 2)$
7. $x(x + 4)$
8. $d(d - 3)$
9. $(5 + m)m$
10. $(6 - n)n$
11. $(r + 7)r$
12. $(t - 4)8$

Find the GCF of each pair of numbers. (Pages 115-117)

13. 4, 8
14. 6, 12
15. 8, 12
16. 4, 10
17. 6, 9
18. 12, 15

Rational Numbers: Multiplication/Division **363**

12-5 Factoring

You have used the Distributive Property to express a product as a sum or as a difference.

Product	Sum	Product	Difference
$3(4 + 6)$	$3(4) + 3(6)$	$2(8 - 5)$	$2(8) - 2(5)$
$2(x + 4)$	$2x + 8$	$3(y - 2)$	$3y - 6$

The Distributive Property is also used to express a sum or a difference as a product. This is called **factoring**.

Sum or Difference	Product
$2(5) + 2(6)$ ⟶	$2(5 + 6)$
$4(8) - 4(3)$ ⟶	$4(8 - 3)$
$3a + 6$ ⟶	$3(a + 2)$
$5b - 10$ ⟶	$5(b - 2)$

To factor a sum or a difference, find all the number factors and variable factors shared by each term. These **common factors** become one of the factors of the product.

PROCEDURE

To factor using the Distributive Property:

1. Write each term of the sum or difference as the product of factors.
2. Use the common factor(s) to write a product.

EXAMPLE 1 Factor: $5y + 5$

Solution:
1. Write the factors of each term. $5y + 5 = 5 \cdot y + 5 \cdot 1$ ◀ 5 is the common factor.
2. Use the common factor, 5, to write a product. $= 5(y + 1)$ ◀ By the Distributive Property

Self-Check *Factor.*

1. $3a + 3$
2. $7x + 7$
3. $4 + 2p$
4. $15 + 5n$

Example 2 shows how to write a difference as a product.

EXAMPLE 2 Factor: $3p - 6$

[1] $3p - 6 = 3 \cdot p - 3 \cdot 2$ ◀ **3 is the common factor.**
[2] $ = 3(p - 2)$

Self-Check *Factor:* **5.** $3x - 6$ **6.** $5m - 10$ **7.** $12n - 3$

Sometimes the common factor is a variable such as the variable x in Example 3.

EXAMPLE 3 Factor: $x^2 + 5x$

[1] $x^2 + 5x = x \cdot x + x \cdot 5$ ◀ **x is the common factor.**
$ = x(x + 5)$

Self-Check *Factor:* **8.** $t^2 + 3t$ **9.** $p^2 + 5p$ **10.** $4y - y^2$

CLASSROOM EXERCISES

Identify the common factor of each pair of terms.
(Examples 1–3, Step 2)

1. $2x$, 2 **2.** $5m$, 5 **3.** $2t$, $6r$ **4.** $7n$, $21p$
5. 15, $3y$ **6.** t^2, $3t$ **7.** m^2, $6m$ **8.** $6a$, 3

For Exercises 9–32, one factor is given for each product. Write the other factor. (Example 1)

9. $3x + 3 = \underline{}(x + 1)$
10. $5p + 5 = \underline{}(p + 1)$
11. $2m + 10 = \underline{}(m + 5)$
12. $3q + 18 = \underline{}(q + 6)$
13. $5r + 35 = 5(\underline{})$
14. $7p + 49 = 7(\underline{})$
15. $18 + 2t = 2(\underline{})$
16. $21 + 3r = 3(\underline{})$

(Example 2)

17. $3p - 6 = \underline{}(p - 2)$
18. $5r - 10 = \underline{}(r - 2)$
19. $12m - 3 = \underline{}(4m - 1)$
20. $7x - 42 = \underline{}(x - 6)$
21. $2r - 2 = 2(\underline{})$
22. $9m - 3 = 3(\underline{})$
23. $4x - 2 = 2(\underline{})$
24. $15a - 3 = 3(\underline{})$

Rational Numbers: Multiplication/Division

(Example 3)

25. $n^2 + 2n = \underline{\;?\;}(n + 2)$
26. $a^2 + 5a = \underline{\;?\;}(a + 5)$
27. $t^2 + 10t = t(\underline{\;?\;})$
28. $y^2 + 100y = y(\underline{\;?\;})$
29. $5x - x^2 = \underline{\;?\;}(5 - x)$
30. $8t - t^2 = \underline{\;?\;}(8 - t)$
31. $12c - c^2 = c(\underline{\;?\;})$
32. $9m - m^2 = m(\underline{\;?\;})$

WRITTEN EXERCISES

Goal: To factor a sum or difference of a binomial by applying the Distributive Property

For Exercises 1-49, factor.   (Example 1)

1. $2n + 2$
2. $7y + 7$
3. $3r + 3$
4. $5t + 5$
5. $3y + 15$
6. $15 + 5m$
7. $21 + 7c$
8. $18 + 3d$

(Example 2)

9. $7n - 14$
10. $13r - 26$
11. $15a - 5$
12. $35x - 7$
13. $7n - 28$
14. $22t - 11$
15. $3p - 12$
16. $2c - 2$

(Example 3)

17. $y^2 + 5y$
18. $p^2 + 9p$
19. $c^2 - 4c$
20. $t^2 - 5t$
21. $13x + x^2$
22. $15b - b^2$
23. $r^2 + r$
24. $q + q^2$
25. $r^2 - r$
26. $q - q^2$
27. $t^2 + 2t$
28. $5p - p^2$

MIXED PRACTICE

29. $13y + 13$
30. $5p - 35$
31. $7n + 42$
32. $2 - 2y$
33. $3t + 12$
34. $d^2 + 3d$
35. $5a - 20b$
36. $36 - 3x$
37. $x^2 - 12x$
38. $21n + 3p$
39. $9y - y^2$
40. $8t - 8$

MORE CHALLENGING EXERCISES

EXAMPLE: $2x^2 + 14x + 10 = 2(x^2 + 7x + 5)$ ◁ The common factor is 2.

41. $3x^2 + 9x + 15$
42. $4y^2 + 10y + 8$
43. $5a^2 + 10a + 15$
44. $6c^2 + 12c + 24$
45. $7d^2 + 14d + 28$
46. $8n^2 + 20n + 32$
47. $18m^2 + 15m + 27$
48. $10x^2 + 30x + 15$
49. $2x^2 - 2x - 2$

DISTRIBUTIVE PROPERTY

When using a calculator to evaluate an expression of the form $(b + c)a$ or $(b - c)a$, it is quicker to use the product form than the sum/difference form.

EXAMPLE Evaluate: $(876 + 958)0.75$

Solution First find the sum. Then find the product.

8 7 6 [+] 9 5 8 [=] [×] . 7 5 [=] | 1375.5

EXERCISES *Evaluate.*

1. $(4.5 - 3.9)52$
2. $(94 + 133)0.5$
3. $(6.72 + 8.86)76$
4. $(1258 + 9355)2.5$
5. $(24{,}856 - 19{,}072)0.9$
6. $(12{,}804 + 9496)12.9$
7. $(1092.8 - 948.9)109.2$
8. $(6294 - 4807)0.8$
9. $(8.31 - 5.85)29$
10. $(6255 - 4075)1.4$
11. $(12{,}806.3 - 8948.8)436.8$
12. $(12.064 + 6.066)0.25$

REVIEW: SECTIONS 12-4 — 12-5

For Exercises 1–24, divide. (Section 12-4)

1. $-24 \div (-3)$
2. $-35 \div (-5)$
3. $36 \div (-2)$
4. $81 \div (-3)$
5. $\frac{-45}{15}$
6. $\frac{-56}{7}$
7. $\frac{-64}{-4}$
8. $\frac{-65}{-13}$
9. $\frac{-124}{4}$
10. $\frac{-200}{40}$
11. $\frac{125}{-25}$
12. $\frac{84}{-7}$
13. $300 \div (-30)$
14. $-256 \div 16$
15. $-120 \div (-24)$
16. $350 \div (-50)$
17. $-4 \div \frac{1}{4}$
18. $-5 \div \frac{1}{3}$
19. $8 \div \left(-\frac{4}{5}\right)$
20. $10 \div \left(-\frac{2}{5}\right)$
21. $-\frac{12}{5} \div \left(-\frac{8}{5}\right)$
22. $-\frac{8}{9} \div \left(-\frac{4}{3}\right)$
23. $\frac{7}{8} \div \left(-\frac{7}{8}\right)$
24. $-\frac{1}{4} \div \frac{5}{4}$

Factor each expression. (Section 12-5)

25. $2x + 2$
26. $5x - 10$
27. $3x - 3$
28. $5x + 15$
29. $x^2 - 2x$
30. $x^2 + 12x$
31. $2x + 14$
32. $x^2 - 3x$
33. $7x - 42$
34. $15 + 3x$
35. $18 + 6x$
36. $12 + 4x$
37. $16x - x^2$
38. $3x + x^2$
39. $7x + x^2$
40. $13x - x^2$

Rational Numbers: Multiplication/Division

CHAPTER REVIEW

PART 1: VOCABULARY

For Exercises 1–8, choose from the box at the right the word(s) or numerical expression(s) that best corresponds to each description.

1. Commutative Property of Multiplication __?__ (356)
2. Distributive Property of Multiplication over Subtraction __?__ (p. 356)
3. Multiplication Property of Zero __?__ (p. 351)
4. All number and variable factors shared by each term __?__ (p. 364)
5. Associative Property of Multiplication __?__ (p. 356)
6. Multiplication Property of One __?__ (p. 351)
7. To write a sum or a difference as a product __?__ (p. 364)
8. Distributive Property of Multiplication over Addition __?__ (p. 356)

$5 \cdot 0 = 0$
$-7 \cdot 1 = -7$
$(6)(-3) = (-3)(6)$
$(-2 \cdot 5)4 = -2(5 \cdot 4)$
$2(1 + 4) = 2(1) + 2(4)$
$5(6 - 3) = 5(6) - 5(3)$
common factors
factoring

PART 2: SKILLS

For Exercises 9–52, multiply. (Section 12-1)

9. $18(-6)$
10. $15(-9)$
11. $-7(12)$
12. $-5(15)$
13. $17(-23)$
14. $11(-57)$
15. $-20(48)$
16. $-35(64)$
17. $126(-19)$
18. $532(-35)$
19. $-42(108)$
20. $-78(228)$
21. $3.6(-2.9)$
22. $1.07(-6.8)$
23. $\left(-\frac{1}{4}\right)\left(\frac{1}{6}\right)$
24. $\left(-\frac{2}{5}\right)\left(\frac{5}{8}\right)$

(Section 12-2)

25. $(-6)(-4)$
26. $(-3)(-9)$
27. $(-12)(-12)$
28. $(-15)(-22)$
29. $(-34)(-47)$
30. $(-28)(-45)$
31. $(-58)(-63)$
32. $(-85)(-79)$
33. $(-7)(-218)$
34. $(-326)(-9)$
35. $(-24)(-119)$
36. $(-281)(-56)$
37. $(-2.7)(-3.2)$
38. $(-0.6)(-2.15)$
39. $\left(-1\frac{2}{3}\right)\left(-2\frac{1}{4}\right)$
40. $\left(-3\frac{3}{5}\right)\left(-5\frac{1}{8}\right)$

(Section 12-3)

41. $2(x + 3)$
42. $(a + 7)3$
43. $-4(b + 5)$
44. $(-2d + 1)(-3)$
45. $r(r - 6)$
46. $2t(t - 9)$
47. $4(-2n + 4)$
48. $(-3m + 1)5$
49. $-1(-2p + 3)$
50. $(-4q + 5)(-2)$
51. $3y(2y - 2)$
52. $(6c - 5)2c$

Divide. (Section 12-4)

53. $-50 \div (-5)$ **54.** $-42 \div (-6)$ **55.** $-18 \div (-6)$ **56.** $-85 \div (-5)$
57. $\frac{56}{-14}$ **58.** $\frac{81}{-3}$ **59.** $\frac{32}{-8}$ **60.** $\frac{138}{-6}$
61. $\frac{-51}{3}$ **62.** $\frac{-132}{22}$ **63.** $\frac{-65}{5}$ **64.** $\frac{-212}{4}$
65. $-8 \div \frac{1}{2}$ **66.** $-16 \div \frac{2}{3}$ **67.** $-\frac{1}{2} \div \left(-\frac{1}{3}\right)$ **68.** $-\frac{5}{2} \div \left(-\frac{10}{3}\right)$
69. $\frac{1}{4} \div \left(-\frac{3}{8}\right)$ **70.** $\frac{3}{5} \div \left(-\frac{3}{5}\right)$ **71.** $-\frac{1}{4} \div (-5)$ **72.** $\frac{5}{6} \div (-10)$

Factor each expression. (Section 12-5)

73. $2x + 2$ **74.** $3b + 12$ **75.** $k^2 + k$ **76.** $x - x^2$
77. $39r - 13$ **78.** $7y + 7$ **79.** $2 + 2h$ **80.** $t^2 + 4t$
81. $14 - 7f$ **82.** $15 - 3n$ **83.** $13y - y^2$ **84.** $12 - 3d$

PART 3: APPLICATIONS

Use positive and negative numbers to represent each situation by a multiplication problem. Then find the product. (Section 12-1)

85. Two penalties of 10 yards each

86. Three debts of $42 each

87. Twice a drop in temperature of 6 degrees

88. Four times a depth of 18 meters below sea level

89. Five times a debt of $12

90. Twice a loss of 125 feet in altitude

Rational Numbers: Multiplication/Division

CHAPTER TEST

For Exercises 1–8, multiply.

1. $-16(24)$
2. $(-8)(-5)$
3. $9(-34)$
4. $-1.2(5.4)$
5. $(3.1)(-2.5)$
6. $-47(56)$
7. $(-12)(-75)$
8. $28(-67)$

For Exercises 9–14, multiply. Show all work.

9. $2(a+1)$
10. $-3(c+3)$
11. $4y(y-2)$
12. $(t+4)5$
13. $(r+6)(-2)$
14. $(3x-1)(6x)$

Divide.

15. $-36 \div (-4)$
16. $68 \div (-17)$
17. $\frac{-24}{3}$
18. $\frac{80}{-5}$
19. $\frac{3}{4} \div \left(-\frac{9}{4}\right)$
20. $-120 \div (-8)$
21. $125 \div (-25)$
22. $-\frac{1}{2} \div \left(-\frac{1}{5}\right)$

Factor each expression.

23. $5x+5$
24. $3p+12$
25. $2y-16$
26. $14-7a$
27. m^2+2m
28. $x-x^2$
29. $5y-20$
30. $2x+8$

ANSWERS TO REVIEW CAPSULES

Page 353 1. -6 2. -12 3. -15 4. -3 5. 48 6. -40 7. 48 8. 143

Page 355 1. -48 2. -48 3. -300 4. -154 5. 84 6. -192 7. $50y$ 8. $48t$ 9. $81r$ 10. $28k$ 11. $56a^2$ 12. $2x^2$ 13. $18z^2$ 14. $15p^2$

Page 360 1. 9 2. 62 3. 25 4. 9 5. $\frac{1}{12}$ 6. $\frac{1}{18}$

Page 363 1. $2a+8$ 2. $3b+3$ 3. $6x-18$ 4. $5y-25$ 5. $4a-4b$ 6. $7x+14$ 7. x^2+4x 8. d^2-3d 9. $5m+m^2$ 10. $6n-n^2$ 11. r^2+7r 12. $8t-32$ 13. 4 14. 6 15. 4 16. 2 17. 3 18. 3

ENRICHMENT

Multiplying Binomials

You can use the distributive property to multiply two binomials.

$$(x + 2)(x + 4) = x(x + 4) + 2(x + 4)$$
$$= x \cdot x + x \cdot 4 + 2 \cdot x + 2 \cdot 4$$
$$= x^2 + 4x + 2x + 8$$
$$= x^2 + 6x + 8$$

Here is a short way to multiply binomials. This is called the **FOIL** method.

	F Product of First Terms	**O I** Sum of Products of Outer and Inner Terms	**L** Product of Last Terms
$(x + 3)(x + 5) =$	$x \cdot x$ +	$x \cdot 5 + 3 \cdot x$ +	$3 \cdot 5$
$=$	x^2 +	$(5x + 3x)$ +	15
$=$	$x^2 + 8x + 15$		

EXERCISES

Multiply.

1. $(x + 2)(x + 3)$
2. $(a + 1)(a + 2)$
3. $(w + 5)(w + 4)$
4. $(y + 9)(y + 4)$
5. $(m + 8)(m + 7)$
6. $(x + 6)(x + 5)$
7. $(a - 1)(a - 4)$
8. $(n - 2)(n - 3)$
9. $(p - 1)(p - 3)$
10. $(y - 2)(y - 8)$
11. $(w - 7)(w - 4)$
12. $(d - 9)(d - 11)$
13. $(c + 7)(c - 3)$
14. $(t + 2)(t - 6)$
15. $(b + 5)(b - 4)$
16. $(b - 7)(b + 3)$
17. $(y - 4)(y + 2)$
18. $(p - 4)(p + 2)$

19. A rectangular garden has a length of $(t + 12)$ units and a width of $(t + 7)$ units Find the area.

20. Each side of a square lot has a length of $(x + 4)$ units. Find the area of the lot.

Rational Numbers: Multiplication/Division

ADDITIONAL PRACTICE

SKILLS

Multiply. (Pages 350–352)

1. $13(-12)$
2. $21(-35)$
3. $21(-9)$
4. $66(-29)$
5. $-63(0)$
6. $-71(1)$
7. $-42(60)$
8. $(-17)(12)$
9. $-3.5(2.4)$
10. $(-4.9)(1.6)$
11. $\left(-\frac{1}{4}\right)\left(2\frac{1}{2}\right)$
12. $\left(-\frac{2}{3}\right)\left(1\frac{1}{4}\right)$

(Pages 353–355)

13. $(-6)(-4)$
14. $(-9)(-7)$
15. $(-11)(-6)$
16. $(-13)(-5)$
17. $(-13)(-108)$
18. $(-42)(-211)$
19. $(-183)(-35)$
20. $(-308)(-48)$
21. $(-1.4)(-0.8)$
22. $(-5.03)(-4.9)$
23. $\left(-1\frac{1}{2}\right)\left(-1\frac{1}{3}\right)$
24. $\left(-2\frac{3}{4}\right)\left(-\frac{6}{11}\right)$

(Pages 356–358)

25. $5(3x + 4)$
26. $4(5s + 10)$
27. $-10(3b + 5)$
28. $-6(m + 1)$
29. $-4(3n + 3)$
30. $-5(5q + 7)$
31. $y(4y - 2)$
32. $5d(5d - 3)$
33. $7p(4p - 5)$
34. $5c(c - 5)$
35. $(3f - 6)(4f)$
36. $(6a - 4)(3a)$

Divide. (Pages 360–363)

37. $-50 \div (-5)$
38. $-49 \div (-7)$
39. $-44 \div 11$
40. $-56 \div 8$
41. $\frac{36}{-9}$
42. $\frac{45}{-5}$
43. $\frac{-18}{6}$
44. $\frac{-96}{6}$
45. $3 \div \left(-\frac{3}{4}\right)$
46. $4 \div \left(-\frac{4}{5}\right)$
47. $-3 \div \frac{1}{2}$
48. $-30 \div \frac{1}{6}$
49. $-\frac{1}{2} \div \left(-\frac{3}{4}\right)$
50. $-\frac{1}{4} \div \left(-\frac{7}{8}\right)$
51. $-\frac{5}{6} \div \frac{1}{12}$
52. $-\frac{7}{10} \div \frac{14}{5}$

Factor. (Pages 364–366)

53. $14x + 14$
54. $6m + 36$
55. $21m + 7$
56. $7v + 7$
57. $7p - 35$
58. $5t - 15$
59. $4c - 4$
60. $4a - 20$
61. $36 - 4v$
62. $49 - 7y$
63. $x^2 - 6x$
64. $x^2 - 8x$

APPLICATIONS

Use positive and negative numbers to represent each situation by a multiplication problem. Then find the product. (Pages 350–352)

65. Four fines of $15 each
66. A temperature drop of 3°C per hour for five hours

372 Chapter 12

CHAPTER 13 Equations: Multiplication/Division

SECTIONS
- **13-1** Equations: Multiplication/Division
- **13-2** Problem Solving and Applications: Words to Symbols: Multiplication and Division
- **13-3** Combined Operations
- **13-4** Equations with Parentheses
- **13-5** Problem Solving and Applications: More Than One Unknown: Angles in a Triangle
- **13-6** Problem Solving and Applications: More Than One Unknown: Consecutive Numbers
- **13-7** Inequalities: Multiplication/Division
- **13-8** Problem Solving and Applications: Using Inequalities

FEATURES
Computer Application: *Solving Equations*
Enrichment: *Sets*
Common Errors

13-1 Equations: Multiplication/Division

The following properties were used to solve division and multiplication equations in Chapter 7.

> **Multiplication Property for Equations**
>
> Multiplying each side of an equation by the same nonzero number forms an equivalent equation.
>
> $$\frac{m}{2} = -3$$
> $$\frac{m}{2}(2) = -3(2)$$
> $$m = -6$$

EXAMPLE 1 Solve and check: $\frac{x}{-12} = 6$

Solution: To get x alone, multiply each side by -12.

$$\frac{x}{-12} = 6$$
$$\frac{x}{-12}(-12) = 6(-12)$$
$$x = -72$$

Check: $\frac{x}{-12} = 6$
$$\frac{-72}{-12}$$
$$6$$

Self-Check Solve and check.

1. $\frac{a}{-7} = -6$
2. $\frac{c}{5} = -2$
3. $14 = \frac{y}{-8}$
4. $-5 = \frac{b}{-4}$

> **Division Property for Equations**
>
> Dividing each side of an equation by the same nonzero number forms an equivalent equation.
>
> $$-3w = 27$$
> $$\frac{-3w}{-3} = \frac{27}{-3}$$
> $$w = -9$$

EXAMPLE 2 Solve and check: $-4p = 16$

Solution: To get p alone, divide each side by -4.

$$-4p = 16$$
$$\frac{-4p}{-4} = \frac{16}{-4}$$
$$p = -4$$

Check: $-4p = 16$
$$-4(-4)$$
$$16$$

Self-Check Solve and check: **5.** $-2y = 18$ **6.** $21 = -3q$

374 Chapter 13

NOTE: By the Multiplication Property of One,
$$x = 1 \cdot x, \text{ or } \mathbf{1}x.$$
Thus, $\qquad -x = -1 \cdot x, \text{ or } \mathbf{-1}x.$

Therefore, to solve an equation such as $-x = 8$, divide each side of the equation by -1.

$$-x = 8 \qquad \blacktriangleleft \quad -x = -1 \cdot x$$
$$\frac{-1 \cdot x}{-1} = \frac{8}{-1}$$

Since $\frac{-1}{-1} = 1,$ $\qquad x = \frac{8}{-1}.$

$$x = \mathbf{-8}$$

CLASSROOM EXERCISES

For Exercises 1-8, by what number would you multiply each side of the equation in order to solve for x? (Example 1)

1. $\frac{x}{2} = -9$
2. $\frac{x}{5} = -2$
3. $\frac{x}{-7} = 1$
4. $\frac{x}{-3} = 8$
5. $\frac{x}{-3} = -5$
6. $\frac{x}{-6} = -8$
7. $8 = \frac{x}{12}$
8. $-3 = \frac{x}{-15}$

For Exercises 9–16, by what number would you divide each side of the equation in order to solve for n? (Example 2)

9. $-4n = 52$
10. $-6n = 36$
11. $7n = -21$
12. $5n = -15$
13. $-3n = -24$
14. $-10n = -40$
15. $36 = -2n$
16. $-48 = -12n$

For Exercises 17-36, solve and check. (Example 1)

17. $\frac{w}{4} = -9$
18. $\frac{p}{6} = -12$
19. $\frac{x}{-9} = -7$
20. $\frac{y}{-3} = -8$
21. $\frac{a}{-2} = 14$
22. $\frac{d}{-5} = 10$
23. $-11 = \frac{b}{4}$
24. $-15 = \frac{c}{7}$

(Example 2)

25. $-2q = 6$
26. $-8t = 64$
27. $10r = -90$
28. $4a = -144$
29. $-12r = -132$
30. $-7p = -49$
31. $-54 = 9c$
32. $-24 = 4m$
33. $-x = 3$
34. $-x = -5$
35. $-x = -3$
36. $-x = 0$

Equations: Multiplication/Division **375**

WRITTEN EXERCISES

Goal: To solve equations by using the Multiplication and Division Properties for Equations

For Exercises 1–40, solve and check. (Example 1)

1. $\dfrac{p}{12} = -8$
2. $\dfrac{q}{3} = -7$
3. $\dfrac{x}{-5} = 12$
4. $\dfrac{y}{-4} = 16$
5. $\dfrac{x}{-3} = -8$
6. $\dfrac{y}{-5} = -4$
7. $\dfrac{a}{4} = -18$
8. $\dfrac{b}{-25} = -5$
9. $\dfrac{c}{6} = -19$
10. $\dfrac{m}{9} = -4$
11. $12 = \dfrac{d}{-43}$
12. $52 = \dfrac{c}{-4}$

(Example 2)

13. $-4c = 16$
14. $-6y = 18$
15. $12x = -132$
16. $15g = -105$
17. $-3t = -72$
18. $-7x = -28$
19. $15 = -5p$
20. $21 = -7r$
21. $2w = -145$
22. $18w = -18$
23. $-50 = -3x$
24. $-19 = 4m$
25. $-x = 28$
26. $-x = -28$
27. $-c = -29$
28. $-c = 29$

MIXED PRACTICE

29. $\dfrac{x}{-5} = -35$
30. $\dfrac{c}{-2} = -130$
31. $\dfrac{n}{-2} = 70$
32. $\dfrac{t}{-4} = -80$
33. $9a = -72$
34. $-5y = 0$
35. $-7b = 0$
36. $13p = -26$
37. $3x = -4.2$
38. $8 = \dfrac{y}{2.1}$
39. $\dfrac{b}{-2.5} = -5$
40. $-7.2 = -1.2p$

REVIEW CAPSULE FOR SECTION 13–2

Determine whether the solution for the given equation is 5. Answer Yes or No.

1. $4y = 20$
2. $3x = 24$
3. $\dfrac{b}{2} = 10$
4. $\dfrac{c}{5} = 1$

Solve and check.

5. $7y = 21$
6. $2p = 14$
7. $12w = 60$
8. $9r = 81$
9. $\dfrac{b}{6} = 8$
10. $\dfrac{c}{10} = 11$
11. $\dfrac{w}{5} = 10$
12. $\dfrac{t}{13} = 7$

The answers are on page 401.

PROBLEM SOLVING AND APPLICATIONS

13-2 Words to Symbols: Multiplication and Division

Albert and Carmen picked apples last Saturday. Carmen stated that Albert had picked four times the number of apples she had picked.

Albert thought:

Number of apples Carmen picked: a

4 times the number Carmen picked: $4a$

The following table shows some word expressions and the corresponding algebraic expressions.

TABLE

Operation	Word Expression	Algebraic Expression
Multiplication	4 times the number of apples, a	$4a$
	The product of 18 and the number of tables, t	$18t$
	Twice the number of swimmers, s	$2s$
	Double Bill's weight, w	$2w$
	Triple the cost, c	$3c$
	Eight multiplied by the number of gallons, g	$8g$
Division	The quotient of the length, l, and 10	$\dfrac{l}{10}$
	The number of days, d, divided by 3	$\dfrac{d}{3}$
	84 divided by the number of kilometers, k	$\dfrac{84}{k}$

Albert knows that he picked 360 apples. He can use an equation to find how many apples Carmen picked.

Equations: Multiplication/Division

EXAMPLE 1 Four times the number of apples Carmen picked is 360. Find how many apples Carmen picked.

Solution:

1. Choose a variable. Let $a=$ the number of apples Carmen picked.
2. Write an equation. 4 times the number of apples Carmen picked is 360.

$$4a = 360$$

3. Solve the equation.

$$\frac{4a}{4} = \frac{360}{4}$$

$$a = 90$$

4. Check: Does $4(90) = 360$?   Yes ✔   Carmen picked **90 apples.**

Self-Check 1. The product of 22 and the number of chairs per table is 132. Find how many chairs there are per table.

EXAMPLE 2 The number of kilometers traveled by the Phillips family divided by 3 hours is 62. Find the distance traveled.

Solution:

1. Let $d=$ the number of kilometers.
2. Number of kilometers divided by 3 is 62.

$$\frac{d}{3} = 62$$

3.

$$\frac{d}{3}(3) = 62(3)$$

$$d = 186$$

4. Check: Does $\frac{186}{3} = 62$?   Yes ✔   Distance: **186 km**

Self-Check 2. The quotient of t and 8 is 15. Find t.

CLASSROOM EXERCISES

Choose from the box at the right the algebraic expression that represents each word expression.
(Table)

1. The product of t and 5
2. The quotient of t and 5
3. q divided by 4
4. The quotient of x and 3
5. 4 multiplied by q
6. x tripled
7. t doubled
8. t divided by 2

$4q$	$5t$
$\frac{t}{2}$	$\frac{t}{5}$
$\frac{q}{4}$	$\frac{x}{3}$
$2t$	$3x$

Complete. (Examples 1 and 2, step 2)

Word Sentence	Equation
9. Seven times the number of miles, m, is 140.	$\underline{?} = 140$
10. Twice the number of club members, s, is 36.	$\underline{?} = 36$
11. The quotient of the total cost, c, and 12 payments is $25.	$\underline{?}$
12. The number of passengers, p, divided by 6 buses is 42.	$\underline{?}$
13. The quotient of the number of hours, h, and 7 is 6.	$\underline{?}$

WRITTEN EXERCISES

Goals: To write an algebraic equation for a word sentence

To apply the skill of writing algebraic equations to solving word problems

Represent each word expression by an algebraic expression. (Table)

1. The quotient of x and 9
2. q multiplied by 7
3. The product of 2 and q
4. Twice p
5. Three times r
6. t divided by 212
7. a multiplied by (-4)
8. a divided by -4
9. 3 times p
10. p times 5
11. w tripled
12. The quotient of y and 12
13. t divided by 14
14. x doubled

For Exercises 15–22, write an equation for each word sentence. Do not solve the equation. (Example 1, step 2)

15. Four times the number of sailboats, s, is 28.
16. Twice the number of millimeters of rain, m, is 82 millimeters.
17. The product of the number of hours, h, and $6 is $132.
18. Twenty books multiplied by the number of shelves, n, is 240.

(Example 2, Step 2)

19. The number of kilometers, k, divided by 78 is 4.
20. The quotient of the total cost, c, and 24 payments is $30.
21. The quotient of the number of club members, m, and 7 is 10.
22. The number of points, p, divided by 12 questions is 13.

For Exercises 23-34, solve each problem. (Example 1)

23. The number of hours times 52 miles per hour is 156 miles. Find the number of hours.

24. Four times the width of a rectangle is 48 centimeters. Find the width of the rectangle.

25. Double the number of points Pedro scored is 76. How many points did Pedro score?

26. The product of 12 players and the number of teams is 60. Find the number of teams.

(Example 2)

27. The number of dimes saved divided by 50 is 8. Find the number of dimes.

28. The number of cars washed divided by 7 is 105. Find the number of cars.

29. The quotient of the amount of a loan and 15 payments is $45. Find the amount of the loan.

30. The quotient of the number of names in a phone directory and 6 is 72. Find the number of names.

MIXED PRACTICE

31. Twice the number of words Marilyn can type per minute is 142. How many words can she type per minute?

34. Eight times the cost of a concert ticket is $96. Find the cost of a concert ticket.

32. The number of balcony seats in a theater divided by 24 rows is 16. Find the number of balcony seats.

33. The quotient of the number of 9th grade students and 30 is 9. Find the number of 9th grade students.

REVIEW CAPSULE FOR SECTION 13-3

Simplify each expression. (Pages 64-66)

1. $3x + 5 - 5$
2. $4n + 6 - 6$
3. $7b - 1 + 1$
4. $5d - 7 + 7$
5. $\frac{x}{3} - 1 + 1$
6. $\frac{d}{4} - 2 + 2$
7. $\frac{h}{3} + 8 - 8$
8. $\frac{r}{6} + 2 - 2$

Solve and check. (Pages 190-195)

9. $x + 5 = 7$
10. $y + 6 = -4$
11. $b - 3 = 4$
12. $a - 9 = -6$
13. $p - 5 = -10$
14. $t + 7 = -10$
15. $r + 12 = 3$
16. $s - 6 = 14$

The answers are on page 401.

13-3 Combined Operations

This procedure will help you to solve equations that involve more than one operation.

> **PROCEDURE**
>
> **To solve equations with more than one operation:**
> 1. In general, use the Addition or Subtraction Property first.
> 2. Then use the Multiplication or Division Property.

EXAMPLE 1 Solve and check: $3x + 11 = 5$

Solution:

$3x + 11 = 5$

1. Subtract 11 from each side. $3x + 11 - 11 = 5 - 11$

   $3x = -6$

2. Divide each side by 3. $\dfrac{3x}{3} = \dfrac{-6}{3}$

   $x = -2$

Check: $3x + 11 = 5$

$3(-2) + 11$

$-6 + 11$

5

Self-Check *Solve and check.*

1. $7n + 20 = 13$
2. $2y - 5 = 7$
3. $81 = -5a + 6$

Be sure to check your answer in the original equation.

EXAMPLE 2 Solve and check: $\dfrac{n}{3} - 7 = -8$

Solution:

$\dfrac{n}{3} - 7 = -8$

1. Add 7 to each side. $\dfrac{n}{3} - 7 + 7 = -8 + 7$

   $\dfrac{n}{3} = -1$

2. Multiply each side by 3. $(3)\left(\dfrac{n}{3}\right) = -1(3)$

   $n = -3$

Check: $\dfrac{n}{3} - 7 = -8$

$\dfrac{-3}{3} - 7$

$-1 - 7$

-8

Self-Check *Solve and check.*

4. $\dfrac{a}{3} - 6 = -14$
5. $\dfrac{x}{-9} + 5 = 3$
6. $1 = \dfrac{t}{5} + 7$

Equations: Multiplication/Division **381**

CLASSROOM EXERCISES

For Exercises 1–16, write an equivalent equation as in step 1 of Examples 1 and 2. Do not solve the equation. (Example 1, step 1)

1. $3k + 5 = 2$
2. $4n + 6 = 14$
3. $7b + 1 = 15$
4. $5d - 7 = 13$
5. $12 = -2y + 8$
6. $15 = -3t + 3$
7. $4x + 7 = -13$
8. $5m + 1 = -29$

(Example 2, step 1)

9. $\frac{x}{3} - 1 = -2$
10. $\frac{d}{4} - 2 = -5$
11. $\frac{v}{-2} + 3 = 7$
12. $\frac{f}{-5} + 6 = 12$
13. $2 = \frac{y}{2} + 4$
14. $6 = \frac{a}{3} + 13$
15. $\frac{c}{-2} - 3 = 4$
16. $\frac{g}{-5} - 4 = 2$

WRITTEN EXERCISES

Goal: To solve an equation with more than one operation

For Exercises 1–40, solve and check. (Example 1)

1. $2x + 7 = 3$
2. $3y + 9 = 6$
3. $5a - 1 = 9$
4. $6p - 2 = 16$
5. $13 = -2b + 5$
6. $7 = -4g + 3$
7. $8z + 10 = -22$
8. $12d + 7 = -29$
9. $5h - 9 = -34$
10. $7w - 5 = -47$
11. $22 = -4f - 10$
12. $44 = -13k - 21$

(Example 2)

13. $\frac{x}{2} - 1 = -4$
14. $\frac{t}{4} - 3 = -5$
15. $\frac{n}{-2} + 1 = 5$
16. $\frac{a}{-3} + 2 = 7$
17. $10 = \frac{y}{5} + 6$
18. $12 = \frac{w}{4} + 5$
19. $\frac{c}{-6} - 5 = 6$
20. $\frac{g}{-5} - 7 = 9$
21. $\frac{h}{-2} + 8 = -2$
22. $\frac{r}{-3} + 7 = -4$
23. $-11 = \frac{q}{8} + 9$
24. $-10 = \frac{z}{7} + 6$

MIXED PRACTICE

25. $6x + 15 = 3$
26. $4p + 21 = 1$
27. $10a - 23 = 27$
28. $9c - 15 = 30$
29. $\frac{y}{5} - 6 = -13$
30. $\frac{b}{7} - 8 = -14$
31. $\frac{m}{-4} + 3 = 6$
32. $\frac{t}{-7} + 5 = 11$
33. $17 = \frac{u}{6} + 7$
34. $22 = \frac{v}{8} + 14$
35. $\frac{e}{-3} - 8 = -15$
36. $\frac{k}{-6} - 9 = -20$
37. $54 = -12x + 18$
38. $79 = -9s - 11$
39. $-4w - 6 = -18$
40. $-5d + 9 = -16$

REVIEW CAPSULE FOR SECTION 13-4

For Exercises 1-8, multiply. (Pages 356-358)

1. $3(4n + 6)$
2. $4(-7y + 3)$
3. $(p + 7)(5)$
4. $(2x + 3)(-4)$
5. $(z + 11)(-5)$
6. $-7(b + 6)$
7. $(4a - 3)5a$
8. $(2t)(4t - 4)$

Solve and check. (Pages 326-330)

9. $-x + 3 + 2x = 33$
10. $6y - 7 - 5y = 83$
11. $5 + 3p + 7 = 44 + 2p$
12. $4 + 10d - 13 = 7d + 8 + 2d$

The answers are on page 401.

13-4 Equations With Parentheses

Equations sometimes contain parentheses. To remove the parentheses, use the Distributive Property.

PROCEDURE

To solve equations containing parentheses:

[1] Use the Distributive Property to express the product as a sum or difference.

[2] Solve the equation.

EXAMPLE 1 Solve and check: $4(y + 2) = 28$

Solution:

[1] Use the Distributive Property.
$$4(y + 2) = 28$$
$$(4 \cdot y) + (4 \cdot 2) = 28$$

[2] Solve the equation.
$$4y + 8 = 28$$
$$4y + 8 - 8 = 28 - 8$$
$$4y = 20$$
$$\frac{4y}{4} = \frac{20}{4}$$
$$y = 5$$

◀ The check is left for you.

Self-Check *Solve and check.*

1. $5(n + 3) = 25$
2. $6(q - 1) = -12$
3. $2(x + 4) = -4$

EXAMPLE 2 Solve and check: $4(a-3) = -2(a+3)$

Solution: $\boxed{1}$
$$4(a-3) = -2(a+3)$$
$$(4 \cdot a) + (4)(-3) = (-2 \cdot a) + (-2 \cdot 3)$$
$\boxed{2}$
$$4a + (-12) = -2a - 6$$
$$4a + 2a - 12 = -2a + 2a - 6$$
$$6a - 12 = -6$$
$$6a - 12 + 12 = -6 + 12$$
$$6a = 6$$
$$\frac{6a}{6} = \frac{6}{6}$$
$$a = 1$$

Check: $4(a-3) \stackrel{?}{=} -2(a+3)$
$4(1-3) \mid -2(1+3)$
$4(-2) \mid -2(4)$
$-8 \mid -8$

Self-Check Solve and check.

4. $5(v-4) = -2(v+3)$ 　　　　**5.** $3(b+6) = -2(b+6)$

CLASSROOM EXERCISES

For Exercises 1–6, complete. (Example 1, step 1)

1. $3(c+7) = 32$
　　$3c + \underline{\ ?\ } = 32$

2. $7(2+d) = 63$
　　$\underline{\ ?\ } + 7d = 63$

3. $9(e+8) = 100$
　　$\underline{\ ?\ } + \underline{\ ?\ } = 100$

4. $2(-4-q) = 7$
　　$-8 - \underline{\ ?\ } = 7$

5. $6(m-5) = 11$
　　$6m \underline{\ ?\ } 30 = 11$

6. $10(n-5) = 81$
　　$10n \underline{\ ?\ } 50 = 81$

For Exercises 7–10, complete. (Example 2, step 1)

7. $5(z+7) = -3(z+7)$
　　$5z + \underline{\ ?\ } = \underline{\ ?\ } - 21$

8. $2(p+3) = -6(p+7)$
　　$\underline{\ ?\ } + 6 = -6p \underline{\ ?\ } 42$

9. $4(m-3) = 6(m-5)$
　　$4m - \underline{\ ?\ } = 6m \underline{\ ?\ } 30$

10. $-8(k+7) = -3(k+2)$
　　$-8k \underline{\ ?\ } 56 = \underline{\ ?\ } \underline{\ ?\ } 6$

Determine whether the given value of the variable is a solution of each equation. Answer Yes or No. (Examples 1 and 2, Check)

11. $7(v+4) = 119$ when $v = 13$

12. $6(r+2) = 66$ when $r = 9$

13. $3(t-5) = 15$ when $t = 11$

14. $2(n-4) = -4$ when $n = 3$

15. $-4(b+2) = 3(b-5)$ when $b = 1$

16. $5(d-2) = -6(d+2)$ when $d = 3$

17. $8(c+4.5) = 68$ when $c = 4$

18. $3.5(e-4) = -2.5(e+20)$ when $e = 6$

WRITTEN EXERCISES

Goal: To solve equations containing parentheses

For Exercises 1–40, solve and check. (Example 1)

1. $8(x + 2) = 48$
2. $2(y - 6) = 10$
3. $6(z - 3) = 4$
4. $11(f + 3) = 88$
5. $5(u - 2) = -25$
6. $4(r - 5) = -16$
7. $3(a + 3) = -39$
8. $7(c + 5) = -77$
9. $\frac{1}{2}(6t + 8) = 37$
10. $\frac{1}{8}(24m + 40) = 26$
11. $8(g + 1.5) = 92$
12. $4(2.5v + 7.5) = 100$

(Example 2)

13. $5(p - 6) = -3(p + 2)$
14. $8(n - 2) = -2(n + 3)$
15. $7(m - 7) = -5(m + 5)$
16. $6(q - 9) = -4(q + 1)$
17. $-1(c + 40) = -3(c + 10)$
18. $-2(d + 83) = -1(d + 15)$
19. $6(r - 4) = -2(r + 10)$
20. $3(s - 4) = -1(s + 11)$
21. $\frac{1}{3}(6e - 12) = \frac{1}{4}(12e - 4)$
22. $\frac{1}{2}(10f - 8) = \frac{1}{8}(24f + 96)$
23. $6(r - 4.5) = -2(r + 22.5)$
24. $3.5(2v + 4) = 1.5(4v - 10)$

MIXED PRACTICE

25. $3(a + 5) = 60$
26. $10(c - 5) = -3(c - 5)$
27. $12(d + 3) = -2(d + 38)$
28. $7(b + 2) = 77$
29. $4(e - 10) = -6(e + 15)$
30. $2(y - 8) = -44$
31. $5(x - 12) = -100$
32. $9(h + 3) = 3(h - 1)$
33. $10(1.5q + 2.5) = 175$
34. $6(0.5k + 6.5) = -4(1.5k + 10.5)$

MORE CHALLENGING EXERCISES

EXAMPLE: $-(-5 - 3x) = -7$ ◀ $-(-5 - 3x) = -1(-5 - 3x)$
$(-1)(-5) - (-1)(3x) = -7$
$5 + 3x = -7$
$3x = -12$
$x = -4$ ◀ The check is left for you.

35. $-(-10 - 2x) = 16$
36. $-(-2y - 4) = -6$
37. $-3(-6 - 4a) = -6$
38. $-2(-6b - 4) = 44$
39. $-3(x - 4) = -(-2x + 8)$
40. $-2(x - 5) = -4(2x + 2)$

REVIEW: SECTIONS 13-1—13-4

For Exercises 1-8, solve and check. (Section 13-1)

1. $\dfrac{n}{6} = -5$
2. $\dfrac{x}{8} = -7$
3. $36 = \dfrac{c}{-3}$
4. $9 = \dfrac{a}{-5}$
5. $-6p = 42$
6. $-10x = 50$
7. $-24 = -4y$
8. $-21 = -7z$

For Exercises 9-12, solve each problem. (Section 13-2)

9. The quotient of the number of fish caught and 4 is 32. Find the number of fish caught.
10. Five times the number of acres is 75. Find the number of acres.
11. The product of the number of weeks and $320 is $1280. Find the number of weeks.
12. The number of miles divided by 42 is 8. Find the number of miles.

For Exercises 13-26, solve and check. (Section 13-3)

13. $3x + 5 = 8$
14. $2a + 7 = 3$
15. $5w - 4 = -19$
16. $4t - 5 = -13$
17. $\dfrac{y}{2} - 2 = -6$
18. $\dfrac{b}{6} - 3 = -4$
19. $-4 = \dfrac{w}{5} + 2$
20. $10 = \dfrac{p}{6} + 7$

(Section 13-4)

21. $3(a + 6) = 30$
22. $5(y - 7) = -30$
23. $2(x - 4) = 4(x + 8)$
24. $7(b + 1) = -4(b + 12)$
25. $-3(d + 4) = 2(9 + d)$
26. $7(c + 5) = 5(c + 3)$

REVIEW CAPSULE FOR SECTION 13-5

Combine like terms. (Pages 76-80)

1. $7x + 9x + 2$
2. $8y + y - 9$
3. $4a + 2a - a$
4. $8a + 3a - 9$
5. $6t - 5t + 1$
6. $6y - y + 9y$
7. $4r - 9r + 32$
8. $10b - 7 + 15b$
9. $3c + 3c + 10$
10. $5g - 12 - 3g$
11. $6p + 3p + 4p$
12. $9s + 4s + 2s$

The answers are on page 401.

COMPUTER APPLICATIONS
Solving Equations

The following program will solve equations of the form **Ax + B = Cx + D** where $A \neq C$.

To use this program, you must first write the equation in the given form. Then you identify A, B, C, and D.

EXAMPLE Write these equations in the form $Ax + B = Cx + D$. Then identify A, B, C, and D.

  a. $-2 + 7x = 5x - 14$
  b. $8 - 3x - 14 = 3x$

Solutions:
  a. $7x - 2 = 5x - 14$
  $A = 7, b = -2, C = 5, D = -14$
  b. $-3x + 8 = 3x + 14$
  $A = -3, B = 8, C = 3, D = 14$

Program:

```
100 PRINT
110 PRINT "FOR THE EQUATION AX + B = CX + D,"
120 PRINT "WHAT ARE A, B, C, AND D";
130 INPUT A, B, C, D
140 LET X = (D - B) / (A - C)
150 PRINT "X = "; X
160 PRINT
170 PRINT "ANY MORE EQUATIONS (1 = YES, 0 = NO)";
180 INPUT Z
190 IF Z = 1 THEN 100
200 PRINT
210 END
```

NOTE: Solving $Ax + B = Cx + D$ for x yields statement 140.

$$x = \frac{D - B}{A - C}$$

Exercises

Write each equation in the form $Ax + B = Cx + D$. Then use the program to solve the equation.

1. $8x - 13 = 2x + 56$
2. $8x - 20 = -32 - 7x$
3. $11 - 2x - 16 = 8x$
4. $4x + 12 = 3 + 7x$
5. $-16 - 3x = 5x + 16$
6. $-5x + 2 = -9x - 16$
7. $8(x + 2) = 4x - 10$
8. $5x - 13 = 6x - 18$
9. $-3(x - 8) = 2x - 1$

PROBLEM SOLVING AND APPLICATIONS

13-5 More Than One Unknown: Angles in a Triangle

Many problems that relate to geometry can be solved by writing an equation.

Word Rule: The sum of the measures of the angles of a triangle is 180°.

Formula: $A + B + C = 180°$,

where A, B, and C represent the measures of the angles of a triangle.

EXAMPLE Find the measure of each angle of this triangle.

Solution:

⒈ Use the formula. Replace A with $2x$, B with x, and C with $2x + 5$.

$A + B + C = 180°$

$2x + x + 2x + 5 = 180$ ◀ **Combine like terms.**

⒉ Solve the equation.

$5x + 5 = 180$
$5x = 175$
$x = 35$ ◀ **Find $2x$ and $2x + 5$.**

$2x = 2(35) = 70 \qquad 2x + 5 = 2(35) + 5 = 75$

⒊ Check: $35 + 70 + 75 = 180$? Yes ✔

The angle measures are **35°, 70°, and 75°**.

Self-Check 1. The measures of the angles of a triangle can be represented by a, $5a$, and $6a$. Find the measure of each angle.

CLASSROOM EXERCISES

Complete. (Example)

1. $x + 2x + 3x = 180$; $x = \underline{}$; $2x = \underline{}$; $3x = \underline{}$
2. $t + 2t + 2t + 20 = 180$; $t = \underline{}$; $2t = \underline{}$; $2t + 20 = \underline{}$
3. $3s + 5s + 2s - 20 = 180$; $3s = \underline{}$; $5s = \underline{}$; $2s - 20 = \underline{}$

WRITTEN EXERCISES

Find the measure of each angle of the triangle. (Example)

1. Triangle ABC with angles x at A, $x + 20$ at B, $2x$ at C.

2. Triangle DGF with angles $5s$ at D, $5s$ at G, $5s$ at F.

3. Triangle RTS with angles $2y$ at R, y at T, y at S.

4. Triangle PNQ with angles $d + 30$ at P, d at N, d at Q.

5. Triangle ACB with angles $y + 15$ at A, y at C, $y + 15$ at B.

6. Triangle KLM with angles $3t$ at K, $8t$ at L, t at M.

Solve.

7. The measures of the angles of a triangle can be represented by d, $d - 15$, and $d - 30$. Find the measure of each angle of the triangle.

8. The measures of the angles of a triangle can be represented by s, $2s$, and $6s$. Find the measure of each angle of the triangle.

9. The measures of the angles of a triangle can be represented by x, $x + 10$, and $x + 20$. Find the measure of each angle of the triangle.

10. The measures of the angles of a triangle can be represented by c, $5c + 20$, and $10c$. Find the measure of each angle of the triangle.

REVIEW CAPSULE FOR SECTION 13-6

Combine like terms. (Pages 76-80)

1. $m + m + 1$
2. $x + x + 1$
3. $p + p + 1 + p + 2$
4. $y + y + 1 + y + 2$
5. $2t + 2t + 1 + 2t + 2$
6. $3g + 3g + 1 + 3g + 2$

Solve and check. (Pages 326-328)

7. $n + n + 1 = 13$
8. $p + p + 1 = 19$
9. $s + s + 1 + s + 2 = 27$
10. $q + q + 1 + q + 2 = 45$
11. $99 = 2y + 2y + 1 + 2y + 2$
12. $57 = 3b + 3b + 1 + 3b + 2$

The answers are on page 401.

Equations: Multiplication/Division

PROBLEM SOLVING AND APPLICATIONS

13-6 More Than One Unknown: Consecutive Numbers

Michael is one year younger than Janet. Thus, their ages can be written as *consecutive whole numbers*.

Consecutive whole numbers are numbers such as 3, 4, 5, 6, 7, and so on. When you find consecutive whole numbers, list them in order from least to greatest.

EXAMPLE 1 The sum of Michael's and Janet's present ages is 31. Their ages are consecutive whole numbers. Find their present ages.

Solution: ① Choose a variable. Represent the two unknowns.

Let n = the age of Michael.

Then $n + 1$ = the age of Janet.   ◂ Janet is one year older.

Michael's age + Janet's age = 31

② Write an equation. n + $n + 1$ = 31
③ Solve the equation. $2n + 1 = 31$
$2n = 30$
$n = 15$ ◂ Don't forget to find $n + 1$.
$n + 1 = 16$

④ Check: a. Are the two numbers consecutive?

Does $15 + 1 = 16$?   Yes ✓

b. Is the sum of their ages 31?

Does $15 + 16 = 31$?   Yes ✓

Michael is **15 years old** and Janet is **16 years old**.

Self-Check

1. The sum of two consecutive whole numbers is 15. Find the two numbers.

You can find three consecutive whole numbers when given their sum.

EXAMPLE 2 The sum of Lorraine's three diving scores is 24. The scores are consecutive whole numbers. Find each score.

Solution:

1. Choose a variable. Represent the three unknowns.

   Let q = the smallest score.

   Then $q + 1$ = the second score.

   And $q + 2$ = the third score.

2. Write an equation.

   Smallest Score + Second Score + Third Score = 24
   $$q + q + 1 + q + 2 = 24$$

3. Solve the equation.
   $$3q + 3 = 24$$
   $$3q = 21$$
   $$q = 7$$
   $$q + 1 = 8$$
   $$q + 2 = 9$$

   Don't forget to find $q + 1$ and $q + 2$.

4. Check: a. Are the numbers consecutive?

   Does $7 + 1 = 8$ and $8 + 1 = 9$?   Yes ✔

   b. Is the sum of the numbers 24?

   Does $7 + 8 + 9 = 24$?   Yes ✔   **Scores: 7, 8, 9**

Self-Check

2. The sum of three consecutive whole numbers is 27. Find the numbers.

CLASSROOM EXERCISES

For Exercises 1–4, determine whether the numbers listed are consecutive whole numbers. Write Yes or No. (Examples 1 and 2, step 4)

1. 86, 87
2. 56, 57, 58, 60
3. 5, 10, 11, 15
4. 21, 22, 23, 24

Write an equation for each sentence. Let h represent the smallest whole number. (Examples 1 and 2, Step 2)

5. The sum of two consecutive whole numbers is 5.
6. The sum of two consecutive whole numbers is 13.
7. The sum of three consecutive whole numbers is 21.
8. The sum of three consecutive whole numbers is 39.

Equations: Multiplication/Division

WRITTEN EXERCISES

Goal: To solve word problems involving consecutive whole numbers

For Exercises 1–14, solve and check. (Example 1)

1. Louise is one year younger than Reynaldo. The sum of their ages is 29. Find their ages.

2. Kyoshi is one year older than Rosa. The sum of their ages is 37. Find their ages.

3. William's score on a spelling test was one point higher than Juan's score. The sum of their scores was 169. Find their scores.

4. Julia's test score was one point lower than Roberta's score. The sum of their scores was 187. Find their scores.

(Example 2)

5. Carol Soong's scores for 3 games of bowling were consecutive whole numbers. The sum of her scores was 315. Find her scores.

6. The ages of Jason, Amy, and Tasha are consecutive whole numbers. Jason is the oldest and Tasha is the youngest. The sum of their ages is 30. Find their ages.

7. The ages of Morris, Eric, and Maria are consecutive whole numbers. Maria is the oldest and Eric is the youngest. The sum of their ages is 63. Find their ages.

8. William Blackfeather's three math scores are consecutive whole numbers. The sum of his scores is 237. Find his scores.

MIXED PRACTICE

9. Hans is one year older than Rosaline. The sum of their ages is 33. Find their ages.

10. Regina's two scores were consecutive whole numbers. The sum of her scores was 173. Find her scores.

11. The sum of Winnie's three diving scores was 21. The scores were consecutive whole numbers. Find her scores.

12. Stan's, Mark's and Wayne's ages are consecutive whole numbers. Stan is the youngest and Wayne is the oldest. The sum of their ages is 111. Find their ages.

13. The sum of two consecutive whole numbers is 129. Find the numbers.

14. The sum of three consecutive whole numbers is 75. Find the numbers.

REVIEW CAPSULE FOR SECTION 13-7

Replace each __?__ with < or >. (Pages 294–296)

1. −2 __?__ 4 **2.** 2 __?__ −4 **3.** 0 __?__ 5 **4.** 0 __?__ −5 **5.** −6 __?__ 10

Solve each equation. (Pages 374–376)

6. $\frac{x}{3} = 8$ **7.** $\frac{t}{-4} = 2$ **8.** $6m = -42$ **9.** $-5p = 30$ **10.** $-7y = -56$

The answers are on page 401.

13-7 Inequalities: Multiplication/Division

Solving an inequality is similar to solving an equation except when you multiply or divide an inequality by a negative number.

Multiplication Property for Inequalities

1. The direction of the inequality symbol <u>remains unchanged</u> when each side of the inequality is multiplied by the <u>same positive number</u>.

$$\frac{n}{2} < 8$$
$$(2)\frac{n}{2} < 8(2)$$
$$n < 16$$

2. The direction of the inequality symbol is <u>reversed</u> when each side of the inequality is multiplied by the <u>same negative number</u>.

$$3 < \frac{x}{-4}$$
$$(-4)3 > \frac{x}{-4}(-4)$$
$$-12 > x, \text{ or } x < -12$$

EXAMPLE 1 Solve and graph: **a.** $\frac{x}{2} > 3$ **b.** $\frac{y}{-3} < 6$

Solutions: **a.** $\frac{x}{2} > 3$

$(2)\frac{x}{2} > 3(2)$

$x > 6$

b. $\frac{y}{-3} < 6$

$(-3)\frac{y}{-3} > 6(-3)$ ◀ Reverse the inequality symbol.

$y > -18$

The solution is **all numbers greater than 6**.

The solution is **all numbers greater than −18**.

Equations: Multiplication/Division **393**

Self-Check *Solve and graph:* **1.** $\frac{c}{4} < 2$ **2.** $\frac{x}{3} > -1$ **3.** $\frac{d}{-6} < 3$ **4.** $-4 < \frac{r}{-5}$

REMEMBER: The open circle on a graph means that the number circled is <u>not</u> one of the solutions of the inequality.

> **Division Property for Inequalities**
> 1. The direction of the inequality symbol <u>remains unchanged</u> when each side of the inequality is divided by the <u>same positive number</u>.
>
> $5y < 15$
> $\frac{5y}{5} < \frac{15}{5}$
> $y < 3$
>
> 2. The direction of the inequality symbol is <u>reversed</u> when each side of the inequality is divided by the <u>same negative number</u>.
>
> $-3a > 9$
> $\frac{-3a}{-3} < \frac{9}{-3}$
> $a < -3$

EXAMPLE 2 Solve and graph: **a.** $-5y < 40$ **b.** $-24 > 8p$

Solutions:

a. $-5y < 40$
$\frac{-5y}{-5} > \frac{40}{-5}$ ◀ Reverse the inequality symbol.
$y > -8$

The solution is **all numbers greater than -8**.

b. $-24 > 8p$
$\frac{-24}{8} > \frac{8p}{8}$
$-3 > p$ or $p < -3$

The solution is **all numbers less than -3**.

Self-Check *Solve and graph:* **5.** $-3m < 15$ **6.** $-54 > 9b$ **7.** $25 \leq -5q$

CLASSROOM EXERCISES

For Exercises 1–10, by what number would you multiply each side of the inequality in order to solve for x? (Example 1)

1. $\frac{x}{5} > 4$ **2.** $\frac{x}{3} < 6$ **3.** $\frac{x}{-2} < 8$ **4.** $\frac{x}{-4} > 1$ **5.** $\frac{x}{9} < -5$

6. $\frac{x}{6} \geq -10$ **7.** $\frac{x}{7} \leq -5$ **8.** $-2 < \frac{m}{-7}$ **9.** $-9 > \frac{h}{-8}$ **10.** $\frac{x}{-5} > 12$

394 Chapter 13

For Exercises 11-18, by what number would you divide each side of the inequality in order to solve for p? (Example 2)

11. $-3p > 12$ **12.** $-4p < 24$ **13.** $-6p < -48$ **14.** $-3p > -15$

15. $-20 \geq 5p$ **16.** $-18 \leq -2p$ **17.** $21 \leq -7p$ **18.** $22 \geq -11p$

WRITTEN EXERCISES

Goal: To solve and graph inequalities using the Multiplication and Division Properties for Inequalities

For Exercises 1-32, solve and graph each inequality. (Example 1)

1. $\frac{y}{3} < 3$ **2.** $\frac{x}{5} > 1$ **3.** $\frac{a}{7} > -10$ **4.** $\frac{g}{6} \leq -8$

5. $\frac{w}{-2} \geq 3$ **6.** $\frac{t}{-4} < 2$ **7.** $-5 < \frac{k}{-10}$ **8.** $-11 > \frac{n}{-8}$

(Example 2)

9. $-5n < 35$ **10.** $-3y > 27$ **11.** $-10b < -40$ **12.** $-8a > -24$

13. $-20 > 2m$ **14.** $-30 < 6x$ **15.** $20 \leq -4t$ **16.** $12 \geq -3z$

MIXED PRACTICE

17. $\frac{x}{4} > 24$ **18.** $\frac{k}{3} > 3$ **19.** $\frac{a}{5} < -8$ **20.** $\frac{t}{6} > -6$

21. $-12j < -120$ **22.** $-9m > -108$ **23.** $-8n < 16$ **24.** $-4y > 32$

25. $\frac{b}{-5} > 3$ **26.** $\frac{c}{-7} < 3$ **27.** $-8 < \frac{d}{-4}$ **28.** $-9 > \frac{e}{-5}$

29. $22 \leq -2p$ **30.** $14 \geq -14w$ **31.** $-32 \leq 2n$ **32.** $-12 \geq 12x$

REVIEW CAPSULE FOR SECTION 13-8

Write an algebraic expression for each word expression.
(Pages 334-337)

1. Five more than a certain number, y   **2.** Nine less than a certain number, p

3. Eight less than a certain number, t   **4.** Ten more than a certain number, b

Find the average score. (Pages 10-12)

5. Team A: 16, 19, 7, 14, 14   **6.** Team B: 33, 25, 29, 25

The answers are on page 401.

PROBLEM SOLVING AND APPLICATIONS

13-8 Using Inequalities

Marla plans to spend less than $50 for a shirt and sweater. She buys a shirt for $21 and wants to know the greatest amount she can spend for a sweater. Marla can use an inequality to find the amount.

The cost of a sweater plus $21 is less than $50.

$$c + 21 < 50$$

The procedure for solving word problems involving inequalities is similar to the procedure for solving word problems involving equations.

EXAMPLE 1 Find the greatest amount Marla can spend for a sweater.

Solution:
1. Choose a variable. Let c = the cost of the sweater.

   The cost of a sweater and $21 is less than $50.

2. Write an inequality. $c + 21 < 50$

3. Solve the inequality.
$$c + 21 - 21 < 50 - 21$$
$$c < 29$$

The greatest amount Marla can spend for a sweater is **$28.99**.

Self-Check 1. Ted plans to spend less than $60 for a baseball glove and bat. He spends $32 for a glove. Find the greatest amount he can spend for a bat.

Recall that the **average** of two or more scores is the sum of the scores divided by the number of scores.

EXAMPLE 2 Raoul's scores on 4 tests were 80, 77, 81, and 84. He wants his average on **five** tests to be greater than 80. Find the lowest score Raoul can get on the fifth test.

Solution:
1. Let x = Raoul's score on the fifth test.

2. $$\frac{80 + 77 + 81 + 84 + x}{5} > 80 \qquad \text{Average} = \frac{\text{Sum of scores}}{5}.$$

396 Chapter 13

$$\boxed{3} \quad 5\left(\frac{80 + 77 + 81 + 84 + x}{5}\right) > 80(5)$$
$$80 + 77 + 81 + 84 + x > 400$$
$$322 + x > 400$$
$$322 + x - 322 > 400 - 322$$
$$x > 78$$

Raoul's score must be **at least 79.**

Self-Check 2. Lisa's scores on 4 tests were 90, 92, 89, and 85. She wants her average on five tests to be greater than 90. Find the lowest score Lisa can get on the fifth test.

CLASSROOM EXERCISES

Write an inequality for each word sentence. Use b as the variable. Do <u>not</u> solve the inequality. (Example 1, Step 2)

1. The cost of a lamp and $8 is less than $45.
2. The number of persons increased by 5 is less than 290.
3. The sum of the cost of a shirt and $15 is less than $32.
4. The number of hats increased by 12 is less than 72.

Write an inequality for each word sentence. Do <u>not</u> solve the inequality. (Example 2, Step 2)

5. The average of 87, 92, and y is greater than 91.
6. The average of 85, 87, 84, and p is greater than 86.
7. The average of 65, 70, 68 and x is greater than 66.
8. The average of 68, 70, 72, and c is greater than 71.

WRITTEN EXERCISES

Goal: To solve word problems involving inequalities

For Exercises 1–8, solve (Example 1)

1. Bill plans to buy a jacket and sweater for less than $70. He spends $42 for a jacket. Find the greatest amount he can spend for the sweater.
2. A family plans to spend less than $1100 for a new sofa and chair. They spend $800 for a sofa. Find the greatest amount they can spend for a chair.

Equations: Multiplications/Divisions   **397**

3. Carl plans to spend less than $70 for a tennis racket and a pair of shoes. He spends $25 for a pair of shoes. Find the greatest amount he can spend for a tennis racket.

4. The Chin family plans to spend less than $650 a month for rent and food. They spend $430 a month for rent. What is the most they can spend for food a month?

(Example 2)

5. Ava's three math quiz scores were 78, 81, and 93. She wants her average on four quizzes to be greater than 85. Find the lowest score she can get on the fourth quiz.

6. Ronald's grades on four tests are 68, 82, 78, and 80. He would like to have an average of more than 80 on five tests. What is the lowest grade he can get on the fifth test?

7. Rocco's scores for three games of bowling were 76, 103, and 121. In order to have an average greater than 100 for four games, what is the lowest score he can bowl on the fourth game?

8. Lillian's scores for four games of bowling were 76, 85, 104 and 92. She wants her average for five games to be greater than 90. Find the lowest score she can bowl on the fifth game.

REVIEW: SECTIONS 13-5—13-8

Solve and check. (Section 13-6)

1. The sum of two consecutive whole numbers is 17. Find the numbers.

2. The sum of three consecutive whole numbers is 39. Find the numbers.

Solve and graph. (Section 13-7)

3. $\frac{x}{4} < -2$

4. $\frac{c}{-2} > 3$

5. $5n < -10$

6. $-8y > -32$

Solve. (Section 13-8)

7. The measures of the angles of a triangle can be represented by x, $2x$, and $4x + 5$. Find the measure of each angle of the triangle. (Section 13-5)

8. The Luna family plans to spend less than $750 a month for rent and food. They spend $400 each month for rent. What is the most they can spend each month for food? (Section 13-8)

CHAPTER REVIEW

PART 1: VOCABULARY

For Exercises 1–6, choose from the box at the right below the word(s) that best corresponds to each description.

1. Words such as product, times, twice, and tripled suggest the operation of __?__. (p. 377)
2. The sum of the measures of the angles of a __?__ is 180°. (p. 388)
3. Numbers such as 12, 13, 14, 15 and so on are __?__ whole numbers. (p. 390)
4. The words quotient and divided by suggest the operation of __?__. (p. 377)
5. The direction of the symbol of inequality is __?__ when each side is multiplied by the same negative number. (p. 393)
6. The direction of the symbol of inequality is __?__ when each side is multiplied by the same positive number. (p. 393)

triangle
unchanged
multiplication
reversed
division
consecutive

PART 2: SKILLS

Solve and check. (Section 13-1)

7. $\dfrac{x}{5} = -8$
8. $\dfrac{y}{-3} = -12$
9. $\dfrac{b}{4} = 9$
10. $\dfrac{c}{-8} = 22$
11. $3p = 24$
12. $-5q = -45$
13. $-8d = 64$
14. $12t = -60$

Write an equation for each word sentence. (Section 13-2)

15. Three times the number of club members, m, is 42.
16. The number of letters, l, divided by 6 is 14.
17. The number of peanuts, p, divided by 16 is 8.
18. Twice the length, l of a rectangle is 56 centimeters.

Solve and check. (Section 13-3)

19. $3m + 6 = -3$
20. $2p + 4 = -6$
21. $-4t - 8 = 16$
22. $-2y - 9 = -13$
23. $\dfrac{p}{7} + 4 = 5$
24. $\dfrac{x}{3} + 5 = 2$
25. $\dfrac{a}{-4} - 2 = 7$
26. $\dfrac{b}{-5} - 6 = -4$

(Section 13-4)

27. $3(c + 9) = 54$
28. $8(m - 2) = 40$
29. $-4(2n + 3) = -2(3n + 18)$
30. $-5(t + 4) = -7(t + 12)$

Equations: Multiplication/Division **399**

Use the formula $A + B + C = 180°$ to find the measure of each angle of the triangle. (Section 13-5)

31.

32.

33.

Solve each equation. Then complete. (Section 13-6)

34. $n + n + 1 = 13$; $n = \underline{\quad?\quad}$; $n + 1 = \underline{\quad?\quad}$
35. $p + p + 1 = 35$; $p = \underline{\quad?\quad}$; $p + 1 = \underline{\quad?\quad}$
36. $x + x + 1 + x + 2 = 24$; $x = \underline{\quad?\quad}$; $x + 1 = \underline{\quad?\quad}$; $x + 2 = \underline{\quad?\quad}$
37. $y + y + 1 + y + 2 = 39$; $y = \underline{\quad?\quad}$; $y + 1 = \underline{\quad?\quad}$; $y + 2 = \underline{\quad?\quad}$

Solve and graph. (Section 13-7)

38. $\dfrac{x}{3} > -8$
39. $\dfrac{p}{-9} > 15$
40. $\dfrac{a}{-5} < -9$
41. $\dfrac{b}{4} < 12$
42. $6y > -24$
43. $-4t < -44$
44. $-5c < 25$
45. $8q > -16$

PART 3: APPLICATIONS

Solve each problem.

46. Twice the number of fish Steve caught is 14. How many fish did he catch? (Section 13-2)
47. The number of bicycles divided by 6 is 15. Find the number of bicycles. (Section 13-2).
48. The measures of the angles of a triangle can be represented by b, $2b + 20$, and $b + 20$. Find the measure of each angle of the triangle. (Section 13-5)
49. Kermit is one year younger than Vera. Find their present ages if the sum of their ages is 37. (Section 13-6)
50. Glenda plans to spend less than $100 for a lamp and a table. She buys a lamp for $35. Find the greatest amount she can spend on a table. (Section 13-8)

CHAPTER TEST

For Exercises 1–10, solve and check.

1. $\dfrac{x}{15} = -4$
2. $\dfrac{c}{-7} = 3$
3. $-6y = 48$
4. $-15z = -45$
5. $6x + 10 = 52$
6. $3a - 8 = -114$
7. $\dfrac{n}{-5} - 2 = 13$
8. $\dfrac{y}{4} + 7 = -5$
9. $10(e + 2) = 50$
10. $6(x + 5) = -3(x + 2)$

11. Solve and graph: $\dfrac{b}{4} < 2$

Solve each problem.

12. The number of seats in an auditorium divided by 36 rows is 14. Find the number of seats.

13. The measures of the angles of a triangle can be represented by x, $3x$, and $6x$. Find the measure of each angle of the triangle.

14. Kurt and Melissa were born in consecutive years. Kurt is 1 year younger than Melissa. Find their present ages if the sum of their ages is 35.

15. Bert plans to buy a shirt and a pair of slacks for less than $53. He spends $21 for a shirt. Find the greatest amount he can spend for a pair of slacks.

ANSWERS TO REVIEW CAPSULES

Page 376 1. Yes 2. No 3. No 4. Yes 5. 3 6. 7 7. 5 8. 9 9. 48 10. 110 11. 50 12. 91

Page 380 1. $3x$ 2. $4n$ 3. $7b$ 4. $5d$ 5. $\dfrac{x}{3}$ 6. $\dfrac{d}{4}$ 7. $\dfrac{h}{3}$ 8. $\dfrac{r}{6}$ 9. 2 10. -10 11. 7 12. 3 13. -5 14. -17 15. -9 16. 20

Page 383 1. $12n + 18$ 2. $-28y + 12$ 3. $5p + 35$ 4. $-8x - 12$ 5. $-5z - 55$ 6. $-7b - 42$ 7. $20a^2 - 15a$ 8. $8t^2 - 8t$ 9. 30 10. 90 11. 32 12. 17

Page 386 1. $16x + 2$ 2. $9y - 9$ 3. $5a$ 4. $11a - 9$ 5. $t + 1$ 6. $14y$ 7. $-5r + 32$ 8. $25b - 7$ 9. $6c + 10$ 10. $2g - 12$ 11. $13p$ 12. $15s$

Page 389 1. $2m + 1$ 2. $2x + 1$ 3. $3p + 3$ 4. $3y + 3$ 5. $6t + 3$ 6. $9g + 3$ 7. 6 8. 9 9. 8 10. 14 11. 16 12. 6

Page 393 1. $<$ 2. $>$ 3. $<$ 4. $>$ 5. $<$ 6. 24 7. -8 8. -7 9. -6 10. 8

Page 395 1. $y + 5$ 2. $p - 9$ 3. $t - 8$ 4. $b + 10$ 5. 14 6. 28

Equations: Multiplications/Division **401**

ENRICHMENT

Sets

A **set** is a collection of objects called **elements**.

Here are two ways to describe a set.

Rule Method	Roster Method
D = The set of natural numbers less than 6	D = {1, 2, 3, 4, 5}
W = The set of whole numbers	W = {0, 1, 2, 3, ···}

The numbers continue without end.

Symbol or Term	Explanation	Example
∈	Is an element of	6 ∈ {2, 4, 6, 8}
∉	Is not an element of	3 ∉ {6, 7, 8}
=	Is equal to; has the same elements	{1, 4, 6} = {6, 4, 1}
≠	Is not equal to	{2, 3, 5} ≠ {4, 6, 8}
Finite set	The elements can be listed.	{1, 2, 5, 6}
Infinite set	The elements can not be completely listed.	{2, 4, 6, 8, ···}
φ	Empty set; has no elements	The set of whole numbers less than 0 = φ.
Equivalent sets	Sets that have the same number of elements	{1, 2, 3, 4} {6, 7, 8, 9} — One-to-one correspondence

EXERCISES

Determine whether each of the following is a true statement. Answer Yes or No

1. 6 ∈ {1, 2, 3, 4, ···}
2. 15 ∉ {3, 6, 9, ···}
3. {1, 2, 3} = {1, 2, 3, ···}
4. {6, 7, 8, 9} ≠ {9, 8, 7, 6}
5. {0} is a finite set.
6. {2, 4, 6, ···} is an infinite set.

ADDITIONAL PRACTICE

SKILLS

Solve and check. (Pages 374–376)

1. $\dfrac{t}{-5} = -30$
2. $\dfrac{x}{-6} = -36$
3. $\dfrac{y}{3} = -12$
4. $\dfrac{r}{4} = -52$
5. $8c = -64$
6. $4w = -56$
7. $-3x = -9.6$
8. $-3v = -8.1$

(Pages 381–382)

9. $5x + 1 = 51$
10. $10v + 26 = 6$
11. $36 = -6g + 6$
12. $14 = -8g + 6$
13. $\dfrac{y}{5} - 5 = -12$
14. $\dfrac{k}{3} - 7 = -35$
15. $20 = \dfrac{x}{8} + 14$
16. $10 = \dfrac{w}{3} + 4$

(Pages 383–385)

17. $7(x + 3) = 42$
18. $8(a + 10) = 32$
19. $2(4w - 18) = -28$
20. $4(b - 3) = -48$
21. $7(y + 7) = 5(y + 5)$
22. $9(g + 3) = 3(g + 1)$

Solve and graph each inequality. (Pages 393–395)

23. $\dfrac{x}{2} < 4$
24. $\dfrac{n}{-3} > 3$
25. $6x > -24$
26. $-2v < -6$

APPLICATIONS *Represent each word expression by an algebraic equation.* (Pages 377–380)

27. The product of 3 and x is 33.
28. The product of 12 and w is 144.
29. The quotient of w and 4 is 9.
30. The quotient of v and 11 is 4.

Solve each problem.

31. The Hearth family plans to spend less than $1200 for a new table and chairs. The chairs cost $360. Find the greatest amount they can spend for the table. (Pages 396–398)

32. Jeff's three test scores were 79, 84, and 94. He wants his average on four tests to be greater than 86. Find the lowest score he can get on the fourth test. (Pages 396–398)

33. Laura's scores for three games of bowling were consecutive whole numbers. The sum of her scores was 327. Find Laura's scores for each game. (Pages 390–392)

34. The measures of the angles of a triangle can be represented by x, $x + 20$, and $6x$. Find the measure of each angle of the triangle. (Pages 388–389)

Equations: Multiplication/Division   **403**

COMMON ERRORS

Each of these problems contains a common error.

a. Find the correct answer.

b. Find the error.

1. $-9 + (-4) = \underline{}$
$|-9| + |-4| = 9 + 4$
$ = 13$

Thus, $-9 + (-4) = 13$

2. $4 + (-7) = \underline{}$
$|4| + |-7| = -4 + 7$
$ = 3$

Thus, $4 + (-7) = 3$

3. $-9 - 7 = -9 + 7$
$ = -2$

4. $6 - (-3) = 6 - 3$
$ = 3$

Solve.

5. $x - 6 = -3$
$x - 6 = -3 - 6$
$x = -9$

6. $y + 5 = 3$
$y + 5 - 5 = 3 - 5$
$y = 2$

7. $2c = 21 - c$
$2c - c = 21 - c + c$
$c = 21$

8. $q + 6 < 9$
$q + 6 - 6 < 9 - 6$
$q = 3$

9. $\dfrac{x}{-3} = 6$
$(-3)\dfrac{x}{-3} = 6(-3)$
$x = 18$

10. $-4y = -24$
$\dfrac{-4y}{-4} = \dfrac{-24}{-4}$
$y = -6$

11. $2x + 9 = -24$
$\dfrac{2x}{2} + 9 = \dfrac{-24}{2}$
$x + 9 = -12$
$x = -21$

12. $-6w > -30$
$\dfrac{-6w}{-6} > \dfrac{-30}{-6}$
$w > 5$

Multiply.

13. $4(m + 9) = 4(m) + 9$
$ = 4m + 9$

14. $p(p + 5) = p(p) + p(5)$
$ = 2p + 5p = 7p$

404 Chapter 13

CUMULATIVE REVIEW: CHAPTERS 10–13

Choose the correct answer. Choose a, b, c, or d.

1. Which integers are arranged in order from least to greatest?
   a. 4, 7, 13, -44
   b. $-4, -6, -9, -13$
   c. $-15, -11, 0, 8$
   d. $-17, -12, -13, 0$

2. Which integer is the opposite of -2?
   a. 2
   b. -1
   c. -3
   d. 3

3. Add: $-93 + 112$
   a. -21
   b. 21
   c. 19
   d. -19

4. Subtract: $-12 - (-36)$
   a. 48
   b. -48
   c. 24
   d. -24

5. The highest temperature on the planet Mars is about 35° F above zero and the lowest temperature is about 120° F below zero. About how much warmer is the highest temperature than the lowest temperature?
   a. 155° F
   b. 85° F
   c. 95° F
   d. 105° F

6. Add: $-35 + 47 + (-74) + 81$
   a. 21
   b. 19
   c. -19
   d. -21

7. Solve for y: $y - 41 = -135$
   a. -176
   b. 176
   c. 94
   d. -94

8. Solve for m: $-13 = m + 49$
   a. 36
   b. 62
   c. -62
   d. -36

9. Solve for t: $-3t + 8 + 4t = -7 + 12$
   a. -3
   b. 13
   c. 3
   d. -13

10. Solve for p: $4p - 17 = 3p + 4$
    a. -13
    b. 21
    c. 13
    d. -21

11. Write an algebraic expression: Three more than the sum of x and y.
    a. $3 > x + y$
    b. $xy + 3$
    c. $(x + y) + 3$
    d. $3 > xy$

12. Write an algebraic expression: Seven less than some number is 12.
    a. $n - 7 = 12$
    b. $n - 12 = 7$
    c. $7 - n = 12$
    d. $12 - n = 7$

13. Last Wednesday's average temperature plus 2° equals $-7°$. Find last Wednesday's average temperature.
    a. $-14°$
    b. $-5°$
    c. $-9°$
    d. $-11°$

14. Which graph below shows the solution to the inequality $-3 > x + 1$?

   a. [number line with closed circle at -4, shaded right from -4]
   b. [number line with open circle at -4, shaded right]
   c. [number line with open circle at -4, shaded left]
   d. [number line with open circle at -2, shaded right]

15. Multiply: $20(-35)$
   a. 700   b. -55 c. 55   d. -700

16. Multiply: $(-43)(-35)$
   a. 1505   b. -1505 c. -78 d. 78

17. Multiply: $8(2x - 10)$
   a. $16x + 10$ b. $16x + (-80)$ c. $-10x + 2$ d. $10x + (-80)$

18. Divide: $-36 \div 4$
   a. $-\frac{1}{9}$ b. 9   c. -9 d. $\frac{1}{9}$

19. Divide: $-\frac{3}{8} \div (-12)$
   a. $-\frac{1}{32}$ b. $4\frac{1}{2}$ c. $\frac{1}{32}$ d. $-4\frac{1}{2}$

20. Factor: $9x - 54$
   a. $9(x - 5)$ b. $9(x + 6)$ c. $9(x - 6)$ d. $9(x + 5)$

21. Factor: $y^2 - 4y$
   a. $4y(y - 1)$ b. $y(y - 4)$ c. $y(y + 4)$ d. $y(3y - 2)$

22. Solve for a: $\frac{a}{-4} = -16$
   a. -64 b. 64   c. -4 d. 4

23. Solve for w: $11w = -286$
   a. -26 b. -297 c. -275 d. -24

24. Write an algebraic expression for the quotient of 12 and y.
   a. $12 - y$ b. $y \div 12$ c. $12 \div y$ d. $12y$

25. Solve for h: $7(h + 5) = -77$
   a. 16   b. -112 c. 112   d. -16

26. One angle of a triangle is 50°. The measures of the other two angles are equal. Find the measures of the equal angles.
   a. 60°   b. 130°   c. 65°   d. 70°

27. The sum of three consecutive whole numbers is 84. Find the three numbers.
   a. 26, 27, 28   b. 27, 28, 29   c. 26, 28, 30   d. 25, 27, 29

Cumulative Review: Chapters 10-13

CHAPTER 14
Graphing and Equations

SECTIONS
- 14-1 Graphing Ordered Pairs
- 14-2 Graphing Equations
- 14-3 Problem Solving and Applications: Broken Line Graphs
- 14-4 Slope of a Line
- 14-5 Intercepts
- 14-6 Direct and Indirect Variation
- 14-7 Solving a System of Equations by Graphing
- 14-8 Graphing Linear Inequalities

Countries Participating in Winter Olympics

FEATURES
Calculator Application: *Computing y Values for Equations*
Computer Application: *Slope of a Line*
Enrichment: *Venn Diagrams*

14-1 Graphing Ordered Pairs

In the figure at the right, the two intersecting number lines are called **axes**. The horizontal number line is the **x axis**. The vertical number line is the **y axis**. The two axes separate the **coordinate plane** into four **quadrants**.

Note that point A is 4 units to the left of the y axis. It is also 3 units above the x axis.

Coordinate Plane

The ordered pair $(-4, 3)$ gives the location of point A in the coordinate plane. The *first* number, -4, is the **x coordinate**; the *second* number, 3, is the **y coordinate**.

$$A(-4, 3)$$

x coordinate ——— y coordinate

The point with coordinates $(0, 0)$ is called the **origin**. It is the point where the x axis and the y axis intersect.

EXAMPLE 1 Use the graph at the right to give the ordered pair for each point.

	Point	Think	Ordered Pair
a.	Q	Start at the origin. Move 3 units to the <u>right</u>. Then move 4 units <u>down</u>.	$(3, -4)$
b.	S	Start at the origin. Move 4 units to the <u>left</u>. Then move 3 units <u>up</u>.	$(-4, 3)$

Self-Check *Refer to the graph above to give the ordered pair for each point.*

1. M 2. N 3. P 4. R

Note that the points $S(-4, 3)$ and $Q(3, -4)$ were graphed using the same two numbers, -4 and 3. However, they are <u>not</u> the same point. Thus, the <u>order</u> of the numbers is important.

408 Chapter 14

EXAMPLE 2 Graph: **a.** $B(2, 1)$ **b.** $C(-3, 0)$

Solutions: **a.** Start at the origin. Move 2 units to the right. Then move 1 unit up. Label the point B.

b. Start at the origin. Move 3 units to the left. Move neither up nor down. Label the point C.

Self-Check Graph each point: **5.** $D(-1, -4)$ **6.** $F(4, 0)$ **7.** $G(0, -2)$

CLASSROOM EXERCISES

For Exercises 1-12, name the point on the graph that corresponds to each ordered pair. (Example 1)

1. $(-2, 5)$
2. $(3, 2)$
3. $(1, -4)$
4. $(-5, 1)$
5. $(-2, -2)$
6. $(4, 4)$
7. $(1, 1)$
8. $(3, -2)$
9. $(-5, -5)$
10. $(-4, 0)$
11. $(0, 3)$
12. $(-4, 4)$

WRITTEN EXERCISES

Goals: To write the ordered pair for a point on a graph
To graph a point in a coordinate plane

Write the ordered pair for each point graphed below. (Example 1)

1. A
2. B
3. C
4. D
5. E
6. F
7. G
8. H
9. J
10. K
11. P
12. Q

Graph the point for each ordered pair. Label each point. (Example 2)

13. $A(4, 2)$
14. $C(-3, -5)$
15. $D(-2, -4)$
16. $E(4, -5)$
17. $F(6, -3)$
18. $G(-5, 4)$
19. $H(-2, 6)$
20. $J(6, 0)$
21. $K(-4, 0)$
22. $L(0, -5)$

Graphing and Equations

REVIEW CAPSULE FOR SECTION 14-2

Evaluate each expression. (Pages 22–24)

1. $y + 3$ when $y = 4$
2. $n + 3$ when $n = 10$
3. $z - 5$ when $z = 8$
4. $3r - 5$ when $r = 4$
5. $2t + 5$ when $t = -4$
6. $4q - 3$ when $q = -2$

The answers are on page 435.

14-2 Graphing Equations

Jay was watching an approaching thunderstorm. Four seconds after he saw a flash of lightning, he heard the thunder. He wondered how far away the lightning was.

You can use this equation to find out how far away a flash of lightning is.

$d = 1100t$

d = distance in feet
t = time in seconds

EXAMPLE 1 Make a table of ordered pairs for the equation $d = 1100t$.

Solution:

1 Choose at least 3 values for t.

2 Find d.

$d = 1100t$

Let $t = 0.$ → $d = 1100 \cdot 0 = 0$
Let $t = 2.$ → $d = 1100 \cdot 2 = 2200$
Let $t = 4.$ → $d = 1100 \cdot 4 = 4400$

3 Make a table.

t (seconds)	d (feet)
0	0
2	2200
4	4400

Self-Check

1. Make a table of ordered pairs for the equation $d = 1100t$. Use $t = 1$, $t = 3$, and $t = 5$.

You can use the table of ordered pairs to graph a <u>linear equation</u>. A **linear equation** is an equation whose graph is a straight line.

PROCEDURE

To graph a linear equation:

1 Make a table of ordered pairs.

2 Graph the ordered pairs. Connect them with a straight line.

410 Chapter 14

EXAMPLE 2 Graph the equation $d = 1100t$. Use the table of ordered pairs from Example 1.

Solution: ⓵

t	d
0	0
2	2200
4	4400

⓶ Graph showing points (0, 0), (2, 2200), (4, 4400) with d-axis marked 1100, 2200, 3300, 4400, 5500, 6600 and t-axis marked 1 through 8.

Self-Check 2. Graph the equation $d = 1100t$. Use the table of ordered pairs from Self-Check 1.

In Example 2, different scales were used on the horizontal and vertical axes for convenience in graphing. In Example 3, the same scale is used on each axis.

You can graph linear equations such as $y = 2x + 1$ in the coordinate plane.

EXAMPLE 3 Graph the linear equation $y = 2x + 1$.

Solution: ⓵ Make a table.   ⓶ $y = 2x + 1$

x	2x + 1
−2	2(−2) + 1 = −3
0	2(0) + 1 = 1
1	2(1) + 1 = 3

x	y
−2	−3
0	1
1	3

Graph showing points (−2, −3), (0, 1), (1, 3).

Self-Check 3. Graph the equation $y = x − 1$.

CLASSROOM EXERCISES

Make a table of ordered pairs for each equation. Use the three given values for t. (Examples 1 and 3)

1. $d = 80t$
   Let $t = 0$.
   Let $t = 2$.
   Let $t = 4$.

2. $d = 60t$
   Let $t = 1$.
   Let $t = 2$.
   Let $t = 3$.

3. $d = 500t$
   Let $t = 0$.
   Let $t = 1$.
   Let $t = 2$.

4. $d = 1000t$
   Let $t = 1$.
   Let $t = 3$.
   Let $t = 5$.

Graphing and Equations

WRITTEN EXERCISES

Goal: To graph linear equations

For Exercises 1–2, copy the table and the coordinate axes. Then complete the table and draw the graph. (Examples 1 and 2)

1. The equation $d = 50t$ relates distance in miles and time in hours for a car traveling at a rate of 50 miles per hour.

   $d = 50t$

t	d
0	?
2	?
4	?

2. The equation $C = \dfrac{5F - 160}{9}$ relates temperature in degrees Fahrenheit to temperature in degrees Celsius.

   $C = \dfrac{5F - 160}{9}$

°F	°C
−40	?
32	?
68	?

Copy and complete the table of ordered pairs for each equation. (Example 1 and Example 3, Step 1)

3. $y = x$

x	y
1	?
0	?
−2	?

4. $y = -x$

x	y
1	?
0	?
−2	?

5. $y = 3x$

x	y
2	?
0	?
−2	?

6. $y = -3x$

x	y
1	?
0	?
−1	?

Graph each equation. Use the corresponding tables from Exercises 3–6. (Examples 2 and 3, Step 2)

7. $y = x$
8. $y = -x$
9. $y = 3x$
10. $y = -3x$

Graph each equation. (Example 3)

11. $y = 3x + 1$
12. $y = 2x + 5$
13. $y = 4x - 3$
14. $y = 3x - 2$
15. $y = -2x + 4$
16. $y = -3x + 6$
17. $y = 2x - 3$
18. $y = 3x - 4$
19. $y = -x + 1$
20. $y = -x - 1$
21. $y = 2x$
22. $y = -2x$
23. $y = -2x - 1$
24. $y = -2x + 1$
25. $y = \dfrac{1}{2}x$
26. $y = -\dfrac{1}{2}x - 1$

COMPUTING Y VALUES FOR EQUATIONS

You can use a calculator to compute y values for an equation.

EXAMPLE Complete the table of values. $y = 3x + 12$

x	y
2	?
0	?
−2	?

SOLUTION Substitute x values from the table in the equation.

$$y = 3x + 12 \longrightarrow y = 3(2) + 12$$

$3 \; \boxed{\times} \; 2 \; \boxed{+} \; 1 \; 2 \; \boxed{=} \qquad \boxed{18.}$

Repeat the procedure for $x = 0$ and $x = -2$.
Then $y = 12$ and $y = 6$.

EXERCISES

1. $y = 3x + 1$

x	y
0	?
−2	?
−3	?

2. $y = -2x + 2$

x	y
4	?
−4	?
−8	?

3. $y = 4x - 3$

x	y
9	?
−3	?
−6	?

REVIEW CAPSULE FOR SECTION 14-3

Add. (Pages 35–37)

1. $22 + 8$ **2.** $17 + 4$ **3.** $35 + 19$ **4.** $4.7 + 2.4$ **5.** $12.8 + 3.2$

Subtract. (Pages 35–37)

6. $20 - 9$ **7.** $27 - 8$ **8.** $9.2 - 1.8$ **9.** $7.0 - 1.1$ **10.** $87 - 38$

Find the average. (Pages 10–12)

11. 1; 7; 13; 15 **12.** 2.4; 1.5; 3.6 **13.** 10; 12; 9; 15; 14 **14.** 22.1; 19.5; 20.5

The answers are on page 435.

PROBLEM SOLVING AND APPLICATIONS

14-3 Broken Line Graphs

Broken line graphs show the amount of change over a certain period of time. This broken line graph shows the amount of rainfall in New Orleans for each month of the year.

A broken line graph has a horizontal axis and a vertical axis. Thus, each point on the graph can be represented by this ordered pair.

(month, inches of rainfall)

EXAMPLE 1 Use the broken line graph above to answer each question.

a. What is the difference in the amount of rainfall for June and March?
b. What is the total amount of rainfall for the year?
c. What is the average amount of rainfall per month?

Solutions: a. Rainfall for June: 9 inches   Rainfall for March: 2 inches
Difference: 9 inches − 2 inches = **7 inches**

b. Total: 3 + 2 + 2 + 3.5 + 5.5 + 9 + 7.5 + 6.5 + 8.5 + 8 + 3 + 1.5
= 60   The total rainfall is **60 inches**.

c. Average = $\dfrac{\text{Total Yearly Rainfall}}{\text{Number of Months}} = \dfrac{60}{12} =$ **5 inches**

Self-Check 1. Use the broken line graph above to find the difference in the amount of rainfall for the months of July and May.

The procedure for constructing a broken line graph is similar to the procedure for graphing equations.

PROCEDURE
To construct a broken line graph:
1 Use the data table as a table of ordered pairs.
2 Graph the ordered pairs. Connect the points in order.

EXAMPLE 2 Construct a broken line graph for the following data.

Solution:

PUBLIC SCHOOL ATTENDANCE

Year	Pupils (in millions)	Ordered Pair
1940	25	(1940, 25)
1950	25	(1950, 25)
1960	36	(1960, 36)
1970	45	(1970, 45)
1980	41	(1980, 41)

Self-Check 2. Construct a broken line graph for the following data.

PRICE OF A SHARE OF STOCK

Month	JAN	FEB	MAR	APR	MAY	JUN
Price	$10	$8	$9	$16	$20	$18

CLASSROOM EXERCISES

For Exercises 1–4, refer to the broken line graph at the right to answer each question. (Example 1)

1. How many minutes of daylight are there on the first day of June?

2. How many minutes of daylight are there on the first day of January?

3. How many more minutes of daylight are there on the first day of June than on the first day of October?

4. Find the total minutes of daylight for April 1, May 1, June 1, and July 1.

Graphing and Equations 415

WRITTEN EXERCISES

Goals: To read broken line graphs

To construct broken line graphs

The broken line graph at the right shows the number of countries that participated in the Winter Olympics. Refer to this graph for Exercises 1–4. (Example 1)

1. How many countries participated in the Winter Olympics in 1984?
2. How many more countries participated in the Winter Olympics in 1984 than in 1980?
3. Find the percent of increase for the number of countries that participated in 1956 and the number that participated in 1976.
4. Find the average number of countries that participated in the Winter Olympics for the years shown.

For Exercises 5–6, construct a broken line graph for the data. (Example 2)

5. POPULATION OF WASHINGTON, D.C.

Year	Population (in hundred thousands)
1940	6.6
1950	8.0
1960	7.6
1970	7.6
1980	6.4

6. MORNING TEMPERATURES

Time	Temperature
6 A.M.	14
7 A.M.	17
8 A.M.	18
9 A.M.	20
10 A.M.	23
11 A.M.	26

REVIEW CAPSULE FOR SECTION 14–4

Subtract. (Pages 308–309)

1. $3 - 4$
2. $5 - 12$
3. $8 - 10$
4. $0 - 9$
5. $2 - (-3)$
6. $5 - (-6)$
7. $4 - (-4)$
8. $0 - (-2)$
9. $-1 - 6$
10. $-2 - 4$
11. $-8 - 3$
12. $-2 - (-2)$
13. $-1 - (-3)$
14. $-4 - (-4)$
15. $-3 - (-8)$

The answers are on page 435.

14-4 Slope of a Line

When you talk about the grade or steepness of the road shown at the right, you say that it rises a vertical distance of 100 feet over a horizontal distance of 400 feet. That is,

$$\text{steepness} = \frac{\text{rise}}{\text{run}} = \frac{\text{vertical change}}{\text{horizontal change}} = \frac{100}{400} = \frac{1}{4}.$$

The measure of the steepness or slant of a line is called <u>slope</u>.

> The **slope** of a line is the ratio of the vertical change to the horizontal change between any two points on the line.
>
> $$\text{slope} = \frac{\text{vertical change}}{\text{horizontal change}}, \text{ (horizontal change} \neq 0)$$

When you know the coordinates of any two points on a line, you can determine the slope of a line.

> **PROCEDURE**
>
> **To find the slope of a line using two points on a line:**
>
> **1** Subtract the *y* coordinates to find the vertical change.
>
> **2** Subtract the *x* coordinates in the same order as the *y* coordinates to find the horizontal change.
>
> **3** Write this ratio in lowest terms: $\frac{\text{vertical change}}{\text{horizontal change}}$

EXAMPLE 1 Find the slope of the line containing the points (1, 3) and (6, 7).

Solution:
1 $7 - 3 = 4$ ◀ Vertical change

2 $6 - 1 = 5$ ◀ Horizontal change

3 Write the ratio: $\frac{\text{Vertical change}}{\text{Horizontal change}} = \frac{4}{5}$ ◀ Slope

Self-Check *Find the slope of the line containing the given points.*

1. $A(4, 2); B(6, 4)$　　　**2.** $C(5, 8); D(2, 3)$

Graphing and Equations **417**

The slope of a line can also be negative or zero.

EXAMPLE 2 Find the slope of the line containing the given points.

  a. $F(1, 4)$; $G(4, 2)$ **b.** $M(2, 2)$; $N(-2, 2)$

Solutions:

a.
1. $4 - 2 = 2$
2. $1 - 4 = 1 + (-4)$
   $= -3$
3. Slope: $\frac{2}{-3}$ or $-\frac{2}{3}$

b.
1. $2 - 2 = 0$
2. $2 - (-2) = 2 + 2$
   $= 4$
3. Slope: $\frac{0}{4}$ or $\mathbf{0}$

Self-Check Find the slope of the line containing the given points.

3. $R(-1, 4)$; $S(3, 2)$ **4.** $Y(-2, -1)$; $Z(4, 3)$

When computing slope, always be careful to subtract the x coordinates in the <u>same order</u> as you subtracted the y coordinates.

The table below shows how the slope of a line determines its slant.

TABLE

Slope: Positive	Slope: Negative	Slope: 0	Slope: Undefined
The line rises from left to right.	The line falls from left to right.	A horizontal line	A vertical line

CLASSROOM EXERCISES

Classify the slope of each line as positive, negative, equal to zero, or undefined. (Table)

1. **2.** **3.** **4.**

418   Chapter 14

For Exercises 5–11, complete the table.
(Examples 1 and 2)

	Coordinates	Vertical Change	Horizontal Change	Slope
5.	$A(5, 4); B(1, 3)$	$4 - 3 = 1$	$5 - 1 = 4$	?
6.	$T(3, 1); V(2, -2)$	$1 - (-2) = 3$	?	?
7.	$E(-1, -5); W(-3, -8)$	$-5 - (-8) = 3$	?	?
8.	$G(-2, -3); H(-4, -6)$	?	?	?
9.	$L(4, 2); M(3, 5)$	$2 - 5 = -3$	$4 - 3 = 1$	?
10.	$S(-2, -1); T(3, -4)$	?	$-2 - 3 = -5$	?
11.	$D(-2, 4); G(-6, 4)$	?	?	?

WRITTEN EXERCISES

Goal: To find the slope of a line

For Exercises 1–36, find the slope of the line containing the given points. (Example 1)

1. $A(6, 8); B(3, 2)$
2. $C(7, 3); D(3, 1)$
3. $E(4, 2); F(-1, -3)$
4. $G(1, 4); H(-5, -2)$
5. $J(-1, -3); K(-5, -4)$
6. $L(-3, -2); M(-6, -9)$
7. $N(2, -3); P(0, -4)$
8. $Q(-1, 5); R(-7, 0)$
9. $S(2, 3); T(7, 4)$
10. $V(-1, -2); W(6, 3)$
11. $A(2, -4); C(3, -1)$
12. $D(5, 3); F(9, 10)$

(Example 2)

13. $B(3, 4); C(6, 2)$
14. $G(1, 4); H(4, 3)$
15. $K(7, 2); L(3, 5)$
16. $M(4, 3); P(1, 7)$
17. $R(-1, -2); S(-5, 3)$
18. $Q(-3, -7); T(-7, 2)$
19. $V(-4, -1); A(-3, -1)$
20. $E(8, 2); F(-4, 2)$
21. $J(-6, -4); L(-2, -6)$
22. $M(-1, -3); N(4, -5)$
23. $K(-2, 5); P(0, 5)$
24. $A(0, -4); D(-3, -4)$

MIXED PRACTICE

25. $C(4, 8); D(3, 1)$
26. $S(7, 1); T(5, 9)$
27. $A(-4, 3); B(-8, 6)$
28. $Q(5, -2); P(-7, 4)$
29. $G(-5, -1); W(-6, -4)$
30. $T(-6, 3); R(-2, -9)$
31. $H(2, 6); T(-2, -3)$
32. $J(3, -5); K(-3, 9)$
33. $B(-8, 2); C(10, -6)$
34. $P(0, 4); S(-6, 4)$
35. $D(2, -8); G(-1, -8)$
36. $E(-6, -2); F(0, 9)$

Graphing and Equations 419

REVIEW: SECTIONS 14-1 – 14-4

Write the ordered pair for each point shown on the graph. (Section 14-1)

1. A
2. B
3. D
4. G

Draw and label a pair of axes. Then graph the point for each ordered pair. Label each point. (Section 14-1)

5. $P(4, 2)$
6. $Q(-3, 1)$
7. $R(-2, -4)$
8. $S(1, -3)$

For Exercises 9-12, evaluate each equation for the given values of t. Use t = 0, t = 2, and t = 4. (Section 14-2)

9. $d = 40t$
10. $y = t - 1$
11. $y = t + 2$
12. $y = 2t$

Graph each equation. (Section 14-2)

13. $y = x + 4$
14. $y = x - 5$
15. $y = 2x + 3$
16. $y = 3x + 1$

17. Use the table to construct a broken line graph. (Section 14-3)

CALORIES BURNED IN ONE WEEK

Day	Mon	Tue	Wed	Thur	Fri	Sat	Sun
Calories	2000	2500	1500	2400	2100	3000	1200

Find the slope of the line containing the given points. (Section 14-4)

18. $A(4, 2); B(1, 1)$
19. $C(6, 2); D(1, 4)$
20. $F(4, -3); C(-1, -2)$

REVIEW CAPSULE FOR SECTION 14-5

Find the value of y when x = 0. (Pages 64-66)

1. $y = x + 4$
2. $y = x - 6$
3. $y = 2x + 3$
4. $y = 3x - 4$

Find the value of x when y = 0. (Pages 190-195)

5. $y = x + 2$
6. $y = x - 5$
7. $y = 2x + 2$
8. $y = 3x - 3$

The answers are on page 435.

COMPUTER APPLICATIONS Slope of a Line

Given the coordinates of two points on a line, the following program will compute and print the slope of the line. This program includes the possibility that the line is vertical. In this case, the slope is undefined.

Program:

```
100 PRINT "WHAT ARE THE COORDINATES OF POINT 1";
110 INPUT X1, Y1
120 PRINT "WHAT ARE THE COORDINATES OF POINT 2";
130 INPUT X2, Y2
140 IF X1 = X2 THEN 180
150 LET M = (Y2 - Y1) / (X2 - X1)
160 PRINT "SLOPE = ";M
170 GOTO 190
180 PRINT "SLOPE IS UNDEFINED"
190 PRINT
200 PRINT "ANY MORE TO COMPUTE (1 = YES, 0 = NO)";
210 INPUT Z
220 IF Z = 1 THEN 100
230 PRINT
240 END
```

Statement 160 contains the following formula for slope.

$$\text{Slope} = \frac{\text{vertical change}}{\text{horizontal change}}, \text{ or } m = \frac{y_2 - y_1}{x_2 - x_1}$$

where (x_1, y_1) are the coordinates of one point and (x_2, y_2) are the coordinates of a second point.

Since the formula cannot be used if $x_2 - x_1 = 0$, statement 150 first tests whether $x_1 = x_2$. If $x_1 = x_2$, the computer is sent to statement 190 and prints SLOPE IS UNDEFINED

Exercises

Use the program to find the slope of the line containing the given points.

1. (−4,3), (6,−9)
2. (−1,−3), (2,−3)
3. (4,5), (4,9)
4. (1,2), (−3,5)
5. (−5,−1), (−2,−4)
6. (−3,−11), (2,−7)
7. (4,−2), (3,2)
8. (3,−2), (0,0)
9. (6,−9), (−1,−1)

14-5 Intercepts

Point *A* in the figure at the right is on the *y* axis. Therefore, its *x* value is 0. Its *y* value, −3, is called the **y intercept** of the line. This means that the graph of the line crosses the *y* axis at the point (0, −3).

PROCEDURE

To find the y intercept of a linear equation:

① Replace *x* with 0.   ② Solve for *y*.

EXAMPLE 1 Find the *y* intercept of $y = 2x + 6$.

① Replace *x* with 0. ⟶ $y = 2x + 6$
$y = 2(0) + 6$

② Solve for *y*. ⟶ $y = 0 + 6$
$y = 6$ ⟵ y intercept

Self-Check *Find the y intercept:* **1.** $y = 3x + 4$ **2.** $y = -2x - 1$

Point *B* in the figure above is on the *x* axis. Therefore, its *y* value is 0. Its *x* value, 2, is called the **x intercept** of the line. This means that the graph of the line crosses the *x* axis at the point (2, 0).

PROCEDURE

To find the x intercept of a linear equation:

① Replace *y* with 0.   ② Solve for *x*.

EXAMPLE 2 Find the *x* intercept of $y = x - 4$.

Solution: ① Replace *y* with 0. $y = x - 4$
$0 = x - 4$

② Solve for *x*. $0 + 4 = x - 4 + 4$

$4 = x$ ⟵ x intercept

Self-Check *Find the x intercept:* **3.** $y = x - 2$ **4.** $y = 2x - 6$

422   Chapter 14

EXAMPLE 3 Graph $y = 3x + 6$. Use the x and y intercepts.

Solution:
① Find the y intercept. $y = 6$ ◀ See Example 1.

② Find the x intercept. $x = -2$ ◀ See Example 2.

Since the y intercept is 6 and the x intercept is -2, graph the points $(0, 6)$ and $(-2, 0)$.

Draw a line through the two points.

Self-Check Graph each linear equation. Use the x and y intercepts.

5. $y = 2x + 4$
6. $y = 5x - 10$
7. $y = -3x + 9$

CLASSROOM EXERCISES

Find the x and y intercept of each equation. (Examples 1 and 2)

1. $y = 4x - 8$
2. $y = 3x + 9$
3. $y = -x + 1$
4. $y = -2x + 6$

WRITTEN EXERCISES

Goals: To find the x and y intercepts of a linear equation

To graph a linear equation using the x and y intercepts

Find the y intercept of each equation. (Example 1)

1. $y = 3x + 2$
2. $y = 6x + 1$
3. $y = 4x + 7$
4. $y = 4x + 3$
5. $y = x - 3$
6. $y = x - 6$
7. $y = -x + 4$
8. $y = 3x$

Find the x intercept of each equation. (Example 2)

9. $y = x - 3$
10. $y = x - 10$
11. $y = x + 6$
12. $y = x + 8$
13. $y = 3x - 9$
14. $y = 5x - 10$
15. $y = 4x + 12$
16. $y = 6x + 24$

Graph each equation. Use the x and y intercepts. (Example 3)

17. $y = x + 3$
18. $y = x + 5$
19. $y = x - 7$
20. $y = x - 2$
21. $y = 2x + 4$
22. $y = 3x + 6$
23. $y = 4x - 4$
24. $y = 2x - 4$
25. $y = -2x + 8$
26. $y = -5x + 10$
27. $y = -3x - 6$
28. $y = -4x - 8$

Graphing and Equations

REVIEW CAPSULE FOR SECTION 14-6

Multiply. (Pages 135–137)

1. $8\left(13\frac{1}{2}\right)$
2. $6\left(10\frac{2}{3}\right)$
3. $12\left(20\frac{1}{4}\right)$
4. $24\left(15\frac{1}{6}\right)$
5. $30\left(15\frac{1}{2}\right)$
6. $25\left(25\frac{1}{5}\right)$

Solve each equation. (Pages 200–202)

7. $8t = 48$
8. $12x = 132$
9. $9p = 126$
10. $24w = 180$

Round to the nearest whole number. (Pages 47–49)

11. 12.35
12. 27.91
13. 9.55
14. $10.\overline{33}$
15. $52.\overline{45}$
16. $16.\overline{66}$

The answers are on page 435.

14-6 Direct and Indirect Variation

April Connors works during the summer to help pay her college expenses. She earns $10 per hour painting houses. April made a table and a graph to show the total amount she earns.

h	T
1	10
2	20
3	30
4	40
5	50
6	60

h = hours worked
T = earnings

T increases as h increases.

From the table: $\dfrac{T}{h} = \dfrac{10}{1} = \dfrac{20}{2} = \dfrac{30}{3} = \dfrac{40}{4} = \dfrac{50}{5} = \dfrac{60}{6} = 10$

The variables T and h are in **direct variation** because the ratio $\dfrac{T}{h}$ is always the same. We say that **T varies directly** as h.

EXAMPLE 1   a. Use April's table of values to write an equation for the variation.
b. Use the equation to find how much April earns in $22\frac{1}{2}$ hours.

Solution: a. Think: T equals 10 times h.
Equation: $T = 10h$

b. $T = 10h$
$T = 10\left(22\frac{1}{2}\right)$
$T = 225$ April earns $225 in $22\frac{1}{2}$ hours.

424 Chapter 14

Self-Check   1. Refer to Example 1 to find how much April earns in 47 hours.

Two variables may also be related by indirect variation. We say that two variables, *x* and *y*, are indirectly or inversely related, when their **product is always the same.** That is, when *x* increases, *y* decreases. When *x* decreases, *y* increases.

EXAMPLE 2   The table below shows that for a trip of 480 kilometers, the rate, *r*, and the time, *t*, vary inversely.

r	80	60	40	30	20	10
t	6	8	12	16	24	48

As *r* decreases, *t* increases.

a. Refer to the table to write an equation for the variation.

b. Use the equation to find *t* when $r = 88$ kilometers per hour. Round the answer to the nearest hour.

Solution: a. Think: For **every ordered pair** in the table, $r \cdot t = 480$.

Equation: $rt = 480$

b. $rt = 480$ Replace *r* with 88.
$88t = 480$
$t = 5.\overline{45}$ Thus, $t = 5$ hours.

Self-Check   2. Use the inverse variation equation in Example 2 to find *r* when $t = 7$ hours. Round the answer to the nearest tenth.

CLASSROOM EXERCISES

Write an equation for the direct variation.   (Example 1)

1.

n	1	2	3	4	5
c	5	10	15	20	25

2.

x	1	2	3	4	5
y	3	6	9	12	15

Write an equation for the inverse variation.   (Example 2)

3.

r	10	15	20	25	30
t	30	20	15	12	10

4.

l	2	5	8	10	20	40
w	20	8	5	4	2	1

Graphing and Equations

WRITTEN EXERCISES

Solve each problem. (Example 1)

1. The cost, c, in cents, of operating a TV varies directly as the number of hours, n, it is in operation.

   a. Refer to the table below to write an equation for the operation.

n	1	2	5	6	7
c	2	4	10	12	14

   b. Use the equation to find the cost in dollars of operating the TV 5 hours a day for 30 days.

2. The volume, V, in liters, of water in a tank varies directly as the time, t, in minutes, that an inlet valve is open.

   a. Refer to the table below to write an equation for the variation.

t	1	2	3	4
V	3	6	9	12

   b. Use the equation to find how many liters are in the tanks after 15 minutes.

(Example 2)

3. The length, l, of a rectangle varies inversely as the width, w.

   a. Refer to the table below to write an equation for the variation.

l	5	20	50	100	200
w	80	40	8	4	2

   b. Use the equation to find the length when $w = 125$.

4. The number of days, d, needed to complete a job varies inversely as the number of persons working, n.

   a. Refer to the table to write an equation for the variation.

n	1	2	3	4	5
d	24	12	8	6	4.8

   b. Use the equation to find how many workers would be needed to complete the job in $3\frac{1}{2}$ days.

REVIEW CAPSULE FOR SECTION 14-7

Find the value of y for each given value of x. (Pages 22–24)

1. $y = x + 4$; $x = -6$
2. $y = -4x + 1$; $x = -3$
3. $y = x - 2$; $x = -1$

Write Yes or No to indicate whether the given ordered pair is a solution of the given equation. (Pages 410–413)

4. $y = x + 5$; $(1, 6)$
5. $y = 2x - 3$; $(4, 0)$
6. $y = -3x + 4$; $(-1, 7)$

The answers are on page 435.

14-7 Solving a System of Equations by Graphing

The Modern Microchip Company plans to produce a new pocket calculator. When the income from sales equals the cost of producing the calculator, the company is said to **break-even**. Any further sales represent a profit.

A pair of linear equations is a **system of equations**. To solve a system of equations means to find the coordinates of the point where the graphs of the equations intersect.

PROCEDURE

To solve a system of equations:

1. Make a table of ordered pairs for each equation.
2. Graph each equation on the same coordinate plane.
3. Write the solution as an ordered pair.

EXAMPLE 1 Find the number of pocket calculators, x, that must be sold in order for the Modern Microchip Company to break-even. Each calculator will sell for \$40. Use the following system of equations.

$$\begin{cases} y = 20x + 2000 \\ y = 40x \end{cases}$$ ← Cost of producing x calculators
← Amount made for selling x calculators

[1] $y = 20x + 2000$

x	20x + 2000
0	20(0) + 2000 = 2000
50	20(50) + 2000 = 3000
100	20(100) + 2000 = 4000

x	y
0	2000
50	3000
100	4000

$y = 40x$

x	40x
0	40(0) = 0
50	40(50) = 2000
150	40(150) = 6000

x	y
0	0
50	2000
150	6000

[2] *(graph showing Production Costs line through (0, 2000), (50, 3000), (100, 4000) and Sales Income line through (0, 0), (50, 2000), (150, 6000), intersecting at (100, 4000))*

[3] The graphs meet at **(100, 4000)**.

One hundred calculators must be sold for the company to break-even.

Check: The check is left for you.

Graphing and Equations **427**

Self-Check 1. Find the number of calculators, x, that must be sold for the Modern Microchip Company to break-even. Each calculator will sell for $60. Use the equations $y = 20x + 2000$ and $y = 60x$.

NOTE: As shown in Example 1, you check the solution of a system of equations by substituting the ordered pair in both equations of the system.

EXAMPLE 2 Graph this system of equations in the coordinate plane. Determine the point of intersection.

$$\begin{cases} y = 2x + 3 \\ y = x + 2 \end{cases}$$

Solution: ⓵ $y = 2x + 3$ $y = x + 2$ ⓶

x	y
−2	−1
−1	1
0	3

x	y
−1	1
0	2
2	4

⓷ The graphs meet at $(-1, 1)$.

Check: $y = 2x + 3$ $y = x + 2$
$1 = 2(-1) + 3$ $1 = -1 + 2$
$ -2 + 3$ $ 1$
$ 1$

Self-Check Graph each system of equations. Determine the point of intersection.

2. $\begin{cases} y = 2x + 2 \\ y = x - 1 \end{cases}$
3. $\begin{cases} y = -x + 1 \\ y = x + 3 \end{cases}$
4. $\begin{cases} y = -2x \\ y = x \end{cases}$

CLASSROOM EXERCISES

Determine whether the given ordered pair is a solution of the system of equations. Answer Yes or No. (Example 2, step 3)

1. $\begin{cases} y = 2x - 3 \\ y = x - 4 \end{cases}$; $(-1, -5)$
2. $\begin{cases} y = 3x - 1 \\ y = x - 1 \end{cases}$; $(1, 0)$
3. $\begin{cases} y = -x + 2 \\ y = x - 6 \end{cases}$; $(4, -2)$
4. $\begin{cases} y = -x - 1 \\ y = x + 1 \end{cases}$; $(0, -1)$
5. $\begin{cases} y = -4x \\ y = x \end{cases}$; $(0, 0)$
6. $\begin{cases} y = -x \\ y = 5x \end{cases}$; $(0, 0)$

WRITTEN EXERCISES

Goal: To solve a system of equations by graphing

For Exercises 1–2:

a. *Complete the tables. Then copy the coordinate axes on a piece of graph paper and draw the graph.*

b. *Write the solution.* (Example 1)

1. Find the number of office chairs, x, that must be sold in order for the Break-Time Office Supplies Company to break-even. Each office chair sells for $75.

   $y = 25x + 2000$ $y = 75x$

x	y
0	?
20	?
80	?

x	y
0	?
20	?
60	?

2. Find the number of camera cases, x, that must be sold in order for the Cogswell Camera Company to break-even. Each camera case sells for $10.

   $y = 5x + 2000$ $y = 10x$

x	y
0	?
100	?
200	?

x	y
0	?
150	?
300	?

Solve by graphing. (Example 2)

3. $\begin{cases} y = 2x + 3 \\ y = x + 1 \end{cases}$

4. $\begin{cases} y = 2x + 1 \\ y = x + 4 \end{cases}$

5. $\begin{cases} y = 3x + 1 \\ y = x - 1 \end{cases}$

6. $\begin{cases} y = 3x + 2 \\ y = x - 2 \end{cases}$

7. $\begin{cases} y = -x + 3 \\ y = x + 3 \end{cases}$

8. $\begin{cases} y = -x + 6 \\ y = x + 2 \end{cases}$

9. $\begin{cases} y = -x - 2 \\ y = x - 4 \end{cases}$

10. $\begin{cases} y = -x - 1 \\ y = x - 3 \end{cases}$

REVIEW CAPSULE FOR SECTION 14–8

Tell whether each of the following is a solution for $x - 1 < 4$. Write Yes *or* No. *(Pages 338–341)*   The answers are on page 435.

1. 1   **2.** 5   **3.** 0   **4.** -4 **5.** -3 **6.** 6   **7.** -1 **8.** 8

14-8 Graphing Linear Inequalities

Mathematical sentences such as $y > \frac{1}{3}x$ and $y < \frac{1}{3}x$ are called **linear inequalities**. To graph a linear inequality, first draw the graph of the related linear equation as a dashed line.

> **PROCEDURE**
>
> To graph a linear inequality, $y < ax + b$ or $y > ax + b$:
>
> **1** Graph the related linear equation, $y = ax + b$.
>
> **2** Test an ordered pair from each side of the graph of the equation to determine which point is a solution of the inequality.
>
> **3** Shade the part of the coordinate plane containing the solution.

EXAMPLE Graph $y < x + 1$

Solution: **1** $y = x + 1$ ◀ Replace < with =.

Draw the graph of $y = x + 1$ as a dashed line.

2 Choose two ordered pairs.
Choose one ordered pair on one side of the line.
Choose another ordered pair on the other side of the line.

a. Test $(0, 2)$.
$y < x + 1$
$2 \stackrel{?}{<} 0 + 1$
$2 \stackrel{?}{<} 1$ No

Thus, $(0, 2)$ is not a solution.

b. Test $(3, -2)$.
$y < x + 1$
$-2 \stackrel{?}{<} 3 + 1$
$-2 \stackrel{?}{<} 4$ Yes

Thus, $(3, -2)$ is a solution.

3 Shade the part of the coordinate plane containing $(3, -2)$. **All points in the shaded part of the coordinate plane are solutions of $y < x + 1$.**

Self-Check Graph each inequality: **1.** $y < x + 3$ **2.** $y > x$

NOTE: The points on the dashed line, $y = x + 1$, are **not** solutions of $y < x + 1$ or $y > x + 1$.

CLASSROOM EXERCISES

Determine whether the given point is a solution of the given inequality. Write Yes or No. (Example, step 2)

1. $y > x + 1$; $(-3, 2)$ **2.** $y < 4x$; $(1, 1)$ **3.** $y > x$; $(0, 0)$
4. $y < 3x - 4$; $(2, 3)$ **5.** $y > -x + 1$; $(3, 4)$ **6.** $y < -2x - 1$; $(1, -5)$

For Exercises 7–10, choose a graph from a–d that matches each inequality. (Example, step 2)

7. $y > \frac{1}{2}x$ **8.** $y < \frac{1}{2}x$ **9.** $y > 4x$ **10.** $y < 4x$

WRITTEN EXERCISES

Goal: To graph linear inequalities in the coordinate plane

Determine whether each point is a solution of $y > x - 2$. Write Yes or No. (Example)

1. A **2.** B **3.** C
4. D **5.** E **6.** F
7. G **8.** H **9.** J

For Exercises 10–21, graph each inequality in the coordinate plane. (Example)

10. $y > x + 2$ **11.** $y < x - 1$ **12.** $y < 2x$ **13.** $y > 3x$
14. $y > 2x - 1$ **15.** $y < 3x - 1$ **16.** $y > -x$ **17.** $y < -3x$
18. $y < -x + 5$ **19.** $y > -x - 2$ **20.** $y < -2x - 4$ **21.** $y > -4x + 3$

Graphing and Equations **431**

REVIEW: SECTIONS 14-5 — 14-8

For Exercises 1-4:
a. *Find the y intercept of each linear equation.*
b. *Find the x intercept of each linear equation.* (Section 14-5)

1. $y = x + 2$ **2.** $y = x - 5$ **3.** $y = 2x - 4$ **4.** $y = 3x + 6$

For Exercises 5-8:
a. *Find the x and y intercepts.*
b. *Use the x and y intercepts to graph the equation.* (Section 14-5)

5. $y = x + 3$ **6.** $y = x - 4$ **7.** $y = 3x + 3$ **8.** $y = 4x - 8$

Solve each problem. (Section 14-6)

9. The actual distance, D, in miles between two towns varies directly as the distance, m, in inches between the towns on a map.

  a. Refer to the table below to write an equation for the variation.

m	2	3	6	8
D	120	180	360	480

  b. Use the equation to find the actual distance when m equals 11 inches.

10. The number of days, d, needed to repair a sidewalk varies inversely as the number of persons, w, working. Assume that all persons work at the same rate.

  a. Refer to the table below to write an equation for the variation.

w	2	3	4	12
d	12	8	6	2

  b. Use the equation to find how many workers are needed to complete the job in 4 days.

Graph each system of equations in a coordinate plane. Determine the point of intersection. (Section 14-7)

11. $\begin{cases} y = x + 2 \\ y = 2x + 1 \end{cases}$ **12.** $\begin{cases} y = x - 1 \\ y = 3x - 3 \end{cases}$ **13.** $\begin{cases} y = x \\ y = 2x - 1 \end{cases}$ **14.** $\begin{cases} y = 2x \\ y = -3x \end{cases}$

Graph each inequality in the coordinate plane. (Section 14-8)

15. $y < x + 3$ **16.** $y > x - 4$ **17.** $y > -3x$ **18.** $y < 2x - 4$

CHAPTER REVIEW

PART 1: VOCABULARY

For Exercises 1–15, choose from the box at the right the word(s) that best corresponds to each description.

1. The point where the x axis and y axis cross __?__ (p. 408)
2. Where the graph of a line crosses the y axis __?__ (p. 422)
3. The vertical number line in the plane __?__ (p. 408)
4. The second number in an ordered pair __?__ (p. 408)
5. A graph that shows the amount of change over a certain period of time __?__ (p. 414)
6. A pair of linear equations __?__ (p. 427)
7. When the income from sales equals the cost of production, a company is said to __?__. (p. 427)
8. $rt = 24$ is an example of __?__ (p. 425)
9. The ratio of the vertical change to the horizontal change between any two points on a line __?__ (p. 416)
10. An equation whose graph is a straight line __?__ (p. 410)
11. $T = 10h$ is an example of __?__ (p. 424)
12. The horizontal number line in the plane __?__ (p. 408)
13. The first number in an ordered pair __?__ (p. 408)
14. For a line, the x value of its point on the x axis __?__ (p. 422)
15. The solution of a system of equations __?__ (p. 427)

> x axis
> y axis
> origin
> x coordinate
> y coordinate
> slope
> x intercept
> y intercept
> linear equation
> line graph
> direct variation
> indirect variation
> system of equations
> point of intersection
> break-even

PART 2: SKILLS

For Exercises 16–27, give the ordered pair for the points graphed at the right. (Section 14-1)

16. A
17. B
18. C
19. D
20. E
21. F
22. G
23. H
24. I
25. J
26. K
27. L

Draw and label a pair of axes. Then graph the point for each ordered pair. Label each point. (Section 14-1)

28. $P(-3, 2)$
29. $Q(4, -4)$
30. $R(1, 2)$
31. $S(3, -2)$
32. $T(0, -3)$
33. $V(5, 0)$
34. $U(-3, -4)$
35. $W(5, 4)$

Graphing and Equations 433

Graph each equation. (Section 14–2)

36. $y = x + 2$ **37.** $y = 2x$ **38.** $y = 3x + 1$ **39.** $y = -x + 4$

This line graph shows the number of car sales for one year. Use this graph for Exercises 40–42. (Section 14–3)

40. How many cars were sold in June?

41. How many more cars were sold in June than in January?

42. What is the difference between the number of cars sold in September and the number sold in November?

CAR SALES FOR ONE YEAR

Find the slope of the line containing the given points. (Section 14–4)

43. $A(3, 5)$; $B(1, 4)$ **44.** $F(5, -1)$; $G(3, 4)$ **45.** $H(-2, -3)$; $J(-5, 2)$
46. $K(7, -2)$; $L(3, -2)$ **47.** $P(-3, 6)$; $Q(4, 0)$ **48.** $R(0, -1)$; $S(-3, 6)$

Find the y intercept of each linear equation. (Section 14–5)

49. $y = x + 7$ **50.** $y = x - 12$ **51.** $y = 2x + 3$ **52.** $y = 3x - 5$

Find the x intercept of each linear equation. (Section 14–5)

53. $y = x + 6$ **54.** $y = -x + 3$ **55.** $y = 2x + 8$ **56.** $y = 4x - 12$

Solve each problem. (Section 14–6)

57. The toll, T, on a bridge varies directly with the number of axles, a, of a vehicle.

  a. Refer to the table below to write an equation for the variation.

a	2	4	8
T	1.50	3.00	6.00

  b. Use the equation to find the amount of the toll for a 6-axle truck.

58. The number of tomato plants, p, in a row in a garden varies inversely as the space, s, between them.

  a. Refer to the table below to write an equation for the variation.

s	15	20	30
p	60	45	30

  b. Use the equation to find how many plants can fit in a row if he places them 25 centimeters apart.

434   Chapter 14

Graph each system of equations in a coordinate plane. Determine the point of intersection. (Section 14-7)

59. $\begin{cases} y = x - 1 \\ y = 2x - 2 \end{cases}$
60. $\begin{cases} y = x + 2 \\ y = -x + 4 \end{cases}$
61. $\begin{cases} y = x \\ y = 2x + 1 \end{cases}$
62. $\begin{cases} y = 2x \\ y = 3x - 1 \end{cases}$

Graph each inequality in the coordinate plane. (Section 14-8)

63. $y < -x + 3$
64. $y > x - 2$
65. $y > 2x + 1$
66. $y < 3x - 1$

PART 3: APPLICATIONS

67. Use the information in the table below to construct a broken line graph. (Section 14-3)

CALORIE INTAKE FOR ONE WEEK

Day	MON.	TUES.	WED.	THURS.	FRI.	SAT.	SUN.
Calories	2500	2400	3000	2200	1900	2500	2000

68. Find the number of phones that must be sold in order for the Dial-A-Tone Company to break-even. Each phone sells for $50. (HINT: Copy the tables and the coordinate axes. Then complete the table and draw the graph.) (Section 14-8)

$y = 25x + 1000$

x	y
0	?
40	?
80	?

$y = 50x$

x	y
0	?
50	?
100	?

ANSWERS TO REVIEW CAPSULES

Page 410   **1.** 7   **2.** 13   **3.** 3   **4.** 7   **5.** −3   **6.** −11

Page 413   **1.** 30   **2.** 21   **3.** 54   **4.** 7.1   **5.** 16.0   **6.** 11   **7.** 19   **8.** 7.4   **9.** 5.9   **10.** 49   **11.** 9   **12.** 2.5   **13.** 12   **14.** 20.7

Page 416   **1.** −1   **2.** −7   **3.** −2   **4.** −9   **5.** 5   **6.** 11   **7.** 8   **8.** 2   **9.** −7   **10.** −6   **11.** −11   **12.** 0   **13.** 2   **14.** 0   **15.** 5

Page 420   **1.** 4   **2.** −6   **3.** 3   **4.** −4   **5.** −2   **6.** 5   **7.** −1   **8.** 1

Page 424   **1.** 108   **2.** 64   **3.** 243   **4.** 364   **5.** 465   **6.** 630   **7.** 6   **8.** 11   **9.** 14   **10.** $7\frac{1}{2}$ **11.** 12   **12.** 28   **13.** 10   **14.** 10   **15.** 52   **16.** 17

Page 426   **1.** −2   **2.** 13   **3.** −3   **4.** Yes   **5.** No   **6.** Yes

Page 429   **1.** Yes   **2.** No   **3.** Yes   **4.** Yes   **5.** Yes   **6.** No   **7.** Yes   **8.** No

Graphing and Equations

CHAPTER TEST

For Exercises 1–4, give the ordered pair for each point graphed at the right.

1. C
2. F
3. H
4. M

Graph each equation.

5. $y = x - 3$
6. $y = 2x + 1$

Find the slope of the line containing the given points.

7. $A(3, 4); B(1, 1)$
8. $C(2, -1); D(4, -5)$

Find the x and y intercepts of each equation. Then graph the equation.

9. $y = x + 2$
10. $y = 3x - 3$

Graph each inequality in the coordinate plane.

11. $y < x + 5$
12. $y > x - 3$

13. Solve by graphing: $\begin{cases} y = 3x \\ y = x + 6 \end{cases}$

14. Use the table below to construct a broken line graph.

POPULARITY OF A CERTAIN RECORD

Rating (1: low – 5: high)	Number of Students
1	1
2	2
3	8
4	12
5	7

15. The speed, v, in miles per hour of the blade tips of a windmill varies directly with the speed, w, of the wind.

   a. Refer to the table below to write an equation for the variation.

w	20	10	8	2
v	180	90	72	18

   b. Use the equation to find the speed of the blade tips in a 5 mile per hour wind.

ENRICHMENT

Venn Diagrams

You can use **Venn diagrams** to show how sets are related. The Venn diagram at the right shows that every element in set B is also in set A. Thus, B is a **subset** of A. Every element of A, however, is not in B. Thus, A is not a subset of B.

A = {whole numbers}

B ⊂ A — Read: "B is a subset of A."

A ⊄ B — Read: "A is not a subset of B."

The union of sets B and C (in symbols: B ∪ C) is shown by the shaded portion of the Venn diagram at the right. The **union** of two sets is the set that contains all the elements in both sets. Thus,

B ∪ C = {1, 2, 3} ∪ {3, 4, 5} = {1, 2, 3, 4, 5}

A = {whole numbers}

The "3" is not repeated.

The intersection of sets B and C (in symbols: B ∩ C) is shown by the shaded portion of the Venn diagram at the right. The **intersection** of two sets is the set that contains all the elements common to both sets. Thus,

B ∩ C = {1, 2, 3} ∩ {3, 4, 5} = {3}

A = {whole numbers}

EXERCISES

Write the elements of each set.

1. B ∪ D
2. A ∩ B
3. D ∪ F
4. B ∪ F
5. B ∩ D
6. A ∩ F

Replace each __?__ with ⊂ or ⊄.

7. A __?__ D
8. F __?__ D
9. F __?__ A
10. B __?__ A
11. {3} __?__ A
12. {6} __?__ B

A = {whole numbers}

Graphing and Equations 437

ADDITIONAL PRACTICE

SKILLS *Graph each ordered pair in the same coordinate plane.* (Pages 408–409)

1. $A(1, 1)$
2. $B(-1, 4)$
3. $F(-2, -2)$
4. $E(3, -2)$

Graph each equation in the coordinate plane. (Pages 410–412)

5. $y = x$
6. $y = 2x$
7. $y = x + 2$
8. $y = 3x + 1$

Find the slope of the line containing the given points. (Pages 417–419)

9. $A(1, 3)$; $B(4, 7)$
10. $C(-1, 2)$; $D(4, -3)$
11. $G(-4, -3)$; $H(5, 3)$

Find the x and y intercepts. (Pages 422–423)

12. $y = x + 2$
13. $y = 3x + 6$
14. $y = -2x + 4$
15. $y = -3x + 9$

16. Write an equation for the direct variation. (Pages 426–428)

x	2	5	7	10	20
y	6	15	21	30	60

17. Write an equation for the indirect variation. (Pages 426–428)

x	2	4	5	8	40
y	20	10	8	5	1

Solve each system of equations by graphing. (Pages 427–429)

18. $\begin{cases} y = x + 1 \\ y = 2x + 2 \end{cases}$
19. $\begin{cases} y = 2x + 3 \\ y = 3x + 3 \end{cases}$
20. $\begin{cases} y = -x - 5 \\ y = -3x + 1 \end{cases}$

Graph each inequality in the coordinate plane. (Pages 430–431)

21. $y > \frac{1}{4}x$
22. $y < x + 2$
23. $y < 2x + 1$

APPLICATIONS

24. Use the information at the right to construct a broken line graph. (Pages 414–416)

CASES OF VEGETABLES SOLD

Month	JAN	FEB	MAR	APR	MAY	JUN
Cases Sold	200	100	100	300	400	600

25. There are four quarts in a gallon. The equation $Q = 4G$ describes this relation. Complete the table. Then draw the graph of the equation. (Pages 410–412)

G	Q
1	?
3	?
5	?

CHAPTER 15 Special Triangles

FEATURES
Calculator Application: *Tangent Values*

Career Application: *Meteorologist*

Calculator/Computer Application: *Area of Triangles*

Enrichment: *Arithmetic Method for Finding $\sqrt{N}$*

Common Errors

SECTIONS
15-1 Meaning of Square Root

15-2 Irrational Numbers

15-3 Table of Squares and Square Roots

15-4 Problem Solving and Applications: Formulas: Using Square Roots

15-5 Pythagorean Theorem and Applications

15-6 Tangent Ratio and Applications

15-1 Meaning of Square Root

You learned in Chapter 3 that the area of a square equals the product of the lengths of any two sides. For the figure at the right,

$$A = 12 \cdot 12 = 144 \text{ ft}^2.$$

Since the lengths of the sides are the same, you are multiplying a number by itself, or finding the **square** of the number.

Kitchen Layout — 12 ft by 12 ft

$A = 12^2$ ◄ Read: "12 <u>squared</u>."
$ = 12 \cdot 12$
$ = 144$ ◄ 144 is the <u>square</u> of 12.

> **PROCEDURE**
>
> To square a number, multiply the number by itself.

EXAMPLE 1 Square each number.

    a. 8   **b.** 15   **c.** 24

Solutions:
a. $8^2 = 8 \cdot 8$
$ = 64$

b. $15^2 = 15 \cdot 15$
$ = 225$

c. $24^2 = 24 \cdot 24$
$ = 576$

Self-Check *Square each number.*

1. 3   **2.** 13   **3.** 18   **4.** 31

The inverse operation of squaring a number is taking its **square root**.

Since $8^2 = 64$, $\sqrt{64} = 8$ ◄ Read: "The square root of 64 is 8."

Since $8^2 = 64$ and $(-8)^2 = 64$, the number 64 has two square roots, 8 and -8.

To avoid confusion, we define

$\sqrt{64} = 8$ ◄ Principal square root

and $-\sqrt{64} = -8$.

EXAMPLE 2 Find each answer.

        a. $\sqrt{25}$ **b.** $\sqrt{121}$ **c.** $-\sqrt{36}$

Solutions: **a. Think:** $25 = 5 \cdot 5.$ So $\sqrt{25} = 5.$

         b. Think: $121 = 11 \cdot 11.$ So $\sqrt{121} = 11.$

         c. Think: $36 = 6 \cdot 6.$ So $-\sqrt{36} = -6$

Self-Check Find each answer.

    5. $\sqrt{49}$ **6.** $\sqrt{9}$ **7.** $-\sqrt{16}$ **8.** $-\sqrt{81}$

CLASSROOM EXERCISES

Complete. (Example 1)

1. $4^2 = 4 \cdot 4$ **2.** $1^2 = \underline{\ ?\ } \cdot \underline{\ ?\ }$ **3.** $9^2 = 9 \cdot \underline{\ ?\ }$ **4.** $11^2 = \underline{\ ?\ } \cdot \underline{\ ?\ }$

    $= \underline{\ ?\ }$ $= \underline{\ ?\ }$ $= \underline{\ ?\ }$ $= \underline{\ ?\ }$

(Example 2)

5. $16 = 4 \cdot 4.$ So $\sqrt{16} = \underline{\ ?\ }.$ **6** $9 = 3 \cdot 3.$ So $\sqrt{9} = \underline{\ ?\ }.$

7. $49 = \underline{\ ?\ } \cdot 7.$ So $\sqrt{49} = \underline{\ ?\ }.$ **8** $81 = \underline{\ ?\ } \cdot 9.$ So $\sqrt{81} = \underline{\ ?\ }.$

WRITTEN EXERCISES

Goals: To square a number

          To find the square root of a number

Square each number. (Example 1)

1. 2	**2.** 3	**3.** 6	**4.** 9	**5.** 10	**6.** 1
7. 4	**8.** 11	**9.** 15	**10.** 25	**11.** 30	**12.** 40
13. 21	**14.** 51	**15.** 13	**16.** 19	**17.** 18	**18.** 14
19. 60	**20.** 80	**21.** 24	**22.** 36	**23.** 100	**24.** 200

Find each answer. (Example 2)

25. $\sqrt{4}$	**26.** $\sqrt{1}$	**27.** $\sqrt{49}$	**28.** $\sqrt{64}$	**29.** $\sqrt{9}$	**30.** $\sqrt{81}$
31. $\sqrt{36}$	**32.** $\sqrt{16}$	**33.** $\sqrt{100}$	**34.** $\sqrt{400}$	**35.** $\sqrt{169}$	**36.** $\sqrt{225}$
37. $-\sqrt{9}$	**38.** $-\sqrt{1}$	**39.** $-\sqrt{49}$	**40.** $-\sqrt{121}$	**41.** $-\sqrt{36}$	**42.** $-\sqrt{100}$
43. $\sqrt{196}$	**44.** $\sqrt{484}$	**45.** $\sqrt{1600}$	**46.** $\sqrt{2500}$	**47.** $\sqrt{3600}$	**48.** $\sqrt{4900}$

REVIEW CAPSULE FOR SECTION 15-2

Divide. Round each answer to the nearest hundredth. (Pages 32-34)

1. $2\overline{)1}$
2. $3\overline{)2}$
3. $5\overline{)3}$
4. $4\overline{)1}$
5. $6\overline{)7}$
6. $4\overline{)9}$

Write a decimal for each fraction. Use a bar to indicate any digit or digits that repeat. (Pages 173-175)

7. $\frac{1}{5}$
8. $\frac{3}{4}$
9. $\frac{1}{3}$
10. $\frac{5}{6}$
11. $\frac{3}{20}$
12. $\frac{5}{8}$
13. $\frac{7}{8}$
14. $\frac{9}{20}$
15. $\frac{10}{9}$
16. $\frac{19}{5}$
17. $\frac{7}{4}$
18. $\frac{11}{3}$

The answers are on page 464.

15-2 Irrational Numbers

Rational numbers such as the following are <u>perfect</u> <u>squares</u>.

$$16 \qquad 144 \qquad \frac{1}{4} \qquad 0.81$$

A rational number is a **perfect square** if it is the square of a rational number.	$16 = 4 \cdot 4$ $144 = 12 \cdot 12$ $\frac{1}{4} = \frac{1}{2} \cdot \frac{1}{2}$ $0.81 = (0.9)(0.9)$

A rational number that is not a perfect square has an **irrational square root**. Thus, the following numbers are irrational.

$$\sqrt{2} \qquad \sqrt{5} \qquad \sqrt{17} \qquad \sqrt{31}$$

EXAMPLE 1 Identify each number as rational or irrational. Give a reason.
  a. $\sqrt{15}$
  b. $\sqrt{49}$
  c. $\sqrt{3}$

Solutions:
  a. $\sqrt{15}$ is **irrational** since 15 is <u>not</u> a perfect square.
  b. $\sqrt{49}$ is **rational** since 49 is a perfect square.
  c. $\sqrt{3}$ is **irrational** since 3 is <u>not</u> a perfect square.

Self-Check *Identify each number as rational or irrational.*

1. $\sqrt{64}$
2. $\sqrt{33}$
3. $\sqrt{114}$
4. $\sqrt{100}$

442 Chapter 15

Every rational number can be written either as a **terminating** or as a **repeating decimal**.

$$2\tfrac{1}{2} = \tfrac{5}{2} = 2.5$$ ◀ **Terminating decimal**

$$\tfrac{2}{3} = 0.6666\cdots = 0.\overline{6}$$ ◀ **Repeating decimal**

The decimal equivalents of irrational numbers are **non-terminating** and **non-repeating**.

$$\sqrt{2} = 1.141421\cdots \quad \sqrt{3} = 1.73205\cdots \quad \sqrt{5} = 2.23606\cdots$$
$$\sqrt{6} = 2.44948\cdots \quad \sqrt{7} = 2.64575\cdots \quad \sqrt{8} = 2.82842\cdots$$

◀ There is no repeating pattern.

EXAMPLE 2 Identify each number as rational or irrational. Give a reason for each answer.

    a. 4.2   **b.** $3.021\overline{616}$ **c.** $5.292292229\cdots$

Solutions:

Number	Rational or Irrational?	Reason
a. 4.2	Rational	It is a terminating decimal.
b. $3.021\overline{616}$	Rational	It is a repeating decimal.
c. $5.292292229\cdots$	Irrational	It is neither a terminating nor a repeating decimal.

Self-Check   *Identify each number as rational or irrational.*

5. $5.4772\cdots$ **6.** $7.\overline{6}$ **7.** 3.4215 **8.** $7.07106\cdots$

CLASSROOM EXERCISES

Tell whether each of the following is a perfect square. Write Yes or No. (Definition)

1. 12   **2.** 18   **3.** 16   **4.** 25   **5.** 30   **6.** 36
7. 49   **8.** 80   **9.** 75   **10.** 81   **11.** 125   **12.** 144

Identify each number as rational or irrational (Example 1)

13. $\sqrt{10}$ **14.** $\sqrt{4}$ **15.** $\sqrt{64}$ **16.** $\sqrt{23}$ **17.** $\sqrt{5}$ **18.** $\sqrt{49}$
19. $\sqrt{81}$ **20.** $\sqrt{7}$ **21.** $\sqrt{9}$ **22.** $\sqrt{21}$ **23.** $\sqrt{15}$ **24.** $\sqrt{144}$

Special Triangles

Identify each number as rational or irrational: (Example 2)

25. 3.5   **26.** 2.828423 · · ·   **27.** 3.31662 · · ·   **28.** 2.$\overline{142857}$
29. 6.4512   **30.** 9.$\overline{1}$ **31.** 5.41621 · · ·   **32.** 9.02$\overline{6}$

WRITTEN EXERCISES

Goal: To identify a number as rational of irrational

For Exercises 1-42, identify each number as rational or irrational. Give a reason for each answer. (Example 1)

1. $\sqrt{25}$ **2.** $\sqrt{3}$ **3.** $\sqrt{100}$ **4.** $\sqrt{400}$ **5.** $\sqrt{18}$ **6.** $\sqrt{96}$
7. $\sqrt{22}$ **8.** $\sqrt{121}$ **9.** $\sqrt{7}$ **10.** $\sqrt{1}$ **11.** $\sqrt{36}$ **12.** $\sqrt{50}$
13. $\sqrt{112}$ **14.** $\sqrt{500}$ **15.** $\sqrt{1600}$ **16.** $\sqrt{2500}$ **17.** $\sqrt{1000}$ **18.** $\sqrt{900}$

(Example 2)

19. 0.52   **20.** 9.05   **21.** 2.6457 · · ·   **22.** 3.8729 · · ·
23. 1.$\overline{378}$ **24.** 4.$\overline{6}$ **25.** 1.$\overline{857142}$ **26.** 2.052
27. 3.06006 · · ·   **28.** 9.$\overline{16}$ **29.** 3.506   **30.** 1.414114 · · ·

MIXED PRACTICE

31. $\sqrt{19}$ **32.** $\sqrt{100}$ **33.** 2.95   **34.** 3.8$\overline{3}$
35. 4.5825 · · ·   **36.** 8.$\overline{6}$ **37.** 3.$\overline{142857}$ **38.** $\sqrt{32}$
39. $\sqrt{61}$ **40.** $\sqrt{169}$ **41.** 14.001   **42.** 9.11043 · · ·

MORE CHALLENGING EXERCISES

Choose one or more letters below to describe each number.
a. whole number   **b.** integer   **c.** rational number   **d.** irrational number

43. −6   **44.** $\sqrt{3}$ **45.** $\frac{1}{8}$ **46.** 3.$\overline{3}$ **47.** 2.23606 · · ·   **48.** −$\sqrt{21}$

REVIEW CAPSULE FOR SECTION 15-3

Evaluate. Round each answer to the nearest whole number.
(Pages 32-34)

1. 6^2 **2.** 15^2 **3.** 1.414^2 **4.** 10.1^2 **5.** 42^2 **6.** 2.2^2
7. 7.5^2 **8.** 58^2 **9.** 3.6^2 **10.** 9.4^2 **11.** 90^2 **12.** 118^2

The answers are on page 464.

15-3 Table of Squares and Square Roots

You can use the table on page 447 to find the squares and square roots of numbers from 1 to 150.

> **PROCEDURE**
>
> **To use a table of squares and square roots:**
>
> 1 Find the number in the **NUMBER** column.
>
> 2 Look directly to the right.
>
>   a. Read the square of the number in the **SQUARE** column, or
>   b. Read the square root of the number in the **SQUARE ROOT** column.

EXAMPLE 1 Find each answer.

a. 24^2 b. $\sqrt{23}$

Solutions: a. $24^2 = 576$ b. $\sqrt{23} = 4.796$

To check your answer in **b**, use a calculator to multiply 4.796 by 4.796.

4 . 7 9 6 [×] 4 . 7 9 6 [=]

[23.001616]

Thus, $\sqrt{23}$ is about **4.796**.

Number	Square	Square Root
21	441	4.583
22	484	4.690
23 →	529 →	4.796
24 →	576	4.899
25	625	5.000

Self-Check *Find each answer.*

1. 35^2 2. 91^2 3. $\sqrt{21}$ 4. $\sqrt{84}$

NOTE: Most numbers in the square root column in the table are approximations (correct to thousandths) for irrational numbers. Thus, when you find the square root of a number in the table, you may be asked to round your answer to the nearest tenth, or to the nearest hundredth, and so on.

You can use the table to find the square roots of some numbers greater than 150. First, find the number in the **SQUARE** column. Then read the square root in the **NUMBER** column.

Special Triangles

EXAMPLE 2 $\sqrt{4761} = $ __?__

Solution: ⓵ Find 4761 in the **SQUARE** column.

⓶ Look directly to the left.

$\sqrt{4761} = $ **69**

Number	Square
66	4356
67	4489
68	4624
69 ←	4761
70	4900

Self-Check *Refer to the table on page 447 to find each answer.*

5. $\sqrt{841}$ **6.** $\sqrt{7225}$ **7.** $\sqrt{400}$ **8.** $\sqrt{9801}$

CLASSROOM EXERCISES

Complete. Refer to the table on page 447. (Example 1)

	Number	Square	Square Root
1.	14	196	?
3.	20	?	?
5.	18	?	?
7.	55	?	?

	Number	Square	Square Root
2.	?	361	?
4.	?	6400	?
6.	?	1225	?
8.	?	529	?

WRITTEN EXERCISES

Goal: To use a table of squares and square roots

Refer to the table on page 447 to find each answer. (Example 1)

1. 13^2 **2.** 17^2 **3.** 21^2 **4.** 27^2 **5.** 32^2 **6.** 45^2
7. 81^2 **8.** 101^2 **9.** 120^2 **10.** 117^2 **11.** 130^2 **12.** 150^2
13. $\sqrt{8}$ **14.** $\sqrt{15}$ **15.** $\sqrt{21}$ **16.** $\sqrt{32}$ **17.** $\sqrt{56}$ **18.** $\sqrt{41}$
19. $\sqrt{121}$ **20.** $\sqrt{144}$ **21.** $\sqrt{45}$ **22.** $\sqrt{27}$ **23.** $\sqrt{149}$ **24.** $\sqrt{130}$

(Example 2)

25. $\sqrt{225}$ **26.** $\sqrt{256}$ **27.** $\sqrt{676}$ **28.** $\sqrt{841}$ **29.** $\sqrt{1521}$ **30.** $\sqrt{1024}$
31. $\sqrt{6561}$ **32.** $\sqrt{7225}$ **33.** $\sqrt{3600}$ **34.** $\sqrt{9025}$ **35.** $\sqrt{3025}$ **36.** $\sqrt{17,956}$

Table of Squares and Square Roots

Number	Square	Square Root	Number	Square	Square Root	Number	Square	Square Root
1	1	1.000	51	2601	7.141	101	10,201	10.050
2	4	1.414	52	2704	7.211	102	10,404	10.100
3	9	1.732	53	2809	7.280	103	10,609	10.149
4	16	2.000	54	2916	7.348	104	10,816	10.198
5	25	2.236	55	3025	7.416	105	11,025	10.247
6	36	2.449	56	3136	7.483	106	11,236	10.296
7	49	2.646	57	3249	7.550	107	11,449	10.344
8	64	2.828	58	3364	7.616	108	11,664	10.392
9	81	3.000	59	3481	7.681	109	11,881	10.440
10	100	3.162	60	3600	7.746	110	12,100	10.488
11	121	3.317	61	3721	7.810	111	12,321	10.536
12	144	3.464	62	3844	7.874	112	12,544	10.583
13	169	3.606	63	3969	7.937	113	12,769	10.630
14	196	3.742	64	4096	8.000	114	12,996	10.677
15	225	3.873	65	4225	8.062	115	13,225	10.724
16	256	4.000	66	4356	8.124	116	13,456	10.770
17	289	4.123	67	4489	8.185	117	13,689	10.817
18	324	4.243	68	4624	8.246	118	13,924	10.863
19	361	4.359	69	4761	8.307	119	14,161	10.909
20	400	4.472	70	4900	8.367	120	14,400	10.954
21	441	4.583	71	5041	8.426	121	14,641	11.000
22	484	4.690	72	5184	8.485	122	14,884	11.045
23	529	4.796	73	5329	8.544	123	15,129	11.091
24	576	4.899	74	5476	8.602	124	15,376	11.136
25	625	5.000	75	5625	8.660	125	15,625	11.180
26	676	5.099	76	5776	8.718	126	15,876	11.225
27	729	5.196	77	5929	8.775	127	16,129	11.269
28	784	5.292	78	6084	8.832	128	16,384	11.314
29	841	5.385	79	6241	8.888	129	16,641	11.358
30	900	5.477	80	6400	8.944	130	16,900	11.402
31	961	5.568	81	6561	9.000	131	17,161	11.446
32	1024	5.657	82	6724	9.055	132	17,424	11.489
33	1089	5.745	83	6889	9.110	133	17,689	11.533
34	1156	5.831	84	7056	9.165	134	17,956	11.576
35	1225	5.916	85	7225	9.220	135	18,225	11.619
36	1296	6.000	86	7396	9.274	136	18,496	11.662
37	1369	6.083	87	7569	9.327	137	18,769	11.705
38	1444	6.164	88	7744	9.381	138	19,044	11.747
39	1521	6.245	89	7921	9.434	139	19,321	11.790
40	1600	6.325	90	8100	9.487	140	19,600	11.832
41	1681	6.403	91	8281	9.539	141	19,881	11.874
42	1764	6.481	92	8464	9.592	142	20,164	11.916
43	1849	6.557	93	8649	9.644	143	20,449	11.958
44	1936	6.633	94	8836	9.695	144	20,736	12.000
45	2025	6.708	95	9025	9.747	145	21,025	12.042
46	2116	6.782	96	9216	9.798	146	21,316	12.083
47	2209	6.856	97	9409	9.849	147	21,609	12.124
48	2304	6.928	98	9604	9.899	148	21,904	12.166
49	2401	7.000	99	9801	9.950	149	22,201	12.207
50	2500	7.071	100	10,000	10.000	150	22,500	12.247

CAREER APPLICATIONS
Meteorologist

Meteorology is the study of the atmosphere. A meteorologist must be able to observe, analyze, and forecast the weather.

One statistic that is often given in winter weather reports in cold climates is **wind chill temperature.** Meteorologists can obtain the wind chill temperature from tables or they can use a formula such as the one below.

$$C = 33 - \frac{(10\sqrt{r} + 10.45 - r)(33 - t)}{22.1}$$

C = wind chill temperature in degrees Celsius.
t = air temperature in degrees Celsius.
r = wind speed in meters per second.

EXAMPLE On a winter day, the air temperature is 2°C. The wind speed is 6 meters per second. Find the wind chill temperature to the nearest tenth.

SOLUTION

$$C = 33 - \frac{(10\sqrt{r} + 10.45 - r)(33 - t)}{22.1}$$

$r = 6$
$t = 2$

$$C = 33 - \frac{(10\sqrt{6} + 10.45 - 6)(33 - 2)}{22.1}$$

$$C = 33 - \frac{(10\sqrt{6} + 10.45 - 6)(31)}{22.1}$$

Use a calculator.

6 √ × 1 0 + 1 0 . 4 5 −

6 = × 3 1 ÷ 2 2 . 1 =

`40.60144`

Round 40.60144 to the nearest tenth: 40.6
Thus, $C = 33 - 40.6 = $ **−7.6°C.**

Exercises

Find each wind chill temperature to the nearest tenth. The air temperature is given in degrees Celsius (°C) and the wind speed is given in meters per second (m/s).

1. Air temperature: 4°C
   Wind speed: 15 m/s

2. Air temperature: −15°C
   Wind speed: 6 m/s

3. Air Temperature: −5°C
   Wind speed: 4.5 m/s

4. Air temperature: 7.2°C
   Wind speed: 4.5 m/s

CALCULATOR/COMPUTER APPLICATIONS
Area of Triangles

If you know the lengths of three sides of a triangle, you can use **Hero's Formula** to find the area.

$$A = \sqrt{s(s-a)(s-b)(s-c)}, \text{ where } s = \frac{a+b+c}{2}$$

Calculator: The sides of a triangle are 212, 341, and 285 centimeters long. Find the area to the nearest whole number.

Solution: First find s, $s - a$, $s - b$, and $s - c$.

$s =$ 212 [+] 341 [+] 285 [÷] 2 [=] $\boxed{419.}$

$s - a = 419 - 212 = 207$ $s - b = 419 - 341 = 78$ $s - c = 419 - 285 = 134$

$$A = \sqrt{419(207)(79)(139)}$$

$A =$ 419 [×] 207 [×] 79 [×] 139 [=] [√] $\boxed{30861.22281}$

The area is **30,861 cm²** to the nearest whole number.

Computer: The following program also uses Hero's Formula.

Program:
```
10 PRINT "ENTER THE LENGTHS OF THE SIDES."
20 INPUT A, B, C
30 LET S = .5 * (A + B + C)
40 LET X = SQR(S * (S - A) * (S - B) * (S - C))
50 PRINT "AREA = "; X
60 PRINT "ANY MORE TRIANGLES (1 = YES, 0 = NO)";
70 INPUT Y
80 IF Y = 1 THEN 10
90 END
```

Exercises

Find the area of each triangle by using a calculator or the program. Round each answer to the nearest whole number.

1. $a = 6$; $b = 25$; $c = 29$
2. $a = 19$; $b = 23$; $c = 14$
3. $a = 62$; $b = 48$; $c = 36$
4. $a = 24$; $b = 36$; $c = 18$
5. $a = 17$; $b = 18$; $c = 19$
6. $a = 16$; $b = 21$; $c = 33$

REVIEW: SECTIONS 15-1—15-3

Square each number. (Section 15-1)

1. 4 **2.** 11 **3.** 19 **4.** 8 **5.** 24 **6.** 30

Find each answer. (Section 15-1)

7. $\sqrt{49}$ **8.** $\sqrt{4}$ **9.** $\sqrt{100}$ **10.** $\sqrt{1}$ **11.** $-\sqrt{36}$ **12.** $-\sqrt{81}$
13. $-\sqrt{16}$ **14.** $\sqrt{25}$ **15.** $-\sqrt{64}$ **16.** $-\sqrt{144}$ **17.** $\sqrt{9}$ **18.** $\sqrt{121}$

Identify each number as rational or irrational. (Section 15-2)

19. $\sqrt{21}$ **20.** $\sqrt{36}$ **21.** 4.84 **22.** 2.3132⋯ **23.** $2.\overline{6}$ **24.** $\sqrt{50}$
25. $\sqrt{14}$ **26.** $\sqrt{64}$ **27.** 3.75 **28.** $12.\overline{326}$ **29.** $-\sqrt{9}$ **30.** $\sqrt{55}$

Refer to the table on page 447 to find each answer. (Section 15-3)

31. 18^2 **32.** 22^2 **33.** 112^2 **34.** 65^2 **35.** $\sqrt{32}$ **36.** $\sqrt{75}$
37. $\sqrt{125}$ **38.** $\sqrt{139}$ **39.** $\sqrt{961}$ **40.** $\sqrt{9409}$ **41.** $\sqrt{16,900}$ **42.** $\sqrt{3844}$
43. $\sqrt{841}$ **44.** $\sqrt{9216}$ **45.** $\sqrt{21,316}$ **46.** $\sqrt{5041}$ **47.** $\sqrt{18,225}$ **48.** $\sqrt{22,201}$

REVIEW CAPSULE FOR SECTION 15-4

Round each number to the nearest tenth. (Pages 32-34)

1. 43.24 **2.** 16.71 **3.** 27.83 **4.** 14.01 **5.** 82.87 **6.** 93.14
7. 12.521 **8.** 39.765 **9.** 55.543 **10.** 68.331 **11.** 10.897 **12.** 50.112

Evaluate each expression. (Pages 38-41)

13. $3.5p$ when $p = 1.214$ **14.** $4.2x$ when $x = 4.561$
15. $2.5t$ when $t = 3.125$ **16.** $1.6y$ when $y = 2.496$

Round each of the following to the nearest whole number. (Pages 32-34)

17. 38.3 **18.** 27.6 **19.** 42.1 **20.** 16.9 **21.** 72.84 **22.** 24.62
23. 6.27 **24.** 18.39 **25.** 9.064 **26.** 2.307 **27.** 8.099 **28.** 123.724

The answers are on page 464.

PROBLEM SOLVING AND APPLICATIONS

15-4 Formulas: Using Square Roots

The area of the square park at the right is 2 square kilometers. You can use the formula $A = s^2$ and the Table of Squares and Square Roots on page 447 to find the length of one side of the park.

EXAMPLE 1 Find the length of one side of the park shown above. Round the answer to the nearest tenth.

Solution:
1. Formula: $A = s^2$ ← Known value: $A = 2$
2. $2 = s^2$, or $s^2 = 2$
   $s = \sqrt{2}$ ← Since $s^2 = 2$, $s = \sqrt{2}$.
   $s = 1.414$ ← From the Table on page 447.

The length of one side of the park is about **1.4 kilometers.**

Self-Check 1. The area of a square is 144 square feet. Find the length of one side.

You can use the formula below to find the distance <u>in miles</u> to the horizon from an aircraft above the ground.

Formula $D = 1.2\sqrt{A}$

D: distance in mile
A: number of feet from the ground

EXAMPLE 2 Find the distance to the horizon from a plane flying at a height of 4096 feet. Round the answer to the nearest mile.

Solution:
1. Formula: $D = 1.2\sqrt{A}$ ← Known value: $A = 4096$
2. $D = 1.2\sqrt{4096}$
   $D = 1.2(64)$
   $D = 76.8$ The distance is about **77 miles.**

Self-Check 2. Find the distance to the horizon from a plane flying at a height of 3136 feet. Round the answer to the nearest mile.

Special Triangles

CLASSROOM EXERCISES

For Exercises 1–8, use the formula $A = s^2$ to find the length, s, of one side of a square with the given area, A. For Exercises 1–4, round each answer to the nearest tenth. (Example 1)

1. $A = 14$ m²
   $s = $?
2. $A = 30$ mm²
   $s = $?
3. $A = 3$ km²
   $s = $?
4. $A = 24$ cm²
   $s = $?
5. $A = 36$ in²
   $s = $?
6. $A = 81$ ft²
   $s = $?
7. $A = 144$ yd²
   $s = $?
8. $A = 4$ mi²
   $s = $?

For Exercises 9–16, use the formula $D = 1.2\sqrt{A}$ to find the distance, D, (in miles) to the horizon from an aircraft that is the given number of feet, A, from the ground. Round each answer to the nearest mile. (Example 2)

	A	$\sqrt{A}$	D
9.	2500 ft	50	?
11.	5041 ft	?	?
13.	9604 ft	?	?
15.	12,996 ft	?	?

	A	$\sqrt{A}$	D
10.	4900 ft	70	?
12.	2209 ft	?	?
14.	6084 ft	?	?
16.	16,900 ft	?	?

WRITTEN EXERCISES

Goal: To use a formula to solve word problems involving square roots

For Exercises 1–12, use the formula $A = s^2$ to find the length, s, of one side of a square with the given area, A. Round each answer to the nearest tenth. (Example 1)

	A	s
1.	23 m²	?
4.	110 mm²	?
7.	42 m²	?
10.	100 mi²	?

	A	s
2.	8 km²	?
5.	136 cm²	?
8.	78 mm²	?
11.	121 in²	?

	A	s
3.	35 cm²	?
6.	10 km²	?
9.	36 ft²	?
12.	16 yd²	?

Solve. Round each answer to the nearest tenth. (Example 1)

13. A square swimming pool has an area of 324 square feet. Find the length of one side of the pool.

14. The area of the top of a square table is 130 square centimeters. Find the length of one side of the table.

For Exercises 15–18, use the formula $D = 1.2\sqrt{A}$ to find the distance, D, in miles to the horizon from an aircraft that is the given number of feet, A, from the ground. Round each answer to the nearest mile. (Example 2)

	Feet from the Ground	Distance to the Horizon			Feet from the Ground	Distance to the Horizon
15.	3025	?		16.	7225	?
17.	14,884	?		18.	10,816	?

Solve. Round each answer to the nearest mile. (Example 2)

19. The height of a hot-air balloon above the ground is 3844 feet. Find the distance to the horizon.

20. A plane is flying at a height of 19,321 feet above the ground. Find the distance to the horizon.

21. A plane is flying at a height of 10,404 feet. Find the distance to the horizon.

22. The height of a plane above the ground is 20,736 feet. Find the distance to the horizon.

REVIEW CAPSULE FOR SECTION 15–5

Evaluate. (Pages 440–441)

1. 3^2
2. 8^2
3. 10^2
4. 20^2
5. 36^2
6. 19^2

Solve and check. (Pages 190–192)

7. $n + 6 = 10$
8. $p + 17 = 30$
9. $x + 3 = 12$
10. $b + 12 = 25$
11. $t + 15 = 40$
12. $y + 30 = 75$
13. $a + 4 = 21$
14. $c + 9 = 27$

Round each number to the nearest tenth. (Pages 32–34)

15. 12.35
16. 21.72
17. 18.09
18. 32.15
19. 49.53
20. 62.43

The answers are on page 464.

Special Triangles 453

15-5 Pythagorean Theorem and Applications

A **right triangle** is a triangle with one right angle (90°).

When you know the lengths of two sides of a right triangle, you can use the **Pythagorean Theorem** to find the length of the other side.

hypotenuse: c
leg: b
leg: a
90°

Pythagorean Theorem: $(\text{hypotenuse})^2 = (\text{leg})^2 + (\text{leg})^2$

$$c^2 = a^2 + b^2$$

NOTE: The **hypotenuse** of a right triangle is the side opposite the right angle. The hypotenuse is the longest side of the right triangle.

PROCEDURE

To find the measure of a side of a right triangle:

1. Draw the triangle. Label the sides with the given measures.
2. Write the Pythagorean Theorem. Replace two letters in the formula by the corresponding given measures.
3. Solve for the third measure.

EXAMPLE 1 The sail on this boat is 5 yards high and 3 yards wide. Find the length of the third side. Round your answer to the nearest tenth.

Solution: Use the Pythagorean Theorem to find c.

2 $c^2 = a^2 + b^2$ Replace a with 5 and b with 3.
$c^2 = 5^2 + 3^2$
3 $c^2 = 25 + 9$
$c^2 = 34$
$c = \sqrt{34}$ Use the table on page 447.
$c = 5.831$ The length is about **5.8 yards**.

454 Chapter 15

Self-Check   *Find the length of the hypotenuse, c, of each right triangle with legs a and b.*

    1. $a = 5$ in; $b = 7$ in   **2.** $a = 7$ m; $b = 8$ m

When you know the lengths of the hypotenuse and one leg of a right triangle, you can use the Pythagorean Theorem to find the length of the other leg.

EXAMPLE 2   A 7-meter ladder is leaning against a shed. If the base of the ladder is 3 meters from the wall, how high up wall does the ladder reach? Round your answer to the nearest tenth.

Solution: Use the Pythagorean Theorem to find b.

    ☐2 $c^2 = a^2 + b^2$ ◀ **Replace c with 7 and a with 3.**

            $7^2 = 3^2 + b^2$

    ☐3 $49 = 9 + b^2$

        $49 - 9 = 9 - 9 + b^2$

           $40 = b^2$

           $\sqrt{40} = b$ ◀ **Use the table on page 447.**

         $6.325 = b$ The ladder reaches about **6.3 meters** above the ground.

Self-Check   *Refer to this right triangle with hypotenuse c and legs a and b. Find each unknown length.*

    3. $b = 4$ m; $c = 5$ m; $a =$ __?__

    4. $a = 5$ ft; $c = 13$ ft; $b =$ __?__

CLASSROOM EXERCISES

In Exercises 1–2, find the length of the hypotenuse, c, of each right triangle. Use the Table of Squares and Square Roots on page 447. Round each answer to the nearest tenth. (Example 1)

1. $a = 6$, $b = 8$, $c =$ __?__   **2.** $a = 8$, $b = 8$, $c =$ __?__

Special Triangles   **455**

For Exercises 3–5, the length of the hypotenuse, c, and the length of one leg of the right triangle are given. Find the unknown length. Refer to the Table of Squares and Square Roots on page 447. Round each answer to the nearest tenth. (Example 2)

3. $a = 30$, $c = 34$, $b = ?$

4. $c = 53$, $b = 45$, $a = ?$

5. $c = 91$, $a = 84$, $b = ?$

WRITTEN EXERCISES

Goals: To use the Pythagorean Theorem to find the measure of the hypotenuse of a right triangle

To use the Pythagorean Theorem to find the measure of one leg of a right triangle

To apply the skill of finding the measure of a side of a right triangle to solving word problems

In Exercises 1–6, use the Pythagorean Theorem to solve each problem. Use the Table of Squares and Square Roots on page 447. (Example 1)

1. In the figure below, a wire supports a pole that is 5 meters high. The distance from the foot of the pole to the point where the wire is fastened to a stake in the ground is 7 meters. How long is the wire? Round the answer to the nearest meter.

2. The figure below shows the escalator between two floors of a department store. How many yards are you carried as you travel on the escalator from one floor to the next? Round the answer to the nearest yard.

456   Chapter 15

3. A ladder is placed against the side of the house as shown in the figure at the right. Find the length of the ladder.

(Example 2)

4. A 18-meter wire staked to the ground supports a pole that is 15 meters high. How far from the pole is the stake?

5. The front wall of an A-frame house is as shown in the figure at the right. How tall is the front of the house?

6. A diagonal wall brace is 14 feet long. The wall is 8 feet high. Find the width of the wall.

REVIEW CAPSULE FOR SECTION 15-6

Write a two-place decimal for each fraction. (Pages 173-175)

1. $\frac{1}{2}$ 2. $\frac{9}{10}$ 3. $\frac{5}{2}$ 4. $\frac{6}{5}$ 5. $\frac{9}{4}$ 6. $\frac{7}{20}$

Round to the nearest whole number. (Pages 32-34)

7. 12.1 8. 18.7 9. 37.32 10. 18.08 11. 100.49 12. 5.98

Solve and check. (Pages 203-205)

13. $\frac{n}{10} = 7$ 14. $\frac{x}{8} = 4$ 15. $15 = \frac{b}{100}$ 16. $9 = \frac{y}{9}$ 17. $\frac{c}{50} = 2$

18. $\frac{y}{25} = 5$ 19. $\frac{p}{10} = 1.5$ 20. $\frac{t}{7} = 6.8$ 21. $0.2 = \frac{y}{9}$ 22. $0.9 = \frac{x}{4}$

The answers are on page 464.

Special Triangles 457

15-6 Tangent Ratio and Applications

Another way to find the lengths of the legs of a right triangle is to use the **tangent ratio**.

The abbreviation for the "tangent of angle A" is "tan A."

$$\tan A = \frac{\text{length of the side opposite angle } A}{\text{length of the side adjacent to angle } A}$$

EXAMPLE 1 Find tan A (nearest hundredth).

Solution: $\tan A = \dfrac{\text{length of side opposite angle } A}{\text{length of side adjacent to angle } A}$

$\tan A = \dfrac{3}{4}$

$\tan A = \mathbf{0.75}$

Self-Check Find tan A: **1.** **2.**

When you know the measure of an acute angle of a right triangle such as angle A, you can use a table (see page 459) to find tan A.

EXAMPLE 2 Find the height of this flagpole. Round your answer to the nearest meter.

Solution: Use the tangent ratio.

[2] $\tan A = \dfrac{\text{length of side opposite angle } A}{\text{length of side adjacent to angle } A}$ [1]

$\tan 50° = \dfrac{h}{12}$ ◀ **Find tan 50° in the Tangent Table.**

[3] $1.19 = \dfrac{h}{12}$ ◀ **Multiply each side by 12.**

$1.19 \times 12 = \dfrac{h}{12} \times \dfrac{12}{1}$

$14.28 = h$ The flagpole is about **14 meters** high.

Self-Check   *Solve for x. Round your answer to the nearest whole number.*

3. $\tan 60° = \dfrac{x}{6}$ **4.** $\tan 5° = \dfrac{x}{15}$ **5.** $\tan 30° = \dfrac{x}{11}$

TANGENT TABLE

Angles	Tangents	Angles	Tangents
5°	0.087	50°	1.19
10°	0.176	55°	1.43
15°	0.268	60°	1.73
20°	0.364	65°	2.14
25°	0.466	70°	2.75
30°	0.577	75°	3.73
35°	0.700	80°	5.67
40°	0.839	85°	11.43
45°	1.00		

◀ tan 65° = 2.14

◀ tan 80° = 5.67

When you know the measure of angle A and the length of one leg of a right triangle, you can use the table and the definition of tan A to find the length of the other leg.

PROCEDURE

To use the tangent ratio to find a leg of a right triangle:

1. Draw and label a right triangle.
2. Use the known angle, the known leg, and the unknown leg to write the tangent ratio.
3. Solve for the length of the unknown leg.

CLASSROOM EXERCISES

In Exercises 1–5, use the Table of Tangents to find each of the following. (Table)

1. tan 10°   **2.** tan 85°   **3.** tan 25°   **4.** tan 75°   **5.** tan 45°

In Exercises 6–9, use the Table of Tangents to solve each problem. Round each answer to the nearest whole number. (Example 2)

6. $\tan 10° = \dfrac{x}{20}$ **7.** $\tan 35° = \dfrac{n}{9}$ **8.** $\tan 20° = \dfrac{h}{15}$ **9.** $\tan 85° = \dfrac{y}{32}$

Special Triangles

WRITTEN EXERCISES

Goals: To use the tangent ratio to find the tangent of an acute angle of a right triangle

To use the tangent ratio and the tangent table to find the measure of a leg of a right triangle

Find tan A. Write a two-place decimal for your answer. (Example 1)

1. Triangle with B at top, C at bottom-right angle, A at bottom-right. BC = 4, CA = 4.

2. Triangle with A at top, right angle at C. AC = 3, BC = 10.

3. Triangle with A at top, right angle at C. AC = 3, BC = 2.

4. Triangle with B at top, right angle at C. BC = 4, CA = 5.

Solve. Use the tangent ratio and the Table of Tangents on page 459. Round each answer to the nearest whole number. (Example 2)

5. Find the distance, d.

(Triangle: angle A = 40°, AC = 80 m, BC = d)

6. Find the width, w.

(Triangle: angle A = 40°, AC = 240 m, BC = w)

7. Find the height, h, of the plane above the ground.

$\left(\text{HINT: } \tan 15° = \dfrac{h}{?}\right)$

(Radar distance, Radar-position angle 15°, AC = 30 km)

8. Find the height, h, of the television tower.

$\left(\text{HINT: } \tan \underline{\ ?\ } = \dfrac{h}{40}\right)$

(angle A = 55°, AC = 40 m)

9. Find the height, h, of the mountain shown at the right.

$\left(\text{HINT: } \tan 30° = \dfrac{h}{?}\right)$

(angle = 30°, base = 8 km)

460   Chapter 15

TANGENT VALUES

You can use a scientific calculator instead of a table to find tangent values, given the angle measures.

EXAMPLE Find tan 42°.

Solution   4 2 [tan]   `0.900404`

EXERCISES

Use a scientific calculator to find each of the following.

1. tan 3°   **2.** tan 16°   **3.** tan 27°   **4.** tan 46°   **5.** tan 88°

6. Use a scientific calculator to check your answers to Exercises 1-5 on page 459.

REVIEW: SECTIONS 15-4 — 15-6

Use the formula $A = s^2$ to find the length, s, of one side of a square with the given area, A. Round each answer to the nearest tenth. (Section 15-4)

1. $A = 12$ square meters; $s =$ __?__

2. $A = 37$ square millimeters; $s =$ __?__

Solve. (Section 15-5)

3. The base of a ramp is 12 meters long and 5 meters high. Find the length of the ramp. Use the Tables of Squares and Square Roots on page 447.

Solve. Use the Table of Tangents on page 459. Round each answer to the nearest whole number. (Section 15-6)

4. Find the height of the mountain shown below.

$$\left(\text{HINT: } \tan 25° = \frac{h}{?}\right)$$

5. Find the height of the building shown below.

$$\left(\text{HINT: } \tan 30° = \frac{?}{42}\right)$$

Special Triangles

CHAPTER REVIEW

PART 1: VOCABULARY

For Exercises 1–6, choose from the box at the right the word(s) that best corresponds to each description.

1. A triangle with one right angle __?__ (p. 454)
2. The square of a rational number __?__ (p. 442)
3. The decimal equivalents of __?__ numbers are non-terminating and non-repeating. (p. 443)
4. The product of a number and itself is the __?__ of the number. (p. 440)
5. The longest side of a right triangle __?__ (p. 454)
6. The inverse operation of squaring a number is taking its __?__. (p. 440)

> square
> irrational
> right triangle
> square root
> hypotenuse
> perfect square

PART 2: SKILLS

Square each of the following. (Section 15-1)

7. 9 8. 15 9. 18 10. 24 11. 42 12. 17

Find each answer. (Section 15-1)

13. $\sqrt{16}$ 14. $\sqrt{1}$ 15. $\sqrt{81}$ 16. $\sqrt{36}$ 17. $\sqrt{100}$ 18. $\sqrt{64}$
19. $\sqrt{121}$ 20. $\sqrt{400}$ 21. $-\sqrt{4}$ 22. $-\sqrt{25}$ 23. $-\sqrt{144}$ 24. $-\sqrt{49}$

Identify each number as rational or irrational. (Section 15-2)

25. 4.5 26. $\sqrt{25}$ 27. $\sqrt{15}$ 28. 3.214··· 29. $9.\overline{6}$ 30. $\sqrt{120}$
31. 8.562··· 32. $3.\overline{13}$ 33. $\sqrt{9}$ 34. $\sqrt{90}$ 35. $4.51\overline{6}$ 36. $\sqrt{75}$

Refer to the table on page 447 to find each answer. (Section 15-3)

37. 18^2 38. 15^2 39. 21^2 40. 84^2 41. 110^2 42. 64^2
43. $\sqrt{75}$ 44. $\sqrt{961}$ 45. $\sqrt{114}$ 46. $\sqrt{19,044}$ 47. $\sqrt{95}$ 48. $\sqrt{729}$

For Exercises 49–54, use $A = s^2$ to find s for the given values of A. Round each answer to the nearest tenth. (Section 15-4)

49. $A = 12$ 50. $A = 5$ 51. $A = 27$ 52. $A = 125$ 53. $A = 8$ 54. $A = 63$

For Exercises 55–59, use $D = 1.2\sqrt{A}$ to find D for the given values of A. Round each answer to the nearest whole number. (Section 15-4)

55. $A = 9604$ **56.** $A = 10{,}609$ **57.** $A = 6241$ **58.** $A = 18{,}496$ **59.** $A = 21{,}904$

For Exercises 60–63 find the unknown length for each right triangle. Round each answer to the nearest tenth. (Section 15-5)

60. $a = 12$, $b = 16$, $c = ?$

61. $a = 5$, $b = 10$, $c = ?$

62. $a = ?$, $b = 25$, $c = 27$

63. $a = 40$, $b = ?$, $c = 50$

For Exercises 64–67, find tan A. Write a two-place decimal for each answer. (Section 15-6)

64. Triangle with B, C, A; $BC = 4$, $CA = 4$.

65. Triangle with A, B, C; $AB = 4$, $BC = 7$.

66. Triangle with B, C, A; $BC = 4$, $CA = 8$.

67. Triangle with B, C, A; $BC = 7$, $CA = 9$.

PART 3: APPLICATIONS

68. A ladder is placed against the side of a house as shown below. Find the length of the ladder. Use the table on page 447. Round the answer to the nearest tenth. (Section 15-5)

69. Find the height, h, of the telephone pole shown below. Use the Table of Tangents on page 459. Round the answer to the nearest whole number.

$\left(\text{HINT: } \tan 45° = \dfrac{h}{11}\right)$ (Section 15-6)

Special Triangles 463

CHAPTER TEST

Find each answer.

1. 5^2
2. 9^2
3. $\sqrt{49}$
4. $\sqrt{36}$
5. $-\sqrt{144}$

Identify each number as rational or irrational.

6. $\sqrt{24}$
7. $\sqrt{81}$
8. $4.2\overline{6}$
9. $1.4132\cdots$
10. 0.18

Use the Table of Squares and Square Roots on page 447 to find each answer.

11. 64^2
12. 37^2
13. $\sqrt{95}$
14. $\sqrt{102}$
15. $\sqrt{4356}$

For Exercises 16–18, find the unknown length of each right triangle with hypotenuse c and legs a and b. Refer to the Table of Squares and Square Roots on page 447. Round each answer to the nearest tenth.

16. $a = 9; b = 12; c = \underline{}$
17. $b = 16; c = 34; a = \underline{}$
18. $a = 5; c = 13; b = \underline{}$

19. A ladder is placed against a shed as shown below. Find the length of the ladder. Use the Table of Squares and Square Roots on page 447. Round the answer to the nearest tenth.

20. Find the height, h, of the smockstack shown below. Use the Table of Tangents on page 459. Round the answer to the nearest whole number.

$$\left(\text{HINT: } \tan 30° = \frac{h}{75}\right)$$

ANSWERS TO REVIEW CAPSULES

Page 442 1. 0.5 2. 0.67 3. 0.6 4. 0.25 5. 1.17 6. 2.25 7. 0.2 8. 0.75 9. $0.\overline{3}$ 10. $0.8\overline{3}$ 11. 0.15 12. 0.625 13. 0.875 14. 0.45 15. $1.\overline{1}$ 16. 3.8 17. 1.75 18. $3.\overline{6}$

Page 444 1. 36 2. 225 3. 2 4. 102 5. 1764 6. 5 7. 56 8. 3364 9. 13 10. 88 11. 8100 12. 13.924

Page 450 1. 43.2 2. 16.7 3. 27.8 4. 14.0 5. 82.9 6. 93.1 7. 12.5 8. 39.8 9. 55.5 10. 68.3 11. 10.9 12. 50.1 13. 4.249 14. 19.1562 15. 7.8125 16. 3.9936 17. 38 18. 28 19. 42 20. 17 21. 73 22. 25 23. 6 24. 18 25. 9 26. 2 27. 8 28. 124

Page 453 1. 9 2. 64 3. 100 4. 400 5. 1296 6. 361 7. 4 8. 13 9. 9 10. 13 11. 25 12. 45 13. 17 14. 18 15. 12.4 16. 21.7 17. 18.1 18. 32.2 19. 49.5 20. 62.4

Page 457 1. 0.50 2. 0.90 3. 2.50 4. 1.20 5. 2.25 6. 0.35 7. 12 8. 19 9. 37 10. 18 11. 100 12. 6 13. 70 14. 32 15. 1500 16. 81 17. 100 18. 125 19. 15 20. 47.6 21. 1.8 22. 3.6

464 Chapter 15

ENRICHMENT

Arithmetic Method for Finding $\sqrt{N}$

The flowchart at the left below outlines a procedure for approximating the square root of an irrational number N to a given number of decimal places. The Example at the right shows how to follow these steps to determine $\sqrt{38}$ to the nearest tenth.

Flowchart

1. DETERMINE TWO CONSECUTIVE INTEGERS A AND B SUCH THAT $A < \sqrt{N} < B$. THEN FIND THE AVERAGE OF A AND B.
2. DIVIDE N BY THE APPROXIMATION.
3. FIND THE AVERAGE OF THE QUOTIENT AND DIVISOR.
4. DO YOU WANT A MORE ACCURATE APPROXIMATION? Yes → (back to step 2) / No ↓
5. $\sqrt{N} \approx ?$

Example

[1] $6 < \sqrt{38} < 7$ ◄ Find the average of 6 and 7.

First approximation: $\dfrac{6+7}{2} = 6.5$

[2] Divide N (38) by the approximation.

$6.5 \overline{)38.00}$ = 5.8 ◄ **First** division: Divide to **one** decimal place.

[3] Second approximation:

$\dfrac{5.8 + 6.5}{2} = 6.15$ ◄ Average of the quotient and divisor

[4] One loop is complete. To approximate $\sqrt{38}$ to the nearest tenth, you divide to two decimal places. Return to Step 2.

[2] $6.15 \overline{)38.0000}$ = 6.17 ◄ **Second** division: Divide to **two** decimal places.

[3] Third approximation:

$\dfrac{6.17 + 6.15}{2} = 6.16$ ◄ Average of the quotient and divisor

[4] Two loops are complete. Round 6.16 to the nearest tenth.

[5] $\sqrt{N} \approx \mathbf{6.2}$

EXERCISES

Follow the steps of the flowchart to approximate the square root of each number.

1. 6 (nearest tenth)
2. 8 (nearest tenth)
3. 15 (nearest tenth)
4. 5 (nearest hundredth)
5. 17 (nearest hundredth)
6. 60 (nearest hundredth)

Special Triangles

ADDITIONAL PRACTICE

SKILLS

Square each number. (Pages 440-441)   *Find each answer.* (Pages 440-441)

1. 7
2. 11
3. 13
4. $\sqrt{25}$
5. $\sqrt{81}$
6. $\sqrt{121}$

Identify each number as rational or irrational. (Pages 442-444)

7. $\sqrt{64}$
8. $\sqrt{3}$
9. 4.12
10. $4.\overline{25}$
11. $4.\overline{6}$
12. 3.24135 . . .

Refer to the table on page 447 to find each answer. (Pages 445-446)

13. 26^2
14. 36^2
15. 47^2
16. $\sqrt{5041}$
17. $\sqrt{729}$
18. $\sqrt{8281}$

Use the Pythagorean Theorem to solve each problem. (Pages 454-457)

19. Triangle with $c = ?$, $a = 5$, $b = 12$
20. Triangle with $c = ?$, $a = 2$, $b = 5$
21. Triangle with $c = 5$, $a = 3$, $b = ?$

22. Find tan A for each triangle of Exercises 19–21. Write a ratio for tan A. (Pages 458-460)

Use the Table of Tangents on page 459 to find x.

23. $\tan 45° = \dfrac{x}{40}$
24. $\tan 35° = \dfrac{x}{16}$
25. $\tan 75° = \dfrac{x}{25}$
26. $\tan 15° = \dfrac{x}{20}$

APPLICATIONS

27. The Ty family plans to build a square garage. The area of the floor will be 529 square meters. Find the length of one side of the garage. (Pages 451-453)

28. Find the height, h, of this radio tower. $\left(\text{HINT: } \tan \underline{} = \dfrac{h}{90}\right)$ (Pages 458-460)

466 Chapter 15

COMMON ERRORS

Each of these problems contains a common error.

a. Find the correct answer.

b. Find the error.

Solve.

1. To graph the ordered pair $P(-3, -6)$, start at the origin.
   Move 3 units to the *right*.
   Then move 6 units *down*.

2. Complete the table of ordered pairs for the equation $y = 2x - 1$.

x	$2x - 1$	y
2	$2(2) - 1 = 3$	3
0	$2(0) - 1 = -1$	-1
-1	$2(-1) - 1 = 1$	1

3. Find the slope of the line containing the points $(3, 5)$ and $(2, 1)$.
   Vertical change: $3 - 2 = 1$
   Horizontal change: $5 - 1 = 4$
   Slope: $\frac{1}{4}$

4. Find the y intercept of $y = 3x + 6$.
   $y = 3x + 6$
   $0 = 3x + 6$
   $-6 = 3x$
   $-2 = x$ ◀ y intercept: -2

5. The area of a square is 10 meters. Find the length of one side.
   $A = s^2$
   $10 = s^2$
   $100 = s$
   The length of one side is **100 m.**

6. Use the Pythagorean Theorem to find the length of the hypotenuse of this right triangle.
   $a^2 + b^2 = c^2$
   $9^2 + 6^2 = c^2$
   $81 + 36 = c^2$
   $117 = c^2$
   $\sqrt{117} = c$
   $13{,}689 = c$ ◀ The unknown length

7. Find $\tan A$. Refer to the figure at the right.
   $\tan A = \dfrac{\text{length of side opposite angle } A}{\text{length of side adjacent to angle } A}$
   $= \dfrac{12}{9} = \dfrac{4}{3} = \mathbf{1.75}$

Special Triangles

CUMULATIVE REVIEW: CHAPTERS 14–15

Choose the correct answer.
Choose a, b, c, or d.

1. Name the point on the graph at the right that corresponds to the ordered pair $(-1, 2)$.
   a. A
   b. B
   c. C
   d. D

2. Find the slope of the line containing the points $A(3, 4)$ and $B(-1, 2)$.
   a. 1
   b. 2
   c. $\frac{1}{3}$
   d. $\frac{1}{2}$

3. Find the y-intercept of the linear equation $y = 4x + 16$.
   a. $y = 0$
   b. $y = 16$
   c. $y = 4$
   d. $y = -4$

4. Which ordered pair is a solution of this system? $\begin{cases} y = 2x + 6 \\ y = 3x + 1 \end{cases}$
   a. $(5, 16)$
   b. $(-3, 0)$
   c. $(3, 12)$
   d. $(2, 7)$

5. Which ordered pair is a solution to the inequality $y > -2x + 1$?
   a. $(4, -6)$
   b. $(-3, 6)$
   c. $(-100, 194)$
   d. $(0, 1)$

6. Which number is a perfect square?
   a. 24
   b. 81
   c. 18
   d. 28

7. Which of the following numbers is irrational?
   a. 3.1
   b. $3.\overline{3}$
   c. $\sqrt{25}$
   d. $\sqrt{17}$

8. Find the length in meters of a side of the square that has an area of 36 square meters.
   a. 9
   b. 1296
   c. 24
   d. 6

9. Use the Pythagorean Theorem to find the length in feet of the ladder shown below.
   a. 14
   b. 9
   c. 13
   d. 10

10. Use the tangent ratio to find the height in meters of the wall. (HINT: $\tan 15° = 0.27$)
    a. 2962
    b. 216
    c. 21.6
    d. 292.2

468 Cumulative Review: Chapters 14–15

CHAPTER 16
Statistics and Probability

SECTIONS
16-1 Mean, Median, and Mode
16-2 Histograms
16-3 Probability
16-4 Probability and Tables
16-5 Tree Diagrams
16-6 Multiplying Probabilities

FEATURES
Calculator Application: *Multiplying Probabilities*
Career Application: *Statistical Clerk*
Enrichment: *Sampling*

16-1 Mean, Median, and Mode

George Blackwater records the amount of rainfall throughout the year for the Weather Service. He uses the <u>data</u> to keep records and spot trends. Every month, George calculates the *mean* rainfall.

> **PROCEDURE**
>
> To find the mean, divide the sum of the given measures by the number of measures.
>
> $$\text{Mean} = \frac{\text{Sum of Measures}}{\text{Number of Measures}}$$
>
> ◂ Mean is another word for <u>average</u>.

EXAMPLE 1 This table shows the rainfall for Louisville, Kentucky from May through September of a certain year. Find the mean rainfall for this five-month period.

Rainfall for Louisville, KY	
Month	Rainfall (in)
May	4.2
June	4.1
July	3.8
August	3.0
September	2.9

Solution:

$$\text{Mean} = \frac{4.2 + 4.1 + 3.8 + 3.0 + 2.9}{5}$$

Mean = 3.6   The mean rainfall is **3.6 inches**.

Self-Check   1. Find the mean amount of savings.

Month	Jan.	Feb.	Mar.	Apr.	May	June	July	Aug.	Sept.
Amount	$20	$25	$17	$25	$21	$25	$17	$20	$19

In a listing of data, the **median** is the middle measure or score. When there are an <u>even number</u> of measures, the median is the <u>mean of the two middle measures</u>.

The **mode** is the measure that occurs most often. There may be more than one mode.

> **PROCEDURE**
>
> **To find the median of a list of data:**
>
> 1. Arrange the data in ascending or descending order.
> 2. a. For an odd number of items, the median is the middle number listed.
>   b. For an even number of items, the median is the mean of the two middle numbers.

EXAMPLE 2 The table at the right shows the number of fiction books checked out of a library for 6 days.

a. Find the median number of books checked out.

b. Find the mode.

Number of Fiction Books Checked Out During Six Days

Day	Number
Monday	30
Tuesday	45
Wednesday	19
Thursday	30
Friday	36
Saturday	50

Solutions: a. 1 Arrange the numbers in order.

19 30 **30**
36 45 50

◄ There are two middle numbers.

2 Find the mean of 30 and 36.

$$\frac{30 + 36}{2} = \frac{66}{2} = 33$$

The median number of fiction books checked out is **33**.

b. Since the number of fiction books checked out most often is 30, the mode is **30**.

Self-Check 2. Find the median and mode of the prices for the television sets.

Brand	A	B	C	D	E	F
Price	$275	$672	$195	$450	$857	$672

CLASSROOM EXERCISES

For Exercises 1–2, find the mean. (Example 1)

1. **Passengers on Eight Buses**

42	45	41	38
40	42	32	40

2. **Prices of Six Trucks**

$8439	$7680	$6755
$7521	$6350	$5945

Statistics and Probability

For Exercises 3–4, find the median and mode. (Example 2)

3. Math Scores for Ten Tests

85	86	89	89	70
74	88	85	100	85

4. Ages of Members of Golf Team

16	15	15	17	16
17	14	16	18	14

WRITTEN EXERCISES

Goal: To find the mean, median, and mode of a list of data

For Exercises 1–4, find the mean. (Example 1)

1. Heights in Feet of Six Mountains in the United States

Mountain	Height
Blackburn	16,390
Churchill	15,638
McKinley	20,320
Pike's Peak	14,110
Rainier	14,410
San Luis	14,010

2. Number of Telephones in Five Cities

City	Number
Albany, NY	170,643
Cambridge, MA	111,498
Lexington, KY	163,114
Lubbock, TX	161,860
San Francisco, CA	815,520

3. Earnings Per Hour for Eight Workers

$6.50	$5.75	$5.25	$4.95
$6.25	$5.00	$8.15	$7.35

4. Number of Hours of Sleep for Ten Adults

8	6	6	8	7
7	7	8	5	8

For Exercises 5–8, find the median and mode. (Example 2)

5. Ages of Twenty People at a Party

22	31	27	26	19
23	26	28	26	21
24	27	31	27	29
25	22	28	30	26

6. Yearly Salaries for Five Employees

Manager	$40,000
Assistant	$25,000
Clerk	$12,500
Clerk	$12,500
Clerk	$12,500

7.

Rainfall for Omaha, Nebraska

Month	Rainfall (in)
September	3.3
October	1.9
November	1.1
December	0.8
January	0.8

8.

Enrollment Per Grade

Grade	Enrollment
7	189
8	198
9	205
10	215
11	198
12	200

MORE CHALLENGING EXERCISES

9. Before ordering dresses for the summer, the owner of a dress shop examined the records of the number and sizes of dresses sold the previous summer. Would the mean, median, or mode be the most useful guide in determining the sizes to order this summer? Give a reason for your answer.

Mary and Bill own a small business. Of the twelve employees, one earns $16,500, one earns $14,000, two earn $9500 each, five earn $8100 each, and three earn $6000 each. Mary and Bill pay themselves $20,000 each.

10. Find the mean of the fourteen salaries.

11. Find the median and the mode of the fourteen salaries.

12. Which of these measures, the mean, median, or mode, gives the highest "average" salary?

13. Which of these measures, the mean, median, or mode, gives the lowest "average" salary?

REVIEW CAPSULE FOR SECTION 16-2

1. Construct a vertical bar graph for the following data. (Pages 270-273)

Personal Income for the First Six Months of a Certain Year

Month	January	February	March	April	May	June
Trillions of Dollars	2045	2055	2070	2070	2060	2080

The answers are on page 494.

Statistics and Probability   **473**

16-2 Histograms

Carol Davis runs the 100-yard dash for the school track team. At each meet, the coach records her time in seconds and lists the data in a table. Grouping the data in intervals makes the table shorter and easier to read.

> **PROCEDURE**
>
> **To show grouped data in a table:**
>
> 1. List the intervals.
> 2. Tally the number of measures in each interval. Count the measures.

EXAMPLE 1 Carol's times in seconds for the 100-yard dash are given at the right. List the data in a table.

Times in Seconds for the 100-yard Dash

11.6 11.2 11.5 12.0 12.0
11.2 11.5 11.6 11.8 11.4
12.0 11.4 11.2 11.6 11.4

Solution:

Intervals	Tally	Count
11.2–11.4	⦀⦀⦀	6
11.5–11.7	⦀⦀	5
11.8–12.0	⦀⦀⦀⦀	4

Self-Check

1. Show the following data in a table. Use the given intervals.

Tractor Output Per Day for 24 Factories

31 40 35 67 85 90 32 39 68 87 55 50
33 43 64 78 82 70 54 60 41 85 32 68

Intervals: 31–40; 41–50; 51–60; 61–70; 71–80; 81–90

A **histogram** is a special kind of bar graph. The width of the bars of a histogram represent the size of the intervals.

> **PROCEDURE**
>
> **To draw a histogram:**
>
> 1. List the data in equal intervals in a table.
> 2. Draw a rectangle to represent the count for each interval.

EXAMPLE 2 Pulse rates for 60 people are listed below. Draw a histogram to show the data.

Pulse Rates of 60 People					
68	70	82	72	74	81
77	73	78	73	77	72
72	75	74	73	76	80
77	74	73	72	84	73
75	79	79	81	69	77
79	69	70	71	84	85
69	74	75	79	75	70
69	70	76	76	75	73
79	81	74	72	76	78
79	71	80	77	79	80

Solution:

1 List the data by intervals.

Pulse Rates	Count
65–69	5
70–74	22
75–79	23
80–84	9
85–89	1

2 Draw a rectangle for each interval.

Pulse Rates of 60 People

(histogram with Number of People on y-axis, Beats Per Minute on x-axis: 67, 72, 77, 82, 87)

Self-Check **2.** Draw a histogram to show the data. Use the given intervals.

Cars	A	B	C	D	E	F	G	H	I	J	K	L	M	N
Gas Mileage (mi/gal)	28	32	25	33	30	24	35	26	31	34	27	38	36	37

Intervals: 21-25; 26-30; 31-35; 36-40

NOTE: In a histogram, the middle number of each interval is at the center of the base of the corresponding bar on the graph.

Statistics and Probability **475**

CLASSROOM EXERCISES

Use the given information to complete each table. (Example 1)

1. **Capacity of 16 College Football Stadiums**

59,000	52,000	32,000	30,000
66,000	35,000	72,000	50,000
50,000	66,000	42,000	58,000
70,000	46,000	47,000	52,000

Capacity	Tally	Count
30,000–39,000	?	?
40,000–49,000	?	?
50,000–59,000	?	?
60,000–69,000	?	?
70,000–79,000	?	?

2. **Average Class Size in Twenty Schools** (Example 2)

19	32	28	38	24
25	39	20	35	37
29	26	25	33	38
32	34	22	18	31

Class Size	Count
15–20	?
21–25	?
26–30	?
31–35	?
36–40	?

3. Conduct an opinion poll of about 40 students on a topic of your choice. Construct a frequency table to show the data. Use the frequency table to draw a histogram.

WRITTEN EXERCISES

Goals: To list data by using equal intervals in a table

To use a histogram to show data

For Exercises 1–2, use the given information to make a table. The intervals are also given. (Example 1)

1. **Heights in Feet of Twenty Major U.S. Dams**

121	340	250	325	519
585	121	250	160	210
550	564	402	637	625
245	382	610	154	239

Intervals: 120–219; 220–319; 320–419; 420–519; 520–619; 620–719

2. **Qualifying Speeds (mph) for a Stock Car Race**

175	186	183	171	182
190	184	187	188	178
191	173	182	186	177
185	193	186	182	189

Intervals: 170–174; 175–179; 180–184; 185–189; 190–194

For Exercises 3–6, make a histogram to show the given data. Use the given intervals. (Example 2)

3. **Diameters of Twenty Trees (cm)**

25	32	36	28	20
26	35	40	27	30
36	23	29	31	24
29	32	43	25	30

Intervals: 17–23; 24–30; 31–37; 38–44

4. **Bowling Scores for One Week**

162	125	146	150
170	190	155	175
135	144	170	165
155	185	162	184

Intervals: 121–140; 141–160; 161–180; 181–200

5. **Miles Bicycled Per Day for Triathlon Training**

25	30	22	32	37
34	15	25	39	24
32	18	30	35	32
20	25	31	28	30

Intervals: 15–19; 20–24; 25–29; 30–34; 35–39

6. **Hours Per Week of Evening Television Viewing by Teens**

4	6	9	11	8
1	12	9	4	11
15	2	9	5	13
4	9	18	4	9

Intervals: 0–4; 5–9; 10–14; 15–19

REVIEW: SECTIONS 16-1 – 16-2

For Exercises 1–4, find the mean, median, and mode. (Section 16-1)

1. **Science Test Scores (%)**

92	85	94	81
83	92	89	92

2. **Weight of a Wrestling Team (lbs)**

160	177	122	151
143	156	160	147

3. **Denver's High Temperatures for a Week**

| 70 | 73 | 72 | 68 | 72 | 70 | 72 |

4. **Ages of Golf Team Members**

19	22	21	19	19	21
18	24	23	26	19	

Make a histogram to show the given data. Use the given intervals. (Section 16-2)

5. **Price of One Ounce of Gold Over One Year**

$525	$500	$400	$425	$440	$510
$530	$500	$560	$530	$515	$495

Intervals: $400–$439; $440–$479; $480–$519; $520–$559; $560–$599

Statistics and Probability

CAREER APPLICATIONS
Statistical Clerk

A **statistical clerk** often helps to collect and analyze data such as test scores. One measure that is computed is the **standard deviation.** The standard deviation shows how data spreads out or varies from the mean of the data.

EXAMPLE Find the standard deviation for these test scores.

80, 84, 87, 89, 90

SOLUTION

1 Find the mean.

$$\frac{80 + 84 + 87 + 89 + 90}{5} = 86$$

2 Find the difference between the mean and each score.

$$86 - 80 = 6$$
$$86 - 84 = 2$$
$$86 - 87 = -1$$
$$86 - 89 = -3$$
$$86 - 90 = -4$$

3 Square the differences.

$$6^2 = 36$$
$$2^2 = 4$$
$$(-1)^2 = 1$$
$$(-3)^2 = 9$$
$$(-4)^2 = 16$$

4 Find the average of the squares.

5 The standard deviation is the square root of the average. Use a calculator.

$$\frac{36 + 4 + 1 + 9 + 16}{5} = 13.2 \quad 13.2 \; \boxed{\sqrt{}} \quad \boxed{3.6331804}$$

The standard deviation is about **3.6.**

Exercises

Find the standard deviation to the nearest tenth.

1. **Science Test Scores**

   75 78 84 91

2. **Heights of Players in Inches**

   76 74 77 70 73

3. **Miles Per Gallon For Six Cars**

   22 30 18 16 25 15

4. **Number of Hits in Six Baseball Games**

   1 1 3 5 6 2

REVIEW CAPSULE FOR SECTION 16-3

Write each fraction in lowest terms. (Pages 124–125)

1. $\frac{4}{6}$
2. $\frac{20}{24}$
3. $\frac{6}{18}$
4. $\frac{54}{60}$
5. $\frac{15}{40}$
6. $\frac{21}{36}$

Write a percent for each fraction. (Pages 227–230)

7. $\frac{1}{4}$
8. $\frac{1}{10}$
9. $\frac{3}{4}$
10. $\frac{4}{5}$
11. $\frac{37}{50}$
12. $\frac{9}{20}$

The answers are on page 494.

16-3 Probability

Jason is one of 20 guests at a party. All the guests write their names on a card and place it in a box. The cards are shuffled and one card is drawn at random to choose a winner for a prize. Since each card is <u>as equally likely</u> to be drawn as any other, Jason can find the *probability* of winning the prize.

PROCEDURE

To find the **probability** of an event with equally likely outcomes, use this ratio.

$$P(\text{probability}) = \frac{\text{Number of Favorable Outcomes}}{\text{Number of Possible Outcomes}}$$

EXAMPLE 1 Find the probability that Jason will win.

Solution: Jason wins only if his card is drawn.

$\frac{\text{Number of Favorable Outcomes}}{\text{Number of Possible Outcomes}} = \frac{1}{20}$ ← Jason's card
← 20 cards in all

The probability that Jason will win the prize is $\frac{1}{20}$.

Self-Check

Each of the 11 letters of the word "MATHEMATICS" is written on a separate card. The cards are placed face down and shuffled. A card is chosen at random. What is the probability that it will show each of the following?

1. The letter M
2. The letter E
3. A vowel

Statistics and Probability 479

You can write a fraction or a percent for a probability.

EXAMPLE 2 Stan and Susan are playing a board game. Susan spins the spinner at the right once during her turn. Find the probability of the spinner stopping on each of the following.

　　a. A three　　　　　　b. A two
　　c. A number greater than three
　　d. A number less than four

Solutions:

a. $\dfrac{\text{Number of Favorable Outcomes}}{\text{Number of Possible Outcomes}} = \dfrac{3}{6} = \dfrac{1}{2}$　　b. $\dfrac{\text{Number of Favorable Outcomes}}{\text{Number of Possible Outcomes}} = \dfrac{2}{6} = \dfrac{1}{3}$

　$P = \dfrac{1}{2}$, or **50%**.　　　　　　　　　　　$P = \dfrac{1}{3}$, or **33$\dfrac{1}{3}$%**.

c. $\dfrac{\text{Number of Favorable Outcomes}}{\text{Number of Possible Outcomes}} = \dfrac{0}{6}$　　d. $\dfrac{\text{Number of Favorable Outcomes}}{\text{Number of Possible Outcomes}} = \dfrac{6}{6}$

　$P = 0$, or **0%**.　　　　　　　　　　　　$P = 1$, or **100%**.

Self-Check *There are 3 red pencils, 2 blue pencils, 1 green pencil, and 4 brown pencils in a desk drawer. A pencil is chosen at random. Find the probability of choosing each of the following.*

　　4. A red pencil　　　**5.** A brown pencil　　　**6.** A purple pencil

The probability of an event that **cannot** happen is **0**.
The probability of an event that is **certain** to happen is **1**.

CLASSROOM EXERCISES

Suppose that there are 12 boys and 18 girls in a mathematics class. The teacher chooses a student at random to read the answers to the homework. Complete the table. (Example 1)

Event	Number of Favorable Outcomes	Number of Possible Outcomes	Probability
1. Choosing a girl	18	30	?
2. Choosing a boy	?	30	?
3. Not choosing a boy	?	?	?
4. Not choosing a girl	?	?	?

480　Chapter 16

There are 4 red marker pens, 3 blue marker pens, 2 brown marker pens, and 1 green marker pen in a box. Felipe chooses a pen at random. Find the probability of each of the following. (Example 2)

5. Choosing a green pen
6. Choosing a brown pen
7. Choosing a red pen
8. Choosing a blue pen
9. Not choosing a red pen
10. Not choosing a green pen
11. Choosing a yellow pen
12. Choosing a colored marker pen

WRITTEN EXERCISES

Goal: To find the probability of an event

These ten names are written on slips of paper and placed in a hat. One name is drawn at random. Find each probability. (Example 1)

| Alice | Jennifer | Ronald | Anthony | Joyce |
| Philip | Anna | Sylvia | Carlos | Michael |

1. Drawing a name that begins with the letter "A"
2. Drawing a name that begins with the letter "J"
3. Drawing a girl's name
4. Drawing a boy's name
5. Drawing a name that ends with the letter "e"
6. Drawing the name Rosa

A coin purse contains two pennies, a nickel, a dime, and a quarter. One coin is drawn at random. Find each probability. Write each answer as a fraction in lowest terms and as a percent. (Example 2)

7. Drawing a dime
8. Drawing a penny
9. Drawing a coin worth more than 1¢
10. Drawing a coin worth more than 5¢
11. Drawing a coin worth less than 25¢
12. Drawing a coin worth more than 25¢

Find the probability of the spinner at the right stopping on each of the following. Write each answer as a fraction in lowest terms and as a percent. (Example 2)

13. An even number
14. A three
15. An odd number
16. A number less than three
17. A number greater than five
18. A number less than nine

Statistics and Probability 481

REVIEW CAPSULE FOR SECTION 16-4

Write each fraction in lowest terms. (Pages 124-125)

1. $\frac{10}{14}$
2. $\frac{6}{10}$
3. $\frac{13}{26}$
4. $\frac{11}{33}$
5. $\frac{15}{45}$
6. $\frac{63}{72}$

Add. (Pages 158-160)

7. $\frac{1}{3} + \frac{1}{3}$
8. $\frac{2}{5} + \frac{2}{5}$
9. $\frac{3}{12} + \frac{4}{12}$
10. $\frac{5}{36} + \frac{2}{36}$
11. $\frac{5}{16} + \frac{7}{16}$
12. $\frac{5}{10} + \frac{3}{10}$

The answers are on page 494.

16-4 Probability and Tables

Eric and Rosa are playing a baseball board game. The player at bat tosses a pair of dice and records the sum of the dice. The play for each sum is given in the table below.

Sum of Two Dice	Play
2	Homerun
3, 4, 5	Single
6, 7, 8	Out
9, 10, 11	Double
12	Triple

Sum of Two Dice

	⚀	⚁	⚂	⚃	⚄	⚅
⚀	2	3	4	5	6	7
⚁	3	4	5	6	7	8
⚂	4	5	6	7	8	9
⚃	5	6	7	8	9	10
⚄	6	7	8	9	10	11
⚅	7	8	9	10	11	12

(Second Die across top, First Die down side)

TABLE 1

Table 1 shows the <u>sum</u> for each of the 36 possible ways that a pair of dice can land. You can get certain sums in more than one way. For example, there are 4 ways to get a sum of 9.

6 and 3   5 and 4   4 and 5   3 and 6

EXAMPLE 1 Rosa tosses the pair of dice. Use Table 1 to find the probability that the sum is 4 (a single). Write a fraction for the probability.

Solution: There are 3 ways to get a sum of 4.

$$P(\text{sum of } 4) = \frac{\text{Number of Favorable Outcomes}}{\text{Number of Possible Outcomes}} = \frac{3}{36} = \frac{1}{12}$$

Thus, the probability that the sum is 4 is $\frac{1}{12}$.

Chapter 16

Self-Check Use Table 1 to find each probability when a pair of dice is tossed. Write each answer as a fraction in lowest terms.

1. Getting a sum of 5 **2.** Getting a sum of 12

Events that cannot happen at the same time are said to be **mutually exclusive.**

EXAMPLE 2 Mark tosses a pair of dice once. Use Table 1 to find the probability that the sum is 2 or 12 (homerun or triple). Write a fraction for the probability.

Solution: *Think:* Ways of getting a sum of 2: 1
Ways of getting a sum of 12: 1

P(sum of 2 or 12) = P(sum of 2) + P(sum of 12) ◀ **Sum of the probabilities**

$$= \frac{1}{36} + \frac{1}{36} = \frac{2}{36}, \text{ or } \frac{1}{18}$$

Self-Check A pair of dice is tossed. Find each probability. Write each answer as a fraction in lowest terms.

3. Getting a sum of 2 or 11 **4.** Getting a sum of 3 or 5

CLASSROOM EXERCISES

A pair of dice is tossed. Refer to Table 1 on page 482 to write the number of favorable outcomes for each given sum of the two dice. (Example 1)

1. 1 **2.** 2 **3.** 3 **4.** 4 **5.** 5 **6.** 6
7. 7 **8.** 8 **9.** 9 **10.** 10 **11.** 11 **12.** 12

A pair of dice is tossed. Refer to Table 1 on page 482 to find each probability. Write each answer as a fraction in lowest terms. (Example 2)

13. P(sum of 2 or 4) = $\frac{1}{36} + \frac{?}{36} =$ ___?___   **14.** P(sum of 3 or 8) = $\frac{2}{36} + \frac{?}{36} =$ ___?___

15. P(sum of 10 or 11) = $\frac{?}{36} + \frac{?}{36} =$ ___?___   **16.** P(sum of 6 or 12) = $\frac{?}{36} + \frac{?}{36} =$ ___?___

17. P(sum of 2 or 3 or 4) = $\frac{1}{36} + \frac{?}{36} + \frac{?}{36} =$ ___?___

18. P(sum of 10 or 11 or 12) = $\frac{?}{36} + \frac{?}{36} + \frac{?}{36} =$ ___?___

Statistics and Probability

WRITTEN EXERCISES

Goals: To use a table to find the probability of an event
To find the probability of mutually exclusive events

A pair of dice is tossed once. Use Table 1 on page 482 to find each probability. Write each answer as a fraction in lowest terms.
(Example 1)

1. What is the probability of getting a sum of 3?
2. What is the probability of getting a sum of 6?
3. What is the probability of getting a sum of 7?
4. What is the probability of getting a sum of 9?
5. What is the probability of getting a sum of 10?
6. What is the probability of getting a sum of 8?

(Example 2)

7. What is the probability of getting a sum of 2 or 9?
8. What is the probability of getting a sum of 10 or 11?
9. What is the probability of getting a sum of 5 or 6?
10. What is the probability of getting a sum of 4 or 5?
11. What is the probability of getting a sum of 3 or 7?
12. What is the probability of getting a sum of 7 or 8?

MORE CHALLENGING EXERCISES

13. What is the probability of getting a single?
14. What is the probability of getting an out?
15. What is the probability of getting a double or a triple?
16. What is the probability of not getting an out?

REVIEW CAPSULE FOR SECTION 16-5

Write each fraction in lowest terms. (Pages 124–125)

1. $\frac{4}{36}$
2. $\frac{12}{16}$
3. $\frac{24}{60}$
4. $\frac{18}{64}$
5. $\frac{18}{32}$
6. $\frac{15}{20}$

Write a percent for each fraction. (Pages 227–230)

7. $\frac{3}{5}$
8. $\frac{1}{4}$
9. $\frac{2}{3}$
10. $\frac{1}{6}$
11. $\frac{7}{8}$
12. $\frac{1}{12}$

The answers are on page 494.

16-5 Tree Diagrams

There are two empty seats next to each other on a school bus. Two boys and two girls get on the bus. You can draw a **tree diagram** to show the number of ways the students can be seated.

CHOOSING TWO SEATS

First Seat	Second Seat	Possible Outcomes
G	G	GG
	B	GB
	B	GB
B	B	BB
	G	BG
	G	BG

PROCEDURE

To use a tree diagram to find the probability of an event:

1. Draw a tree diagram that shows all the possible ways the event can happen.
2. Use the tree diagram to find the probability.

EXAMPLE Two girls and two boys get on a school bus. There are only two empty seats next to each other. Refer to the tree diagram above to find the probability that a girl sits in the first seat and a boy sits in the second seat.

Solution: Think: There are six possible outcomes. In two of the six outcomes, a girl sits in the first seat and a boy sits in the second seat.

$$P(G, B) = \frac{\text{Number of Favorable Outcomes}}{\text{Number of Possible Outcomes}} = \frac{2}{6} = \frac{1}{3}, \text{ or } 33\frac{1}{3}\%$$

Self-Check 1. What is the probability that two boys will sit in the empty seats?

Statistics and Probability

CLASSROOM EXERCISES

For Exercises 1–4, refer to the tree diagram on page 485 to find each probability. Write a fraction in lowest terms and a percent for each probability. (Example)

1. What is the probability that a boy sits in the first seat and a girl sits in the second?

2. What is the probability that a boy and girl sit next to each other?

3. What is the probability that both seats are occupied by girls?

4. What is the probability that both seats are occupied by boys?

WRITTEN EXERCISES

Goals: To draw a tree diagram
To use tree diagrams to find probabilities

For a trip from Miami to Chicago, Acme Tours offers a choice of three routes (Coastal, Central, or Overland) and two routes (Direct and Scenic) from Chicago to San Francisco.

1. Complete this tree diagram to show all possible choices of a route.

CHOOSING A ROUTE

Miami–Chicago	Chicago–San Francisco	Possible Outcomes
Coastal	Direct	Coastal, ?
	Scenic	? , Scenic
Central	Direct	? , Direct
	Scenic	Central, ?
Overland	Direct	Overland, ?
	Scenic	? , ?

Miami →

For Exercises 2–7, refer to the tree diagram above to find each probability. Write a fraction in lowest terms and a percent for each probability. (Example)

2. What is the probability of choosing the coastal scenic route?

3. What is the probability of choosing the overland direct route?

486 Chapter 16

4. What is the probability of choosing a scenic route?

5. What is the probability of choosing a central route?

6. What is the probability of choosing a coastal or central route?

7. What is the probability of choosing the coastal direct or the overland scenic route?

MORE CHALLENGING EXERCISES

On a certain day, the menu at a school cafeteria offers a choice of two main courses – chicken or fish. Each of three students chooses one main course.

Lunch Menu
— Main Courses —
CHICKEN
BROILED FISH

8. Draw a tree diagram that shows all the possible ways in which each of three students can order a main course.

For Exercises 9–14, refer to the tree diagram of Exercise 8 to find each probability. Write a fraction in lowest terms and a percent for each probability. (Example)

9. What is the probability that each student chooses chicken?

10. What is the probability that exactly two students choose fish?

11. What is the probability that only one student chooses chicken?

12. What is the probability that all three students chose fish?

13. What is the probability that at least two students choose chicken?

14. What is the probability that the first two students each choose fish?

REVIEW CAPSULE FOR SECTION 16–6

Write a percent for each fraction. (Pages 227–230)

1. $\frac{3}{25}$ 2. $\frac{13}{50}$ 3. $\frac{5}{40}$ 4. $\frac{8}{40}$ 5. $\frac{8}{8}$ 6. $\frac{0}{9}$

Multiply. (Pages 129–131)

7. $\frac{1}{4} \times \frac{1}{4}$ 8. $\frac{1}{8} \times \frac{1}{8}$ 9. $\frac{1}{3} \times \frac{2}{5}$ 10. $\frac{3}{4} \times \frac{5}{8}$

11. $\frac{1}{2} \times \frac{6}{7}$ 12. $\frac{3}{5} \times \frac{5}{6}$ 13. $\frac{1}{3} \times \frac{2}{3}$ 14. $\frac{3}{8} \times \frac{3}{4}$

The answers are on page 494.

16-6 Multiplying Probabilities

A student from each class at Smalltown High School is chosen by a random drawing for a community internship program. One name is drawn from each of four boxes labeled by class.

Smalltown High School

Class	Number of Boys	Number of Girls
Freshman	35	35
Sophomore	27	33
Junior	24	32
Senior	20	22

Since the outcome of the draw from one box **does not** affect the outcome of the draw from any other box, the events are **independent.**

You can find the probability of two or more independent events occurring by <u>multiplying</u> the probabilities of each event.

EXAMPLE 1 Find the probability of choosing a freshman boy <u>and</u> a junior girl.

Solution: P(freshman boy <u>and</u> junior girl) = P(freshman boy) × P(junior girl)

P(freshman boy) = $\frac{35}{70}$, or $\frac{1}{2}$ P(junior girl) = $\frac{32}{56} = \frac{4}{7}$

P(freshman boy <u>and</u> junior girl) = $\frac{1}{2} \times \frac{4}{7} = \frac{4}{14}$, or $\frac{2}{7}$

Self-Check 1. Find the probability of choosing a sophomore girl <u>and</u> a senior boy.

Sometimes what happens in the first event **does affect** what happens in following events. Events which affect each other are called **dependent events.**

EXAMPLE 2 A drawer contains 6 black socks and 4 brown socks. Two socks are drawn at random, one after the other, without replacement. Find the probability of drawing two black socks in a row.

Solution: P(black, then black) = P(black) × P(black)

First draw: P(black) = $\frac{6}{10}$, or $\frac{3}{5}$ ◀ The first sock is not replaced.

Second draw: P(black) = $\frac{5}{9}$ ◀ There are now 5 black socks out of 9 socks.

P(black, then black) = $\frac{3}{5} \times \frac{5}{9} = \frac{1}{3}$ ◀ The probability is $\frac{1}{3}$.

Self-Check 2. Find the probability of drawing two brown socks in a row without replacement (see Example 2).

CLASSROOM EXERCISES

Refer to the table on page 488 to find each probability. Write each answer as a fraction in lowest terms. (Example 1)

1. P(freshman girl and junior boy)
   P(freshman girl) = __?__
   P(junior boy) = __?__
   P(freshman girl and junior boy) = __?__

2. P(sophomore boy and senior girl)
   P(sophomore boy) = __?__
   P(senior girl) = __?__
   P(sophomore boy and senior girl) = __?__

A drawer contains 8 black socks, 6 grey socks, and 2 white socks. Two socks are drawn at random, one after the other, without replacement. Complete to find each probability. (Example 2)

3. P(black, then black)
   First draw: P(black) = __?__
   Second draw: P(black) = __?__
   P(black, then black) = __?__

4. P(grey, then white)
   First draw: P(grey) = __?__
   Second draw: P(white) = __?__
   P(grey, then white) =

WRITTEN EXERCISES

Goals: To find the probability of two or more independent events
To find the probability of two or more dependent events

Connie has six scarves. Three are striped, one is blue, and two are white. She also has ten blouses. Three are blue, two are yellow, and five are plaid. She chooses a scarf and a blouse at random. For Exercises 1–6, find each probability. Write each probability as a fraction in lowest terms. (Example 1)

1. P(white scarf and plaid blouse)
2. P(blue scarf and yellow blouse)
3. P(striped scarf and blue blouse)
4. P(striped scarf and plaid blouse)
5. P(blue scarf and blue blouse)
6. P(white scarf and yellow blouse)

Statistics and Probability

A box has 4 quarters, 3 dimes, 2 nickels, and 1 penny. Two coins are drawn at random, one after the other, without replacement. Write each probability as a fraction in lowest terms. (Example 2)

7. P(quarter, then quarter)
8. P(dime, then dime)
9. P(nickel, then penny)
10. P(penny, then dime)
11. P(dime, then quarter)
12. P(quarter, then nickel)

MORE CHALLENGING EXERCISES

In a bag there are 4 blue marbles, 3 yellow marbles, 2 white marbles, and 1 black marble. For Exercises 13–16, three marbles are drawn at random, one after the other, with replacement. Find each probability. Write each probability as a fraction in lowest terms.

13. P(blue and blue and yellow)
14. P(yellow and yellow and white)
15. P(blue and white and black)
16. P(yellow and blue and black)

For Exercises 17–20, three marbles are drawn at random, one after the other, without replacement. Find each probability. Write each probability as a fraction in lowest terms.

17. P(blue, then blue, then yellow)
18. P(yellow, then yellow, then white)
19. P(blue, then white, then black)
20. P(yellow, then blue, then black)

MULTIPLYING PROBABILITIES

A calculator with the (and) keys enables you to multiply two or more probabilities written as fractions to obtain a probability written as a percent.

EXAMPLE Multiply: $\frac{7}{14} \cdot \frac{3}{13} \cdot \frac{5}{12}$ ▶ **The probabilities for 3 dependent events occurring in a row**

Solution

(7 × 3 × 5) ÷ (1 4 × 1 3 × 1 2) = × 1 0 0 =   *4.8076923*

The answer rounded to the nearest tenth is **4.8%**.

NOTE: Pressing the) key completes the multiplication.

EXERCISES *Multiply. Round each answer to the nearest tenth.*

1. $\frac{5}{14} \cdot \frac{4}{13} \cdot \frac{1}{12}$
2. $\frac{9}{10} \cdot \frac{5}{9} \cdot \frac{2}{8}$
3. $\frac{4}{7} \cdot \frac{5}{6} \cdot \frac{3}{5}$
4. $\frac{8}{13} \cdot \frac{2}{12} \cdot \frac{6}{11}$
5. $\frac{11}{15} \cdot \frac{9}{14} \cdot \frac{7}{13}$
6. $\frac{15}{17} \cdot \frac{12}{16} \cdot \frac{4}{15}$

REVIEW: SECTIONS 16-3 – 16-6

Each of the 6 letters of the word "CHANCE" is written on a separate card. The cards are placed face down and shuffled. A card is chosen at random. What is the probability that it will contain each of the following? (Section 16-3)

1. The letter N?
2. The letter C?
3. A consonant?
4. A vowel?

A pair of dice is tossed once. For Exercises 5-8, refer to the table on page 482 to find each probability. Write each probability as a fraction in lowest terms. (Section 16-4)

5. What is the probability of getting a sum of 11?
6. What is the probability of getting a sum of 7?
7. What is the probability of getting a sum of 4 or 8?
8. What is the probability of getting a sum of 9 or 12?

9. Two coins are tossed at the same time. To find all the possible ways the coins can land, copy and complete the tree diagram below. (Section 16-5)

TOSSING TWO COINS

First Coin	Second Coin	Possible Outcomes
H	H	H H
	?	? ?
T	?	? ?
	?	? ?

For Exercises 10-13, refer to the tree diagram in Exercise 9 to find each probability. Write a fraction in lowest terms and a percent for each probability. (Section 16-5)

10. Getting two heads
11. Getting two tails
12. Getting a head and a tail
13. Getting a tail on the second coin

A box contains 12 eggs, two of which are cracked. (Section 16-6)

14. One egg is picked from the box and replaced. Then another egg is picked. Find P(good egg, then cracked egg.)
15. One egg is picked from the box and not replaced. Another egg is picked. Find P(good egg, then good egg).

Statistics and Probability   491

CHAPTER REVIEW

PART 1: VOCABULARY

For Exercises 1-9, choose from the box at the right below the word(s) or number(s) that best corresponds to each description.

> mean
> mode
> median
> data
> 0
> 1
> mutually exclusive events
> independent events
> dependent events

1. The middle number in a list of data arranged in order of size __?__ (p. 470)
2. Another word for information __?__ (p. 470)
3. The probability of an event that cannot happen __?__ (p. 480)
4. Events which affect each other __?__ (p. 488)
5. The quotient of the sum of the given measures and the number of measures __?__ (p. 470)
6. Events that cannot happen at the same time __?__ (p. 483)
7. Events which do not affect each other __?__ (p. 488)
8. The probability of an event that is certain to happen __?__ (p. 480)
9. The number that occurs most often in a set of data __?__ (p. 470)

PART 2: SKILLS AND APPLICATIONS

For Exercises 10-11, find the mean, median, and mode. (Section 16-1)

10. **Cost of Lettuce at Seven Stores**

    $.80 $.59 $.68 $.80
    $.96 $.97 $.59

11. **Miles Driven Per Day on a Trip**

    250   375   132   200
    132   400   182   225

12. Make a histogram to show the number of words typed per minute by 20 students. Use the data in the table. (Section 16-2)

Number of Words	26-34	35-43	44-52	53-61
Count	5	6	5	4

Each of the 16 letters of the words "MASSACHUSETTS BAY" is written on a separate card. A card is chosen at random. Find each probability. Write each probability as a fraction in lowest terms and as a percent. (Section 16-3)

13. What is the probability of drawing an M?
14. What is the probability of drawing an A?
15. What is the probability of drawing an S?
16. What is the probability of drawing a T?

This table shows the possible outcomes for a game. In the game, each player spins the arrow twice and adds the scores.

	Second Spin				
+	2	4	6	8	10
2	4	6	8	10	12
4	6	8	10	12	14
6	8	10	12	14	16
8	10	12	14	16	18
10	12	14	16	18	20

(First Spin labels rows)

For Exercises 17–22, refer to the table above to find each probability. Write each answer as a fraction in lowest terms. (Section 16-4)

17. What is the probability of getting a sum of 6?

18. What is the probability of getting a sum of 10?

19. What is the probability of getting a sum of 20?

20. What is the probability of getting a sum of 12?

21. What is the probability of getting a sum of 6 or 8?

22. What is the probability of getting a sum of 10 or 12?

23. Three coins are tossed at the same time. Draw a tree diagram that shows all the possible ways for three coins to land. (Section 16-5)

For Exercises 24–27, refer to the tree diagram of Exercise 23 to find each probability. Write each probability as a fraction in lowest terms and as a percent. (Section 16-5).

24. What is the probability of getting two tails and a head?

25. What is the probability of getting three heads?

26. What is the probability of getting three tails?

27. What is the probability of getting a tail on the first coin?

A bag contains 3 green marbles, 5 white marbles, and 7 blue marbles. For Exercises 28–30, one marble is drawn and replaced. Then a second marble is drawn. Find each probability. Write each probability as a fraction in lowest terms and as a percent. (Section 16-6)

28. P(green, then green)   **29.** P(green, then white)   **30.** P(white, then blue)

For Exercises 31–33, one marble is drawn and not replaced. Then a second marble is drawn. Find each probability. Write each probability as a fraction in lowest terms and as a percent. (Section 16-6)

31. P(green, then green)   **32.** P(green, then white)   **33.** P(white, then blue)

Statistics and Probability

CHAPTER TEST

1. Find the mean, median, and mode.

   Number of Weeks on the Chart for the Top Ten Records

14	6	8	14	15
9	7	3	7	7

2. Draw a histogram to show this data.

   Heights in Meters of Ten Trees

23	14	17	25	16
23	19	18	12	11

   Intervals: 11–14; 15–18; 19–22; 23–26

Each of the eleven letters of the word "MISSISSIPPI" is written on a separate card. The cards are placed face down and shuffled. A card is chosen at random. What is the probability that the card will show each of the following?

3. The letter S
4. A vowel
5. Not the letter S

6. There are two Yes–No questions on a test. Draw a tree diagram to show all the possible ways to answer the questions.

A box contains twenty light bulbs, four of which are broken. One light bulb is drawn from the box. Then a second bulb is drawn. For Exercises 7–8, find each probability if the first light bulb is replaced. Write each answer as a fraction in lowest terms.

7. What is the probability of drawing a broken bulb and a good bulb?
8. What is the probability of drawing a good bulb and a good bulb?

For Exercises 9–10, find each probability if the first light bulb is not replaced. Write each answer as a fraction in lowest terms.

9. What is the probability of drawing a broken bulb, then a good bulb?
10. What is the probability of drawing a good bulb, then a good bulb?

ANSWERS TO REVIEW CAPSULES

Page 473 1. Let 10 cm represent the scale from 2000 to 2100. Jan: 4.5 cm; Feb: 5.5 cm; Mar: 7.0 cm; Apr: 7.0 cm; May: 6.0 cm; Jun: 8.0 cm

Page 479 1. $\frac{2}{3}$ 2. $\frac{5}{6}$ 3. $\frac{1}{3}$ 4. $\frac{9}{10}$ 5. $\frac{3}{8}$ 6. $\frac{7}{12}$ 7. 25% 8. 10% 9. 75% 10. 80% 11. 74% 12. 45%

Page 482 1. $\frac{5}{7}$ 2. $\frac{3}{5}$ 3. $\frac{1}{2}$ 4. $\frac{1}{3}$ 5. $\frac{1}{3}$ 6. $\frac{7}{8}$ 7. $\frac{2}{3}$ 8. $\frac{4}{5}$ 9. $\frac{7}{12}$ 10. $\frac{7}{36}$ 11. $\frac{3}{4}$ 12. $\frac{4}{5}$

Page 484 1. $\frac{1}{9}$ 2. $\frac{3}{4}$ 3. $\frac{2}{5}$ 4. $\frac{9}{32}$ 5. $\frac{9}{16}$ 6. $\frac{3}{4}$ 7. 60% 8. 25% 9. $66\frac{2}{3}$% 10. $16\frac{2}{3}$% 11. $87\frac{1}{2}$% 12. $8\frac{1}{3}$%

Page 487 1. 12% 2. 26% 3. $12\frac{1}{2}$% 4. 20% 5. 100% 6. 0% 7. $\frac{1}{16}$ 8. $\frac{1}{64}$ 9. $\frac{2}{15}$ 10. $\frac{15}{32}$ 11. $\frac{3}{7}$ 12. $\frac{1}{2}$ 13. $\frac{2}{9}$ 14. $\frac{9}{32}$

ENRICHMENT

Sampling

To find how many defective batteries there are in a large shipment, a quality control technician inspects and tests a certain number of batteries. The batteries that are actually tested are called a **sample**.

EXAMPLE A technician found that 6 out of every 120 batteries tested were defective. Find the probable number of defective batteries in a shipment of 80,000.

Solution:

1. Find the percent of defective batteries in the sample.

    6 out of 120 = $\frac{6}{120}$ = $\frac{1}{20}$ = 0.05 = **5%** ◀ **5% are defective.**

2. Since 5% of the batteries in the sample are defective, it is probable that 5% of the batteries in the shipment are also defective. Find 5% of 80,000.

    5% of 80,000 = 0.05 × 80,000 = **4000** ◀ **Probable number of defective batteries**

EXERCISES

For Exercises 1–4, complete the table.

Product	Defective Items in Sample	Percent of Defective Items in Sample	Total Number of Items in Shipment	Probable Number of Defective Items in Shipment
1. Calculators	4 out of 100	?	800	?
2. Light bulbs	6 out of 100	?	13,400	?
3. Stereos	2 out of 50	?	4000	?
4. Watches	3 out of 60	?	6800	?

5. A sample of 400 cameras contains 20 that are defective. Find the probable number of defective cameras in a shipment of 26,000.

6. In a shipment of 950 radios, the probability that any one is defective is 2%. Find the probable number of defective radios in the shipment.

Statistics and Probability

ADDITIONAL PRACTICE

SKILLS

1. Find the mean, median, and mode. (Pages 470–473)

Number of Passengers in Eight Taxis

6	5	3	6
5	4	3	3

2. Construct a histogram to show the data. Use the given intervals. (Pages 474–477)

Miles Per Day on a Bicycle Trip

16	30	33	34	37
34	39	26	21	38
32	36	35	24	33

Intervals: 16–20; 21–25; 26–30; 31–35; 36–40

A box contains ten neckties. Three are green, two are white, one is black, and four are blue. One necktie is chosen at random.

Find each probability. Write each answer as a fraction in lowest terms and as a percent. (Pages 479–481)

3. Choosing a blue necktie

4. Choosing a green necktie

5. Not choosing a white necktie

6. Not choosing a black necktie

For Exercises 7–10, refer to Table 1 on page 482 to find each probability. Write each answer as a fraction in lowest terms. (Pages 482–484)

7. Getting a sum of 5 <u>or</u> 10

8. Getting a sum of 4 <u>or</u> 8

9. Getting a sum of 1 <u>or</u> 7

10. Getting a sum of 2 <u>or</u> 12

11. Four coins are tossed at the same time. Draw a tree diagram to find all possible ways the coins can land. (Pages 485–487)

APPLICATIONS

12. A basket contains 8 oranges and 2 grapefruit. One piece of fruit is picked from the basket and replaced. Then another piece of fruit is picked. What is the probability of picking one orange, then one grapefruit? (Pages 488–490)

13. In the basket of fruit for Exercise 12, one piece of fruit is picked and not replaced. Then another piece of fruit is picked. What is the probability of picking one orange, then one orange? (Pages 488–490)

CHAPTER 17 Geometry

SECTIONS
- **17-1** Introduction to Geometry
- **17-2** Angles
- **17-3** Properties of Triangles
- **17-4** Problem Solving and Applications: Using Similar Triangles
- **17-5** Perpendicular and Parallel Lines
- **17-6** Properties of Quadrilaterals
- **17-7** Problem Solving and Applications: Surface Area and Rectangular Prisms
- **17-8** Volume: Cylinders
- **17-9** Volume: Pyramids and Cones

FEATURES
Calculator Application: *Similar Polygons*
Computer Application: *Angle Measures*
Enrichment: *Spheres*
Common Errors

17-1 Introduction to Geometry

Geometry is the study of points, lines, planes, and the figures they form. This table shows some basic geometric figures.

TABLE

Figure	Name	Symbol	Description
•A	Point A	A	Shows location. All points are named with capital letters.
A———B	Line AB	$\overleftrightarrow{AB}$	Extends indefinitely in opposite directions.
A———B	Line segment AB	$\overline{AB}$	Part of a line that is named by its endpoints.
A———B	Ray AB (In this figure, A is the endpoint)	$\overrightarrow{AB}$	Extends indefinitely in one direction from its endpoint.
Angle with vertex B, rays to A and C	Angle ABC, angle CBA, or angle B (In this figure, B is the vertex)	$\angle ABC$, $\angle CBA$, or $\angle B$	A pair of rays that have a common endpoint called the vertex. The vertex (plural: vertices) is the middle point named.
Parallelogram ABCD labeled I	Plane **I**, or Plane ABCD	Plane **I**, or Plane ABCD	A flat surface that extends indefinitely in all directions. A plane is represented by a four-sided figure. Points, lines, rays, and angles all lie in planes.

Use the symbols to name lines, line segments, rays, and angles.

EXAMPLE Use symbols and the given line to name each of the following.

    a. Three different segments   **b.** Three different rays
    c. Ray AB in two other ways   **d.** The line in three ways

    ←——•——————•——————•—•——→
        A   B   C D

Solutions: **a.** $\overline{AB}$, $\overline{AC}$, $\overline{BC}$ **b.** $\overrightarrow{AD}$, $\overrightarrow{DA}$, $\overrightarrow{BD}$
               c. $\overrightarrow{AC}$, $\overrightarrow{AD}$ **d.** $\overleftrightarrow{AD}$, $\overleftrightarrow{BC}$, $\overleftrightarrow{BD}$

Self-Check 1. Use symbols and line AB in the Example to name two different segments and two different rays not already named.

NOTE: $\overline{AD}$ and $\overline{DA}$ are the same segment. However, $\overrightarrow{AD}$ and $\overrightarrow{DA}$ are different rays.

CLASSROOM EXERCISES

Replace the __?__ with the missing symbol or name. (Table)

1. Name: Ray TS
   Symbol: __?__

2. Name: __?__
   Symbol: $\overleftrightarrow{MN}$

3. Name: __?__
   Symbol: $\angle DEF$

4. Name: Point X
   Symbol: __?__

5. Name: Line segment QR
   Symbol: __?__

6. Name: Angle H
   Symbol: __?__

For Exercises 7–16, refer to lines MR and HK which intersect at E. Complete each statement. (Table)

7. Point E is between points M and __?__.
8. Point E is between points H and __?__.
9. Point E is on lines __?__ and __?__.
10. Line segment EM has endpoints __?__ and __?__.
11. Line EM has __?__ endpoints.
12. The endpoint of $\overrightarrow{HE}$ is __?__.
13. Rays EH and ER form angle __?__.
14. Rays EM and EK form angle __?__.
15. Points M, E, and R are on the __?__ line.
16. Points K, E, and H are on the __?__ line.

Geometry 499

WRITTEN EXERCISES

Goals: To identify points, lines, segments, rays, angles, and planes

Match one or more names in Column A with each figure in Exercises 1–6. (Table)

COLUMN A

a. Plane II
b. $\overline{RS}$
c. $\overrightarrow{TV}$
d. $\angle HBJ$
e. $\overrightarrow{WX}$
f. $\overleftrightarrow{PQ}$
g. $\angle B$
h. $\overline{SR}$
i. $\overleftrightarrow{WY}$
j. Plane $DEFG$
k. $\angle JBH$
l. $\overleftrightarrow{QP}$

For Exercises 7–14, refer to the figure at the right. (Example)

7. Name line AF in three different ways.
8. Name four rays that have B as an endpoint.
9. Name $\overrightarrow{BE}$ in a different way.
10. Name three different line segments on line HE.
11. Name angle HGI in one other way.
12. Name four angles with G as a vertex.
13. Name two lines that contain point G.
14. Name the point where $\overleftrightarrow{IG}$ and $\overleftrightarrow{FE}$ intersect.

REVIEW CAPSULE FOR SECTION 17–2

Replace the __?__ with > or <. (Pages 294–296)

1. 32 __?__ 90
2. 110 __?__ 90
3. 90 __?__ 60
4. 90 __?__ 15
5. 160 __?__ 90
6. 25 __?__ 90
7. 90 __?__ 89
8. 90 __?__ 179

The answers are on page 533.

17-2 Angles

A ski jumper can increase the length of a jump by keeping the skis at the proper angle. If the angle is either too large or too small, the distance covered by the jump will be short.

The degree (symbol: "°") is the standard unit for measuring an angle. You can use a protractor to measure an angle.

PROCEDURE

To use a protractor to measure an angle:

1. Place the center of the base of the protractor at the vertex of the angle.

2. Place the protractor so that one ray of the angle passes through "0" on the protractor.

3. Read the measure of the angle from the second ray.

EXAMPLE 1 Use a protractor to find the measure of the given angles.

Solutions: a. ∠CBD: 30°   b. ∠CBE: 90°   c. ∠CBF: 130°   d. ∠ABD: 150°

Self-Check Use the protractor in Example 1 to find the measure of each angle.

1. Angle CBH
2. Angle CBG
3. Angle ABH

Geometry 501

Angles can be classified according to their measures.

TABLE

Right Angle	Acute Angle	Obtuse Angle	Straight Angle
Equals 90°	Less than 90°	Greater than 90°	Equals 180°

This symbol means that the angle is a right angle.

A protractor can also be used to draw angles.

EXAMPLE 2 Draw ∠RST so that its measure is 150.

Solution:

1. Label a point S and a point T. Then draw $\vec{ST}$. Place the center of the protractor at S.

2. Use the scale that begins with 0 on $\vec{ST}$. Find the 150 mark and label point R. Draw $\vec{SR}$.

The measure of ∠**RST** is **150°**.

Self-Check *Use a protractor to draw angles having the given measures.*

4. 45°   5. 90°   6. 115°   7. 135°

CLASSROOM EXERCISES

For Exercises 1–6, choose the angle that is closest to the given measure. Choose a or b. (Example 1)

1. 30°
   a.
   b.

2. 120°
   a.
   b.

3. 60°
   a.
   b.

502   Chapter 17

4. 170°
a.
b.

5. 90°
a.
b.

6. 45°
a.
b.

WRITTEN EXERCISES

Goals: To use a protractor to measure a given angle
To classify an angle according to its measure
To use a protractor to draw a given angle

For Exercises 1–9:

a. Use the protractor to find the measure of each angle.

b. Identify each angle as right, acute, obtuse, or straight.
(Example 1 and Table)

1. $\angle CBD$ 2. $\angle CBE$ 3. $\angle CBF$
4. $\angle CBG$ 5. $\angle CBH$ 6. $\angle CBA$
7. $\angle ABH$ 8. $\angle ABD$ 9. $\angle EBA$

For each of Exercises 10–19, use a protractor to draw an angle with the given measure. (Example 2)

10. 32°   11. 47°   12. 85°   13. 90°   14. 55°
15. 105° 16. 122° 17. 137° 18. 155° 19. 175°

MORE CHALLENGING EXERCISES

Use the protractor above to find the measure of each angle.

20. $\angle DBE$ 21. $\angle FBG$ 22. $\angle EBF$ 23. $\angle GBH$ 24. $\angle DBF$

Geometry **503**

17-3 Properties of Triangles

The figures formed by the wood beams on the gable at the right are **closed plane figures**. Closed plane figures formed by line segments are called **polygons**.

A **triangle** is a polygon with three sides and three angles. The gable at the right is in the shape of a triangle. The triangle formed by the gable can be named as follows.

Gable

△ABC, △BCA, or △CAB

Read: "Triangle ABC, triangle BCA, or triangle CAB."

Triangles can be classified by the lengths of their sides.

Sides with the same number of marks have the same length.

TABLE 1

Triangle	Name	Description
△ABC	Equilateral	All sides have the same length.
△DEF	Isosceles	Two sides have the same length.
△GHT	Scalene	No sides have the same length.

Triangles can also be classified by the measures of their angles.

TABLE 2

Right Triangle	Acute Triangle	Obtuse Triangle
One angle is a right angle.	All three angles are acute angles.	One angle is an obtuse angle.

504 Chapter 17

Congruent triangles are triangles that have the same shape and size.

Triangles ABC and DEF shown on the gate at the right are congruent triangles.

In symbols,
△ABC ≅ △DEF

Read: "△**ABC** is congruent to △**DEF**."

In congruent triangles, **corresponding angles** have the same measure and **corresponding sides** have the same length. Note that corresponding congruent sides are **opposite** corresponding congruent angles.

TABLE 3

Corresponding Congruent Angles	Corresponding Congruent Sides
∠A ≅ ∠D	$\overline{BC} \cong \overline{FE}$
∠B ≅ ∠E	$\overline{AC} \cong \overline{FD}$
∠C ≅ ∠F	$\overline{AB} \cong \overline{DE}$

To show that two triangles are congruent, it is *not* necessary to know that all six corresponding parts are congruent.

Side–Angle–Side (SAS)

If two corresponding sides and their included angles are congruent, then the triangles are congruent.

Side–Side–Side (SSS)

If three corresponding sides are congruent, then the triangles are congruent.

Angle–Side–Angle (ASA)

If two corresponding angles and their included side are congruent, then the triangles are congruent.

NOTE: Tick marks show corresponding congruent parts.

Geometry

EXAMPLE 1 *Determine whether each pair of triangles is congruent by SAS, SSS, or ASA. Give a reason for each answer.*

Solutions: a. SSS, because $\overline{AB} \cong \overline{ED}$, $\overline{BC} \cong \overline{EF}$, and $\overline{CA} \cong \overline{FD}$.

b. SAS, because $\overline{KP} \cong \overline{HR}$, $\angle P \cong \angle R$, and $\overline{PJ} \cong \overline{RG}$.

c. ASA, because $\angle A \cong \angle M$, $\overline{AH} \cong \overline{MB}$, and $\angle H \cong \angle B$.

Self-Check *Draw a pair of triangles to illustrate each kind of congruence. Mark the corresponding congruent parts.*

1. SSS   2. SAS   3. ASA

CLASSROOM EXERCISES

The sides of a triangle have the lengths shown. Classify each triangle as equilateral, isosceles, or scalene. (Table 1)

1. 4 in; 5 in; 4 in
2. 6 cm; 6 cm; 6 cm
3. 7 in; 9 in; 12 in
4. 8 cm; 10 cm; 8 cm

The angles of a triangle have the measures shown. Classify each triangle as acute, right, or obtuse. (Table 2)

5. 89°; 43°; 48°
6. 25°; 20°; 135°
7. 30°; 60°; 90°
8. 60°; 60°; 60°

For Exercises 9–14, it is given that triangles RST and MPQ are congruent. Complete each statement. (Table 3)

9. $\angle R \cong$ ___?___
10. $\overline{RS} \cong$ ___?___
11. $\overline{ST} \cong$ ___?___
12. $\angle S \cong$ ___?___
13. ___?___ $\cong \angle Q$
14. ___?___ $\cong \overline{QM}$

506   Chapter 17

WRITTEN EXERCISES

Goals: To classify triangles according to the lengths of their sides
To classify triangles according to the measures of their angles
To determine whether triangles are congruent by SSS, SAS, or ASA

Classify each triangle as equilateral, isosceles, or scalene. (Table 1)

1. Triangle CED with sides 8, 8, 8

2. Triangle TRK with angles 92, 57, 92

3. Triangle SWY with sides 15, 15, 22

4. Triangle NMP with sides 15, 8, 17

Determine whether each pair of triangles is congruent by SSS, SAS, or ASA. Name the corresponding congruent angles and sides. (Example 1)

5. Coat Hangers

6. Bicycle Frames

7. Tent Flaps

8. Antenna

9. Lean-to

10. Table Legs

Geometry **507**

REVIEW CAPSULE FOR SECTION 17-4

Solve each proportion. (Pages 219–221)

1. $\dfrac{n}{18} = \dfrac{15}{20}$ 2. $\dfrac{10}{9} \times \dfrac{40}{x}$ 3. $\dfrac{16}{p} = \dfrac{4}{2}$ 4. $\dfrac{a}{12} = \dfrac{2}{1}$ 5. $\dfrac{11}{8} = \dfrac{b}{24}$ 6. $\dfrac{25}{t} = \dfrac{5}{3}$

Round each number to the nearest tenth. (Pages 32–34)

7. 9.83 8. 11.88 9. 11.58 10. 12.85 11. 9.05 12. 11.51

The answers are on page 533.

PROBLEM SOLVING AND APPLICATIONS

17-4 Using Similar Triangles

Similar triangles are triangles that have the same shape.

> In similar triangles, angles that correspond have **equal measures**.
>
> In similar triangles, sides that correspond are **opposite equal angles**.
>
> Sides that correspond have **equivalent ratios**.

When you know that triangles are similar, you can write a proportion to find the length of an unknown side.

TABLE	SIMILAR TRIANGLES	EQUIVALENT RATIOS
	Triangle 1: 8 m, 10 m, 15 m (108°, 42°, 30°) Triangle 2: 16 m, 20 m, 30 m (108°, 42°, 30°)	Triangle 1 — Triangle 2 $\dfrac{8}{15} = \dfrac{16}{30}$ $\dfrac{8}{10} = \dfrac{16}{20}$ $\dfrac{10}{15} = \dfrac{20}{30}$

PROCEDURE

To find the length of an unknown side:

1. Use the unknown side and a known side in one triangle to write a ratio.
2. Use sides that correspond in the second triangle to write another ratio.
3. Use the two ratios to write a proportion.
4. Solve the proportion.

EXAMPLE Jill is 1.8 meters tall. On a sunny day, her shadow was 2 meters long at the same time that the shadow of a water tower was 40 meters long. Find the height of the tower.

Solution:

1. Use the triangle with the unknown side to write a ratio.

   $\dfrac{h}{40}$ ← Tower's height
   ← Shadow length

2. Use sides that correspond in the second triangle to write another ratio.

   $\dfrac{1.8}{2}$ ← Jill's height
   ← Shadow length

3. Use the ratios to write a proportion.

   $\dfrac{h}{40} = \dfrac{1.8}{2}$

4. Solve the proportion.

   $h \times 2 = 40 \times 1.8$
   $2h = 72$
   $h = \mathbf{36}$ ◀ The tower is 36 meters high.

Self-Check

1. Bill is 6 feet tall. On a sunny day, Bill's shadow is 8 feet long and the shadow of a high diving platform is 52 feet long. Find the height of the diving platform.

CLASSROOM EXERCISES

For Exercises 1–4, each pair of triangles is similar. Complete the proportion. (Table)

1.
   $\dfrac{n}{10} = \dfrac{?}{20}$

2.
   $\dfrac{n}{56} = \dfrac{20}{?}$

3. (Example)

$$\frac{n}{4} = \frac{?}{?}$$

4.

$$\frac{n}{85} = \frac{?}{?}$$

WRITTEN EXERCISES

Goals: To use a proportion to solve problems involving similar triangles
To apply this skill to solving word problems

Each pair of triangles is similar. Write a proportion to find the unknown length. Then solve the proportion. When necessary, round your answer to the nearest tenth. (Example)

1.

2.

3.

4.

For Exercises 5–7, use a proportion to solve each problem. (Example)

5. Norman is 1.8 meters tall. On a sunny day, Patty measured Norman's shadow and the shadow of the school. Use the similar triangles shown below to find the height of the school. Round the answer to the nearest tenth.

510 Chapter 17

6. Jaimie placed a mirror on the sidewalk. Then he walked backwards until he saw the top of the streetlight in the mirror. The figure below shows how he formed similar triangles. Find the height of the street lamp. Round your answer to the nearest tenth of a meter.

7. Find the distance across this lake. Round your answer to the nearest tenth of a meter.

SIMILAR POLYGONS

You can use a scientific calculator and the ratio of corresponding sides of two similar triangles to find the lengths of unknown sides.

EXAMPLE The ratio of corresponding sides of similar triangles *ABC* and *DEF* is $\frac{11}{8}$. Find *EF* and *DE* when *CB* = 14 and *AB* = 17.

Solution Use the constant and multiplication keys ([K] and [×]) to store the ratio $\frac{11}{8}$. Then enter each known value and press

1 1 [÷] 8 [=] [×] [K] 1 4 [=]   19.25

1 7 [=]   23.375

EXERCISES

Use a scientific calculator to check the solutions to Exercises 1–4 on page 510.

Geometry **511**

COMPUTER APPLICATIONS: Angle Measures

If you are given the measures of two angles greater than 0 and less than 180, you can use the following program to determine if they are supplementary.

Program 1:

```
100 PRINT "ENTER THE ANGLE MEASURES."
110 INPUT A, B          ← These are the angle measures.
120 IF A + B = 180 THEN 150   ← If A + B ≠ 180, then 130.
130 PRINT "ANGLES A AND B ARE NOT SUPPLEMENTARY."
140 GOTO 160
150 PRINT "ANGLES A AND B ARE SUPPLEMENTARY."
160 PRINT "ANY MORE ANGLES (1 = YES, 0 = NO)";
170 INPUT X             ← X = 1 or X = 0
180 IF X = 1 THEN 100
190 END
```

If you know the measure of the vertex angle of an isosceles triangle, you can use the following program to compute the measure of each base angle.

Program 2:

```
10 PRINT "ENTER THE MEASURE OF THE VERTEX ANGLE."
20 INPUT V              ← V is the measure of the vertex angle.
30 LET B = (180 - V) / 2   ← B is the measure of a base angle.
40 PRINT "THE MEASURE OF EACH BASE ANGLE IS"; B
50 PRINT "ANY MORE ANGLES (1 = YES, 0 = NO)";
60 INPUT X
70 IF X = 1 THEN 10
80 END
```

NOTE: Statement 30 gives the formula for finding the measure of the base angle. It stores this value in memory location B.

Exercises

Run Program 1 for the following values.

1. $A = 40$; $B = 140$
2. $A = 60$; $B = 120$
3. $A = 82$; $B = 96$
4. $A = 95$; $B = 85$
5. $A = 62.3$; $B = 97.6$
6. $A = 42.5$; $B = 138.5$

Run Program 2 for the following values of V.

7. 40
8. 70
9. 120
10. 32.5
11. 15.8
12. 92.3

REVIEW: SECTIONS 17-1—17-4

For Exercises 1-3, refer to the figure at the right. (Section 17-1)

1. Name three different rays that have *E* as an endpoint.
2. Name two lines that contain point *C*.
3. Name four angles with *C* as a vertex.

Draw an angle with the given measure. (Section 17-2)

4. 30°   **5.** 87°   **6.** 112°   **7.** 90°   **8.** 45°

Determine whether each pair of triangles is congruent by SSS, SAS, or ASA. Name the corresponding congruent parts. (Section 17-3)

9.   **10.**   **11.**

12. On a sunny day, the shadow of a tree was 12 meters long. Use the similar triangles at the right to find the height of the tree. Round your answer to the nearest tenth of a meter.

REVIEW CAPSULE FOR SECTION 17-5

Match one or more symbols in Column A with each figure in Exercises 1-4. (Pages 498-500)

The answers are on page 533.

1.
2.
3.
4.

COLUMN A

a. $\angle ABC$ e. $\overrightarrow{XY}$
b. $\overleftrightarrow{RS}$ f. $\overleftrightarrow{DF}$
c. $\overleftrightarrow{FD}$ g. $\overrightarrow{SR}$
d. $\angle CBA$ h. $\overleftrightarrow{XZ}$

Geometry 513

17-5 Perpendicular and Parallel Lines

In the figure at the right, boats 1 and 2 are sailing on parallel courses (lines). **Parallel lines** are lines in a plane that are always the same distance apart.

$\overleftrightarrow{AC} \parallel \overleftrightarrow{DF}$ ◀ Read: "Line *AC* is parallel to line *DF*."

The course of boat 3 meets (intersects) the courses of boats 1 and 2 at right angles. In a plane, **intersecting lines** have exactly one point in common. When two intersecting lines form a right angle, the lines are **perpendicular lines.**

$\overleftrightarrow{DF} \perp \overleftrightarrow{BE}$ ◀ Read: "Line *DF* is perpendicular to line *BE*."

EXAMPLE 1 Identify each of the following. Refer to the figure at the right.
   a. A pair of parallel lines
   b. A pair of intersecting lines that are not perpendicular
   c. A pair of perpendicular lines

Solutions: a. $\overleftrightarrow{AB} \parallel \overleftrightarrow{CD}$.   b. $\overleftrightarrow{BC}$ intersects $\overleftrightarrow{AB}$ at *B*.   c. $\overleftrightarrow{AB} \perp \overleftrightarrow{AC}$.

Self-Check *Identify each of the following. Refer to the figure of Example 1.*
1. A second pair of intersecting lines
2. A second pair of perpendicular lines

A **transversal** is a line that intersects two other lines. When parallel lines are cut by a transversal, pairs of **congruent** angles are formed.

Recall that congruent angles have equal measures.

$\angle 1 \cong \angle 3$ ◀ Read: "Angle 1 is congruent to angle 3."

The following table classifies the pairs of congruent angles formed when a transversal intersects two parallel lines. Refer to the figure at the bottom of page 514.

TABLE

Pairs of Congruent Angles	Name
$\angle 1 \cong \angle 3, \angle 2 \cong \angle 4, \angle 5 \cong \angle 7, \angle 6 \cong \angle 8$	Vertical Angles
$\angle 3 \cong \angle 5, \angle 4 \cong \angle 6$	Alternate interior angles
$\angle 1 \cong \angle 7, \angle 2 \cong \angle 8$	Alternate exterior angles
$\angle 1 \cong \angle 5, \angle 2 \cong \angle 6, \angle 3 \cong \angle 7, \angle 4 \cong \angle 8$	Corresponding angles

NOTE: A line can be also named with a single letter. Thus, in the figure at the bottom of page 514, the transversal is line t and lines p and q are parallel.

EXAMPLE 2 In the figure at the right, lines m, n, and p are parallel to each other. Line t is a transversal. Classify each pair of congruent angles as vertical, alternate interior, alternate exterior, or corresponding.

    a. $\angle 1$ and $\angle 3$ **b.** $\angle 3$ and $\angle 5$

    c. $\angle 5$ and $\angle 9$ **d.** $\angle 5$ and $\angle 11$

Solutions: **a.** Vertical   **b.** Alternate interior

                c. Corresponding   **d.** Alternate exterior

Self-Check   *Classify each pair of congruent angles in Example 2 as corresponding, alternate interior, alternate exterior, or vertical.*

    3. $\angle 4$ and $\angle 2$ **4.** $\angle 7$ and $\angle 9$ **5.** $\angle 6$ and $\angle 4$ **6.** $\angle 9$ and $\angle 11$

CLASSROOM EXERCISES

For Exercises 1–6, identify each pair of lines as intersecting, parallel, or perpendicular. Some exercises may have more than one answer. (Example 1)

1. $\overleftrightarrow{AB}$ and $\overleftrightarrow{AD}$ **2.** $\overleftrightarrow{BE}$ and $\overleftrightarrow{DE}$ **3.** $\overleftrightarrow{AD}$ and $\overleftrightarrow{AC}$

4. $\overleftrightarrow{AD}$ and $\overleftrightarrow{DE}$ **5.** $\overleftrightarrow{AB}$ and $\overleftrightarrow{DE}$ **6.** $\overleftrightarrow{AD}$ and $\overleftrightarrow{CE}$

For Exercises 7–12, parallel lines AB and CD intersect transversals EF and GH. Match each pair of congruent angles with the correct name. Choose a, b, c, or d. Example 2)

7. ∠3 and ∠5
8. ∠10 and ∠12
9. ∠5 and ∠7
10. ∠2 and ∠8
11. ∠13 and ∠11
12. ∠14 and ∠10

a. Vertical angles
b. Alternate interior angles
c. Alternate exterior angles
d. Corresponding angles

WRITTEN EXERCISES

Goals: To identify intersecting, perpendicular, and parallel lines

To identify the angle relationships that occur when parallel lines are cut by a transversal

Identify each of the following. Refer to the figure at the right.
(Example 1)

1. Four pairs of perpendicular lines
2. Two pairs of parallel lines
3. Four pairs of intersecting lines that are not perpendicular

In the figure at the right, $\overleftrightarrow{AB}$ is parallel to $\overleftrightarrow{CD}$ and $\overleftrightarrow{EF}$ is parallel to $\overleftrightarrow{GH}$. Classify each pair of congruent angles as corresponding, alternate interior, alternate exterior, or vertical (Example 2)

4. ∠1 and ∠13
5. ∠13 and ∠15
6. ∠15 and ∠3
7. ∠3 and ∠5
8. ∠5 and ∠7
9. ∠7 and ∠11
10. ∠11 and ∠13
11. ∠2 and ∠16
12. ∠13 and ∠3

516 Chapter 17

17-6 Properties of Quadrilaterals

The wings of the model plane at the right are in the shape of quadrilaterals. A **quadrilateral** is a polygon having four sides. The following table lists some quadrilaterals having special names and properties.

Quadrilateral	Figure	Definition
Trapezoid ABCD		Quadrilateral with exactly one pair of opposite sides parallel ($\overline{DC}$ and $\overline{AB}$ in this figure)
Parallelogram EFGH		Quadrilateral with both pairs of of opposite sides parallel
Rectangle KLIJ		Parallelogram with four right angles
Rhombus MPQN		Parallelogram with four congruent sides
Square WXYZ		Parallelogram with four congruent sides and four right angles

NOTE: Rectangles, rhombuses, and squares are all parallelograms. In all parallelograms, opposite sides are equal and opposite angles are equal.

Geometry

EXAMPLE 1 In the figure at the right, line segments *AD*, *GE*, and *MH* are parallel, segments *MA*, *KG*, *JF*, and *HD* are parallel, and segments *MB*, *LC*, and *KD* are parallel. Refer to the figure to classify each figure as a trapezoid, parallelogram, rectangle, rhombus, or square.

    a. FEHJ   **b.** ACLM   **c.** GFJK   **d.** CDFG

Solutions: Refer to the definitions in the Table on page 517 as needed.

	Figure	Trapezoid	Parallelogram	Rectangle	Rhombus	Square
a.	FEHJ	No	Yes	Yes	No	No
b.	ACLM	Yes	No	No	No	No
c.	GFJK	No	Yes	Yes	Yes	Yes
d.	CDFG	No	Yes	No	No	No

Self-Check *Refer to the figure in Example 1 to classify each figure as a trapezoid, parallelogram, rectangle, rhombus, or square.*

    1. *GFJL*   **2.** *AMHD*   **3.** *GEHK*

CLASSROOM EXERCISES

Complete. Answer Yes or No. (Table)

Property	Trapezoid	Parallelogram	Rectangle	Rhombus	Square
1. Opposite sides parallel	No	Yes	Yes	?	?
2. All sides congruent	?	?	?	?	?
3. Opposite sides congruent	?	?	?	?	?
4. Opposite angles congruent	?	?	?	?	?
5. Four right angles	?	?	?	?	?
6. Only one pair of parallel sides	?	?	?	?	?

WRITTEN EXERCISES

Goal: To identify the properties of quadrilaterals

Determine the unknown measure. (Table)

1. **2.** **3.** **4.**

In the figure below, line segments LA, HB, NK, and PC are parallel, segments EJ, RF, and AC are parallel, and segments PL and MG are parallel. Refer to the figure to classify each given quadrilateral as a trapezoid, parallelogram, rectangle, rhombus, or square. Some figures may have more than one name. (Example 2)

5. *BCFD*   **6.** *BCJH*   **7.** *DFJH*   **8.** *GMPL*   **9.** *ACFD*   **10.** *ACPL*
11. *KMPN*   **12.** *ACJE*   **13.** *JELP*   **14.** *AJMG*   **15.** *ABHE*   **16.** *ABDR*

MORE CHALLENGING EXERCISES

Complete. Use the words parallelogram, rhombus, rectangle, or square.

17. Every rectangle is also a __?__ .   **18.** Every rhombus is also a __?__ .
19. Every square is also a __?__ , a __?__ , and a __?__ .
20. A parallelogram with four congruent angles is a __?__ or a __?__ .
21. A parallelogram with four congruent sides is a __?__ or a __?__ .

Geometry 519

REVIEW CAPSULE FOR SECTION 17-7

Evaluate each expression. (Pages 18–21)

1. $2 \cdot 5 + 2 \cdot 8$
2. $3 \cdot 5 + 2 \cdot 9$
3. $2(3.4) + 2(6.5)$
4. $4(1.8) + 3(2.2)$
5. $2 \cdot \frac{1}{2} + 2 \cdot \frac{1}{3}$
6. $3 \cdot 1\frac{1}{2} + 2 \cdot 1\frac{1}{4}$

Find the area of each rectangle. (Pages 70–73)

7. $l = 5$ m; $w = 3$ m
8. $l = 1.4$ cm; $w = 0.5$ cm
9. $l = 13$ ft; $w = 8$ ft

Find the area of each square. (Pages 70–73)

10. $s = 7$ in
11. $s = 8\frac{1}{2}$ ft
12. $s = 12.5$ m

The answers are on page 533.

PROBLEM SOLVING AND APPLICATIONS

17-7 Surface Area and Rectangular Prisms

Each of the six sides or **faces** of a rectangular prism is a rectangle. The surface area is the combined areas of these six rectangles.

Word Rule: The **surface area** of a rectangular prism is the sum of the areas of the six faces.

The opposite faces of a rectangular prism, such as the top and bottom, have equal areas.

$2\ell w \quad + \quad 2\ell h \quad + \quad 2wh$

520 Chapter 17

Thus, the surface area of a rectangular prism can be found as follows.

$$\text{Surface Area} = \underbrace{\text{Area of Top and Bottom}} + \underbrace{\text{Area of Front and Back}} + \underbrace{\text{Area of Left and Right Sides}}$$

Formula: $\quad S \quad = \quad 2lw \quad + \quad 2lh \quad + \quad 2wh$

EXAMPLE Find the surface area of the box of paper clips shown at the right.

Solution:
1. Formula: $S = 2lw + 2lh + 2wh$
2. Known values: $l = 6$ cm; $w = 4$ cm; $h = 3$ cm
3. Replace l with 6, w with 4, and h with 3. $S = 2(6)(4) + 2(6)(3) + 2(4)(3)$
4. Multiply. $S = 48 + 36 + 24$
   $S = 108$

The surface area is **108 cm²**.

Self-Check

1. Find the surface area of a rectangular prism with a length of 6 centimeters, a width of 3 centimeters, and a height of 2 centimeters.

A **cube** is a special kind of rectangular prism. The length, width, and height are the same. The formula for the surface area of a cube can be written as follows.

$S = 6 \cdot s \cdot s$
$S = 6s^2$

s = length of any side

Cube

CLASSROOM EXERCISES

Complete. (Example)

	Length	Width	Height	2lw	2lh	2wh	Surface Area: 2lw + 2lh + 2wh
1.	4 m	3 m	5 m	?	?	?	?
2.	2.5 cm	4 cm	1 cm	?	?	?	?
3.	3 ft	4 ft	2 ft	?	?	?	?
4.	8 in	6 in	4 in	?	?	?	?

Geometry **521**

WRITTEN EXERCISES

Goals: To find the surface area of a rectangular solid

To apply the skill of finding surface area to solving word problems

METRIC MEASURES

Find the surface area of each rectangular solid. (Example)

1. Book

45 cm, 3.5 cm, 20 cm

2. Door

1 m, 0.05 m, 2.5 m

3. Package

3 cm, 3 cm, 3 cm

	Length	Width	Height
4.	3 m	4 m	6 m
6.	40 cm	29 cm	10.2 cm

	Length	Width	Height
5.	12 cm	8 cm	8 cm
7.	1.1 m	2.4 m	6 m

CUSTOMARY MEASURES

8. Cereal Box

8 in, 3 in, 10 in

9. Refrigerator

3 ft, 2 ft, 6 ft

10. Radio

10 in, $2\frac{1}{2}$ in, 7 in

	Length	Width	Height
11.	5 ft	10 ft	16 ft
13.	$6\frac{2}{3}$ in	8 in	$10\frac{1}{2}$ in

	Length	Width	Height
12.	10 in	15 in	8 in
14.	$1\frac{1}{2}$ yd	$1\frac{1}{2}$ yd	$1\frac{1}{2}$ yd

15. Howard is wrapping a present. The box he is using is a rectangular prism with a length of 24 inches, a width of 15 inches, and a height of 4 inches. Find how many square inches of paper he needs to wrap the entire box. Add 368 square inches to the answer to allow for overlapping the paper.

522 Chapter 17

16. Lucy is painting a toy chest that is a rectangular prism. The chest is 4 feet long, 2 feet wide, and $1\frac{1}{2}$ feet high. What is the surface area of the chest?

17. A certain music box has the shape of a cube. Each side of the music box is 12 centimeters long. What is the surface area of the music box?

MORE CHALLENGING EXERCISES

Find the cost of roofing each building when the cost of shingles is $0.35 per square foot. All sides of the roof have the same shape.

18.

19.

20. The roofs of 6 houses in a development are shaped as shown in the figure. Both sides of the roof are the same. The builder pays $0.27 per square foot for shingles and $8.50 per hour for labor. Five roofers shingle the 6 roofs in 40 hours. Find the total cost to the builder.

REVIEW CAPSULE FOR SECTION 17-8

Find the volume of each rectangular prism. (Pages 145-147)

1. $l = 4$ cm; $w = 6$ cm; $h = 3$ cm
2. $l = 5$ m; $w = 2.5$ m; $h = 7$ m
3. $l = 3$ ft; $w = 5$ ft; $h = 2$ ft
4. $l = 10$ in; $w = 8$ in; $h = 9$ in

Evaluate each expression. (Pages 176-179)

5. $3p$ when $p = 3.14$
6. $9t$ when $t = 3.14$
7. r^2 when $r = 0.6$
8. a^2 when $a = 5\frac{1}{2}$
9. $\frac{22}{7}p$ when $p = 5$
10. $\frac{22}{7}t$ when $t = 2\frac{1}{3}$

Round each decimal to the nearest whole number. (Pages 32-34)

11. 0.62   **12.** 1.98   **13.** 25.31   **14.** 36.81   **15.** 46.54   **16.** 99.5

The answers are on page 533.

Geometry   **523**

17-8 Volume: Cylinders

Howard Lopez built a new silo for his farm. The silo has the shape of a cylinder. Howard wants to know how much corn the silo will hold: that is, he wants to know the **volume** of the silo.

Finding the volume of a cylinder is similar to finding the volume of a rectangular prism. Volume is measured in **cubic units**.

Rectangular Prism

◀ The base is a rectangle.

Cylinder

◀ The base is a circle.

Volume = Area of Base × Height

$V = l \times w \times h$
$V = lwh$

Volume = Area of Base × Height

$V = \pi \times r^2 \times h$
$V = \pi r^2 h$

PROCEDURE

To find the volume of a cylinder, multiply the area of the base and the height.

EXAMPLE Howard's silo has a radius of 6.5 meters. The height is 18 meters. Find the volume of the silo. Use 3.14 for π. Round the answer to the nearest meter.

Solution: $V = \pi r^2 h$ ⟵ r^2 means $r \cdot r$.

$V = 3.14(6.5)(6.5)(18)$

$V = 2387.97$ m^3 The volume is about **2388 m^3**.

Self-Check

1. Find the volume of a cylinder with a radius of 5.2 centimeters and a height of 10 centimeters. Use 3.14 for π. Round the answer to the nearest whole number.

CLASSROOM EXERCISES

METRIC MEASURES

For Exercises 1–4, find the area of the base and the volume of each cylinder. Use 3.14 for π. Round each answer to the nearest whole number. (Example)

	Radius	Height	Area of Base	Volume of Cylinder
1.	3 m	6 m	28.26 m²	__?__ m³
2.	6 cm	8 cm	__?__ cm²	__?__ cm³
3.	2.4 m	5 m	__?__ m²	__?__ m³
4.	25 mm	8.5 mm	__?__ mm²	__?__ mm³

CUSTOMARY MEASURES

For Exercises 5–8, find the area of the base and the volume of each cylinder. Use $\frac{22}{7}$ for π. Round each answer to the nearest whole number. (Example)

	Radius	Height	Area of Base	Volume of Cylinder
5.	7 ft	10 ft	154 ft²	__?__ ft³
6.	14 yd	8 yd	__?__ yd²	__?__ yd³
7.	$2\frac{1}{3}$ ft	4 ft	__?__ ft²	__?__ ft³
8.	42 in	$3\frac{1}{3}$ in	__?__ in²	__?__ in³

WRITTEN EXERCISES

Goal: To find the volume of a cylinder

METRIC MEASURES

For Exercises 1–7, find the volume of each cylinder. Use 3.14 for π. Round each answer to the nearest whole number. (Example)

1. Mug

$r = 4$ cm; $h = 8$ cm

2. Water Tower

$r = 5$ m; $h = 16.6$ m

3. Trash Can

$r = 23.5$ cm; $h = 56$ cm

Geometry **525**

	Radius	Height
4.	5 cm	18 cm
6.	0.8 m	1 m

	Radius	Height
5.	3 cm	6 cm
7.	4.5 cm	10.4 cm

CUSTOMARY MEASURES

For Exercises 8–14, find the volume of each cylinder. Use $\frac{22}{7}$ for π. Round each answer to the nearest whole number. (Example)

8. Can of Varnish

$r = 3\frac{1}{2}$ in; $h = 8$ in

9. Glass

$r = 2$ in; $h = 5\frac{1}{2}$ in

10. Wading Pool

$r = 10\frac{1}{2}$ ft; $h = 4$ ft

	Radius	Height
11.	14 in	8 in
13.	$10\frac{1}{2}$ in	13 in

	Radius	Height
12.	6 in	15 in
14.	3 ft	$7\frac{1}{2}$ ft

APPLICATIONS: USING VOLUME OF A CYLINDER

Solve. Round each answer to the nearest whole number. (Example)

15. The radius of a hot water tank is 0.4 meters and its height is 1.5 meters. Find the volume of the tank. Use 3.14 for π.

16. A box of salt has the shape of a cylinder. The radius of the box is $1\frac{1}{2}$ inches and the height is 5 inches. Find the volume of the box. Use $\frac{22}{7}$ for π.

17. A cylinder-shaped water pitcher has a radius of 4 inches and a height of $12\frac{1}{2}$ inches. Find the volume of the water pitcher. Use $\frac{22}{7}$ for π.

18. The radius of a can of soup is 3.2 centimeters and the height is 10 centimeters. Find the volume of the can. Use 3.14 for π.

REVIEW CAPSULE FOR SECTION 17-9

Evaluate each expression. (Pages 22–24, 129–131)

1. $\frac{1}{3}y$ when $y = 18$
2. $\frac{1}{3}c$ when $c = 7.2$
3. $\frac{1}{3}mp$ when $m = 3.14$ and $p = 6$
4. $\frac{1}{3}nq$ when $n = \frac{22}{7}$ and $q = 36$

The answers are on page 533.

17-9 Volume: Pyramids and Cones

The pyramid at the right is a **rectangular pyramid**. The base of a pyramid determines its name.

When a pyramid and a rectangular prism have equal heights and bases that are equal in area, the volume of the pyramid is $\frac{1}{3}$ of the volume of the rectangular prism.

PROCEDURE

To find the volume of a pyramid, use this formula.

$$V = \frac{1}{3}Bh$$

B = area of base
h = height

EXAMPLE 1 The Pyramid of the Sun in Mexico has a rectangular base. The base is 229 meters long and 137 meters wide. The pyramid is 60 meters high. Find the volume.

Solution: Since the base is a rectangle,
$B = (229)(137) = 31,373$ m².

$V = \frac{1}{3}Bh$ Replace *B* with 31,373 and *h* with 60.

$V = \frac{1}{3}(31,373)(60)$

$V = 627,460$ The volume is **627,460 m³**.

Self-Check *Find the volume of each rectangular pyramid.*

1. $l = 3$ m; $w = 3$ m; $h = 6$ m
2. $l = 6$ ft; $w = 9$ ft; $h = 10$ ft

Geometry **527**

The base of a cylinder is a circle. The base of a cone is also a circle. When a cone and a cylinder have equal radii (plural of radius) and equal heights, the volume of the cone is $\frac{1}{3}$ of the volume of the cylinder.

Cylinder
$V = \pi r^2 h$

Cone
$V = \frac{1}{3}\pi r^2 h$

PROCEDURE

To find the volume of a cone, use this formula.

$V = \frac{1}{3} \cdot B\text{(area of base)} \cdot h$ ◀ The base is a circle.

$V = \frac{1}{3} \cdot \pi \cdot r^2 \cdot h$, or $V = \frac{1}{3}\pi r^2 h$

EXAMPLE 2 Find the volume of this paper cup. Use 3.14 for π. Round your answer to the nearest whole number.

$r = 3$ cm
$h = 8$ cm

Solution: $V = \frac{1}{3}\pi r^2 h$ $r = 3$
$h = 8$

$V = \frac{1}{3}(3.14)(3)(3)(8)$

$V = 75.36$ The volume is about **75 cm³**.

Self-Check *Find the volume of each cone. Round each answer to the nearest whole number.*

3. $r = 4$ m; $h = 10$ m; $\pi = 3.14$
4. $r = 6$ ft; $h = 15$ ft; $\pi = \frac{22}{7}$

CLASSROOM EXERCISES

Complete to find the volume of each rectangular pyramid. (Example 1)

1. $l = 5$ ft; $w = 5$ ft; $h = 12$ ft
$V = \frac{1}{3}Bh = \frac{1}{3}(5)(5)(12) =$ __?__

2. $l = 2.4$ m; $w = 1.8$ m; $h = 6$ m
$V = \frac{1}{3}Bh = \frac{1}{3}(2.4)(1.8)(6) =$ __?__

Complete to find the volume of each cone. Round each answer to the nearest whole number. (Example 2)

3. $r = 6$ m; $h = 20$ m; $\pi = 3.14$
$V = \frac{1}{3}\pi r^2 h = \frac{1}{3}(3.14)(6)(6)(20) =$ __?__

4. $r = 5$ ft; $h = 12$ ft; $\pi = \frac{22}{7}$
$V = \frac{1}{3}\pi r^2 h = \frac{1}{3}\left(\frac{22}{7}\right)(5)(5)(12) =$ __?__

WRITTEN EXERCISES

Goals: To find the volume of a pyramid
To find the volume of a cone

METRIC MEASURES

Find each volume. Use $\pi = 3.14$. When necessary, round the answer to the nearest whole number. (Examples 1 and 2)

1. 4 m, 2 m, 3 m
2. 12 cm, 5 cm, 10 cm
3. 10 cm, 30 cm
4. 3 mm, 15 mm

CUSTOMARY MEASURES

Find each volume. Use $\pi = \frac{22}{7}$. When necessary, round the answer to the nearest whole number. (Examples 1 and 2)

5. 15 ft, 6 ft, 10 ft
6. 4 yd, 3 yd, 3 yd
7. 2 ft, 9 ft
8. 4.5 in, 10 in

APPLICATIONS: USING VOLUME

Round each answer to the nearest whole number.

9. Martha and Cherie pitched a tent that has the shape of a pyramid. The base of the tent is a rectangle that is 2.1 meters wide and 2.4 meters long. The tent is 2 meters high. Find the volume.

10. A pyramid-shaped container has a rectangular base. The base is 13 centimeters long and 9 centimeters wide. The container is 22.5 centimeters high. Find the volume.

11. A cone-shaped hanging planter has a radius of 7 centimeters and a height of 15 centimeters. Find the volume. Use 3.14 for π.

12. A cone-shaped papercup has a radius of 4 inches and a height of 8 inches. Find the volume. Use $\frac{22}{7}$ for π. Round your answer to the nearest whole number.

Geometry 529

REVIEW: SECTIONS 17-5–17-9

Identify each of the following. Refer to the figure at the right. (Section 17-5)

1. Four pairs of perpendicular lines.
2. Two pairs of parallel lines.

For Exercises 3-6, find each missing measure. (Section 17-6)

3.
4.
5.
6.

Find the surface area of each rectangular solid. (Section 17-7)

7. $l = 2$ m; $w = 5$ m; $h = 8$ m
8. $l = 2\frac{1}{2}$ ft; $w = 8$ ft; $h = 3\frac{1}{4}$ ft

Find the volume of each cylinder. Round each answer to the nearest whole number. (Section 17-8)

9. $r = 4$ cm; $h = 12$ cm; $\pi = 3.14$
10. $r = 1\frac{1}{2}$ in; $h = 7$ in; $\pi = \frac{22}{7}$

Find the volume of each pyramid. (Section 17-9)

11. $l = 2$ m; $w = 4$ m; $h = 9$ m
12. $l = 6$ cm; $w = 3.5$ cm; $h = 4.5$ cm

Find the volume of each cone. Round each answer to the nearest whole number. (Section 17-9)

13. $r = 7$ m; $h = 9$ m; $\pi = 3.14$
14. $r = 9$ cm; $h = 21$ cm; $\pi = 3.14$
15. $r = 12$ ft; $h = 14$ ft; $\pi = \frac{22}{7}$
16. $r = 3\frac{1}{2}$ yd; $h = 16$ yd; $\pi = \frac{22}{7}$

17. Each side of a cube is 22 inches long. What is the surface area of the cube? (Section 17-7)

18. A cone-shaped candle has a radius of 2 inches and a height of 7 inches. Find the volume. Use $\frac{22}{7}$ for π. (Section 17-9)

19. A can of frozen juice has the shape of a cylinder. The radius of the cylinder is 2 centimeters and the height is 9 centimeters. Find the volume. Use 3.14 for π. (Section 17-8)

20. A parking lot has pyramid-shaped markers with rectangular bases. The markers are 0.3 meters wide, 0.4 meters long, and 0.8 meters high. Find the volume of one marker. (Section 17-9)

530 Chapter 17

CHAPTER REVIEW

PART 1: VOCABULARY

For Exercises 1–18, choose from the box at the right the word(s) that best corresponds to each description.

1. The standard unit for measuring an angle __?__ (p. 501)
2. A polygon with three sides __?__ (p. 504)
3. A __?__ extends indefinitely in opposite directions. (p. 498)
4. A rectangular prism in which the length, width, and height are the same __?__ (p. 521)
5. A closed plane figure formed by line segments __?__ (p. 504)
6. Triangles that have the same size and shape __?__ (p. 505)
7. A polygon with four sides __?__ (p. 517)
8. A __?__ extends indefinitely in one direction from its endpoint. (p. 498)
9. An angle that has a measure of 90° __?__ (p. 502)
10. Volume is measured in __?__. (p. 524)
11. Triangles that have the same shape but not the same size __?__ (p. 508)
12. A pair of rays having a common endpoint __?__ (p. 498)
13. An angle having a measure less than 90° __?__ (p. 502)
14. A triangle having no sides the same length __?__ (p. 504)
15. An angle having a measure greater than 90° __?__ (p. 502)
16. A triangle having two sides the same length __?__ (p. 504)
17. A part of a line that is named by its endpoints __?__ (p. 498)
18. A triangle having all sides the same length __?__ (p. 504)

line
line segment
ray
angle
degree
right angle
acute angle
obtuse angle
polygon
triangle
equilateral triangle
isosceles triangle
scalene triangle
congruent triangles
similar triangles
quadrilateral
cube
cubic units

PART 2: SKILLS

For Exercises 19–22, refer to the figure at the right. (Section 17-1)

19. Name line *BE* in three different ways.
20. Name $\overrightarrow{EB}$ in a different way.
21. Name four angles with *B* as a vertex.
22. Name two lines that contain point *E*.

Geometry 531

For each of Exercises 23–26, use a protractor to draw an angle with the given measure. (Section 17–2)

23. 20°   **24.** 50°   **25.** 90°   **26.** 165°

Classify each triangle as equilateral, isosceles, or scalene. (Section 17–3)

27. (triangle with sides 97, 62, 97)

28. (triangle with sides 6, 6, 6)

29. (triangle with sides 26, 19, 28)

30. (triangle with sides 18, 18, 25)

For Exercises 31–32, each pair of triangles is similar. Use a proportion to find the unknown length. (Section 17–4)

31. (triangles with sides 45, 36 and 20, n)

32. (triangles with sides 13, 10, 84° and 19.5, n, 84°)

In the figure at the right, $\overleftrightarrow{EF}$ is parallel to $\overleftrightarrow{GH}$, and $\overleftrightarrow{IJ}$ is a transversal. Classify each pair of angles as alternate exterior, alternate interior, corresponding, or vertical. (Section 17–5)

33. ∠6 ≅ ∠4   **34.** ∠1 ≅ ∠7   **35.** ∠5 ≅ ∠7   **36.** ∠2 ≅ ∠6

For Exercises 37–39, classify each quadrilateral as a trapezoid, parallelogram, rectangle, rhombus, or square. Some figures may have more than one name. (Section 17–6)

37.   **38.**   **39.**

532 Chapter 17

Find the surface area of each rectangular solid. (Section 17-7)

	Length	Width	Height
40.	$4\frac{1}{2}$ ft	6 ft	10 ft
41.	12 m	10.5 m	20.2 m
42.	$1\frac{1}{2}$ yd	8 yd	$5\frac{1}{5}$ yd

Find the volume of each cylinder. Round each answer to the nearest whole number. (Section 17-8)

	Radius	Height	π
43.	6 m	12 m	3.14
44.	$5\frac{1}{4}$ ft	9 ft	$\frac{22}{7}$
45.	14.4 cm	10.5 cm	3.14

PART 3: APPLICATIONS

46. Sue is 1.6 meters tall. On a sunny day, her shadow is 2 meters long at the same time as the shadow of a TV tower is 70 meters long. Find the height of the tower. (Section 17-4)

47. The radius of a cylindrical silo is 7 meters and its height is 28.5 meters. Find the volume of the silo. Use 3.14 for π. Round the answer to the nearest whole number. (Section 17-8)

48. Ramona is buying screen for a hamster cage that has the shape of a rectangular prism. The cage is 2 feet wide, $2\frac{1}{2}$ feet long, and $1\frac{1}{2}$ feet high. Find how many square feet of screen she needs to cover the entire cage. (Section 17-7)

49. The Great Pyramid of Egypt has a square base. Each side of the base is 230 meters long. The height of the pyramid is 147 meters. Find the volume. (Section 17-9)

ANSWERS TO REVIEW CAPSULES

Page 500 **1.** < **2.** > **3.** > **4.** > **5.** > **6.** < **7.** > **8.** <

Page 508 **1.** $13\frac{1}{2}$ **2.** 36 **3.** 8 **4.** 24 **5.** 33 **6.** 15 **7.** 9.8 **8.** 11.9 **9.** 11.6 **10.** 12.9 **11.** 9.1 **12.** 11.5

Page 513 **1.** b; g **2.** a; d **3.** e **4.** c; f

Page 520 **1.** 26 **2.** 33 **3.** 19.8 **4.** 13.8 **5.** $1\frac{2}{3}$ **6.** 7 **7.** 15 m² **8.** 0.7 cm² **9.** 104 ft² **10.** 49 in² **11.** $72\frac{1}{4}$ ft² **12.** 156.25 m²

Page 523 **1.** 72 cm³ **2.** 87.5 m³ **3.** 30 ft³ **4.** 720 in³ **5.** 9.42 **6.** 28.26 **7.** 0.36 **8.** $30\frac{1}{4}$ **9.** $15\frac{5}{7}$ **10.** $7\frac{1}{3}$ **11.** 1 **12.** 2 **13.** 25 **14.** 37 **15.** 47 **16.** 100

Page 527 **1.** 6 **2.** 2.4 **3.** 6.28 **4.** $37\frac{5}{7}$

Geometry 533

CHAPTER TEST

For Exercises 1–7, refer to the figure at the right below.

1. Name three different rays that have *G* as an endpoint.
2. Name four angles having *E* as a vertex.
3. Is angle *ABJ* a right angle, an obtuse angle, or an acute angle?
4. Name a pair of intersecting lines.
5. Name a pair of perpendicular lines.
6. Lines *HD* and *AC* are not parallel. Which type of quadrilateral is quadrilateral *BCEG*?
7. Quadrilateral *BCFJ* is a parallelogram. List two properties of this quadrilateral.

Write whether each pair of triangles is congruent by SAS, SSS, or ASA. Then name the corresponding congruent sides and angles.

8.

9.

For Exercises 10–11, each pair of triangles is similar. Use a proportion to find the unknown length.

10.

11.

12. Eduardo is painting a box that is a rectangular prism. The box is 3 feet wide, 4 feet long, and 2 feet high. What is the surface area?

13. A cylinder-shaped water tower has a radius of 8.5 meters and a height of 20 meters. Find the volume. Use 3.14 for π.

14. The Step Pyramid in Egypt has a rectangular base, 120 meters long and 108 meters wide. The height is 60 meters. Find the volume.

15. A cone-shaped cup has a radius of 3 inches and a height of 7 inches. Find the volume. Use $\frac{22}{7}$ for π.

ENRICHMENT

Spheres

You can use this formula to find the volume of a sphere.

$$V = \frac{4}{3}\pi r^3 \qquad r = \text{radius}$$

EXAMPLE The radius of a basketball is 12 centimeters. Find the volume of the basketball. Use 3.14 for π. Round your answer to the nearest whole number.

Solution: Use paper and pencil or use a calculator.

$$V = \frac{4}{3}\pi r^3 \qquad r^3 = r \times r \times r$$

$$V = \frac{4}{3} \times 3.14 \times 12 \times 12 \times 12$$

$V = 4 \boxed{\div} 3 \boxed{\times} 3.14 \boxed{\times} 12 \boxed{\times} 12 \boxed{\times} 12 \boxed{=}$

`7234.56`

The volume is about **7235 cm³**. — Nearest whole number

EXERCISES

For Exercises 1–6, find the volume of each sphere. Round each answer to the nearest whole number.

Sphere	Radius	π
1. Golf ball	2.1 cm	3.14
3. Volleyball	4.2 in	$\frac{22}{7}$
5. Beachball	10.2 cm	3.14

Sphere	Radius	π
2. Storage tank	7.1 m	3.14
4. Tennis ball	1.5 in	$\frac{22}{7}$
6. Marble	0.6 cm	3.14

7. Large aluminum spheres are used to transport liquid natural gas on tankers. One of these spheres has a diameter of 36 meters. Find its volume. (Use $\pi = 3.14$)

8. How much greater is the volume of a grapefruit with a radius of 4 inches than the volume of an orange with a radius of 2 inches? (Use $\pi = \frac{22}{7}$)

Geometry

ADDITIONAL PRACTICE

SKILLS   *Use symbols and the given line to name each of the following.* (Pages 498–500)

1. Three different rays.
2. Three different segments.
3. The line in three ways.

Identify each angle as right, acute, obtuse, or straight. (Pages 501–503)

4. 180°
5. 45°
6. 90°
7. 105°
8. 85°
9. 165°

10. Refer to the triangles at the right to determine whether they are congruent by SSS, SAS, or ASA. (Pages 504–507)

In the figure at the right, $\overleftrightarrow{AB}$ is parallel to $\overleftrightarrow{CD}$. Identify each pair of angles as congruent (C) or not congruent (NC). (Pages 514–516)

11. $\angle 6$ and $\angle 8$
12. $\angle 2$ and $\angle 6$
13. $\angle 3$ and $\angle 6$
14. $\angle 2$ and $\angle 8$

Identify each quadrilateral as a parallelogram, square, rectangle, or trapezoid. (Pages 517–519)

15. Exactly one pair of opposite sides parallel
16. Parallelogram with four congruent sides and four right angles
17. Both pairs of opposite sides parallel
18. Parallelogram with four right angles

APPLICATIONS

19. A fence post is 2 meters tall. On a sunny day, the shadow of the post was 2.2 meters long at the same time that the shadow of a light pole was 22 meters long. Find the height of the light pole. (Pages 508–511)

20. A planter shaped like a pyramid has a rectangular base that is 9 inches wide and 12.3 inches long. The planter is 8 inches high. Find the volume of the planter. (Pages 527–529)

21. The radius of a paint can is 7 centimeters and its height is 10 centimeters. Find the volume of the can. Use 3.14 for π. (Pages 524–526)

22. Find the surface area of a rectangular prism that has a length of 5 feet, a width of 4 feet, and a height of 3 feet. (Pages 520–523)

COMMON ERRORS

Each of these problems contains a common error.

a. Find the correct answer.

b. Find the error.

For Exercises 1–2, refer to the table at the right.

Scores on Six Tests

| 85 | 82 | 90 | 76 | 87 | 90 |

1. Find the median.

Middle Scores: 90 and 76

$\dfrac{90 + 76}{2} = \dfrac{166}{2} = \mathbf{83}$

2. Find the mean.

$\dfrac{85 + 82 + 90 + 76 + 87}{5} = \mathbf{84}$

3. Janet is one of 25 guests at a party. A drawing for one prize is held. What is the probability that she will win the prize?

$P = \dfrac{\text{Number of Favorable Outcomes}}{\text{Number of Possible Outcomes}}$

$P(\text{winning}) = \dfrac{1}{24}$

4. A bag contains 5 red marbles, 3 green marbles, and one black marble. Two marbles are drawn at random, one after the other, without replacement. Find the probability of drawing a red marble, then a green marble.

$P(\text{red, then green}) = \dfrac{5}{9} + \dfrac{3}{9} = \dfrac{8}{9}$

5. These triangles are similar. Use a proportion to find the unknown length.

$\dfrac{6}{8} = \dfrac{n}{4}$

$8n = 24$

$n = 3$

6. Find the surface area of this rectangular prism.

$l = 4$ m; $w = 3$ m; $h = 2$ m

$S = 2lw + 2lh + 2wh$

$= 4 \cdot 3 + 4 \cdot 2 + 3 \cdot 2$

$= 12 + 8 + 6$

$= \mathbf{26}$ **m³**

7. Find the volume of this cylinder.

$r = 5$ cm; $h = 6$ cm; $\pi = 3.14$

$V = \pi r^2 h$

$= 3.14(2)(5)(10)$

$= 3.14(100)$

$= \mathbf{314}$ **cm³**

8. Find the volume of this cone.

$r = 3$ in; $h = 7$ in; $\pi = \dfrac{22}{7}$

$V = \dfrac{1}{3}\pi r^2 h$

$= \dfrac{22}{7} \cdot 3 \cdot 3 \cdot 7$

$= \mathbf{198}$ **in³**

Geometry

CUMULATIVE REVIEW: CHAPTERS 16–17

Choose the correct answer. Choose a, b, c, or d.
Refer to the table at the right below for Exercises 1–3.

Points Scored by a Basketball Team

65	42	81	53	64
66	72	51	64	79

1. Find the mean.
   a. 63.7   b. 64   c. 64.5   d. 63

2. Find the median.
   a. 63.7   b. 64   c. 64.5   d. 63

3. Find the mode.
   a. 63.7   b. 64   c. 64.5   d. 63

4. A box contains six quarters, eight dimes, and six nickels. One coin is chosen at random. What is the probability of choosing a quarter?
   a. $\frac{1}{3}$ b. $\frac{3}{10}$ c. $\frac{1}{2}$ d. $\frac{2}{5}$

5. The numbers 1 through 5 are written on separate slips of paper and placed into a hat. A number is chosen at random. Find the probability of choosing an odd number.
   a. 60%   b. 50%   c. 40%   d. 30%

6. A pair of dice is tossed. What is the probability that the sum is 2 or 12?
   a. 0   b. $\frac{1}{12}$ c. $\frac{1}{9}$ d. $\frac{1}{18}$

7. The tree diagram at the right shows all the possible ways two coins can land when tossed at the same time. What is the probability that both coins will be heads?
   a. $\frac{1}{2}$ b. $\frac{1}{6}$
   c. $\frac{1}{4}$ d. $\frac{2}{6}$

8. Three nickels and two dimes are placed into a coin purse. One coin is drawn at random and replaced. Then a second coin is drawn. What is the probability that both coins are nickels?
   a. $\frac{6}{5}$ b. $\frac{3}{5}$ c. $\frac{6}{25}$ d. $\frac{9}{25}$

9. Classify an angle of 95° as one of the following.
   a. acute angle   b. obtuse angle   c. right angle   d. straight angle

10. Determine whether the pair of triangles shown at the right are congruent by SSS, SAS, ASA, or AAA.
    a. SAS
    b. SSS
    c. ASA
    d. AAA

11. Two similar triangles are shown in the figure at the right. Use a proportion to find the height of the larger triangle.
    a. 16.2 m
    b. 18.2 m
    c. 162 m
    d. 182 m

12. In the figure at the right lines AB and CD are parallel. Which of the following name a pair of corresponding angles?
    a. ∠2 and ∠5
    b. ∠4 and ∠5
    b. ∠1 and ∠3
    d. ∠8 and ∠2

13. Refer to the parallelogram at the right to find the measure of the unknown angle.
    a. 70°
    b. 180°
    c. 110°
    d. 40°

14. Find the surface area of a rectangular prism with a length of 20 centimeters, a width of 20 centimeters, and a height of 25 centimeters.
    a. 1400 cm^2
    b. 10,000 cm^2
    c. 2800 cm^2
    d. 65 cm^2

15. A cylinder-shaped salt container has a diameter of 8 inches and a height of 14 inches. Find the volume of the container. Use $\frac{22}{7}$ for π.
    a. 176 in^3
    b. 1408 in^3
    c. 2816 in^3
    d. 704 in^3

16. The pyramid in the figure at the right has a base 300 feet long and 300 feet wide. The height is 150 feet. Find the volume in cubic feet.
    a. 13,500,000
    b. 4,500,000
    c. 45,000
    d. 6,750,000

Cumulative Review: Chapters 16–17 **539**

GLOSSARY

The following definitions and statements reflect the usage of terms in this textbook.

Absolute Value To indicate distance, but not direction, from zero, you use *absolute value*. (Page 304)

Acute Angle An *acute angle* is an angle whose measure is less than 90°. (Page 502)

Algebraic Expression An *algebraic expression* contains at least one variable. (Page 22)

Angle An *angle* is a pair of rays that have a common endpoint. (Page 498)

Area The *area* is the number of square units needed to cover a surface. (Page 70)

Average The *average* or *mean* of two or more numbers or measures is the sum of the measures divided by the number of measures. (Page 10)

Binomial A *binomial* is the sum or difference of two monomials. (Page 315)

Circumference The *circumference* of a circle is the distance around the circle. (Page 176)

Combining Like Terms When you add or subtract like terms, you are *combining like terms*. (Page 76)

Composite Number A *composite number* is a counting number greater than 1 that is not a prime number. (Page 103)

Congruent Triangles Triangles that have the same size and shape are *congruent triangles*. (Page 505)

Coordinate Plane The *coordinate plane* has four regions, or quadrants, which are separated by the x and y axes. (Page 408)

Discount The amount that an item is reduced in price is the *discount*. (Page 244)

Divisible A number is *divisible* by another number if the remainder is 0. (Page 96)

Equation An *equation* is a sentence that contains the equality symbol "=." (Page 190)

Equivalent Equations Equations that have the same solution(s) are *equivalent equations*. (Page 193)

Equivalent Fractions *Equivalent fractions* name the same number. (Page 161)

Evaluate *Evaluate* means to find the value. (Page 18)

Even Number A number that is divisible by 2 is an *even number*. (Page 100)

Exponent The second power of 3 is written as 3^2. The raised two is called an *exponent*. The exponent indicates how many times a number is multiplied by itself. (Page 8)

Factor Since 24 is divisible by 8, 8 is a *factor* of 24. (Page 96)

Factoring Writing a sum or difference as a product is called *factoring*. (Page 364)

Formula A *formula* is a shorthand way of writing a word rule. In a formula, variables and symbols are used to represent words. (Page 2)

Greatest Common Factor The *greatest common factor* (GCF) of two or more counting numbers is the greatest number that is a factor of each number. (Page 115)

Hypotenuse The *hypotenuse* of a right triangle is the side opposite the right angle. (Page 454)

Integers *Integers* are made up of the positive integers, zero, and the negative integers. (Page 294)

Interest *Interest* is the money charged for the use of borrowed money. (Page 240)

Interest Rate Interest is usually expressed as a percent of the principal. This percent is called the *interest rate*. (Page 240)

Least Common Multiple The *least common multiple* (LCM) of two or more counting numbers is the smallest counting number that is divisible by the given numbers. (Page 111)

Like Fractions *Like fractions* have a like denominator. (Page 158)

Like Terms Two or more *like terms* have the same variables and the same powers of these variables. (Page 76)

Linear Equation A *linear equation* is an equation whose graph is a straight line. (Page 410)

List Price The regular price of an article is called the *list price*. (Page 244)

Median In a listing of data, the *median* is the middle measure or score. (Page 470)

Mode In a listing of data, the *mode* is the measure that occurs most often. (Page 470)

Monomial A monomial over the set of rational numbers in one variable is an expression of the form ax^n, where a is a rational number and n is any natural number. (Page 315)

Multiple A *multiple* of a given number is a product that has the given number as one of its factors. (Page 96)

Net Price The price of an article after a discount is deducted is the *net price*. (Page 245)

Numerical Expression A *numerical expression* includes at least one of the operations of addition, subtraction, multiplication, or division. (Page 18)

Obtuse Angle An *obtuse angle* is an angle whose measure is greater than 90°. (Page 502)

Odd Number A number that is not divisible by 2 is an *odd number*. (Page 100)

Opposites Two numbers that are the same distance from 0 and are in opposite directions from 0 are *opposites*. (Page 297)

Ordered Pair An *ordered pair* is a pair of numbers one of which is designated as the first and the other is the second. (Page 408)

Origin In the coordinate plane, the point with coordinates (0, 0) is called the *origin*. (Page 408)

Parallel Lines *Parallel lines* are lines in a plane that are always the same distance apart. (Page 514)

Parallelogram A *parallelogram* is a quadrilateral with both pairs of opposite sides parallel. (Page 517)

Percent A *percent* is a ratio in which the second term is 100. (Page 227)

Perfect Square A rational number is a *perfect square* if it is the square of a rational number. (Page 442)

Perimeter The *perimeter* of a figure is the distance around it. To find perimeter, add the lengths of the sides. (Page 81)

Perpendicular Lines When two intersecting lines form a right angle, the lines are *perpendicular lines*. (Page 514)

Polygon A *polygon* is a closed plane figure formed by line segments. (Page 504)

Polynomial A *polynomial* is a monomial or the sum or difference of two or more monomials. (Page 315)

Prime Factorization Every composite number can be expressed as a product of prime factors. Such a product is called the *prime factorization* of the number. (Page 108)

Prime Number A *prime number* is a counting number greater than 1 that has exactly two counting number factors, 1 and the number itself. (Page 103)

Principal The amount of money borrowed on which interest is paid is called the *principal*. (Page 240)

Probability The ratio of the number of favorable outcomes to the number of possible outcomes is the *probability* that an event will occur. (Page 479)

Proportion A *proportion* is an equation that states that two ratios are equal. (Page 219)

Pythagorean Theorem In any right triangle, the square of the length of the hypotenuse is equal to the sum of the squares of the lengths of the legs, or $c^2 = a^2 + b^2$. (Page 454)

Ratio A *ratio* is a quotient that compares two numbers. (Page 216)

Rational Numbers A number that can be written as a fraction is a *rational number*. (Page 297)

Reciprocals Two numbers having 1 as their product are called *reciprocals*. (Page 148)

Rectangular prism A space figure having two equal rectangles as bases. The bases are parallel.

Retail Price The public pays the *retail price* for items sold in stores. (Page 261)

Right Angle A *right angle* is an angle whose measure is 90°. (Page 502)

Similar Triangles Triangles that have the same shape are *similar triangles*. (Page 508)

Slope The *slope* of a line is the ratio of the vertical change to the horizontal change between any two points on a line. (Page 417)

Solve an Equation To *solve an equation* means to find the numbers that can replace the variable to make the equation true. (Page 190)

System of Equations Two equations in the same two variables form a *system of equations*. (Page 427)

Tangent Ratio In a right triangle, the *tangent* of an acute angle, A, is the ratio

$$\frac{\text{length of side opposite angle } A}{\text{length of side adjacent to angle } A}$$

(Page 458)

Triangle A *triangle* is a polygon with three sides. (Page 504)

Trinomial A *trinomial* is the sum or difference of three monomials. (Page 315)

Variable A *variable* is a letter representing one or more numbers. (Page 22)

Volume The number of cubic units contained in a solid is the *volume* of the solid. (Page 145)

Wholesale Price *Wholesale price* is the price that a store pays for items it will sell. (Page 261)

X Intercept The *x intercept* of a line is the *x* value of its point on the *x* axis. (Page 422)

Y Intercept The *y intercept* of a line is the *y* value of its point on the *y* axis. (Page 422)

INDEX

Boldface numerals indicate the pages that contain formal or informal definitions.

Absolute value, **304**-305
Acute angle, **502**-504
Addition
   Associative Property of, **79**, 312
   Commutative Property of, **79**, 312
   of decimals, 35
   of fractions having unlike denominators, 161
   of like fractions, 158
   of mixed numerals, 158
   of mixed numerals with unlike denominators, 165
   of polynomials, 315
   of positive and negative numbers, 300, 304-305
   of rational numbers, 304-305
   on a number line, 300
   Property for Equations, **324**, 326, 329
   Property for Inequalities, **340**
   Property of Opposites, **312**
   Property of Zero, **64**, 312
   to solve equations, 193-194
Algebraic expression, **22**, 334, 377
Angle(s), **498**, 501-502
   congruent, 505, **514**-515
   of a triangle, 388
Annual percentage rate, 332
Applications
   addition and subtraction in equations, 196
   algebraic expressions, 334-335, 377-378
   angles in a triangle, 388
   area, 70-71, 140-141
   broken line graphs, 414-415
   career, 138, 311, 448, 478
   circumference and area, 176-178
   computer, 43, 75, 168, 226, 333, 387, 421, 449, 512
   consecutive numbers, 390-391
   consumer, 13, 43, 107, 139, 199, 269, 303, 332, 359
   discount, 244-245
   estimating with decimals, 47-48
   estimating with whole numbers, 14, 15
   inequalities, 396-397
   multiplication and division in equations, 206-207
   of scientific notation, 55-56
   percent, 257-258, 264-266, 274-275, 279-280
   perimeter, 81-82
   proportion, 222-223
   ratio, 222, 417, 424, 479
   simple interest, 240-241
   square root, **440**, 442, 445, 451
   surface area and rectangular prisms, 520-521
   temperature, 132-133
   time cards, 169
   using similar triangles, 508-509
   volume, 145
Area, **70**
   of a circle, 177
   of a parallelogram, 140, 177
   of a rectangle, 70
   of a square, 70
   of a triangle, 141
Associative Property
   of Addition, **79**, 312
   of Multiplication, **67**, 356
Average, **10**, 470

Bar graph, **270**
Batting average, **207**
Binomial, **315**
Break-even, **427**
Broken line graph, **414**-415

Calculator Applications
   adding rational numbers, 314
   annual percentage rate, 332
   area of triangles, 449
   checking equations, 331
   checking inequalities, 342
   computing y values for equations, 413
   cost of wallpaper, 199
   distributive property, 367
   estimation, 50
   evaluating powers, 168
   finding discount, 282
   greatest safe load, 138
   multiplying probabilities, 490
   multiplying rational numbers, 355
   Order of Operations, 21
   prime number/exponents, 106
   safe stopping distance, 303
   similar polygons, 511
   simple interest, 243
   standard deviation, 478
   tangent values, 461
   wind chill temperature, 448
Career Applications
   Engineering Technician, 138
   Meteorologist, 448
   Statistical Clerk, 478
   Travel Agent, 311
Celsius, **132**
Centigram, 52
Centiliter, 52
Centimeter, **51**
Circumference, **176**, 177
Clock arithmetic, 321
Commission, **236**
Common factors, **364**
Commutative Property
   of Addition, **79**, 312
   of Multiplication, **67**, 356
Composite number, **103**, 108
Computer Applications
   Adding Fractions, 333
   Angle Measures, 512
   Area of Triangles, 449
   Evaluating Expressions, 75
   Evaluating Powers, 168
   Problem Solving, 226
   Slope of a Line, 421
   Solving Equations, 387
   Unit Price, 43
Congruent angles, 505, **514**-515
Congruent triangles, **505**-506
Consecutive number problems, 390-391
Consumer Applications
   Banking, 332
   Conserving Energy, 359
   Driving Safety, 303
   Nutrition, 269
   Painting and Estimation, 107
   Saving $ on Cooling Costs, 139
   Saving $ on Housing Costs, 13
   Unit Price, 43
   Wallpaper and Estimation, 199
Coordinate plane, **408**
Counting numbers, **96**, 97
Cube, **521**
Cubic units, **524**
Customary measures, 126-127

Data, **270**, 474

Index   543

Decigram, 52
Deciliter, 52
Decimals, 32
   as percent, 231-232
   comparing, 33
   non-repeating, 443
   place value, 32
   repeating, **174**, 213, 443
   terminating, **173**, 443
Decimeter, 51
Degree, **501**
Dekagram, 52
Dekaliter, 52
Dekameter, 51
Dependent events, **488**
Diameter of a circle, **176**
Digits, **99**
Direct variation, **424**
Discount, **244**, 245, 282
Distributive Property, **356**
Divisible by, **96**, 99-100
Division
   of decimals, **44**
   of fractions, 148
   of mixed numerals, 149
   of rational numbers, 360-361
   Property for Equations, 374-375
   Property for Inequalities, **394**
   to solve equations, 200-201, 374
Equation(s), 190-191
   Addition Property for, **324**, 326, 329
   Division Property for, **374**-375
   equivalent, **193**, 201, 203, 324
   for finding percent of a number, 235
   linear, **410**
   Multiplication Property for, **374**
   Solving, 190, 193, 196, 200, 203
   Subtraction Property for, **324**, 326, 329
   to solve word problems, 335
   with like terms, 326
   with more than one operation, 326, 381
   with parentheses, 383
   with variable on both sides, 329
Equilateral triangle, **504**
Equivalent equations, **193**, 201, 203, 324
Equivalent fractions, **161**, 235
Equivalent inequalities, **340**
Equivalent ratios, **216**-217, 508
Estimation, 14, 47, 107, 139, 176, 184, 199, 279, 332, 458
Evaluate, **18**, 22

Even number, **100**
Exponent(s), **8**

Factor, **96**, 97, 108, 124
Factoring, **364**-365
Fahrenheit, **132**
Formula, **2**
   for area of a circle, 177
   for area of a parallelogram, 140, 177
   for area of a rectangle, 70
   for area of a square, 70
   for area of a triangle, 141
   for compound interest, 251
   for converting temperatures, 132, 133
   for discount, 244-245
   for distance in feet an object will fall, 8
   for distance in miles to horizon from aircraft above the ground, 451
   for distance/rate/time, 7, 206
   for finding airspeed in miles per hour, 196
   for finding an average, 10
   for finding angles of a triangle, 388
   for finding batting average, 207
   for finding certain prime numbers, 106
   for finding kilowatt-hours, 5
   for finding net pay, 196
   for finding sticker price, 2-3
   for greatest safe load, 138
   for heat transfer, 359
   for length of one side of a square, 451
   for perimeter of a rectangle, 81
   for perimeter of a square, 82
   for perimeter of a triangle, 82
   for prime numbers, 106, 117
   for sale price, 245
   for simple interest, 240
   for stopping distance, 303
   for sum of n numbers, 86
   for surface area of a cube, 521
   for surface area of a rectangular prism, 521
   for tangent ratio, 458
   for volume of a cone, 528
   for volume of a cylinder, 524
   for volume of a pyramid, 527
   for volume of a rectangular prism, 145
   for volume of a sphere, 535

Pythagorean Theorem, 454-455
Fraction(s), **124**
   as a decimal, 173-174
   as percent, 231
   equivalent, **161**

Geometry, **498**
Gram, **52**
Graph(s)
   of an ordered pair, 408-409
   of integers, 294, 300
   of linear equations, 410-411
   of linear inequalities, 430
   on a number line, 294, 300, 338, 340
   to show data as percents, 274
Graphical method of solving a system of equations, 427-428
Greatest common factor, **115**, 124
Gross earnings, **196**
Groundspeed, **196**

Head wind, **196**
Hecto-, **51**, 52
Hectogram, 52
Hectoliter, 52
Hectometer, 51
Hidden question, **3**, 170
Histogram, **474**-475
Hypotenuse, **454**

Independent events, **488**
Indirect variation, **425**
Inequalities, **295**, 338, 340
   Addition Property for, **340**
   applications of, 396-397
   Compound, **347**
   Division Property for, **394**
   equivalent, 340
   Multiplication Property for, **393**
   on a number line, 338-339
   Subtraction Property for, **340**
Integers, **294**
   graphs of, 294, 300
   negative, 294
   positive, 294
Intercepts of a graph, 422-423
Interest, **240**
   compound, 251
   simple, 240, 243
Interest rate, **240**
Inverse operation, **190**
   addition as inverse of subtraction, 193
   division as inverse of multiplication, 200

544   *Index*

multiplication as inverse of division, 203
subtraction as inverse of addition, 190, 193
Irrational square root, **442**
Isosceles triangle, **504**

Kilo-, 51, 52
Kilogram, **52**
Kiloliter, **52**
Kilometer, **51**
Kilowatt-hour, **5**, 38

Least common multiple, 111
Like denominators, 158
Like fractions, **158**
Like terms, **76**, 326
  combining, 76, 326
  in solving equations, 326
Line(s), **498**
  parallel, **514**
  perpendicular, **514**
Linear equation, **410**
Linear inequalities, **430**
  graph of, 430
Line segment, **498**
List price, **244**
Liter, **52**

Mathematical sentences, 64
Mean, **10**, 470
Median, **470**-471
Meter, **51**
Metric ton, **52**
Milli-, **51**, 52
Milligram, **52**
Milliliter, **52**
Millimeter, **51**
Mixed numeral(s), **126**, 135, 158, 165
  multiplication of, 135-136
  division of, 149
Mode, **470**
Monomial, **315**
Multiple, **96**
Multiplication
  Associative Property of, **67**, 356
  Commutative Property of, **67**, 356
  of decimals, 38-39
  of fractions, 129
  of mixed numerals, 135-136
  of numbers with like signs, 353
  of probabilities, 488, 491
  of rational numbers, 353, 356

of two numbers with unlike signs, 350
Property for Equations, **374**
Property for Inequalities, **393**
Property of One, **64**, 351
Property of Zero, **64**, 351
to solve equations, 203-204, 374
Multiplicative inverses, **148**
Mutually exclusive events, **483**

Negative integers, 294
Negative rational numbers, **297**
Net pay, **196**
Net price, **245**
Non-repeating decimal, 443
Number(s)
  composite, 103, 108
  counting, 96, 97
  even, **100**
  geometric, 29
  odd, **100**
  prime, 103, 108
  rational, 297-298, 304
  whole, **99**
Number line, 294, 300
  inequalities, 338, 340
Number sequences, **288**
Numerical expressions, **18**

Obtuse angle, 502, 504
Odd number, **100**
Operation(s)
  inverse, 190-191, 193-194, 200-201, 203-204
  order of, 18, 19, 21
Opposites, 297-298, 308
Order of Operations, 18, 19, 21
Ordered pair, 408
Origin, **408**

Parallel lines, 514
Percent, **227**
  applications, 257-279
  and equivalent fractions, 235
  as a decimal, 231-232
  as a fraction, 231
  in circle graphs, 274
  of decrease, **258**
  of increase, **257**
  of what one number is of another, 254
  to find a number, 261
Perfect square, **442**
Perimeter, 22, 81
  of a rectangle, 81

of a square, 22, 82
of a triangle, 82
Perpendicular lines, **514**
Pi, **176**, 177
Place name, 32
Plane, **498**
Point, **498**
Point of intersection, 427-428
Polygon, **504**
Polynomial, **315**
Positive integers, 294
Positive rational numbers, 297, 304-305, 308
Price
  list, **244**
  net, **245**
  retail, 261
  sale, **245**
  wholesale, 261
Prime factorization, **108**-109, 111 115, 124
Prime number, **103**, 108
Principal, **240**
Probability
  multiplication of, 488, 491
  ratio, 479-480
  tables, 482-483
  using tree diagram, 485
Problem Solving and Applications (see Applications)
Property
  Addition, for Equations, **324**, 326, 329
  Addition, for Inequalities, **340**
  Addition, of Opposites, **312**
  Addition, of Zero, **64**, 312
  Associative, of Addition, 79, 312
  Associative, of Multiplication, 67, 356
  Commutative, of Addition, 79, 312
  Commutative, of Multiplication, 67, 356
  Distributive, **356**
  Division, for Equations, 374-375
  Division, for Inequalities, **394**
  Multiplication, for Equations, **374**
  Multiplication, for Inequalities, **393**
  Multiplication, of One, **64**, 351
  Multiplication, of Zero, **64**, 351
  Subtraction, for Equations, 324, 326, 329

Index **545**

Subtraction, for Inequalities, 340
Proportion(s), **219**-220
   applications of, 222-223
Pythagorean Theorem, **454**-455

Quadrilateral, 517-518

Radius of a circle, 176-177
Raised to a power, **8**
Ratio, **216**-217
   applications of, **222**, **417**, **424**, **479**
   equivalent, **216**-217, 508
   probability, **479**-480
   tangent, **458**-459
Rational numbers, **297**-298
   addition of, **304**-305
   as fractions, 297
   as perfect square, 442
   negative, **297**-298, **304**-305
   positive, **297**, **304**-305
   subtraction of, **308**-309
Ray, **498**
Reciprocals, **148**
Rectangle, **70**, **81**, 140
Rectangular prism, **145**, 520-521
Repeating decimal, **174**, 213, 443
Retail price, 261
Right angle, **502**

Sale price, 245
Sample, **495**
Scale, **270**
Scalene triangle, **504**
Scientific notation, 55-56
Set(s), **402**
   element of, **402**, 437
   empty, **402**
   intersection of, **437**
   subset of, **437**
Similar triangles, **508**-509

union of, **437**
Slope, **417**-418
Square, **70**, **82**, **517**
Square of a number, **440**-441
Square root, **440**-441
   irrational, **442**
Square units, 70
Standard deviation, **478**
Sticker price, 2, 3
Straight angle, 502
Subtraction
   of decimals, 35
   of fractions having unlike denominators, 161
   of like fractions, 158
   of mixed numerals, 158
   of mixed numerals with unlike denominators, 165
   of rational numbers, 308-309
   Property for Equations, **324**, 326, 329
   Property for Inequalities, 340
   to solve equations, 190-191
System(s) of equations, **427**
   solved by graphing, 427-428

Table of Squares and Square Roots, 445-446, 447
Table of Tangents, 459
Tangent ratio, **458**-459
Tangent value, 461
Temperature
   Celsius, **132**
   conversions, 132-133
   Fahrenheit, **132**
Terminating decimal, **173**, 443
Total cost of the options, 3
Transversal, **514**
Tree diagram, **485**
Triangle(s)
   congruent, **505**-506

   equilateral, **504**
   isosceles, **504**
   scalene, **504**
   similar, **508**-509
   right, **454**, **458**, 504
Trinomial, **315**

Unlike denominators, 161-162, 165

Value of numerical expression, 18
Variable, **22**, 235
   on both sides of equation, 329
Variation
   direct, **424**
   indirect, **425**
Venn diagram(s), **437**
Vertex, **498**
Volume, **145**
   of a cone, 528
   of a cylinder, 524
   of a pyramid, 527
   of a rectangular prism, **145**
   of a sphere, 535

Whole numbers, 99
Wholesale price, **261**
Word expression, 334, 377

x axis, 408
x coordinate, **408**
x intercept, **422**-423

y axis, 408
y coordinate, **408**
y intercept, **422**-423

Zero, Addition Property of, 64, 312
Zero, Multiplication Property of, **64**, 351

ANSWERS TO SELECTED EXERCISES

ANSWERS TO SELECTED EXERCISES

CHAPTER 1

Pages 2-3 Self-Check 1. $9875 2. $8308

Page 3 Classroom Exercises 1. 20 3. 743 5. 3875 7. 8409

Page 4 Written Exercises 1. $10,235 3. $9135 5. $10,640 7. $9880 9. $8828 11. $9667

Page 5 Self-Check 1. 176

Page 5 Classroom Exercises 1. 1212 3. 8317 5. 2526 7. 1249

Page 6 Written Exercises 1. 817 3. 4204 5. 878 7. 56 9. 8196 11. 14,543

Pages 7-8 Self-Check 1. 568 miles 2. 2000 meters 3. 144 feet

Page 8 Classroom Exercises 1. 15 3. 522 5. 81 7. 64 9. 1

Page 9 Written Exercises 1. 2400 feet 3. 375 kilometers 5. 465 feet 7. 2058 meters 9. 4 11. 16 13. 1 15. 9 17. 36 19. 1000 21. 576 feet 23. 256 feet

Page 10 Self-Check 1. 143 2. 122 3. 5°

Page 11 Classroom Exercises 1. 12 3. 42 5. 123 7. 45 9. 7

Page 11 Written Exercises 1. 143 3. 159 5. 147 7. 45 kilograms 9. 83 11. $150,600 13. 90 15. 22 17. 76,329 19. 25 21. Yes

Page 13 Special Topic 1. 140 3. 1 5. 4 7. 50 9. $86

Pages 14-15 Self-Check 1. 2300 kilometers 2. 300 kilometers 3. 900 rentals 4. 8 hours 5. 26 hours

Page 16 Classroom Exercises 1. a 3. c

Page 16 Written Exercises 1. a 3. a 5. a 7. b 9. 5000 meters 11. 2700 seats

Page 17 Review: Sections 1-1–1-5 1. $7653 2. $13,792 3. 1084 kilowatt-hours 4. 1628 kilowatt-hours 5. 186 meters 6. 784 feet 7. 59 8. 85 9. 130 seconds 10. 20 points

Page 19 Classroom Exercises 1. Rule 1 3. Rule 3 5. Rule 4 7. Rule 1 9. Rule 2 11. Rule 3 13. Rule 3 15. Rule 2

Page 20 Written Exercises 1. addition 3. multiplication 5. division 7. multiplication 9. 8 11. 27 13. 28 15. 8 17. 24 19. 54 21. 100 23. 4 25. 4 27. 47 29. 60 31. 27 33. 36 35. 96 37. 57 39. 50 41. 35 43. 3 45. 19 47. 23 49. 18 51. 56 53. 4 55. 90 57. 21 59. 1 61. 2 63. 24 65. 38 67. 49 69. 10 71. 8 73. 14 75. 19 77. 0 79. 25 81. 6 + 7 · 3 = 27 83. (9 + 9) · 4 = 72 85. 4 · 4 · 4 = 64 87. 8 · 9 + 4 − 9 + 6 = 73 89. 8 · 2 + 9 · 3 = 43 91. (12 + 3) · (21 ÷ 7) = 45 93. (6 · 9) − 14 = 40 95. 18 − (35 ÷ 7) = 13 97. (9 − 9) · (33 ÷ 11) = 0

Page 21 Calculator Exercises 1. 51 3. 5 5. 35 7. 35 9. 13

Pages 22-23 Self-Check 1. 108 2. 5 3. 1 4. 19 5. 51 6. 10 7. 3

Page 23 Classroom Exercises 1. 15 · 5 + 32 3. 6 · 4 + 5 · 7 5. 24 · 19 + 18 · 15 − 30 7. 2 · 3 · 4 − 4 9. 2 11. 3 13. 4 15. 9 17. 4 19. 6 21. 8 23. 22 25. 12 27. 120 29. 46 31. 216

Page 24 Written Exercises 1. 5 3. 8 5. 4 7. 4 9. 7 11. 9 13. 36 15. 51 17. 46 19. 12 21. 29 23. 100 25. 42 27. 4 29. 42 31. 15 33. 48

Page 24 Review: Sections 1-6—1-7 1. 6 2. 14 3. 10 4. 16 5. 43 6. 3 7. 5 8. 4 9. 4 10. 42 11. 13 12. 54 13. 0 14. 28 15. 22 16. 25

Page 25 Chapter Review 1. variable 3. formula 5. kilowatt-hour 7. evaluate 9. $5382 11. $9104 13. 219 15. 667 17. 160 hours 19. 90 meters 21. 1 23. 1 25. 49 27. 64 29. 400 31. 3600 33. 14° 35. 83° 37. c 39. b 41. 12 43. 20 45. 90 47. 30 49. 31 51. 21 53. 8 55. 1 57. 4 59. $10,166 61. 3531 miles 63. 140 cases

Page 28 Chapter Test 1. $9100 3. 1440 meters 5. 1936 feet 7. b 9. b 11. 20 13. 7 15. 18 17. 35 19. $5528

Page 29 Enrichment 1. **a.** A triangle with 1 dot in the first row, 2 dots in the second row, 3 dots in the third row, and so on. The last (fifth) row has 5 dots. **b.** A triangle with 1 dot in the first row, 2 dots in the second row, and so on. The last (sixth) row has 6 dots. **c.** A triangle with 1 dot in the first row, 2 dots in the second row, and so on. The last (seventh) row has 7 dots. 3. **a.** A square having 5 rows and 5 columns; **b.** A square having 6 rows and 6 columns; **c.** A square having 7 rows and 7 columns.

Page 30 Additional Practice 1. $9810 3. 1527 kilowatt-hours 5. 1425 feet 7. 576 feet 9. 2304 feet 11. 78 13. 374 15. a 17. a 19. 11 21. 40 23. 48 25. 203 meters

CHAPTER 2

Pages 32-33 Self-Check 1. 50 2. 0.03 3. 0.4 4. 0 5. > 6. < 7. >

Page 33 Classroom Exercises 1. ones 3. thousands
5. thousandths 7. > 9. >

Page 34 Written Exercises 1. 0.5 3. 5 5. 20
7. 200 9. 2000 11. 0.007 13. 400 15. 0.3
17. 0.07 19. > 21. < 23. < 25. > 27. >
29. < 31. Jane 33. 22.00124

Page 35 Self-Check 1. 31.88 2. 225.653

Page 36 Classroom Exercises 1. 0.58 3. 19.94
5. 214.3 7. 115.263 9. 133.706 11. 2.32
13. 2.043 15. 0.40 17. 5.884 19. 6.8 21. 23.7

Page 36 Written Exercises 1. 24.92 3. 38.542
5. 242.921 7. 27.74 9. 946.66 11. 2.593
13. 0.83 15. 10.5 17. 8.952 19. 7.17
21. 135.44 23. 25.3 25. 71.21 27. 116 seconds
29. 0.09 meters

Pages 38-39 Self-Check 1. 73¢ 2. 2¢

Page 39 Classroom Exercises 1. 20.0 3. 20.0
5. 0.200 7. 0.00090 9. 0.000090 11. 27.0
13. 15.73 15. 18.200 17. 0.014 19. 0.002848

Page 40 Written Exercises 1. 9.6 3. 97.20
5. 13.6890 7. 28.084 9. 0.58357 11. 1.0224
13. 0.087528 15. 0.018 17. 0.00800 19. 0.0244
21. 0.0456 23. 0.00175 25. 0.00728 27. $3.08,
$19.25, $1.54, $3.96 29. $5.47, $34.16, $2.73, $7.03
31. $.42, $2.63, $.21, $.54 33. $8.53, $53.34, $4.27,
$10.97 35. $.02 37. $.05 39. $6.45
41. $11.96 43. $10.95

Page 42 Review: Sections 2-1—2-3 1. hundredths
2. tens 3. < 4. < 5. < 6. > 7. 19.94
8. 192.7 9. 2.19 inches 10. $3.47 11. $.06
12. 2.05 inches

Page 43 Special Topic 1. $0.045 3. $0.11
5. $0.017 7. $0.075 9. $0.15

Pages 44-45 Self-Check 1. 26.2 miles 2. 35
3. 730 4. 2900

Page 45 Classroom Exercises 1. 3.48 3. 0.716
5. $32 \overline{)896}$ 7. $36 \overline{)7200}$ 9. 9.4 11. 5470

Page 45 Written Exercises 1. 2.84 3. 9.81
5. 0.065 7. 8.07 9. 1.47 11. 0.056 13. 6.2
15. 89 17. 0.35 19. 7060 21. 3670 23. 8620
25. 16.7 miles 27. 84.5 miles 29. first car

Pages 47-48 Self-Check 1. $13.00 2. 50 pounds

Page 48 Classroom Exercises 1. c 3. a 5. b
7. b 9. b 11. c

Page 49 Written Exercises 1. b 3. b 5. c 7. b
9. $11 11. $66 13. 50 pennants

Page 50 Calculator Exercises 1. 280, 281
3. 12,000, 12,139 5. 70, 68.6

Pages 52-53 Self-Check 1. 25,000 m 2. 1.5 cm

3. 790 cm 4. 3600 cg 5. 0.350 g 6. 53.1 dg
7. 0.022 L 8. 37,000 mL 9. 792,000 L

Page 53 Classroom Exercises 1. divide 3. multiply
5. divide 7. 10 9. 1000

Page 53 Written Exercises 1. 30 mm 3. 54,000 m
5. 3500 m, 350,000 cm 7. 0.256 Km, 25,600 cm
9. 5.100 Kg 11. 900 Kg 13. 3500 g, 3,500,000 mg
15. 0.72 Kg, 720,000 mg 17. 1000 L 19. 450 ml
21. 23,000 ml 23. 620 L 25. 4200 m
27. 0.946 ml 29. 1000 mg

Pages 55-56 Self-Check 1. 2.6×10^3 2. 864,000

Page 56 Classroom Exercises 1. 1 3. 3 5. 4
7. 4×10 9. 4×10^2 11. 4×10^3 13. 160
15. 38,000

Page 56 Written Exercises 1. 3.4×10^3 3. 9×10^4
5. 1.05×10^6 7. 8.75×10^2 9. 8×10^7
11. 7.6×10^4 13. 3200 15. 251,000
17. 9,123,000 19. 52,000,000 21. 14.2 23. 2×10^6

Page 57 Review: Sections 2-4—2-7 1. 0.7 m
2. 18.4 miles 3. b 4. a 5. c 6. a 7. 0.0043
8. 400 9. 0.3204 10. 3 11. 5 12. 8.25
13. 4.5

Page 58 Chapter Review 1. liter 3. deka-
5. meter 7. metric ton 9. scientific notation
11. gram 13. 0.4 15. 0.04 17. 0.005 19. >
21. < 23. > 25. 10.84 27. 1.669 29. 23.595
31. 105.792 33. 40.81 35. $14.94, $11.43, $7.16
37. 5.89 39. 122 41. c 43. b 45. 20
47. 4510 49. 550 51. 2.5×10^2 53. 5×10^6
55. c 57. $299.64

Page 60 Chapter Test 1. hundredths 3. > 5. <
7. 191.83 9. 37.35 11. 476.37 13. 3.26
15. 0.002832 17. b 19. 2600 21. 3 23. 6
25. 28.2 miles

Page 61 Enrichment 1. 500 sec 3. a. 16,292 h
b. 679 days

Page 62 Additional Practice 1. 0.8 3. 60
5. 0.0003 7. < 9. > 11. 82.993 13. 1.26
15. 0.000912 17. 13.2 19. 660 21. b 23. b
25. 4,300,000 27. 14,000 29. 7.7×10^4
31. 1.3×10^2 33. 310 35. 35,000,000 37. $.01

CHAPTER 3

Pages 64-65 Self-Check 1. $t = 0$ 2. $s = 0$
3. $r = 58$ 4. $r = 21$ 5. $r = 15$ 6. $r = 21$

Page 65 Classroom Exercises 1. Add. Prop. of Zero
3. Mult. Prop. of Zero 5. $x = 0$ 7. $c = 7$ 9. $p = 1$
11. $t = 0$ 13. $m = 24$ 15. $k = 0.3$ 17. $x = 7$
19. $a = 2.5$

Answers to Selected Exercises 549

Page 65 Written Exercises 1. $n = 0$ 3. $x = 0.45$ 5. $y = 0$ 7. $m = 0$ 9. $n = 1$ 11. $k = 4.5$ 13. 13. $x = 13$ 15. $z = 19$ 17. $n = 21.3$ 19. $p = 23$ 21. $c = 3$ 23. $m = 5$ 25. $n = 1$ 27. $p = 0$ 29. $t = 0$ 31. $c = 1$ 33. $n = 16$ 35. $x = 0$ 37. $n = 3$ 39. $n = 7$ 41. $n = 3$ 43. $x = 13$ 45. $b = 5$ 47. $z = 0$ 49. $n = 6$ 51. $r = 36$ 53. $n = 5$

Pages 67-68 Self-Check 1. $k = 2$ 2. $d = 7$ 3. $30x$ 4. $62xy$ 5. $33cz$ 6. $11xy$ 7. $50c^2n$ 8. $30a^3b$ 9. $48xy^3$

Page 68 Classroom Exercises 1. $x = 5$ 3. $n = 30$ 5. $p = 2.5$ 7. $y = 7$ 9. $t = 3$ 11. $48t$ 13. $5.5k$ 15. $8ab$ 17. $8rs$ 19. $6x^2y$ 21. $21p^2q^2r$ 23. $90f^2g^2h$ 25. $49k^2n^3$

Page 69 Written Exercises 1. $a = 7$ 3. $n = 2$ 5. $n = 6.2$ 7. $k = 5$ 9. $x = 3$ 11. $24y$ 13. $90b$ 15. $8.4n$ 17. $96r$ 19. $96st$ 21. $8.4pq$ 23. $19.2ab$ 25. $6.08rst$ 27. $84y^2$ 29. $36k^2t$ 31. $42m^3n$ 33. $6st^3$ 35. $80m^2n^3$ 37. $81a^2d^2p^2$ 39. $72mn$ 41. $297p$ 43. $6m^2n^2$ 45. $5.6a$ 47. $24abc$ 49. $6m^3np^2$ 51. $24x^2y^2$; 96 53. x^2y^5; 4 55. $90x^3y^5$; 720

Pages 70-71 Self-Check 1. $1.25 m^2$ 2. $9 yd^2$

Page 71 Classroom Exercises 1. $50 cm^2$ 3. $5 km^2$ 5. $21 in^2$ 7. $64 cm^2$ 9. $0.25 m^2$ 11. $225 in^2$ 13. $625 ft^2$

Page 71 Written Exercises 1. $1134 m^2$ 3. $12 m^2$ 5. $3 m^2$ 7. $36.4 mm^2$ 9. $21 m^2$ 11. $308 yd^2$ 13. $104 in^2$ 15. $48 yd^2$ 17. $3 mi^2$ 19. $961 m^2$ 21. $1.21 m^2$ 23. $100 m^2$ 25. $190.44 mm^2$ 27. $4 mi^2$ 29. $9 ft^2$ 31. $256 yd^2$ 33. $16 mi^2$ 35. $12,500 cm^2$ 37. $13,120 ft^2$ 39. $225 in^2$ 41. $20,908.4 m^2$ 43. $8.99 m^2$

Page 74 Review: Sections 3-1–3-3 1. Mult. Prop. of Zero 2. Add. Prop. of Zero 3. Mult. Prop. of One 4. $t = 3.8$ 5. $r = 32$ 6. $x = 0$ 7. Assoc. Prop. of Mult. 8. Comm. Prop. of Mult. 9. Assoc. Prop. of Mult. 10. $x = 8$ 11. $g = 6$ 12. $y = 9$ 13. $24y$ 14. $24t$ 15. $12a^2b$ 16. $5670.09 m^2$ 17. $1008 in^2$ 18. $2914 cm^2$ 19. $324 in^2$

Page 75 Special Topic 1. 9 3. 15 5. 27.5 7. 7 9. 20 11. 36 13. 84 15. 108

Pages 76-77 Self-Check 1. yes 2. yes 3. no 4. no 5. $10y$ 6. $13rs$ 7. $9.5z^2$ 8. $6p$ 9. $12r$ 10. $4.1t^2$

Page 77 Classroom Exercises 1. like; same variables 3. unlike; different exponents 5. like; same variables 7. unlike; different variables 9. $7 + 15 = 22$ 11. $0.9 - 0.4 = 0.5$ 13. $5 - 4 = 1$ 15. $32 - 19 = 13$ 17. $11x$ 19. $0.9d$ 21. $21r^2$ 23. $34ad$ 25. $5y$ 27. $25b$ 29. $3.3m^2$ 31. $20st$

Page 78 Written Exercises 1. H 3. D, R 5. O 7. C 9. Q 11. $16a$ 13. $13b$ 15. $39y$ 17. $3.9x$ 19. $12m^2$ 21. $14xy$ 23. $7s$ 25. $5a$ 27. $5.9r$ 29. $3.8a$ 31. $2x^2$ 33. $3ab$ 35. $2n$ 37. $45st$ 39. $8.2f^2$ 41. $16.02 m$

Pages 79-80 Self-Check 1. $q = 11.5$ 2. $t = 3$ 3. $17a + 7$ 4. $11t$ 5. $3.9x + 13$

Page 80 Classroom Exercises 1. $r = 11$, Comm. and Assoc. Prop. of Add. 3. $x = 5$, Comm. Prop. of Add. 5. $52r + 10$ 7. $42st + 25$ 9. $1.5t + 1.8$ 11. $87n^2 + 9$

Page 80 Written Exercises 1. $r = 2.7$ 3. $x = 4$ 5. $x = 6$ 7. $5a + 4$ 9. $29a + 15$ 11. $3.9m + 4.2$ 13. $55k + 7$ 15. $18x$ 17. $9.5t$ 19. $13a^2 + 4$ 21. $10xy + 4$ 23. $3.9p + 1.9$ 25. $15p + 16$ 27. $91r$ 29. $190 + 254p$

Pages 81-82 Self-Check 1. $82 m$ 2. $74 ft$ 3. $20 m$ 4. $13 yd$

Page 82 Classroom Exercises 1. $22 cm$ 3. $54 m$ 5. $24 in$ 7. $19 cm$ 9. $33.8 cm$ 11. $22 in$

Page 83 Written Exercises 1. $130 cm$ 3. $7.2 m$ 5. $69.8 m$ 7. $10.2 cm$ 9. $64 mm$ 11. $16 ft$ 13. $42 in$ 15. $24 yd$ 17. $20 ft$ 19. $2.9 m$ 21. $25.1 m$ 23. $9.8 m$ 25. $32 cm$ 27. $9 m$ 29. $36 ft$ 31. $12 ft$ 33. $15 yd$ 35. $49 in$ 37. $2256 ft$ 39. $104 yd$ 41. $8b$ 43. $3a$ 45. $2a + 2b + c$ 47. $12x$

Page 86 Review: Sections 3-4—3-6 1. yes 2. no 3. yes 4. $5x$ 5. $7.2b$ 6. $1.9ef$ 7. $8n^2$ 8. $a = 3.4$ 9. $c = 4$ 10. $m = 3$ 11. $12b$ 12. $14g + 17$ 13. $11.3t + 10$ 14. $8.2x^2 + 2.5$ 15. $180 mi$ 16. $18.4 km$

Page 86 Calculator Exercises 1. 153 3. 2346 5. 183,315

Page 87 Chapter Review 1. area 3. $4 + 3 = 3 + 4$ 5. square 7. $25 \cdot 0 = 0$ 9. $5 \cdot 12 = 12 \cdot 5$ 11. perimeter 13. Mult. Prop. of Zero 15. Comm. Prop. of Mult. 17. Assoc. Prop. of Mult. 19. $y = 1.9$ 21. $x = 3$ 23. $k = 4$ 25. $25.2m$ 27. $17rt$ 29. $45a^2$ 31. $68p^2q$ 33. $100a^2b^3$ 35. $54x^2y^3$ 37. $70.56 cm^2$ 39. $441 m^2$ 41. like 43. unlike 45. unlike 47. $13n$ 49. $11a$ 51. $12.1t$ 53. $2.8n^2$ 55. $y = 3$ 57. $6.7a + 3$ 59. $13ab + 5$ 61. $9t + 8$ 63. $9x^2$ 65. $84 in$ 67. $29.5 cm$ 69. $108 ft^2$ 71. $136.6 m$

Page 89 Chapter Test 1. $x = 5.2$ 3. $a = 1$ 5. $14.76x$ 7. $36x^2$ 9. $11.56 m^2$ 11. $328 in^2$ 13. $15x$ 15. $14a + 3$ 17. $40.4 m$ 19. $27 cm$ 21. $111 ft$ 23. $21.9 cm$ 25. $8.3 m$

Page 90 Enrichment 1. 190 3. 21,000 5. 37,100 7. 162 9. 500 11. 16,100

Page 91 Additional Practice 1. $x = 0$ 3. $x = 1$

Answers to Selected Exercises

5. $x = 6$ **7.** $y = 13$ **9.** $36x$ **11.** $20p^3$
13. $20x^3y^3$ **15.** 84.64 cm^2 **17.** 137.64 m^2
19. $3cd$ **21.** $36.7x^2y^2$ **23.** $5xy + 3$ **25.** $12.8x + 4$
27. $18.3m^2n + 13$ **29.** 208 yd **31.** 72 cm
33. square; 1 yd greater **35.** 23.32 yd^2

Page 92 Common Errors **1.** 9 **3.** 11 **5.** 21.39
7. 4000 mm **9.** $9b$ **11.** 32 cm

Page 93 Cumulative Review: Chapters 1-3 **1.** c
3. c **5.** a **7.** b **9.** b **11.** a **13.** b **15.** c
17. d **19.** d **21.** c **23.** d **25.** d

CHAPTER 4

Pages 96-97 Self-Check **1.** False; since $5 \cdot 7 = 35$
2. True; since $28 \div 4 = 7$ **3.** 1, 2, 3, 4, 6, 12 **4.** 1, 2, 3, 4, 6, 8, 12, 16, 24, 48 **5.** 1, 3 **6.** 1, 11 **7.** 1, 2, 4, 19, 38, 76 **8.** 1, 3, 5, 7, 15, 21, 35, 105

Page 97 Classroom Exercises **1.** 40 is divisible by 20. **3.** 6 is divisible by 2. **5.** 40 is divisible by 5.
7. 7 is divisible by 7. **9.** 11 is a factor of 55. **11.** 8 is a factor of 40. **13.** 1 is a factor of 7. **15.** 33 is a factor of 33. **17.** 24 is a multiple of 8. **19.** 6 is a multiple of 6. **21.** 9 is a multiple of 1. **23.** 15 is a multiple of 3. **25.** $1 \cdot 13$ **27.** $1 \cdot 18; 2 \cdot 9; 3 \cdot 6$
29. $1 \cdot 30; 2 \cdot 15; 3 \cdot 10; 5 \cdot 6$

Page 98 Written Exercises **1.** True, since $18 \div 6 = 3$
3. False, since $21 \div 8 = 2R5$ **5.** True, since $29 \cdot 1 = 29$
7. True, since $5 \cdot 15 = 75$ **9.** True, since $143 \div 13 = 11$ **11.** 1, 13 **13.** 1, 23 **15.** 1, 31 **17.** 1, 2, 3, 4, 5, 6, 10, 12, 15, 20, 30, 60 **19.** 1, 2, 23, 46
21. 1, 7, 49 **23.** No; No; No; 1, 3, 5, 15 **25.** Yes; Yes; Yes; 1, 2, 4, 7, 8, 14, 28, 56

Page 100 Self-Check **1.** 2, 4 **2.** 2, 4 **3.** 2, 4, 5, 10 **4.** 5 **5.** 2, 4, 5, 10 **6.** 3, 9 **7.** 3 **8.** 3
9. neither **10.** No **11.** 2

Page 101 Classroom Exercises **1.** False; 71 does not end in 2 **3.** True; 16 is divisible by 4 **5.** False; 385 does not end in 0 **7.** True; 4130 ends in 0 **9.** True; 9 is divisible by 3 **11.** False; 19 is not divisible by 9
13. True; 18 is divisible by 9 **15.** True; 18 is divisible by 9 **17.** True; 15 is divisible by 3 **19.** True; 27 is divisible by 9

Page 101 Written Exercises **1.** True; 428 ends in 8
3. True; 95 ends in 5 **5.** False; 17 is not divisible by 4
7. True; 28 is divisible by 4 **9.** False; 926 does not end in 0 **11.** True; 15 is divisible by 3 **13.** False; 10 is not divisible by 3 **15.** True; 21 is divisible by 3
17. False; 21 is not divisible by 9 **19.** True; 27 is divisible by 9 **21.** False; 52,374 does not end in 0
23. True; 18 is divisible by 3 **25.** True; 44 is divisible by 4 **27.** False; 67,237 does not end in 0, 2, 4, 6, or 8
29. True; 675 ends in 5 **31.** False; 16 is not divisible by 9 **33.** No; 8 **35.** No; 2

Page 104 Self-Check **1.** 3 **2.** 2 **3.** 2 **4.** 5
5. 31 **6.** $3^2 \cdot 5^2 \cdot 7^2 \cdot 11$ **7.** $2 \cdot 3^2 \cdot 5^3 \cdot 19$

Page 104 Classroom Exercises **1.** prime **3.** composite **5.** prime **7.** 2 **9.** 3 **11.** 7 **13.** $2 \cdot 3 \cdot 5^2$
15. $2^3 \cdot 3^2 \cdot 7 \cdot 11^2$ **17.** $2^4 \cdot 11^2$ **19.** $3^4 \cdot 4^2 \cdot 7^3$

Page 105 Written Exercises **1.** 2 **3.** 3 **5.** 5
7. 5 **9.** 11 **11.** 13 **13.** 29 **15.** 3 **17.** 11
19. $2^3 \cdot 3^2$ **21.** $2^3 \cdot 3^2 \cdot 5 \cdot 11^2$ **23.** $2^2 \cdot 11^3 \cdot 13^2$
25. $2^3 \cdot 3^3 \cdot 5^2 \cdot 7$ **27.** $2^6 \cdot 3^3$

Page 105 Review: Sections 4-1–4-3 **1.** 1, 3, 11, 33
2. 1, 2, 5, 10 **3.** 1, 2, 4, 5, 8, 10, 20, 40 **4.** 1, 2, 13, 26 **5.** 1, 3, 5, 15 **6.** 1, 2, 3, 4, 6, 8, 12, 24 **7.** 1, 2, 4, 5, 10, 20 **8.** 1, 2, 3, 4, 6, 12 **9.** 1, 7 **10.** 1, 13 **11.** 1, 2, 3, 5, 6, 10, 15, 30 **12.** 1, 2, 4, 8, 16, 32
13. True; 12 is divisible by 3 **14.** True; 24 is divisible by 3 **15.** False; 25 is not divisible by 9 **16.** True; 24 is divisible by 4 **17.** True; 44 is divisible by 4
18. False; 1379 does not end in 0, 2, 4, 6, or 8
19. True; 2160 ends in 0 **20.** True; 13,370 ends in 0
21. 2 **22.** 5 **23.** 3 **24.** 7 **25.** $3^3 \cdot 5^2$
26. $2^3 \cdot 3 \cdot 7^2$ **27.** $2^2 \cdot 3^2 \cdot 11^3$ **28.** $2 \cdot 3^3 \cdot 5^3 \cdot 7^2$
29. $3^3 \cdot 5^3$ **30.** $2^2 \cdot 3^2 \cdot 4^2 \cdot 5^2$ **31.** No; 3
32. No; 1

Page 106 Calculator Exercises **1.** 7 **3.** 1023
5. 32,767

Page 107 Special Topic **1.** 6 L **3.** 5 L

Pages 108-109 Self-Check **1.** $7 \cdot 11$ **2.** $3 \cdot 11$
3. $2 \cdot 3 \cdot 5$ **4.** $2 \cdot 3 \cdot 5 \cdot 7$ **5.** $2^2 \cdot 5$ **6.** $2 \cdot 3^2$
7. $2^2 \cdot 5^2$ **8.** $2 \cdot 11^2$

Page 109 Classroom Exercises **1.** $5 \cdot 7 \cdot 13$
3. $2 \cdot 3 \cdot 5^2$ **5.** $2^2 \cdot 3 \cdot 41$ **7.** $2 \cdot 3^2 \cdot 11$ **9.** $3^2 \cdot 29$ **11.** $2 \cdot 3^2 \cdot 5^2$ **13.** $5 \cdot 7$ **15.** $3 \cdot 5$ **17.** $2 \cdot 5 \cdot 13$ **19.** $2^2 \cdot 3^2$ **21.** $2^2 \cdot 5 \cdot 7$ **23.** $2 \cdot 3^3 \cdot 11$

Page 110 Written Exercises **1.** $2 \cdot 3$ **3.** $2 \cdot 17$
5. $5 \cdot 11$ **7.** $2 \cdot 5 \cdot 11$ **9.** $5 \cdot 17$ **11.** $7 \cdot 13$
13. $2^3 \cdot 7$ **15.** $2 \cdot 5^2$ **17.** $2 \cdot 3 \cdot 13$ **19.** $2^2 \cdot 3 \cdot 7$
21. 2^7 **23.** 2^6 **25.** $2 \cdot 3^3 \cdot 7$ **27.** $2 \cdot 3^4$
29. $2^2 \cdot 3^3 \cdot 5$ **31.** $2^2 \cdot 3^2 \cdot 5$ **33.** $7 \cdot 17$ **35.** $2^2 \cdot 3^2 \cdot 13$ **37.** $11 \cdot 17$ **39.** 2^9 **41.** $2^2 \cdot 11^2$
43. $2^3 \cdot 13 \cdot 17$ **45.** $11 \cdot 13 \cdot 17$ **47.** $17 \cdot 19$
49. $5 \cdot 7^2 \cdot 13$ **51.** $19 \cdot 23$ **53.** $2 \cdot 5^2 \cdot 23$
55. $5 \cdot 7^2 \cdot 19$ **57.** $2^2 \cdot 3^2 \cdot 7 \cdot 13^2$ **59.** $3 \cdot 5^2 \cdot 13^2$

Page 112 Self-Check **1.** 8 **2.** 84 **3.** 63 **4.** 12
5. 54 **6.** 300

Page 113 Classroom Exercises **1.** $2^2 \cdot 3^2$ **3.** $2^3 \cdot 3^3 \cdot 5^2 \cdot 7$ **5.** $2^3 \cdot 3^3 \cdot 5^3$ **7.** $2^3 \cdot 3^2 \cdot 5$ **9.** 20
11. 15 **13.** 189 **15.** 54 **17.** 21 **19.** 75
21. 240 **23.** 240 **25.** 15 **27.** 135 **29.** 27
31. 75 **33.** 42 **35.** 36

Page 113 Written Exercises **1.** 75 **3.** 60 **5.** 33
7. 280 **9.** 168 **11.** 135 **13.** 294 **15.** 672
17. 168 **19.** 288 **21.** 120 **23.** 240 **25.** 2025
27. 840 **29.** 200 **31.** 180 **33.** 2100 **35.** 216

Answers to Selected Exercises 551

37. 187 **39.** 319 **41.** 120 **43.** 630 **45.** 144 **47.** 210 **49.** twelfth day **51.** 360 in

Page 116 Self-Check **1.** 2 **2.** 4 **3.** 3

Page 116 Classroom Exercises **1.** 2 **3.** 2 **5.** 2·3·3;2

Page 116 Written Exercises **1.** 3;5;2 **3.** 3;3;6 **5.** 2 **7.** 3 **9.** 8 **11.** 4 **13.** 15 **15.** 15 **17.** 6 **19.** 5 **21.** 4 **23.** 9

Page 117 Review: Sections 4-4–4-6 **1.** $2^4 \cdot 3$ **2.** $2^3 \cdot 5$ **3.** $2^2 \cdot 3^2 \cdot 5$ **4.** $2^3 \cdot 5^2$ **5.** 5^3 **6.** $2 \cdot 3 \cdot 11$ **7.** $3 \cdot 5^2$ **8.** $2^2 \cdot 5 \cdot 17$ **9.** $5 \cdot 19$ **10.** $3 \cdot 11^2$ **11.** $2^5 \cdot 3^2$ **12.** $2 \cdot 5^2$ **13.** 40 **14.** 360 **15.** 336 **16.** 300 **17.** 1260 **18.** 210 **19.** 247 **20.** 140 **21.** 12 days **22.** 24 days **23.** 18 **24.** 8 **25.** 12 **26.** 11 **27.** 3 **28.** 5 **29.** 4 **30.** 6

Page 117 Calculator Exercises **1.** 131 **3.** 691 **5.** 1447

Page 118 Chapter Review **1.** digits **3.** least common multiple **5.** composite numbers **7.** even number **9.** counting numbers **11.** multiple **13.** 1, 2, 7, 14 **15.** 1, 2, 3, 4, 6, 9, 12, 18, 36 **17.** 1, 2, 3, 6, 9, 18 **19.** 1, 2, 5, 10 **21.** 1, 17 **23.** 1, 2, 3, 4, 6, 8, 9, 12, 18, 24, 36, 72 **25.** True; 435 ends in 5 **27.** False; 11 is not divisible by 9 **29.** False; 531 does not end in 0, 2, 4, 6, or 8 **31.** $2 \cdot 3^4 \cdot 5^2 \cdot 7 \cdot 11$ **33.** $3 \cdot 5^4 \cdot 27^2$ **35.** $3 \cdot 5 \cdot 7$ **37.** $2^2 \cdot 3 \cdot 5^2$ **39.** $2^3 \cdot 11$ **41.** $2 \cdot 3^2 \cdot 5^3 \cdot 13^2$ **43.** $2^4 \cdot 3^4 \cdot 5^3 \cdot 7 \cdot 11$ **45.** 120 **47.** 252 **49.** 7 **51.** 3 **53.** 5 **55.** 3 **57.** No; 2 **59.** 360

Page 120 Chapter Test **1.** 1, 2, 3, 5, 6, 10, 15, 30 **3.** 1, 2, 3, 4, 6, 8, 12, 24 **5.** True; 27 is divisible by 3 **7.** False; 13,324 is not divisible by 15 **9.** 7 **11.** $2^4 \cdot 3^3 \cdot 5^2$ **13.** $5^5 \cdot 7^2 \cdot 11 \cdot 13$ **15.** $2 \cdot 3^2 \cdot 5$ **17.** $2^3 \cdot 3^2 \cdot 5$ **19.** 1 **21.** No; 1

Page 121 Enrichment **1.** $2 \times 3 \times 7^2$ **3.** $2^2 \times 3^2$ **5.** $2^2 \times 5^2$ **7.** $2 \times 3 \times 5^2$ **9.** $2^3 \times 5^3$ **11.** $3^4 \times 5^2$

Page 122 Additional Practice **1.** 1, 3, 9 **3.** 1, 2, 3, 4, 6, 8, 12, 16, 24, 48 **5.** 1, 2, 4, 8, 16, 32 **7.** True; 9 is divisible by 9 **9.** True; since $6 \cdot 11 = 66$ **11.** True; since 135 ends in 5 **13.** False; 1342 does not end in 0 **15.** 37 **17.** 3 **19.** 3 **21.** $2^3 \cdot 3^2 \cdot 5$ **23.** $2^3 \cdot 3^2 \cdot 11^2 \cdot 17$ **25.** $5 \cdot 13$ **27.** $2 \cdot 7 \cdot 13$ **29.** $2^3 \cdot 3^3$ **31.** 450; 2 **33.** 630; 1 **35.** No; 2

CHAPTER 5

Page 125 Classroom Exercises **1.** No **3.** Yes **5.** No

Page 125 Written Exercises **1.** $\frac{2}{3}$ **3.** $\frac{4}{7}$ **5.** $\frac{1}{8}$ **7.** $\frac{7}{8}$ **9.** $\frac{2}{15}$ **11.** $\frac{1}{3}$ **13.** $\frac{9}{16}$ **15.** $\frac{4}{9}$ **17.** $\frac{19}{30}$ **19.** $\frac{2}{5}$ **21.** $\frac{t}{2}$ **23.** $\frac{2}{3y}$ **25.** $\frac{2}{3n}$ **27.** $\frac{3}{7y}$ **29.** $\frac{3n}{5p}$ **31.** $\frac{9c}{20f}$

Page 126-127 Self-Check **1.** $3\frac{3}{4}$ lb **2.** $4\frac{2}{3}$ ft **3.** $2\frac{2}{3}$ yd **4.** $8\frac{1}{3}$ yd **5.** $4\frac{7}{8}$ lb **6.** $4\frac{1}{2}$ gal

Page 127 Classroom Exercises **1.** $4\frac{1}{2}$ gal **3.** $5\frac{1}{2}$ qt **5.** $6\frac{1}{2}$ lb **7.** $3\frac{1}{2}$ qt **9.** $6\frac{2}{3}$ ft **11.** $21\frac{1}{4}$ lb **13.** $4\frac{5}{6}$ ft **15.** $3\frac{2}{3}$ yd

Page 127 Written Exercises **1.** $6\frac{3}{4}$ gal **3.** $6\frac{2}{3}$ yd **5.** $4\frac{3}{4}$ gal **7.** $2\frac{1}{2}$ pt **9.** $1\frac{1}{10}$ mi **11.** $2\frac{2}{3}$ yd **13.** $3\frac{11}{12}$ ft **15.** $10\frac{1}{3}$ yd **17.** $11\frac{3}{4}$ gal **19.** $13\frac{1}{4}$ lb **21.** $2\frac{1}{4}$ ft **23.** $6\frac{2}{3}$ yd **25.** $7\frac{1}{4}$ gal **27.** $4\frac{1}{2}$ lb **29.** $3\frac{1}{4}$ gal **31.** $1\frac{7}{12}$

Pages 129-130 Self-Check **1.** $\frac{9}{40}$ **2.** $\frac{8}{15}$ **3.** $\frac{11}{18}$ **4.** $\frac{1}{5}$ **5.** $\frac{4}{5}$ **6.** $\frac{2}{3}$ **7.** $\frac{3}{5}$ **8.** $\frac{1}{4}$

Page 130 Classroom Exercises **1.** $\frac{1}{14}$ **3.** $\frac{3}{16}$ **5.** $\frac{16}{45}$ **7.** $\frac{5}{21}$ **9.** $\frac{15}{7}$ **11.** $\frac{6}{7}$ **13.** $\frac{4}{7}$ **15.** $\frac{2}{5}$ **17.** $\frac{1}{5}$ **19.** $\frac{2}{5}$ **21.** $\frac{7}{18}$ **23.** $\frac{3}{10}$

Page 130 Written Exercises **1.** $\frac{1}{8}$ **3.** $\frac{3}{10}$ **5.** $\frac{8}{45}$ **7.** $\frac{4}{5}$ **9.** $\frac{1}{2}$ **11.** $\frac{8}{15}$ **13.** $\frac{7}{15}$ **15.** $\frac{5}{21}$ **17.** $\frac{3}{10}$ **19.** $\frac{1}{4}$ **21.** $\frac{3}{4}$ **23.** $\frac{1}{6}$ **25.** $\frac{11}{40}$ **27.** $\frac{4}{9}$ **29.** 1 **31.** $\frac{3}{4}$ **33.** 1 **35.** $\frac{8}{15}$ **37.** $\frac{5}{9}$ hr **39.** $\frac{3}{10}$ gal **41.** $\frac{1}{3}$ **43.** $\frac{3a}{5}$ **45.** st **47.** $\frac{6}{b}$ **49.** $\frac{3}{m}$

Pages 132-133 Self-Check **1.** $41°F$ **2.** $221°F$ **3.** $1472°F$ **4.** $2192°F$ **5.** $5°C$ **6.** $105°C$ **7.** $530°C$ **8.** $85°C$

Page 133 Classroom Exercises **1.** $122°F$ **3.** $59°F$ **5.** $392°F$ **7.** $20°C$ **9.** $60°C$ **11.** $55°C$

Page 133 Written Exercises **1.** $50°F$ **3.** $212°F$ **5.** $437°F$ **7.** $10°C$ **9.** $45°C$ **11.** $240°C$ **13.** $120°C$ **15.** $9977°F$

Page 134 Review: Sections 5-1–5-4 **1.** $\frac{3}{4}$ **2.** $\frac{5}{8}$ **3.** $\frac{6}{7}$ **4.** $\frac{2}{5}$ **5.** $\frac{4}{9}$ **6.** $\frac{2}{3}$ **7.** $8\frac{11}{12}$ ft **8.** $19\frac{5}{16}$ lb **9.** $3\frac{2}{3}$ yd **10.** $2\frac{3}{4}$ gal **11.** $\frac{5}{12}$ **12.** $\frac{1}{3}$ **13.** $\frac{6}{7}$ **14.** $\frac{2}{15}$ **15.** $20°C$ **16.** $122°F$ **17.** $180°C$ **18.** $932°F$ **19.** $1\frac{7}{12}$ ft **20.** $\frac{3}{4}$ hr

Pages 135-136 Self-Check **1.** $3\frac{2}{7}$ **2.** $8\frac{1}{5}$ **3.** $5\frac{1}{3}$ **4.** $13\frac{1}{2}$ **5.** 10 **6.** 60 **7.** $5\frac{1}{2}$ **8.** $5\frac{11}{16}$

Page 136 Classroom Exercises **1.** $4\frac{1}{2}$ **3.** $5\frac{2}{5}$ **5.** $6\frac{4}{7}$ **7.** $1\frac{3}{5}$ **9.** $\frac{9}{4}$ **11.** $\frac{75}{8}$ **13.** $\frac{55}{16}$ **15.** $17\frac{1}{2}$ **17.** 4 **19.** 8

Page 136 Written Exercises **1.** $1\frac{3}{7}$ **3.** $2\frac{2}{5}$ **5.** $1\frac{7}{8}$ **7.** $2\frac{7}{10}$ **9.** $\frac{11}{3}$ **11.** $\frac{37}{5}$ **13.** $\frac{31}{9}$ **15.** 7 **17.** 6 **19.** 9 **21.** $7\frac{7}{8}$ **23.** $11\frac{7}{10}$ **25.** $34\frac{5}{6}$ **27.** $18\frac{39}{40}$ **29.** $2\frac{1}{3}$

31. $13\frac{4}{9}$ **33.** $18\frac{1}{3}$ **35.** $116\frac{2}{3}$ mi **37.** $3\frac{3}{4}$ cups
39. $\frac{p}{2}$ **41.** $\frac{9}{2a}$ **43.** $\frac{7m}{9}$ **45.** $\frac{7t}{9}$ **47.** $\frac{3}{2p}$

Page 138 Special Topic **1.** 8975 lb **3.** 34,464 lb
5. 13,350 lb

Page 139 Special Topic **1.** 6500 **3.** 9000 **5.** 3500

Page 141 Self-Check **1.** 22.5 in^2 **2.** 71.68 cm^2
3. $13\frac{1}{2}$ ft^2 **4.** 43 m^2

Page 142 Classroom Exercises **1.** 28 in^2 **3.** 91 yd^2
5. 98.8 m^2 **7.** 6 ft^2 **9.** 21 yd^2 **11.** 36.6 cm^2

Page 142 Written Exercises **1.** 225 in^2 **3.** $1\frac{3}{4}$ ft^2
5. 96 ft^2 **7.** 200 in^2 **9.** 227.5 cm^2 **11.** 12.8 cm^2
13. 84 m^2 **15.** $4\frac{1}{2}$ ft^2 **17.** $26\frac{5}{6}$ in^2 **19.** $91\frac{1}{4}$ ft^2
21. $7\frac{7}{9}$ yd^2 **23.** 5.75 m^2 **25.** 12.6 cm^2
27. 4.55 cm^2 **29.** $67\frac{1}{2}$ ft^2 **31.** $15\frac{3}{4}$ in^2

Page 145 Self-Check **1.** $43\frac{1}{3}$ ft^3 **2.** 299.52 m^3

Page 146 Classroom Exercises **1.** 135 ft^3
3. 0.85 cm^3 **5.** $59\frac{1}{2}$ yd^3 **7.** 754.4 mm^3

Page 146 Written Exercises **1.** 102 ft^3 **3.** $67\frac{1}{2}$ ft^3
5. 297 in^3 **7.** 125 ft^3 **9.** 47,424 mm^3 **11.** 40 cm^3
13. 275.4 m^3 **15.** $13\frac{1}{8}$ ft^3 **17.** 34,290 cm^3
19. 161.4 m^3 **21.** 3630 ft^3

Pages 148-149 Self-Check **1.** $2\frac{1}{4}$ **2.** $\frac{1}{24}$ **3.** $\frac{2}{25}$
4. $\frac{5}{12}$ **5.** $\frac{1}{2}$ **6.** $4\frac{2}{3}$ **7.** $\frac{2}{3}$ **8.** $1\frac{7}{11}$

Page 149 Classroom Exercises **1.** $\frac{3}{5}$ **3.** $\frac{7}{9}$ **5.** $\frac{1}{4}$
7. $\frac{2}{3}$ **9.** $\frac{1}{8}$ **11.** $\frac{6}{7}$ **13.** $\frac{1}{20}$ **15.** $\frac{3}{50}$ **17.** $\frac{1}{20}$
19. $2\frac{8}{25}$ **21.** 1 **23.** $\frac{1}{2}$ **25.** 3 **27.** $6\frac{1}{8}$

Page 150 Written Exercises **1.** 3 **3.** $\frac{7}{4}$ **5.** $\frac{1}{8}$
7. $\frac{1}{12}$ **9.** $\frac{11}{18}$ **11.** $\frac{4}{13}$ **13.** $\frac{9}{7}$ **15.** $\frac{2}{3}$ **17.** $\frac{3}{4}$
19. $\frac{5}{6}$ **21.** $\frac{3}{5}$ **23.** $\frac{20}{33}$ **25.** $\frac{1}{30}$ **27.** 6 **29.** $\frac{2}{15}$
31. $3\frac{7}{25}$ **33.** $\frac{1}{2}$ **35.** $3\frac{1}{2}$ **37.** 4 **39.** $\frac{4}{5}$ **41.** $3\frac{17}{21}$
43. $\frac{31}{32}$ **45.** 12 **47.** $\frac{1}{2}$ **49.** $7\frac{1}{2}$ **51.** $1\frac{7}{8}$ **53.** $\frac{1}{2}$
55. $\frac{6}{11}$ **57.** $1\frac{1}{5}$ **59.** $1\frac{1}{10}$ **61.** $1\frac{7}{20}$ **63.** $\frac{2}{3}$ **65.** 16
67. 9 **69.** $\frac{1}{6}$ **71.** $\frac{1}{t}$ **73.** 2 **75.** $\frac{3}{2p}$ **77.** $\frac{5t}{4}$

Page 151 Review: Sections 5-5—5-8 **1.** 3 **2.** 5
3. $23\frac{1}{3}$ **4.** $13\frac{5}{9}$ **5.** 15 ft^2 **6.** 15.05 m^2 **7.** $12\frac{3}{8}$ in^2
8. 102 yd^3 **9.** 48.1 cm^3 **10.** $1\frac{1}{2}$ **11.** $\frac{1}{8}$ **12.** $1\frac{2}{5}$
13. $\frac{15}{16}$ **14.** $7\frac{7}{8}$ in^2 **15.** $18\frac{3}{4}$ miles

Page 152 Chapter Review **1.** Fahrenheit **3.** mixed numeral **5.** lowest terms **7.** $\frac{1}{2}$ **9.** $\frac{1}{4}$ **11.** $\frac{7}{16}$
13. $3\frac{11}{12}$ ft **15.** $10\frac{1}{2}$ gal **17.** $4\frac{2}{3}$ yd **19.** $6\frac{1}{4}$ gal
21. $\frac{3}{16}$ **23.** $\frac{2}{5}$ **25.** $\frac{2}{3}$ **27.** $\frac{1}{4}$ **29.** $\frac{1}{2}$ **31.** 104°F
33. 50°F **35.** $3\frac{1}{2}$ **37.** $3\frac{3}{4}$ **39.** $2\frac{5}{8}$ **41.** $\frac{5}{4}$

43. $\frac{35}{8}$ **45.** $\frac{18}{7}$ **47.** 39 **49.** $2\frac{3}{4}$ **51.** $2\frac{1}{10}$
53. $16\frac{1}{3}$ **55.** 12 **57.** 10 in^2 **59.** 4.55 cm^2
61. 127.1 m^3 **63.** 455 in^3 **65.** $2\frac{7}{9}$ **67.** $\frac{3}{8}$ **69.** $\frac{1}{2}$
71. $2\frac{1}{2}$ **73.** 17 **75.** 15°C **77.** 490 ft^3

Page 154 Chapter Test **1.** $\frac{1}{2}$ **3.** $\frac{3}{8}$ **5.** $8\frac{3}{4}$ ft
7. $\frac{5}{12}$ **9.** 14 **11.** 130.5 m^2 **13.** $26\frac{2}{3}$ in^2 **15.** 2
17. $6\frac{3}{4}$ **19.** 55°C

Page 155 Enrichment **1.** $32\frac{1}{2}$ m^2 **3.** 66 m^2
5. 28 ft^2

Page 156 Additional Practice **1.** $\frac{1}{2}$ **3.** $\frac{2}{7}$ **5.** $\frac{3}{7}$
7. $28\frac{1}{4}$ lb **9.** $1\frac{3}{4}$ ft **11.** $\frac{1}{4}$ **13.** $\frac{6}{35}$ **15.** $\frac{6}{11}$
17. 95°F **19.** 0°C **21.** $22\frac{1}{2}$ **23.** 6 **25.** 28 cm^2
27. 12 m^2 **29.** 60 m^3 **31.** $1\frac{1}{4}$ **33.** $\frac{9}{20}$ **35.** $5\frac{1}{3}$
37. 7500 m^3

CHAPTER 6

Pages 158-159 Self-Check **1.** $1\frac{1}{3}$ **2.** $\frac{2}{5}$ **3.** $\frac{2}{5}$
4. $1\frac{3}{5}$ **5.** 8 **6.** $3\frac{1}{2}$ **7.** $5\frac{1}{5}$ **8.** $7\frac{4}{5}$

Page 159 Classroom Exercises **1.** $\frac{1}{2}$ **3.** 1 **5.** $1\frac{1}{2}$
7. $2\frac{1}{2}$ **9.** $1\frac{1}{2}$

Page 159 Written Exercises **1.** $\frac{2}{3}$ **3.** $\frac{1}{3}$ **5.** $\frac{1}{2}$
7. 1 **9.** $1\frac{2}{5}$ **11.** $5\frac{2}{7}$ **13.** $9\frac{5}{9}$ **15.** $6\frac{2}{3}$ **17.** $5\frac{2}{3}$
19. $9\frac{5}{7}$ **21.** $10\frac{1}{11}$ **23.** $\frac{3}{5}$ **25.** $\frac{3}{25}$ **27.** $8\frac{2}{5}$ **29.** $2\frac{1}{4}$
31. $14\frac{1}{4}$ **33.** $13\frac{1}{4}$ **35.** $6\frac{1}{2}$ hrs **37.** $\frac{8}{3}$ **39.** $\frac{9}{x}$
41. $\frac{2x}{3}$ **43.** $\frac{2s}{9}$ **45.** $\frac{m}{5}$ **47.** $\frac{3x}{y}$ **49.** $\frac{5a}{b}$ **51.** $\frac{5a}{3f}$

Pages 161-162 Self-Check **1.** $\frac{5}{10}$ **2.** $\frac{8}{20}$ **3.** $\frac{35}{60}$
4. $\frac{20}{64}$ **5.** $1\frac{1}{2}$ **6.** $1\frac{1}{5}$ **7.** $\frac{19}{20}$ **8.** $\frac{19}{24}$ **9.** $\frac{1}{4}$ **10.** $\frac{1}{2}$
11. $\frac{1}{40}$ **12.** $\frac{1}{16}$

Page 162 Classroom Exercises **1.** $\frac{3}{9}$ **3.** $\frac{12}{16}$ **5.** $\frac{15}{35}$
7. $\frac{7}{8}$ **9.** $\frac{7}{18}$ **11.** $\frac{17}{36}$ **13.** $1\frac{3}{14}$ **15.** $1\frac{7}{40}$ **77.** $\frac{1}{2}$
19. $\frac{11}{24}$ **21.** $\frac{11}{20}$ **23.** $\frac{1}{6}$ **25.** $\frac{1}{8}$

Page 163 Written Exercises **1.** $\frac{2}{6}$ **3.** $\frac{25}{40}$ **5.** $\frac{30}{36}$
7. $\frac{18}{42}$ **9.** $\frac{10}{36}$ **11.** $\frac{21}{36}$ **13.** $\frac{7}{8}$ **15.** $\frac{4}{5}$ **17.** $\frac{7}{18}$
19. $\frac{22}{45}$ **21.** $\frac{14}{15}$ **23.** $\frac{23}{24}$ **25.** $1\frac{1}{4}$ **27.** $1\frac{1}{42}$ **29.** $\frac{1}{2}$
31. $\frac{2}{15}$ **33.** $\frac{7}{36}$ **35.** $\frac{11}{18}$ **37.** $\frac{7}{45}$ **39.** $\frac{7}{39}$ **41.** $\frac{1}{8}$
43. $1\frac{1}{3}$ **45.** $\frac{21}{40}$ **47.** $\frac{25}{44}$ **49.** $1\frac{1}{45}$ **51.** $\frac{13}{14}$
53. $1\frac{3}{8}$ in **55.** $\frac{7}{30}$ mi **57.** $\frac{6x}{25}$ **59.** $\frac{a}{8}$ **61.** $\frac{5x}{8}$
63. $\frac{13}{2y}$ **65.** $\frac{7}{3x}$ **67.** $\frac{7a}{2n}$ **69.** $\frac{9y}{16z}$ **71.** $\frac{61s}{15t}$

Page 165 Self-Check **1.** $14\frac{2}{15}$ **2.** $3\frac{1}{4}$ **3.** $2\frac{5}{12}$
4. $8\frac{1}{8}$ **5.** $1\frac{1}{2}$ **6.** $2\frac{1}{2}$ **7.** $5\frac{2}{5}$ **8.** $2\frac{2}{3}$

Page 166 Classroom Exercises **1.** $2\frac{2}{4}$ **3.** $6\frac{16}{20}$

Answers to Selected Exercises   553

5. $6\frac{7}{6}$ 7. $5\frac{4}{6}$

Page 166 Written Exercises 1. $7\frac{3}{4}$ 3. $1\frac{11}{12}$ 5. 8 7. $7\frac{3}{20}$ 9. $3\frac{1}{4}$ 11. $2\frac{1}{12}$ 13. $2\frac{1}{8}$ 15. $6\frac{2}{15}$ 17. $4\frac{2}{3}$ 19. $1\frac{1}{2}$ 21. $10\frac{2}{3}$ 23. $4\frac{47}{60}$ 25. $5\frac{1}{10}$ 27. $3\frac{7}{8}$ 29. $6\frac{11}{16}$ 31. $9\frac{14}{15}$ 33. $16\frac{1}{2}$ 35. $5\frac{7}{8}$ 37. $1\frac{5}{12}$ 39. $12\frac{5}{12}$ 41. $\frac{11}{20}$ min 43. $16\frac{1}{6}$ ft

Page 167 Review: Sections 6-1–6-3 1. $\frac{3}{4}$ 2. $7\frac{2}{3}$ 3. $\frac{3}{13}$ 4. $11\frac{3}{10}$ 5. $10\frac{1}{2}$ 6. $\frac{10}{20}$ 7. $\frac{12}{28}$ 8. $\frac{9}{12}$ 9. $\frac{25}{30}$ 10. $\frac{42}{48}$ 11. $1\frac{1}{6}$ 12. $\frac{1}{2}$ 13. $\frac{17}{24}$ 14. $\frac{1}{2}$ 15. $\frac{1}{3}$ 16. $4\frac{1}{12}$ 17. $9\frac{13}{15}$ 18. $4\frac{1}{4}$ 19. $2\frac{17}{20}$ 20. $10\frac{7}{8}$ 21. $3\frac{4}{5}$ mi 22. $1\frac{5}{12}$ hrs

Page 168 Special Topic 1. 49 3. 4096 5. 1024 7. 19,487,171 9. 62,748,517 11. 4,100,625

Pages 169-170 Self-Check 1. $15\frac{5}{12}$ hrs 2. $63.98

Page 170 Classroom Exercises 1. $4\frac{1}{2}$; 7; 7; $6\frac{1}{2}$; $2\frac{3}{4}$; $27\frac{3}{4}$ hrs

Page 171 Written Exercises 1. $7\frac{1}{4}$; $7\frac{1}{2}$; $8\frac{3}{4}$; $7\frac{1}{2}$; $7\frac{1}{2}$; $38\frac{1}{2}$ hrs 3. $7\frac{1}{2}$; $7\frac{3}{4}$; $3\frac{3}{4}$; $7\frac{1}{4}$; $7\frac{1}{2}$; $34\frac{1}{4}$ hrs 5. $200.20 7. $212.35 9. $93.80 11. $134.20 13. $22.95 15. $12.90 per hour

Pages 173-174 Self-Check 1. 0.5 2. 0.4 3. 0.625 4. 2.5 5. 1.8 6. $0.\overline{333}$ 7. $0.\overline{833}$ 8. $0.\overline{133}$ 9. $0.\overline{555}$ 10. $0.\overline{277}$ 11. < 12. < 13. > 14. =

Page 174 Classroom Exercises 1. = 3. ≠ 5. ≠ 7. = 9. T 11. R 13. T 15. < 17. <

Page 175 Written Exercises 1. 0.55 3. 3.8 5. 0.15 7. 0.625 9. 0.325 11. $0.\overline{3}$ 13. $0.\overline{7}$ 15. $0.\overline{73}$ 17. $1.\overline{1}$ 19. $0.58\overline{3}$ 21. $0.8\overline{3}$ 23. $2.\overline{3}$ 25. $0.\overline{16}$ 27. > 29. < 31. > 33. < 35. < 37. Earl 39. Phyllis

Pages 177-178 Self-Check 1. 19 in 2. 63 ft 3. 16 cm 4. 79 m 5. 314 cm² 6. 39 ft²

Page 178 Classroom Exercises 1. 88 in 3. 94 m 5. 28 yd²

Page 178 Written Exercises 1. 188 cm 3. 20 cm 5. 4 m 7. 76 cm 9. 47 ft 11. 42 in 13. 13 yd 15. 38 in 17. 1256 cm² 19. 2 m² 21. 408 cm² 23. 23 m² 25. 227 in² 27. 16 in² 29. 50 in² 31. 13 in²

Page 180 Review: Sections 6-4–6-6 1. a. $7\frac{1}{4}$; 7; $7\frac{1}{3}$; $7\frac{1}{4}$; $6\frac{2}{3}$ b. $35\frac{1}{2}$ 2. $191.70 3. < 4. > 5. = 6. < 7. 38 in; 113 in² 8. 26 m; 55 m² 9. Tara's 10. $234\frac{1}{7}$ in²

Page 180 Calculator Exercises 31. > 33. < 35. <

Page 181 Chapter Review 1. terminating 3. like 5. unlike 7. $\frac{6}{7}$ 9. $\frac{1}{2}$ 11. $3\frac{2}{3}$ 13. $4\frac{1}{2}$ 15. $\frac{1}{5}$ 17. $\frac{8}{18}$ 19. $\frac{21}{36}$ 21. $\frac{14}{15}$ 23. $\frac{43}{48}$ 25. $\frac{5}{9}$ 27. $\frac{13}{30}$ 29. $\frac{16}{21}$ 31. $3\frac{5}{6}$ 33. $15\frac{20}{21}$ 35. $7\frac{23}{24}$ 37. $5\frac{4}{9}$ 39. $6\frac{7}{24}$ 41. $38\frac{5}{6}$; $186.40 43. 0.125 45. $0.\overline{54}$ 47. 0.15 49. < 51. > 53. 24 m; 45 m² 55. 75 ft; 453 ft² 57. $\frac{1}{5}$ hr 59. $1\frac{17}{20}$ mi 61. 100 m

Page 183 Chapter Test 1. $\frac{8}{10}$ 3. $\frac{20}{48}$ 5. $\frac{1}{2}$ 7. $1\frac{1}{18}$ 9. $3\frac{4}{7}$ 11. $13\frac{5}{24}$ 13. 0.6 15. $0.8\overline{3}$ 17. 14 m; 17 m² 19. $93.10

Page 184 Enrichment 1. .1 cm; ± 0.05 cm; 4.35 cm; 4.25 cm 3. $\frac{1}{2}$ in; ± $\frac{1}{4}$ in; $2\frac{3}{4}$ in; $2\frac{1}{4}$ in

Page 185 Additional Practice 1. $\frac{3}{5}$ 3. $\frac{1}{2}$ 5. $\frac{3}{4}$ 7. $\frac{17}{35}$ 9. $\frac{1}{2}$ 11. $\frac{9}{14}$ 13. $8\frac{1}{2}$ 15. $9\frac{5}{8}$ 17. $3\frac{5}{18}$ 19. $5\frac{13}{15}$ 21. 0.7 23. $0.2\overline{6}$ 25. $0.41\overline{6}$ 27. >; $0.8 > 0.75$ 29. >; $0.4 > 0.375$ 31. >; $0.375 > 0.\overline{3}$ 33. 79 m; 491 m² 35. 95 cm; 725 cm² 37. 13 mi; 13 mi² 39. 57 yd; 255 yd² 41. more computers

Page 186 Common Errors 1. 1, 2, 3, 6, 9, 18 3. 24 5. $\frac{12}{25}$ 7. $2\frac{4}{25}$ 9. $1\frac{5}{12}$ 11. $9\frac{1}{3}$ 13. $2\frac{1}{4}$ yd²

Page 187 Cumulative Review: Chapters 4-6 1. b 3. d 5. b 7. a 9. c 11. c 13. d 15. a 17. a 19. b 21. c 23. d

CHAPTER 7

Pages 190-191 Self-Check 1. 36 2. 3 3. 0 4. 1.8 5. $\frac{3}{8}$ 6. $1\frac{1}{4}$

Page 191 Classroom Exercises 1. 8 3. n 5. x 7. n 9. $x + 5 - 5 = 19 - 5$ 11. $n + 5.7 - 5.7 = 10.3 - 5.7$ 13. $b + \frac{1}{4} - \frac{1}{4} = \frac{1}{2} - \frac{1}{4}$ 15. $6\frac{7}{8} - 2\frac{3}{4} = y + 2\frac{3}{4} - 2\frac{3}{4}$ 17. $3.2 - 3.2 + r = 9 - 3.2$ 19. $6 - 6 + c = 9 - 6$

Page 192 Written Exercises 1. 7 3. 17 5. 24 7. 17 9. 27 11. 49 13. 1.6 15. 3.5 17. 2.9 19. 11.5 21. 0.38 23. 21.74 25. $1\frac{1}{4}$ 27. $2\frac{1}{4}$ 29. $3\frac{1}{2}$ 31. $\frac{1}{3}$ 33. $3\frac{5}{9}$ 35. $2\frac{2}{15}$ 37. 2.54 39. $3\frac{5}{6}$ 41. 213 43. 1.07 45. 7 47. $1\frac{7}{8}$ 49. 1.58 51. 0

Pages 193-194 Self-Check 1. 23 2. 18 3. 5 4. 5.9 5. 1 6. $4\frac{1}{2}$

Page 194 Classroom Exercises 1. a 3. r 5. x 7. a 9. $3\frac{1}{4}$ 11. $a - 12 + 12 = 5 + 12$ 13. $47 +$

83 = t − 83 + 83   **15.** 12.9 + 5.6 = b − 5.6 + 5.6   **17.** $4\frac{1}{4} + 1\frac{7}{8} = n - 1\frac{7}{8} + 1\frac{7}{8}$

Page 194 Written Exercises 1. 58 **3.** 101 **5.** 203 **7.** 392 **9.** 338 **11.** 4107 **13.** 10.1 **15.** 1.13 **17.** 48 **19.** 24.37 **21.** 20.5 **23.** 110.18 **25.** 8 **27.** $14\frac{1}{2}$ **29.** $1\frac{5}{8}$ **31.** $5\frac{1}{2}$ **33.** 14 **35.** $3\frac{7}{8}$ **37.** 127 **39.** 368 **41.** $6\frac{1}{4}$ **43.** $3\frac{11}{16}$ **45.** 3.05 **47.** 334.4 **49.** 8 **51.** 2, 4, 6, 8, 10 **53.** 8

Pages 196-197 Self-Check 1. $336.20 **2.** 655 mi/h

Page 197 Classroom Exercises 1. $233 **3.** $462.11 **5.** 375 mi/h

Page 197 Written Exercises 1. $439.00 **3.** $491.68 **5.** $337.23 **7.** $291.77 **9.** 337 mi/h **11.** 875.4 mi/h **13.** 376.8 mi/h **15.** 475 mi/h **17.** a + t = g

Page 198 Review: Sections 7-1—7-3 1. 15 **2.** 22 **3.** 8.6 **4.** 0.33 **5.** $\frac{1}{3}$ **6.** $2\frac{5}{8}$ **7.** 4.3 **8.** $\frac{1}{6}$ **9.** $4\frac{3}{10}$ **10.** 38 **11.** 130 **12.** 13.6 **13.** 108 **14.** 8 **15.** $2\frac{1}{8}$ **16.** 10.0 **17.** $12\frac{1}{12}$ **18.** $5\frac{17}{20}$ **19.** $257.69 **20.** $148.80 **21.** 569.9 mi/h **22.** 597.5 mi/h

Page 199 Special Topic 1. 11; $63.69 **3.** 11; $87.89

Pages 200-201 Self-Check 1. 14 **2.** 86 **3.** 16 **4.** 32 **5.** 70 **6.** $15\frac{4}{5}$ **7.** 2 **8.** $12\frac{1}{2}$

Page 201 Classroom Exercises 1. x **3.** n **5.** $\frac{4n}{4} = \frac{32}{4}$ **7.** $\frac{44}{11} = \frac{11y}{11}$ **9.** $\frac{2.6a}{2.6} = \frac{33.8}{2.6}$ **11.** $\frac{96}{19.2} = \frac{19.2x}{19.2}$ **13.** No **15.** Yes **17.** Yes **19.** No

Page 202 Written Exercises 1. 9 **3.** 5 **5.** 32 **7.** 5 **9.** 14 **11.** 41 **13.** 20 **15.** 120 **17.** 40 **19.** $9\frac{1}{5}$ **21.** $16\frac{3}{4}$ **23.** $10\frac{5}{8}$ **25.** $\frac{3}{4}$ **27.** 32 **29.** 12 **31.** 53.5 **33.** 125 **35.** $12\frac{1}{2}$ **37.** 24.5 **39.** $\frac{2}{5}$ **41.** 22 **43.** $15\frac{1}{3}$ **45.** $\frac{5}{8}$ **47.** 12 **49.** $\frac{9}{10}$ **51.** $2\frac{1}{4}$ **53.** 28 **55.** $1\frac{1}{2}$

Pages 203-204 Self-Check 1. 27 **2.** 8 **3.** 336 **4.** 108 **5.** 126 **6.** 18 **7.** 10 **8.** 12.3

Page 204 Classroom Exercises 1. q **3.** a **5.** r **7.** p **9.** $\frac{n}{5} \cdot 5 = 6 \cdot 5$ **11.** $\frac{y}{10} \cdot 10 = 3 \cdot 10$ **13.** $\frac{s}{5} \cdot 5 = 1.1(5)$ **15.** $2\frac{1}{2} \cdot 4 = \frac{K}{4} \cdot 4$ **17.** $\frac{b}{17} \cdot 17 = 5 \cdot 17$ **19.** $3\frac{3}{4} \cdot 8 = \frac{d}{8} \cdot 8$ **21.** $\frac{t}{3} \cdot 3 = 4\frac{5}{6} \cdot 3$ **23.** $\frac{r}{7} \cdot 7 = 12 \cdot 7$

Page 205 Written Exercises 1. 136 **3.** 276 **5.** 861 **7.** 4498 **9.** 10 **11.** 16 **13.** 126 **15.** 7.47 **17.** 2.96 **19.** 56.4 **21.** 10.0 **23.** 9.9 **25.** 28 **27.** 70 **29.** 4 **31.** $162\frac{1}{2}$ **33.** 33.6 **35.** 910 **37.** 62 **39.** 4794 **41.** 4.8 **43.** 24 **45.** 9 **47.** $3\frac{1}{5}$ **49.** $\frac{2}{3}$ **51.** $3\frac{2}{3}$

Pages 206-207 Self-Check 1. 255 seconds **2.** 26 hits

Page 207 Classroom Exercises 1. 20 hours **3.** 24 km/h **5.** 28 hits

Page 208 Written Exercises 1. 6 h **3.** 5.24 m/sec **5.** $8\frac{1}{2}$ sec **7.** 1083.75 mi/h **9.** 15 h **11.** 50 hits **13.** 30 hits **15.** 39 hits **17.** 176 hits **19.** 26 hits **21.** $2\frac{3}{5}$ hours **23.** 200 times at bat

Page 209 Review: Sections 7-4—7-6 1. 7 **2.** 6 **3.** 6 **4.** 15 **5.** $4\frac{2}{7}$ **6.** $\frac{3}{8}$ **7.** 26 **8.** 11 **9.** 117 **10.** 100 **11.** 24 **12.** 16.87 **13.** $10\frac{2}{3}$ **14.** $39\frac{2}{3}$ **15.** 1664 **16.** 0.104 **17.** 1460 m/sec **18.** 17 hits

Page 210 Chapter Review 1. net pay **3.** addition **5.** divide **7.** equation **9.** equivalent equations **11.** 34 **13.** 12.8 **15.** $10\frac{3}{4}$ **17.** 22 **19.** 36.4 **21.** $11\frac{3}{8}$ **23.** $224.00 **25.** $239.45 **27.** 440 mi/h **29.** 277.4 mi/h **31.** 9 **33.** $8\frac{3}{4}$ **35.** 182 **37.** $126\frac{2}{3}$ **39.** 60 sec **41.** 52 km/h **43.** 102 hits **45.** $299.45 **47.** 5.1 mi/h

Page 212 Chapter Test 1. 25 **3.** 2.89 **5.** 77 **7.** $1\frac{4}{5}$ **9.** 11 **11.** $10\frac{1}{2}$ **13.** 4112 **15.** 219.24 **17.** $352.11 **19.** 156.25 mi/h

Page 213 Enrichment 1. $\frac{1}{3}$ **3.** $\frac{2}{11}$ **5.** $\frac{8}{11}$ **7.** $\frac{5}{3}$ **9.** $\frac{38}{11}$ **11.** $\frac{34}{37}$ **13.** $\frac{619}{99}$ **15.** $\frac{725}{99}$

Page 214 Additional Practice 1. 305 **3.** 1.12 **5.** $1\frac{3}{8}$ **7.** 89 **9.** 30.2 **11.** $2\frac{9}{16}$ **13.** $515.00 **15.** 75 km/h **17.** 16 **19.** 65 **21.** 12.3 **23.** 1.52 **25.** 448 **27.** 27.6 **29.** 26.88 **31.** 55 **33.** 195 hits **35.** $262.92

CHAPTER 8

Page 216-217 Self-Check 1. $\frac{1}{3}$ **2.** $\frac{4}{3}$ **3.** $\frac{1}{7}$ **4.** $\frac{5}{1}$ **5.** No **6.** Yes **7.** No **8.** No **9.** $\frac{2}{5}$ **10.** $\frac{5}{7}$ **11.** $\frac{1}{4n}$ **12.** $\frac{3}{7p}$

Page 218 Classroom Exercises 1. $\frac{3}{4}$ **3.** $\frac{5}{2}$ **5.** $\frac{4}{7}$ **7.** Yes **9.** No **11.** $\frac{3}{5}$ **13.** $\frac{7p}{2}$ **15.** $\frac{3}{2}$ **17.** $\frac{1}{2q}$ **19.** $\frac{1}{3}$ **21.** $\frac{1}{1}$

Page 218 Written Exercises 1. $\frac{1}{3}$ **3.** $\frac{3}{8}$ **5.** $\frac{1}{10}$ **7.** $\frac{1}{3}$ **9.** $\frac{1}{4}$ **11.** $\frac{1}{3}$ **13.** Yes **15.** Yes **17.** No **19.** No **21.** $\frac{5}{6}$ **23.** $\frac{5r}{2}$ **25.** $\frac{10}{7}$ **27.** $\frac{3}{7}$ **29.** $\frac{5}{12}$ **31.** $\frac{15a}{8}$ **33.** $\frac{1}{1}$ **35.** $\frac{4}{3}$ **37.** $\frac{r}{2}$ **39.** $\frac{r}{1}$ **41.** $\frac{1}{9}$ **43.** $\frac{8}{3}$

Page 220 Self-Check 1. 1 **2.** 1 **3.** 13 **4.** $2\frac{3}{5}$ **5.** $2\frac{2}{5}$ **6.** 4 **7.** 8 **8.** 16

Page 220 Classroom Exercises 1. 16n = 80 **3.** 45 = 7x **5.** 24 = 4m **7.** x = 3 **9.** p = 8 **11.** k = 2

Page 220 Written Exercises 1. 18 **3.** 1 **5.** 2

Answers to Selected Exercises 555

7. 7 **9.** 27 **11.** 16 **13.** 54 **15.** 7 **17.** 18
19. 60 **21.** 11 **23.** 36 **25.** $3\frac{1}{5}$ **27.** $4\frac{1}{2}$ **29.** $11\frac{1}{4}$
31. 2 **33.** 400 **35.** $1\frac{1}{3}$

Page 222-223 Self-Check 1. $21\frac{1}{3}$ quarts **2.** 95 beats

Page 223 Classroom Exercises 1. $\frac{5}{x}$ **3.** $\frac{90}{x}$

Page 224 Written Exercises 1. 585 km **3.** 18
5. $77\frac{7}{9}$ **7.** 10.5 in **9.** 30 in **11.** $762\frac{1}{2}$ bu **13.** 300

Page 225 Review: Sections 8-1–8-3 1. $\frac{4}{5}$ **2.** $\frac{3}{2}$
3. $\frac{8}{5}$ **4.** $\frac{1}{5}$ **5.** $\frac{3}{5}$ **6.** $\frac{13}{5}$ **7.** $\frac{6}{25}$ **8.** $\frac{2k}{3}$ **9.** Yes
10. Yes **11.** No **12.** No **13.** 15 **14.** 6 **15.** 6
16. $16\frac{4}{5}$ **17.** 480 **18.** 50

Page 226 Special Topic
```
10 INPUT H, R       (H = Number of Hours)
20 LET P = H * R    (R = Pay per Hour)
30 PRINT P          (P = Total Pay)
40 END
```

Pages 227-228 Self-Check 1. 25% **2.** 60%
3. 70% **4.** 0% **5.** 45% **6.** 82% **7.** $33\frac{1}{3}$%
8. $83\frac{1}{3}$% **9.** $42\frac{6}{7}$% **10.** $22\frac{2}{9}$% **11.** $41\frac{2}{3}$%
12. $56\frac{1}{4}$% **13.** 100% **14.** 100% **15.** 350%
16. 625% **17.** $216\frac{2}{3}$% **18.** 325%

Page 229 Classroom Exercises 1. 9% **3.** 7.6%
5. $16\frac{2}{3}$% **7.** $\frac{n}{100} = \frac{3}{4}$ **9.** $\frac{n}{100} = \frac{7}{20}$ **11.** $\frac{n}{100} = \frac{1}{2}$
13. $\frac{n}{100} = \frac{2}{3}$ **15.** $\frac{n}{100} = \frac{4}{15}$ **17.** $\frac{n}{100} = \frac{1}{6}$ **19.** $\frac{n}{100} = \frac{16}{5}$ **21.** $\frac{n}{100} = \frac{20}{7}$ **23.** $\frac{n}{100} = \frac{20}{19}$

Page 229 Written Exercises 1. 25% **3.** 20%
5. 8% **7.** 80% **9.** 16% **11.** 85% **13.** $62\frac{1}{2}$%
15. $2\frac{1}{2}$% **17.** $71\frac{3}{7}$% **19.** 2% **21.** $63\frac{7}{11}$%
23. $32\frac{1}{2}$% **25.** 120% **27.** 150% **29.** $131\frac{1}{4}$%
31. 102% **33.** 220% **35.** 340% **37.** 40%
39. $77\frac{7}{9}$% **41.** $146\frac{2}{3}$% **43.** 18% **45.** $63\frac{7}{11}$%
47. $8\frac{1}{3}$% **49.** 610% **51.** 1190% **53.** 70%
55. 40% **57.** 88% **59.** 68% **61.** 62%

Page 232 Self-Check 1. 8% **2.** 63% **3.** 50.1%
4. 940% **5.** 358% **6.** 0.15 **7.** 0.06 **8.** 0.215
9. 0.004 **10.** 1.13

Page 233 Classroom Exercises 1. $\frac{41}{100}$ **3.** $\frac{97}{100}$
5. $\frac{790}{100}$ **7.** 0.21 **9.** 0.007 **11.** 0.063 **13.** 2.00
15. 0.015 **17.** 0.0005

Page 233 Written Exercises 1. 40% **3.** 940%
5. 20.3% **7.** 150% **9.** 26% **11.** 10% **13.** 860%
15. 4% **17.** 362% **19.** 0.18 **21.** 0.661 **23.** 0.03
25. 1.01 **27.** 0.005 **29.** 0.20 **31.** 0.0025
33. 4.00 **35.** 0.045 **37.** 14%; $\frac{7}{50}$ **39.** 0.33; $\frac{33}{100}$
41. 85%; 0.85 **43.** 0.60; $\frac{3}{5}$ **45.** 90%; 0.90

47. 0.47 **49.** 0.05 **51.** 0.025 **53.** 0.025
55. 0.52 **57.** 0.131

Pages 235-236 Self-Check 1. 0.3 **2.** 2.4 **3.** 2
4. 60 **5.** $3900

Page 237 Classroom Exercises 1. $n = 0.07 \cdot 121$
3. $r = 0.15 \cdot 144$ **5.** $w = \frac{3}{5} \cdot 75$ **7.** $r = \frac{1}{3} \cdot 96$
9. $c = \frac{3}{4} \cdot 4$ **11.** $n = \frac{3}{4} \cdot 30$

Page 237 Written Exercises 1. $n = 0.35 \cdot 400$
3. $n = 0.83 \cdot 100$ **5.** $n = \frac{3}{8} \cdot 16$ **7.** 51.6 **9.** 54
11. 0.312 **13.** 108 **15.** 20 **17.** 48 **19.** 93
21. 8 **23.** 34 **25.** 120 **27.** $7500 **29.** $3.00
31. 15 **33.** $255.36 **35.** 12 **37.** $5.45

Pages 240-241 Self-Check 1. $240 **2.** 27.50
3. 781.25

Page 242 Classroom Exercises 1. $i = (950)(0.08)(1)$
3. $i = (1956)(0.05)(\frac{1}{2})$ **5.** $i = (2575)(0.125)(2)$
7. $5.00 **9.** $16.00 **11.** $3.00 **13.** $8.00
15. $100.00 **17.** $19.00

Page 242 Written Exercises 1. $40.00 **3.** $12.00
5. $6.00 **7.** $22.50 **9.** $30.00 **11.** $75.00
13. $144.00 **15.** $2812.50 **17.** $725.00
19. $575.00 **21.** $3120.00

Page 243 Calculator Exercises 1. $76.00
3. $48.90 **5.** $643.75 **7.** $5.00 **9.** $16.00
11. $3.00 **13.** $8.00 **15.** $100.00 **17.** $19.00

Page 245 Self-Check 1. $12.50 **2.** $42.32
3. $12.99 **4.** $99.83

Page 246 Classroom Exercises 1. $6.00 **3.** $27.30
5. $15.80 **7.** $25.13

Page 246 Written Exercises 1. $57.00 **3.** $19.50
5. $52.80 **7.** $323.00 **9.** $13.00 **11.** $2144.00
13. $504.00

Page 247 Review: Sections 8-4–8-8 1. 50%
2. 15% **3.** 54% **4.** 84% **5.** $122\frac{1}{2}$% **6.** 620%
7. 90% **8.** 380% **9.** 17% **10.** 104% **11.** 10%
12. 1% **13.** 0.12 **14.** 0.02 **15.** 0.85 **16.** 0.70
17. 0.095 **18.** 0.153 **19.** 6.6 **20.** 1.28
21. $4130 **22.** $66.00 **23.** $480.00 **24.** $120.00
25. $14.41 **26.** $20.52

Page 248 Chapter Review 1. commission
3. discount **5.** proportion **7.** principal **9.** net price **11.** $\frac{3}{5}$ **13.** $\frac{3}{2}$ **15.** $\frac{12}{7}$ **17.** $\frac{1}{2}$ **19.** $\frac{3}{7}$
21. $\frac{3}{5}$ **23.** No **25.** No **27.** Yes **29.** Yes
31. 4 **33.** 3 **35.** 16 **37.** 12 **39.** $4\frac{4}{5}$ **41.** $6\frac{2}{3}$
43. $\frac{1}{20}$; 5% **45.** $\frac{1}{40}$; 0.025 **47.** 45 **49.** 73.08
51. $1840 **53.** $100 **55.** $8.94 **57.** $216
59. 60¢

Page 250 Chapter Test 1. Yes **3.** No **5.** 24
7. 42 **9.** 60% **11.** 70% **13.** 0.18 **15.** 0.085
17. 21 **19.** 0.81 **21.** 8 in **23.** $127.50
25. $3.75

556 *Answers to Selected Exercises*

Page 251 Enrichment **1.** $1030.41 **3.** $944.78 **5.** $1464.10 **7.** $4207.66

Page 252 Additional Practice **1.** $\frac{1}{5}$ **3.** $\frac{4}{9}$ **5.** $\frac{4}{1}$ **7.** No **9.** Yes **11.** $\frac{2}{5}$ **13.** $\frac{5t}{1}$ **15.** $\frac{6}{25y}$ **17.** $11\frac{2}{3}$ **19.** 36 **21.** $8\frac{3}{4}$ **23.** $36\frac{4}{11}$% **25.** 275% **27.** 14% **29.** 25%; $\frac{1}{4}$ **31.** 0.42; $\frac{21}{50}$ **33.** 85%; 0.85 **35.** 52.91 **37.** $24 **39.** 132 beats

CHAPTER 9

Pages 254-255 Self-Check **1.** 20% **2.** 40% **3.** 8%

Page 255 Classroom Exercises **1.** $p \cdot 36 = 20$ **3.** $p \cdot 108 = 81$ **5.** $120 = p \cdot 12$ **7.** $p \cdot 4000 = 200$

Page 256 Written Exercises **1.** $p \cdot 20 = 15$ **3.** $p \cdot 15 = 9$ **5.** $56 = p \cdot 64$ **7.** $37\frac{1}{2}$% **9.** 250% **11.** $33\frac{1}{3}$% **13.** 20% **15.** 30% **17.** $83\frac{1}{3}$% **19.** $66\frac{2}{3}$% **21.** 60% **23.** 80% **25.** 16%

Pages 257-258 Self-Check **1.** 40% **2.** 15%

Page 258 Classroom Exercises **1.** $19; \frac{19}{100}$; 19% **3.** $16; \frac{16}{80}$; 20% **5.** $12; \frac{12}{80}$; 15% **7.** $9; \frac{9}{150}$; 6%

Page 259 Written Exercises **1.** 10% **3.** 11% **5.** 14% **7.** 19% **9.** 9% **11.** 6% **13.** 18% **15.** 21% **17.** 12% **19.** 12% **21.** 26% **23.** 20%

Pages 261-262 Self-Check **1.** 37.5 **2.** 375 **3.** $53

Page 262 Classroom Exercises **1.** $50 = 0.20n$ **3.** $9 = 0.15t$ **5.** $26 = 0.15p$ **7.** $18 = 0.45n$ **9.** $6 = 0.30n$ **11.** 40; $18,000; 20; 75

Page 263 Written Exercises **1.** 75 **3.** 200 **5.** 168 **7.** 450 **9.** 20 **11.** 97.5 **13.** 120 **15.** 300 **17.** $109 **19.** $1944 **21.** 42; $\frac{1}{6}$ **23.** 640; $\frac{1}{8}$ **25.** 21 min; $\frac{1}{5}$

Page 265 Self-Check **1.** 9.6 **2.** 350% **3.** 20

Page 266 Classroom Exercises **1.** 8% of 21 is n **3.** n% of 85 is 17 **5.** 15% of n is 4 **7.** 100% of k is k; $\frac{100}{300} = \frac{k}{86}$ **9.** 100% of t is t; $\frac{100}{128} = \frac{t}{45}$ **11.** 100% of 126 is 126; $\frac{3}{100} = \frac{h}{126}$ **13.** 100% of a is a; $\frac{100}{15} = \frac{a}{10}$

Page 266 Written Exercises **1.** 7 **3.** 10.8 **5.** 105 **7.** 5.12 **9.** 75% **11.** 25 **13.** 25 **15.** 43.75 **17.** 150% **19.** 50 **21.** 25% **23.** 560 **25.** 340 **27.** 85% **29.** 3.12 oz **31.** $2080 **33.** $2\frac{1}{2}$%

Page 267 Review: Sections 9-1—9-4 **1.** 90% **2.** 120% **3.** $41\frac{2}{3}$% **4.** 140% **5.** 64% **6.** 75% **7.** 6% **8.** 52% **9.** 30% **10.** 9% **11.** 90 **12.** 30 **13.** 80 **14.** 200 **15.** 60 **16.** 62.5 **17.** 6% **18.** $300

Page 269 Special Topic **1.** 150 I.U. **3.** 9mg **5.** 2% **7.** 20mg

Pages 270-271 Self-Check **1.** 40% **2.** Let 4 cm represent 40°C. Phoenix: 3.4 cm; Chicago: 2.4 cm; Boston: 2.2 cm; Los Angeles: 2.2 cm; Cheyenne: 2.0 cm

Page 271 Classroom Exercises **1.** 120 **3.** $8\frac{1}{3}$%

Page 272 Written Exercises **1.** 185,000 **3.** 14% **5.** 86 **7.** 52% **9.** Let 6 cm represent 600 books. January: 1.5 cm; February: 2 cm; March: 3 cm; April: 4 cm; May: 1.75 cm; June: 5 cm **11.** Let 5 cm represent 500 m. Yosemite Lower Falls: 1.0 cm; Yosemite Upper Falls: 4.4 cm; Nevada Falls: 1.8 cm; Great Falls: 2.6 cm; Ribbon Falls: 4.9 cm; Silver Strand Falls: 3.6 cm

Pages 274-275 Self-Check **1.** 25 **2.** New York: 133°; Los Angeles: 68°; San Francisco: 61°; Washington, D.C.: 50°; Orlando: 47°

Page 276 Classroom Exercises **1.** $375 **3.** $n = 0.20(900)$; $180 **5.** 15%; $n = 0.15(900)$; $135 **7.** 90% **9.** 0.18; 65° **11.** 0.21; 76°

Page 277 Written Exercises **1.** $900 **3.** $1200 **5.** 1000 **7.** 2200 **9.** Schools: 126°; Public Works: 54°; Health and Safety: 144°; Other: 36° **11.** Lightning: 122°; Children: 61°; Campers: 40°; Auto Passengers: 36°; Other: 101° **13.** 1976: 79°; 1977: 83°; 1980: 90°; 1985: 108°

Pages 279-280 Self-Check **1.** $12 **2.** $40 **3.** 20%

Page 280 Classroom Exercises **1.** c **3.** b **5.** c **7.** b **9.** a **11.** b **13.** b **15.** b **17.** b

Page 281 Written Exercises **1.** a **3.** a **5.** c **7.** a **9.** b **11.** 25% **13.** 20% **15.** 10% **17.** a **19.** a **21.** 25% **23.** $16 **25.** 50%

Page 282 Calculator Exercises **1.** $200; $208.00 **3.** $22; $21.98 **5.** $2; $2.13

Page 283 Review: Sections 9-5—9-7 **1.** Let 4 cm represent scale 900 — 1100 students. 1976: 2.6 cm; 1977: 2.4 cm; 1978: 3.9 cm; 1979: 3.8 cm; 1980: 3.5 cm **2.** $100 **3.** $300 **4.** $200 **5.** 65 and over: 40°; 45-65: 72°; 20-44: 126°; 19 and under: 122° **6.** a **7.** c **8.** b **9.** c **10.** $33\frac{1}{3}$% **11.** $16\frac{2}{3}$% **12.** $125 **13.** $24

Page 284 Chapter Review **1.** percent of decrease **3.** data **5.** percent of increase **7.** 60% **9.** $85\frac{5}{7}$% **11.** $12\frac{1}{2}$% **13.** 19% increase **15.** 3% decrease **17.** 24% decrease **19.** 41% increase **21.** 120 **23.** 60 **25.** 50 **27.** 189 **29.** 10% **31.** 256 **33.** 25% **35.** Pacific: 166°; Atlantic: 83°; Indian: 72°; Arctic: 14°; Other: 25° **37.** a **39.** c **41.** $33\frac{1}{3}$% **43.** 10% **45.** 10% **47.** $250

Page 287 Chapter Test **1.** 8 **3.** 20 **5.** 14 **7.** 256 **9.** 19% **11.** 30% **13.** $150 **15.** Newspaper: 54°; TV: 198°; Radio: 72°; Other: 36°

Page 288 Enrichment **1.** 9, 11, 13 **3.** 512, 2048,

Answers to Selected Exercises 557

8192 **5.** 16, 8, 4 **7.** 21, 34, 55 **9.** 35, 57, 92

Page 289 Additional Practice 1. 50% **3.** 50% **5.** 50% **7.** 48 **9.** 680 **11.** Suits: 144°; Dresses: 126°; Skirts: 54°; Shoes: 36° **13.** b **15.** 1200

Page 290 Common Errors 1. 11 **3.** 2 **5.** $16.\overline{6}$ **7.** 6 **9.** 300 **11.** 25%

Page 291 Cumulative Review: Chapters 7-9 1. d **3.** c **5.** c **7.** d **9.** b **11.** b **13.** b **15.** a **17.** d **19.** a **21.** a

CHAPTER 10

Pages 294-295 Self-Check 1. +200; or 200 **2.** −120 **3.** −7; −6; −5; 3; 4 **4.** −9; −3; 0; 8; 10 **5.** > **6.** > **7.** < **8.** <

Page 295 Classroom Exercises 1. −8 **3.** −20 **5.** −3; 1; 2; 3; 7 **7.** −8; −1; 0; 2; 11 **9.** −12; 0; 6; 12; 13 **11.** < **13.** >

Page 296 Written Exercises 1. 10 **3.** −2000 **5.** −2 **7.** −3; 0; 3; 6 **9.** −2; 0; 1; 3 **11.** −4; −2; 1; 3; 8 **13.** < **15.** > **17.** < **19.** < **21.** > **23.** < **25.** > **27.** <

Pages 297-298 Self-Check 1. $\frac{4}{1}$ **2.** $\frac{-9}{2}$ **3.** $\frac{38}{10}$ **4.** $\frac{-5}{1}$ **5.** $\frac{9}{4}$ **6.** 15 **7.** $\frac{8}{3}$ **8.** 0 **9.** 2.7 **10.** $\frac{-4}{5}$

Page 298 Classroom Exercises 1. $\frac{4}{1}$ **3.** $-\frac{10}{1}$ **5.** $\frac{32}{10}$ **7.** $-\frac{11}{5}$ **9.** $\frac{9}{4}$ **11.** $-\frac{18}{10}$ **13.** −12 **15.** 6 **17.** −3.2 **19.** 1.5 **21.** $-\frac{1}{4}$ **23.** $1\frac{4}{5}$

Page 298 Written Exercises 1. $\frac{17}{1}$ **3.** $-\frac{7}{1}$ **5.** $\frac{38}{10}$ **7.** $-\frac{10}{3}$ **9.** $\frac{13}{4}$ **11.** $\frac{27}{5}$ **13.** Answers may vary. **15.** $-\frac{93}{10}$ **17.** $\frac{75}{10}$ **19.** −12 **21.** 20 **23.** $-2\frac{1}{3}$ **25.** $\frac{3}{4}$ **27.** $-\frac{2}{5}$ **29.** $8\frac{1}{2}$ **31.** −4.8 **33.** 8.9 **35.** 0 **37.** $\frac{5}{2}$ **39.** $-\frac{326}{10}$ **41.** $\frac{435}{100}$ **43.** $\frac{1325}{10}$ **45.** $-\frac{55}{10}$ **47.** $\frac{125}{10}$ **49.** $-5\frac{1}{5}$; $-2\frac{1}{5}$; $-\frac{1}{4}$; 1; $3\frac{1}{3}$ **51.** $-1\frac{1}{2}$; $-\frac{3}{4}$; $-\frac{2}{3}$; $\frac{4}{3}$; 1; 2 **53.** −1.875; −1.83; −0.687; 0; 1.5; 2.25 **55.** −3; −2.875; −2.5; −1.125; −1; −0.25

Pages 300-301 Self-Check 1. −4 **2.** −5 **3.** −9 **4.** −14 **5.** 1 **6.** −3 **7.** 0 **8.** −7 **9.** −3 **10.** 0 **11.** 3 **12.** 5

Page 301 Classroom Exercises 1. −6 **3.** −10 **5.** −10 **7.** −7 **9.** 3 **11.** 3 **13.** −4 **15.** 3 **17.** −5 **19.** −3 **21.** 1 **23.** 5

Page 301 Written Exercises 1. −9 **3.** −7 **5.** −10 **7.** −4 **9.** −15 **11.** −5 **13.** 3 **15.** 5 **17.** 6 **19.** 4 **21.** −4 **23.** 0 **25.** −2 **27.** 2 **29.** −5 **31.** 5 **33.** −6 **35.** 7 **37.** −3 **39.** −3 **41.** −5 **43.** 6 **45.** −6 **47.** 3 **49.** −5 **51.** loss of 4 yd

Page 302 Review: Sections 10-1–10-3 1. −3; −1; 0; 2; 5 **2.** −8; −4; 1; 7; 9 **3.** −4; −2; 0; 1; 4 **4.** $\frac{15}{1}$

5. $-\frac{8}{1}$ **6.** $-\frac{58}{10}$ **7.** $\frac{17}{8}$ **8.** $-\frac{22}{10}$ **9.** Answers may vary. **10.** −3 **11.** −13 **12.** 8 **13.** −8 **14.** −11 **15.** 2 **16.** 0 **17.** 0 **18.** 2°C

Page 303 Special Topic 1. 121 ft **3.** 162 ft **5.** 101 ft

Page 305 Self-Check 1. −18 **2.** −22 **3.** −13 **4.** 6 **5.** 8 **6.** 7 **7.** −1 **8.** −6 **9.** −6

Page 306 Classroom Exercises 1. 8 **3.** 6 **5.** 15 **7.** −18 **9.** −22 **11.** 29 **13.** −22 **15.** 2 **17.** 11 **19.** 4 **21.** 12 **23.** −3 **25.** −8 **27.** −4 **29.** −15

Page 306 Written Exercises 1. 7 **3.** 5 **5.** 15 **7.** 8 **9.** 32 **11.** 0 **13.** 1.3 **15.** 6.7 **17.** $\frac{1}{3}$ **19.** −16 **21.** 28 **23.** −36 **25.** −18 **27.** −5.5 **29.** −5.2 **31.** $-\frac{2}{3}$ **33.** $-\frac{3}{4}$ **35.** 4 **37.** 8 **39.** 9 **41.** 5 **43.** 0.7 **45.** 3.9 **47.** $\frac{1}{5}$ **49.** $\frac{1}{6}$ **51.** −5 **53.** −12 **55.** −21 **57.** −8 **59.** −1.9 **61.** −1.1 **63.** $-\frac{1}{4}$ **65.** $-\frac{1}{8}$ **67.** 5 **69.** −34 **71.** −5 **73.** 15 **75.** −25 **77.** 6 **79.** −5.7 **81.** 3.9 **83.** $-\frac{3}{5}$ **85.** $\frac{1}{4}$ **87.** 185 m **89.** −8°C

Pages 308-309 Self-Check 1. 4 + (−9) **2.** −10 + (−18) **3.** −1 + 12 **4.** 17 + 21 **5.** −6 **6.** −85 **7.** −2.6 **8.** −5.9 **9.** 29 **10.** 63 **11.** 15 **12.** 7

Page 309 Classroom Exercises 1. (−5) **3.** 14 **5.** (−17) **7.** −3 **9.** −22 **11.** −3 **13.** −25 **15.** 9 **17.** −3 **19.** 21 **21.** 6

Page 309 Written Exercises 1. 11 + (−3) **3.** 14 + 6 **5.** −6 + (−1) **7.** −8 + 5 **9.** −7 + (−7) **11.** 4.7 + (−8.6) **13.** −4 **15.** −41 **17.** −24 **19.** −121 **21.** −0.9 **23.** $-1\frac{1}{2}$ **25.** 4 **27.** 8 **29.** 25 **31.** 57 **33.** 23.4 **35.** $5\frac{1}{2}$ **37.** 0 **39.** 7 **41.** −188 **43.** −351 **45.** 12 **47.** −81 **49.** 11.5 **51.** $-2\frac{7}{8}$ **53.** 56°C **55.** 14,776 ft

Page 311 Special Topic 1. 12:05 P.M.; ZD = 6 **3.** 12:05 P.M.; ZD = 6 **5.** 11:05 P.M.; ZD = −5

Pages 312-313 Self-Check 1. −10 **2.** 12 **3.** 6 **4.** 4

Page 313 Classroom Exercises 1. (−1); 11 **3.** 4; −10 **5.** (−2); 1 **7.** 0 **9.** −3 **11.** 0

Page 313 Written Exercises 1. −5 **3.** 2 **5.** −29 **7.** −24 **9.** 16 **11.** −64 **13.** −10 **15.** 13 **17.** 5 **19.** 0 **21.** −6 **23.** 8 **25.** −9 **27.** −9 **29.** 26 **31.** −4 **33.** −12 **35.** 2 **37.** −10.7 **39.** $-\frac{1}{4}$ **41.** −0.7 **43.** $1\frac{1}{4}$

Page 314 Calculator Exercises 1. −97 **3.** −0.67 **5.** −6.99

Pages 315-316 Self-Check 1. $4x + 3$ **2.** $-5p^2 - 3p$ **3.** $-3y - 2$ **4.** $5n^2 + 3n + 7$

Page 316 Classroom Exercises 1. monomial **3.** binomial **5.** 2 **7.** 3

558 *Answers to Selected Exercises*

Page 316 Written Exercises 1. $7p + 3$ 3. $6c^2 - 6c$ 5. $7d - 2$ 7. $5n^2 - 6n + 5$ 9. $3x^2 - 2x - 2$ 11. $4r^2 + 4r - 2$ 13. $11x$ 15. $-9p^2$ 17. $-6y$ 19. $-9t^2$ 21. $8x^2$ 23. $-2y$

Page 317 Review: Sections 10-4–10-7 1. 6 2. -21 3. -5 4. 11 5. 11 6. 50 7. -14 8. -39 9. 7 10. -24 11. 24 12. -17 13. 39 14. -16 15. 6 16. -12 17. $-9°$F 18. $50°$C 19. -18 20. -26 21. -14 22. 3 23. 3 24. 3 25. $-x + 8$ 26. $-6y^2 - y - 3$ 27. $20t + 1$ 28. $3p^2 + 4p$ 29. $5y^2 - 4y + 2$ 30. $3n^2 + 4n - 7$

Page 318 Chapter Review 1. $4 + (-7) = -7 + 4$ 3. opposites 5. polynomial 7. negative integers 9. $3 + (-3) = 0$ 11. positive integers 13. The points on the numberline from left to right in this order: -5; $1; 2; 4$ 15. The points on the numberline from left to right in this order: $-3; -1; 0; 5$ 17. $>$ 19. $<$ 21. $<$ 23. $>$ 25. $-\frac{5}{1}$ 27. $-\frac{10}{3}$ 29. $-\frac{23}{10}$ 31. 6 33. -8 35. 7 37. 8 39. -5 41. -22 43. 16 45. -33 47. 0 49. -17 51. 5 53. -12 55. 16 57. 13 59. -17 61. -7 63. -23 65. 36 67. $-4x^2 - 3x$ 69. $6b^2 + b - 1$ 71. $-\frac{65}{2}$ 73. $\frac{245}{1}$ 75. gain of 2 yd 77. $130°$F

Page 320 Chapter Test 1. $>$ 3. $>$ 5. 7 7. -72 9. -9 11. -28 13. -25 15. -13 17. -15 19. -20 21. -13 23. 21 25. $34°$F

Page 321 Enrichment 1. 1 3. 6 5. 2 7. 3 9. 6 11. 6 13. 1 15. 1 17. 1

Page 322 Additional Practice 1. $-6; -4; 0; 2$ 3. $-1; 2; 3; 4; 5$ 5. $>$ 7. $<$ 9. $\frac{6}{10}$ 11. $-\frac{35}{10}$ 13. $\frac{10}{3}$ 15. -4 17. 7 19. 2.5 21. -4 23. -6 25. 3 27. 32 29. 0 31. 9 33. -78 35. -7.9 37. 2 39. -14 41. 9 43. $\frac{1}{8}$ 45. $\frac{1}{2}$ 47. 0 49. -1 51. gain of 2 yd

CHAPTER 11

Page 324 Self-Check 1. -4 2. 21 3. -3 4. -15 5. -4 6. -17

Page 325 Classroom Exercises 1. $y + 18 - 18 = 7 - 18$ 3. $a - 5 + 5 = 22 + 5$ 5. $-15 - 23 = 23 + b - 23$ 7. -1 9. 29 11. -2 13. -12 15. 3 17. -7

Page 325 Written Exercises 1. 26 3. 2 5. -13 7. 39 9. 31 11. -1.8 13. 6 15. -9 17. -21 19. -13 21. -35 23. 5.9 25. -25 27. 18 29. 22 31. -102 33. 0 35. -7.5 37. 10.5 39. $-1\frac{2}{5}$

Pages 326-327 Self-Check 1. 4 2. 5 3. 6 4. 3

Page 327 Classroom Exercises 1. $8x = 32$ 3. $c = 19$ 5. $7x - 5 = 16$ 7. $2a + 4 = 6$ 9. 6 11. 8 13. 7 15. 4

Page 327 Written Exercises 1. 9 3. 4 5. 9 7. 3 9. 23 11. 2 13. 4 15. 6 17. 2 19. 5 21. 6 23. 19 25. 33 27. $5\frac{3}{5}$ 29. $3\frac{1}{2}$ 31. 13 33. $6\frac{1}{2}$ 35. 6 37. 8 39. 10 41. 5 43. 10 45. 4 47. 5.5

Page 329 Self-Check 1. 5 2. 3 3. 12 4. 4 5. 4

Page 330 Classroom Exercises 1. $v; v; 2v; 3$ 3. $d; d; 6d; 12$ 5. $5h; 5h; h; 12; 7$

Page 330 Written Exercises 1. 10 3. 7 5. 16 7. $7\frac{1}{2}$ 9. $3\frac{1}{2}$ 11. 50 13. 7 15. 10 17. 8 19. 7 21. 8 23. 5 25. 3 27. 2 29. 30 31. 4

Page 331 Calculator Exercises 1. Does not check 3. Does not check

Page 331 Review: Sections 11-1–11-3 1. 7 3. -24 5. 2 7. -3 9. 1 11. 6 13. 3 15. 8

Page 332 Special Topic 1. 14.7% 3. 17.2% 5. 15.3%

Page 333 Special Topic 1. $\frac{7}{10}$ 3. $\frac{1}{9}$ 5. $-\frac{3}{6}$ 7. $-\frac{12}{5}$

Page 335 Self-Check 1. $53 2. 104 lb

Page 336 Classroom Exercises 1. $y - 6$ 3. $y + 6$ 5. $6 + y; y + 6$ 7. $6 + y; y + 6$ 9. $y - 6$ 11. $c + 12$ 13. $60 - d = 24$ 15. $k - 100 = 425$

Page 336 Written Exercises 1. $25 - n$ 3. $x + 5$, or $5 + x$ 5. $e - 6$ 7. $32 + b$ 9. $85 + c$, or $c + 85$ 11. $d - 14.5$ 13. $y + 25$ 15. $42 - (-p)$ 17. $t + 19 = 42$ 19. $t + 32 = 78$ 21. $24 - t = 15$ 23. $t - 12 = 52$ 25. 682 km 27. $36.75 29. $165 31. 8 33. 9 cm 35. 34

Pages 338-339 Self-Check 1. All points to the right of and not including 2 2. All points to the left of and not including -1 3. All points to the left of and not including 0 4. All points to the left of and including 3 5. All points to the right and including -5 6. All points to the right of and including 0

Page 339 Classroom Exercises 1. No 3. Yes 5. Yes 7. Yes 9. No 11. Yes 13. Yes 15. No 17. Yes 19. Yes 21. No 23. Yes

Page 339 Written Exercises 1. All points to the right of and not including 2 3. All points to the left of and not including -1 5. All points to the left of and not including 4 7. All points to the right of and not including 0 9. All points to the right of and not including 1 11. All points to the right of and including 1 13. All points to the right of and including -5 15. All points to the left of and including 0 17. All points to the left of and including -2 19. All points to the right of and including -2 21. All points to the right of and not including 3 23. All points to the right of and including -3 25. All points to the left of and including -1 27. All points to the right of and includ-

Answers to Selected Exercises 559

ing 5   **29.** All points to the right of and not including −5   **31.** All points to the left of and not including −3   **33.** All points to the left of and including 4   **35.** All points to the right of and including −4

Pages 340-341   Self-Check   1. $p > 7$; All points to the right of and not including 7   **2.** $b < 5$; All points to the left of and not including 5   **3.** $t < 1$; All points to the left of and not including 1   **4.** $x < 3$; All points to the left of and not including 3   **5.** $c > -6$; All points to the right of and not including −6   **6.** $a > -5$; All points to the right of and not including −5

Page 341   Classroom Exercises   1. $a - 5 + 5 > 3 + 5$ **3.** $w - 4 + 4 < 8 + 4$ **5.** $y - 2 + 2 > 6 + 2$ **7.** $m + 12 - 12 \leq 3 - 12$ **9.** $t + 15 - 15 \geq -5 - 15$ **11.** $c + 5 - 5 < 12 - 5$ **13.** $g > 12$; All points to the right of and not including 12   **15.** $w \leq -3$; All points to the left of and including −3   **17.** $b < 5$; All points to the left of and not including 5   **19.** $p < 6$; All points to the left of and not including 6   **21.** $y \geq 4$; All points to the right of and including 4   **23.** $b > 10$; All points to the left of and not including 10

Page 341   Written Exercises   1. $t > 5$; All points to the right of and not including 5   **3.** $p < 2$; All points to the left of and not including 2   **5.** $x > 16$; All points to the right of and not including 16   **7.** $c > 17$; All points to the right of and not including 17   **9.** $b \leq -6$; All points to the left of and including −6   **11.** $z < 1$; All points to the left of and not including 1   **13.** $n < 4$; All points to the left of and not including 4   **15.** $k > -2$; All points to the right of and not including −2   **17.** $w < -6$; All points to the left of and not including −6   **19.** $d \leq 11$; All points to the left of and including 11   **21.** $b \geq -15$; All points to the right of and including −15   **23.** $n < 1$; All points to the left of and not including 1   **25.** $x \geq -7$; All points to the right of and including −7   **27.** $a > 16$; All points to the right of and not including 16   **29.** $t \leq 50$; All points to the left of and including 50   **31.** $x \geq -9$; All points to the right of and including −9   **33.** $y > -1$; All points to the right of and not including −1   **35.** $x < 4$; All points to the left of and not including 4   **37.** $n < -4$; All points to the left of and not including −4   **39.** $r < 2$; All points to the left of and not including 2   **41.** $w < 43$; All points to the left of and not including 43   **43.** $p \geq -9$; All points to the right of and including −9   **45.** $x > -7$; All points to the right of and not including −7   **47.** $c \geq -9$; All points to the right of and including −9

Page 342   Calculator Exercises   1. Checks   **3.** Does not check   **5.** Does not check

Page 343   Review: Sections 11-4–11-6   1. $n + 5$ **2.** $d - 7$ **3.** $w - 5$ **4.** $b + 31$ **5.** $c - 5$ **6.** $y - 27$ **7.** $x + 14$ **8.** $n + 3$ **9.** $c + 8 = 33$ **10.** $12 + p = 56$ **11.** $a - 3 = 13$ **12.** $d - 10 = 36$ **13.** $x + 11 = 34$ **14.** $y - 5 = 19$ **15.** $20 - n = 13$ **16.** $s - 28 = 103$ **17.** $65   **18.** $141   **19.** $48\frac{1}{2}$ in   **20.** 608 km

21. All points to the right of and not including −8   **22.** All points to the left of and not including 5   **23.** All points to the left of and including −3   **24.** All points to the right of and including 6   **25.** $y < 9$; All points to the left of and not including 9   **26.** $c > 2$; All points to the right of and not including 2   **27.** $t > -1$; All points to the right of and not including −1   **28.** $k \geq 4$; All points to the right of and including 4

Page 344   Chapter Review   1. inequalities   **3.** word   **5.** algebraic   **7.** 4   **9.** −44   **11.** −4   **13.** −3   **15.** 9   **17.** 6   **19.** 3   **21.** 4   **23.** 9   **25.** 5   **27.** 1   **29.** $35 + t$ **31.** $c - 6$ **33.** $d + 5$ **35.** $r - 15 = 42$ **37.** $w + 16 = 84$ **39.** All points to the right of and not including 5   **41.** All points to the right of and not including −4   **43.** All points to the left and including −4   **45.** All points to the right of and including 0   **47.** $t < 4$; All points to the left of and not including 4   **49.** $y \geq -3$; All points to the right of and including −3   **51.** $t < -5$; All points to the left of and not including −5   **53.** $k > -3$; All points to the right of and not including −3   **55.** $16   **57.** $88

Page 346   Chapter Test   1. 27   **3.** −15   **5.** 2   **7.** 3   **9.** 4   **11.** $40 - d$ **13.** $11 + n$ **15.** $24.50   **17.** $153.20   **19.** $z > 8$; All points to the right of and not including 8

Page 347   Enrichment   1. All points between 3 and 4 and not including 3 and 4   **3.** All points between −2 and 3 including −2, but not including 3   **5.** All points to the right of, and not including 0, and all points to the left of, and including −3   **7.** $x > 150$ and $x \leq 220$

Page 348   Additional Practice   1. 3   **3.** −93   **5.** −41   **7.** −44.6   **9.** −40.7   **11.** 3   **13.** −13   **15.** 3   **17.** 7   **19.** 7   **21.** 50   **23.** 3   **25.** 19   **27.** 33   **29.** All points to the right of and not including 5   **31.** All points to the right of and including −4   **33.** $n < 3$; All points to the left of and not including 3   **35.** $a > 14$; All points to the right of and not including 14   **37.** $3 + n = 84$ **39.** $c - 24 = 36$ **41.** $c - 55 = 326$; $381

CHAPTER 12

Page 350   Self-Check   1. −70   **2.** −225   **3.** −96   **4.** −36   **5.** −72   **6.** −54

Page 351   Classroom Exercises   1. −16   **3.** −10; −15   **5.** 0   **7.** −65   **9.** −231   **11.** −3456   **13.** −6   **15.** −20   **17.** 0   **19.** −384   **21.** −425   **23.** −1107

Page 351   Written Exercises   1. −30   **3.** −180   **5.** −665   **7.** −2100   **9.** −3.15   **11.** −3   **13.** 0   **15.** −81   **17.** −156   **19.** −840   **21.** −32.8   **23.** $-\frac{1}{12}$ **25.** 0   **27.** −2016   **29.** −6272   **31.** −4402   **33.** −86.8   **35.** $-\frac{4}{5}$ **37.** −3(5); −15   **39.** 6(−15); −90   **41.** 4(−6); −24   **43.** −2(78); −156   **45.** −145   **47.** −102   **49.** −27   **51.** $-50y$ **53.** $-8ab$ **55.** $-56a^2$ **57.** $-6m^3$ **59.** $-2v^5$ **61.** $-30n^3$

560   Answers to Selected Exercises

Page 353 Self-Check **1.** 14 **2.** 130 **3.** 300 **4.** 77

Page 354 Classroom Exercises **1.** 12 **3.** 9; 18

Page 354 Written Exercises **1.** 35 **3.** 9 **5.** 735 **7.** 810 **9.** 625 **11.** 2992 **13.** 1498 **15.** 7812 **17.** 6.6 **19.** 8.28 **21.** 5.7 **23.** 21.294 **25.** $\frac{1}{2}$ **27.** $10\frac{1}{5}$ **29.** $\frac{1}{8}$ **31.** $4\frac{1}{6}$ **33.** 13 **35.** 36 **37.** 52 **39.** 1 **41.** $18a$ **43.** $15x^2$ **45.** $10a^2b^2$ **47.** $12x^3y$

Page 355 Calculator Exercises **1.** $-26,752$ **3.** 5044.48 **5.** -0.1623792 **7.** 51,504 **9.** $-46,818.486$

Pages 356-357 Self-Check **1.** $10n + 8$ **2.** $-24y + 18$ **3.** $4p + 28$ **4.** $-12p - 21$ **5.** $-4x - 48$ **6.** $-8x - 2$ **7.** $24x^2 - 12x$ **8.** $15t^2 - 6t$ **9.** $-6r^2 - 2r$

Page 357 Classroom Exercises **1.** $2x$ **3.** -8 **5.** $6(-3)$; $-6x$ **7.** $2n^2$

Page 357 Written Exercises **1.** $15s + 50$ **3.** $8x + 14$ **5.** $14n + 21$ **7.** $24q + 6$ **9.** $-27c + 18$ **11.** $-2r - 14$ **13.** $6p - 3$ **15.** $4c - 3$ **17.** $-5z + 10$ **19.** $a^2 - a$ **21.** $q^2 - 2q$ **23.** $3y^2 - 12y$ **25.** $6p^2 - 12p$ **27.** $12x^2 - 28x$ **29.** $8x + 20$ **31.** $2t + 18$ **33.** $-3p - 15$ **35.** $-12n - 4$ **37.** $40b + 70$ **39.** $16d^2 - 8d$ **41.** $4n + 12$ **43.** $-b - 6$ **45.** $-14p + 12$ **47.** $m^2 + 3md$

Page 358 Review: Sections 12-1–12-3 **1.** -40 **2.** -60 **3.** -135 **4.** -357 **5.** -672 **6.** -840 **7.** -3150 **8.** $-11,742$ **9.** 32 **10.** 15 **11.** 63 **12.** 512 **13.** 1000 **14.** 4484 **15.** 3696 **16.** 28,743 **17.** $10a + 6$ **18.** $6b + 48$ **19.** $-4x - 8$ **20.** $-15t - 5$ **21.** $y^2 - 5y$ **22.** $8r^2 - 4r$ **23.** $-20c^2 - 15c$ **24.** $-18v^2 + 42v$

Page 359 Special Topic **1.** -678 BTU's **3.** -2236 BTU's

Page 361 Self-Check **1.** 9 **2.** -7 **3.** -6 **4.** -8 **5.** -12 **6.** $-\frac{1}{10}$ **7.** $-3\frac{1}{2}$

Page 362 Classroom Exercises **1.** 8; 6 **3.** 7; -8 **5.** 6 **7.** -9 **9.** -10 **11.** 11 **13.** -6 **15.** -4 **17.** -12 **19.** 10 **21.** $-\frac{1}{5}$ **23.** $1\frac{1}{2}$

Page 362 Written Exercises **1.** 9 **3.** 7 **5.** -3 **7.** -7 **9.** -3 **11.** -18 **13.** 5 **15.** -15 **17.** -8 **19.** -23 **21.** -5 **23.** -7 **25.** -48 **27.** $\frac{1}{27}$ **29.** -6 **31.** $-1\frac{1}{2}$ **33.** $1\frac{2}{9}$ **35.** -10 **37.** $-\frac{a}{8}$; or $-\frac{1}{8}a$ **39.** $\frac{2e}{3}$; or $\frac{2}{3}e$ **41.** $\frac{b}{18}$; or $\frac{1}{18}b$ **43.** $-\frac{5h}{2}$; or $-\frac{5}{2}h$ **45.** $-\frac{1}{x}$ **47.** 1 **49.** $\frac{c}{4}$; or $\frac{1}{4}c$ **51.** $-\frac{100}{f}$ **53.** No; difference may be a negative integer **55.** Yes

Pages 364-365 Self-Check **1.** $3(a + 1)$ **2.** $7(x + 1)$ **3.** $2(2 + p)$ **4.** $5(3 + n)$ **5.** $3(x - 2)$ **6.** $5(m - 2)$ **7.** $3(4n - 1)$ **8.** $t(t + 3)$ **9.** $p(p + 5)$ **10.** $y(4 - y)$

Page 365 Classroom Exercises **1.** 2 **3.** 2 **5.** 3 **7.** m **9.** 3 **11.** 2 **13.** $r + 7$ **15.** $9 + t$ **17.** 3 **19.** 3 **21.** $r - 1$ **23.** $2x - 1$ **25.** n **27.** $t + 10$ **29.** x **31.** $12 - c$

Page 366 Written Exercises **1.** $2(n + 1)$ **3.** $3(r + 1)$ **5.** $3(y + 5)$ **7.** $7(3 + c)$ **9.** $7(n - 2)$ **11.** $5(3a - 1)$ **13.** $7(n - 4)$ **15.** $3(p - 4)$ **17.** $y(y + 5)$ **19.** $c(c - 4)$ **21.** $x(13 + x)$ **23.** $r(r + 1)$ **25.** $r(r - 1)$ **27.** $t(t + 2)$ **29.** $13(y + 1)$ **31.** $7(n + 6)$ **33.** $3(t + 4)$ **35.** $5(a - 4b)$ **37.** $x(x - 12)$ **39.** $y(9 - y)$ **41.** $3(x^2 + 3x + 5)$ **43.** $5(a^2 + 2a + 3)$ **45.** $7(d^2 + 2d + 4)$ **47.** $3(6m^2 + 5m + 9)$ **49.** $2(x^2 - x - 1)$

Page 367 Calculator Exercises **1.** 31.2 **3.** 1184.08 **5.** 5205.6 **7.** 15,713.88 **9.** 71.34 **11.** 1,684,956

Page 367 Review: Sections 12-4–12-5 **1.** 8 **2.** 7 **3.** -18 **4.** -27 **5.** -3 **6.** -8 **7.** 16 **8.** 5 **9.** -31 **10.** -5 **11.** -5 **12.** -12 **13.** -10 **14.** -16 **15.** 5 **16.** -7 **17.** -16 **18.** -15 **19.** -10 **20.** -25 **21.** $1\frac{1}{2}$ **22.** $\frac{2}{3}$ **23.** -1 **24.** $-\frac{1}{5}$ **25.** $2(x + 1)$ **26.** $5(x - 2)$ **27.** $3(x - 1)$ **28.** $5(x + 3)$ **29.** $x(x - 2)$ **30.** $x(x + 12)$ **31.** $2(x + 7)$ **32.** $x(x - 3)$ **33.** $7(x - 6)$ **34.** $3(5 + x)$ **35.** $6(3 + x)$ **36.** $4(3 + x)$ **37.** $x(16 - x)$ **38.** $x(3 + x)$ **39.** $x(7 + x)$ **40.** $x(13 - x)$

Page 368 Chapter Review **1.** $(6)(-3) = (-3)(6)$ **3.** $5 \cdot 0 = 0$ **5.** $(-2 \cdot 5)4 = -2(5 \cdot 4)$ **7.** factoring **9.** -108 **11.** -84 **13.** -391 **15.** -960 **17.** -2394 **19.** -4536 **21.** -10.44 **23.** $-\frac{1}{24}$ **25.** 24 **27.** 144 **29.** 1598 **31.** 3654 **33.** 1526 **35.** 2856 **37.** 8.64 **39.** $3\frac{3}{4}$ **41.** $2x + 6$ **43.** $-4b - 20$ **45.** $r^2 - 6r$ **47.** $-8n + 16$ **49.** $2p - 3$ **51.** $6y^2 - 6y$ **53.** 10 **55.** 3 **57.** -4 **59.** -4 **61.** -17 **63.** -13 **65.** -16 **67.** $1\frac{1}{2}$ **69.** $-\frac{2}{3}$ **71.** $\frac{1}{20}$ **73.** $2(x + 1)$ **75.** $k(k + 1)$ **77.** $13(3r - 1)$ **79.** $2(1 + h)$ **81.** $7(2 - f)$ **83.** $y(13 - y)$ **85.** $2(-10); -20$ **87.** $2(-6); -12$ **89.** $5(-12); -60$

Page 370 Chapter Test **1.** -384 **3.** -306 **5.** -7.75 **7.** 900 **9.** $2a + 2$ **11.** $4y^2 - 8y$ **13.** $-2r - 12$ **15.** 9 **17.** -8 **19.** $-\frac{1}{3}$ **21.** -5 **23.** $5(x + 1)$ **25.** $2(y - 8)$ **27.** $m(m + 2)$ **29.** $5(y - 4)$

Page 371 Enrichment **1.** $x^2 + 5x + 6$ **3.** $w^2 + 9w + 20$ **5.** $m^2 + 15m + 56$ **7.** $a^2 - 5a + 4$ **9.** $p^2 - 4p + 3$ **11.** $w^2 - 11w + 28$ **13.** $c^2 + 4c - 21$ **15.** $b^2 + b - 20$ **17.** $y^2 - 2y - 8$ **19.** $t^2 + 19t + 84$

Page 372 Additional Practice **1.** -156 **3.** -189 **5.** 0 **7.** -2520 **9.** -8.40 **11.** $-\frac{5}{8}$ **13.** 24 **15.** 66 **17.** 1404 **19.** 6405 **21.** 1.12 **23.** 2 **25.** $15x + 20$ **27.** $-30b - 50$ **29.** $-12n - 12$ **31.** $4y^2 - 2y$ **33.** $28p^2 - 35p$ **35.** $12f^2 - 24f$ **37.** 10 **39.** -4 **41.** -4 **43.** -3 **45.** -4 **47.** -6 **49.** $\frac{2}{3}$ **51.** -10 **53.** $14(x + 1)$

Answers to Selected Exercises 561

55. 7(3m + 1) **57.** 7(p − 5) **59.** 4(c − 1)
61. 4(9 − v) **63.** x(x − 6) **65.** 4(−15); −$60

CHAPTER 13

Page 374 Self-Check **1.** 42 **2.** −10 **3.** −112
4. 20 **5.** −9 **6.** −7

Page 375 Classroom Exercises **1.** −18 **3.** −7
5. 15 **7.** 96 **9.** −4 **11.** 7 **13.** −3 **15.** −2
17. −36 **19.** 63 **21.** −28 **23.** −44 **25.** −3
27. −9 **29.** 11 **31.** −6 **33.** −3 **35.** 3

Page 376 Written Exercises **1.** −96 **3.** −60
5. 24 **7.** −72 **9.** −114 **11.** −516 **13.** −4
15. −11 **17.** 24 **19.** −3 **21.** $-72\frac{1}{2}$ **23.** $16\frac{2}{3}$
25. −28 **27.** 29 **29.** 175 **31.** −140 **33.** −8
35. 0 **37.** −1.4 **39.** 12.5

Page 378 Self-Check **1.** 6 **2.** 120

Page 378 Classroom Exercises **1.** 5t **3.** $\frac{q}{4}$ **5.** 4q
7. 2t **9.** 7m **11.** $\frac{c}{12} = 25$ **13.** $\frac{h}{7} = 6$

Page 379 Written Exercises **1.** $\frac{x}{9}$ **3.** 2q **5.** 3r
7. −4a **9.** 3p **11.** 3w **13.** $\frac{t}{14}$ **15.** 4s = 28
17. 6h = 132 **19.** $\frac{k}{78} = 4$ **21.** $\frac{m}{7} = 10$ **23.** 3
25. 38 **27.** 400 **29.** $675 **31.** 71 **33.** 270

Page 381 Self-Check **1.** −1 **2.** 6 **3.** −15
4. −24 **5.** 18 **6.** −30

Page 382 Classroom Exercises **1.** 3k + 5 − 5 = 2 − 5
3. 7b + 1 − 1 = 15 − 1 **5.** 12 − 8 = −2y + 8 − 8
7. 4x + 7 − 7 = −13 − 7 **9.** $\frac{x}{3} - 1 + 1 = -2 + 1$
11. $\frac{v}{-2} + 3 - 3 = 7 - 3$ **13.** $2 - 4 = \frac{y}{2} + 4 - 4$
15. $\frac{c}{-2} - 3 + 3 = 4 + 3$

Page 382 Written Exercises **1.** −2 **3.** 2 **5.** −4
7. −4 **9.** −5 **11.** −8 **13.** −6 **15.** −8 **17.** 20
19. −66 **21.** 20 **23.** −160 **25.** −2 **27.** 5
29. −35 **31.** −12 **33.** 60 **35.** 21 **37.** −3 **39.** 3

Pages 383-384 Self-Check **1.** 2 **2.** −1 **3.** −6
4. 2 **5.** −6

Page 384 Classroom Exercises **1.** 21 **3.** 9e; 72
5. — **7.** 35; −3z **9.** 12; — **11.** Yes **13.** No
15. Yes **17.** Yes

Page 385 Written Exercises **1.** 4 **3.** $3\frac{2}{3}$ **5.** −3
7. −16 **9.** 11 **11.** 10 **13.** 3 **15.** 2 **17.** 5
19. $\frac{1}{2}$ **21.** −3 **23.** $-2\frac{1}{4}$ **25.** 15 **27.** −8
29. −5 **31.** −8 **33.** 10 **35.** 3 **37.** −2 **39.** 4

Page 386 Review: Sections 13-1—13-4 **1.** −30
2. −56 **3.** −108 **4.** −45 **5.** −7 **6.** −5 **7.** 6
8. 3 **9.** 128 **10.** 15 **11.** 4 **12.** 336 **13.** 1
14. −2 **15.** −3 **16.** −2 **17.** −8 **18.** −6
19. −30 **20.** 18 **21.** 4 **22.** 1 **23.** −20 **24.** −5
25. −6 **26.** −10

Page 387 Special Topic **1.** 11.5 **3.** −0.5 **5.** −4
7. −6.5 **9.** 5

Page 388 Self-Check **1.** 15°; 75°; 90°

Page 388 Classroom Exercises **1.** 30; 60; 90
3. 60; 100; 20

Page 389 Written Exercises **1.** x = 40°; 2x = 80°;
x + 20 = 60° **3.** y = 45°; 2y = 90° **5.** y = 50°; y + 15 = 65° **7.** d = 75°; d − 15 = 60°; d − 30 = 45°
9. x = 50°; x + 10 = 60°; x + 20 = 70°

Pages 390-391 Self-Check **1.** 7; 8 **2.** 8; 9; 10

Page 391 Classroom Exercises **1.** Yes **3.** No
5. h + h + 1 = 5 **7.** h + h + 1 + h + 2 = 21

Page 392 Written Exercises **1.** Louise; 14;
Reynaldo: 15 **3.** William: 85; Juan: 84 **5.** 104; 105; 106 **7.** Morris: 21; Eric: 20; Maria: 22 **9.** Hans: 17; Rosaline: 16 **11.** 6; 7; 8 **13.** 64; 65

Page 394 Self-Check **1.** c < 8; All points to the left of and not including 8 **2.** x > −3; All points to the right of and not including −3 **3.** d > −18; All points to the right of and not including −18 **4.** r < 20; All points to the left of and not including 20 **5.** m > −5; All points to the right of and not including −5 **6.** b < −6; All points to the left of and not including −6
7. q ≤ −5; All points to the left of and including −5

Page 394 Classroom Exercises **1.** 5 **3.** −2 **5.** 9
7. 7 **9.** −8 **11.** −3 **13.** −6 **15.** 5 **17.** −7

Page 395 Written Exercises **1.** y < 9; All points to the left of and not including 9 **3.** a > −70; All points to the right of and not including −70 **5.** w ≤ −6; All points to the left of and including −6 **7.** k < 50; All points to the left of and not including 50 **9.** n > −7; All points to the right of and not including −7
11. b > 4; All points to the right of and not including 4
13. m < −10; All points to the left of and not including −10 **15.** t ≤ −5; All points to the left of and including −5 **17.** x > 96; All points to the right of and not including 96 **19.** a < −40; All points to the left of and not including −40 **21.** j > 10; All points to the right of and not including 10 **23.** n > −2; All points to the right of and not including −2 **25.** b < −15; All points to the left of and not including −15 **27.** d < 32; All points to the left of and not including 32 **29.** p ≤ −11; All points to the left of and including −11
31. n ≥ −16; All points to the right of and including −16

Pages 396-397 Self-Check **1.** $27.99 **2.** 95

Page 397 Classroom Exercises **1.** b + 8 < 45
3. b + 15 < 32 **5.** $\frac{87 + 92 + y}{3} > 91$ **7.** $\frac{65 + 70 + 68 + x}{4} > 66$

Page 397 Written Exercises **1.** $27.99 **3.** $44.99
5. 89 **7.** 101

Page 398 Review: Sections 13-5—13-8 **1.** 8; 9
2. 12; 13; 14 **3.** x < −8; All points to the left of and not including −8 **4.** c < −6; All points to the left of and not including −6 **5.** n < −2; All points to the left

562 Answers to Selected Exercises

of and not including −2   **6.** $y < 4$; All points to the left of and not including 4   **7.** $x = 25°$; $2x = 50°$; $4x + 5 = 105°$ **8.** $349.99

Page 399   Chapter Review   **1.** multiplication
3. consecutive   **5.** reversed   **7.** −40   **9.** 36   **11.** 8
13. −8   **15.** $3m = 42$ **17.** $\frac{p}{16} = 8$ **19.** −3   **21.** −6
23. 7   **25.** −36   **27.** 9   **29.** 12   **31.** $x = 60°$
33. $t = 30°$; $4t = 120°$ **35.** 17; 18   **37.** 12; 13; 14
39. $p < -135$; All points to the left of and not including −135   **41.** $b < 48$; All points to the left of and not including 48   **43.** $t > 11$; All points to the right of and not including 11   **45.** $q > -2$; All points to the right of and not including −2   **47.** 90   **49.** Kermit: 18; Vera: 19

Page 401   Chapter Test   **1.** −60   **3.** −8   **5.** 7
7. −75   **9.** 3   **11.** $b < 8$; All points to the left of and not including 8   **13.** $x = 18°$; $3x = 54°$; $6x = 108°$
15. $31.99

Page 402   Enrichment   **1.** Yes   **3.** No   **5.** Yes

Page 403   Additional Practice   **1.** 150   **3.** −36
5. −8   **7.** 3.2   **9.** 10   **11.** −5   **13.** −35   **15.** 48
17. 3   **19.** 1   **21.** −12   **23.** $x < 8$; All points to the left of and not including 8   **25.** $x > -4$; All points to the right of and not including −4   **27.** $3x = 33$
29. $\frac{w}{4} = 9$ **31.** $839.99   **33.** 108; 109; 110

Page 404   Common Errors   **1.** −13   **3.** −16   **5.** 3
7. 7   **9.** −18   **11.** $-16\frac{1}{2}$ **13.** $4m + 36$

Page 405   Cumulative Review: Chapters 10-13   **1.** c
3. c   **5.** a   **7.** d   **9.** a   **11.** c   **13.** c   **15.** d   **17.** ⌣
19. c   **21.** b   **23.** a   **25.** d   **27.** b

CHAPTER 14

Pages 408-409   Self-Check   **1.** (−2, 4)   **2.** (−5, −5)
3. (2, −2)   **4.** (2, 3)   For Exercises 5-7, start at the origin. Then make the indicated moves for each point.
5. Left 1; down 4   **6.** Right 4   **7.** Down 2

Page 409   Classroom Exercises   **1.** N   **3.** P   **5.** Q
7. J   **9.** L   **11.** S

Page 409   Written Exercises   **1.** (2, 3)   **3.** (−3, 3)
5. (−4, 1)   **7.** (−5, −2)   **9.** (4, −3)   **11.** (0, −4)
For Exercises 13-21, start at the origin. Then make the indicated moves for each point.   **13.** Right 4; up 2
15. Left 2; down 4   **17.** Right 6; down 3   **19.** Left 2; up 6   **21.** Left 4

Pages 410-411   Self-Check   **1.** $\begin{array}{c|ccc} t & 1 & 3 & 5 \\ \hline d & 1100 & 3300 & 5500 \end{array}$
2. The graph is a straight line starting at (0, 0) and continuing through (5, 5500).   **3.** The graph is a straight line containing (−2, −3) and (1, 0).

Page 411   Classroom Exercises   **1.** $\begin{array}{c|ccc} t & 0 & 2 & 4 \\ \hline d & 0 & 160 & 320 \end{array}$
3. $\begin{array}{c|ccc} t & 0 & 1 & 2 \\ \hline d & 0 & 500 & 1000 \end{array}$

Page 412   Written Exercises   **1.** $d - 0, 100, 200$; the graph is a straight line starting at (0, 0) and continuing through (4, 200).   **3.** 1; 0; −2   **5.** 6; 0; −6   For Exercises 7-25, each graph is a straight line containing the given points.   **7.** (1, 1); (0, 0); (−2, −2)   **9.** (2, 6); (0, 0); (−2, −6)   **11.** (0, 1); (−1, −2)   **13.** (1, 1); (0, −3)   **15.** (0, 4); (2, 0)   **17.** (0, −3); (2, 1)
19. (−2, 3); (4, −3)   **21.** (2, 4); (−2, −4)   **23.** (−2, 3); (2, −5)   **25.** (−4, −2); (2, 1)

Page 413   Calculator Exercises   **1.** 1; −5; −8   **3.** 33; −15; −27

Pages 414-415   Self-Check   **1.** 2 in   **2.** The graph is a broken line connecting these points in order from left to right: (J, 10); (F, 8); (M, 9); (A, 16); (M, 20); (J, 18)

Page 415   Classroom Exercises   **1.** 900   **3.** 200

Page 416   Written Exercises   **1.** 49   **3.** $12\frac{1}{2}\%$
5. The graph is a broken line connecting these points: (1940, 6.6); (1950, 8); (1960, 7.6); (1970, 7.6); (1980, 6.4)

Pages 417-418   Self-Check   **1.** 1   **2.** $\frac{5}{3}$ **3.** $-\frac{1}{2}$
4. $\frac{2}{3}$

Page 418   Classroom Exercises   **1.** negative   **3.** zero
5. $\frac{1}{4}$ **7.** $-1 - (-3) = 2$; $\frac{3}{2}$ **9.** −3   **11.** $4 - 4 = 0$; $-2 - (-6) = 4$; 0

Page 419   Written Exercises   **1.** 2   **3.** 1   **5.** $\frac{1}{4}$
7. $\frac{1}{2}$ **9.** $\frac{1}{5}$ **11.** 3   **13.** $-\frac{2}{3}$ **15.** $-\frac{3}{4}$ **17.** $-\frac{5}{4}$
19. 0   **21.** $-\frac{1}{2}$ **23.** 0   **25.** 7   **27.** $-\frac{3}{4}$ **29.** 3
31. $\frac{9}{4}$ **33.** $-\frac{4}{9}$ **35.** 0

Page 420   Review: Sections 14-1–14-4   **1.** (−4, 4)
2. (1, 3)   **3.** (0, −1)   **4.** (3, −2)   For Exercises 5-8, start at the origin and make the moves given for each point.   **5.** Right 4; up 2   **6.** Left 3; up 1   **7.** Left 2; down 4   **8.** Right 1; down 3   **9.** 0; 80; 160   **10.** −1; 1; 3   **11.** 2; 4; 6   **12.** 0; 4; 8   For Exercises 13-16, each graph is a straight line containing the given points.
13. (0, 4); (−4, 0)   **14.** (−1, −6); (3, −2)   **15.** (0, 3); (−2, −1)   **16.** (−2, −5); (1,4)   **17.** The graph is a broken line connecting these points: (M, 2000); (T, 2500); (W, 1500); (Th, 2400); (F, 2100); (S, 3000); (S, 1200)   **18.** $\frac{1}{3}$ **19.** $-\frac{2}{5}$ **20.** $-\frac{1}{5}$

Page 421   Special Topic   **1.** −1.2   **3.** SLOPE IS UNDEFINED   **5.** −1   **7.** −4   **9.** −1.14285

Pages 422-423   Self-Check   **1.** 4   **2.** −1   **3.** 2
4. 3   For Exercises 5-7, each graph is a straight line containing the given points.   **5.** (0, 4); (−2, 0)
6. (0, −10); (2, 0)   **7.** (0, 9); (3, 0)

Page 423   Classroom Exercises   **1.** 2; −8   **3.** 1; 1

Page 423   Written Exercises   **1.** 2   **3.** 7   **5.** −3
7. 4   **9.** 3   **11.** −6   **13.** 3   **15.** −3   For Exercises 17-27, each graph is a straight line containing the given points.   **17.** (0, 3); (−3, 0)   **19.** (0, −7); (7, 0)

Answers to Selected Exercises   563

21. (0,4); (−2,0) **23.** (0,−4); (1,0) **25.** (0,8); (4,0) **27.** (0,−6); (−2,0)

Page 425 Self-Check **1.** $470 **2.** 68.6 km/h

Page 425 Classroom Exercises **1.** $c = 5n$ **3.** $rt = 300$

Page 426 Written Exercises **1.** $c = 2n$; $3 **3.** $lw = 400$; $3\frac{1}{5}$

Page 428 Self-Check **1.** 50 For Exercises 2-4, each graph is a pair of straight lines. The solution is given for each exercise. **2.** (−3,−4) **3.** (−1,2) **4.** (0,0)

Page 428 Classroom Exercises **1.** Yes **3.** Yes **5.** Yes

Page 429 Written Exercises **1.** 2000, 2500, 4000; 0, 1500, 4500; the graph is a pair of straight lines. The solution is (40, 3000). For Exercises 3-11, each graph is a pair of straight lines. The solution is given for each exercise. **3.** (−2,−1) **5.** (−1,−2) **7.** (0,3) **9.** (1,−3)

Page 430 Self-Check For Exercises 1-2, each graph is a region that does not include its linear boundary. The location of the region with respect to the boundary is indicated for each exercise. **1.** Below the line containing (0, 3) and (−3, 0) **2.** Above the line containing (−3, −3) and (2, 2)

Page 431 Classroom Exercises **1.** Yes **3.** No **5.** Yes **7.** a **9.** b

Page 431 Written Exercises **1.** Yes **3.** Yes **5.** Yes **7.** No **9.** No For Exercises 11-21, each graph is a region that does not include its linear boundary. The location of the region with respect to the boundary is indicated for each exercise. **11.** Below the line containing (−3, −4) and (3, 2) **13.** To the left of the line containing (−1, −3) and (1, 3) **15.** To the right of the line containing (−1, −4) and (1, 2) **17.** To the left of the line containing (−1, 3) and (1, −3) **19.** Above the line containing (−3, 1) and (2, −4) **21.** To the right of the line containing (0, 3) and (2, −5)

Page 432 Review: Sections 14-5 − 14-8 **1.** 2; −2 **2.** −5; 5 **3.** −4; 2 **4.** 6; −2 For Exercises 5-8, each graph is a straight line containing the given points. **5.** (0,3); (−3,0) **6.** (0,−4); (4,0) **7.** (0,3); (−1,0) **8.** (0,−8); (2,0) **9.** $D = 60m$; 660 **10.** $dw = 24$; 6 For Exercises 11-14, each graph is a pair of straight lines. The solution is given for each exercise **11.** (1, 3) **12.** (1, 0) **13.** (1, 1) **14.** (0, 0) For Exercises 15-18, each graph is a region that does not include its linear boundary. The location of the region with respect to the boundary is indicated for each exercise. **15.** Below the line containing (0, 3) and (−3, 0) **16.** Above the line containing (0, −4) and (4, 0) **17.** To the right of the line containing (−1, 3) and (1, −3) **18.** To the right of the line containing (0, −4) and (2, 0)

Page 433 Chapter Review **1.** origin **3.** y axis **5.** line graph **7.** break-even **9.** slope **11.** indirect variation **13.** x coordinate **15.** point of intersection **17.** (1, −1) **19.** (−4, 2) **21.** (4, 3) **23.** (2, 1) **25.** (−4, −2) **27.** (1, 2) For Exercises 29-35, start at the origin. Then make the indicated moves for each point. **29.** Right 4; down 4 **31.** Right 3; down 2 **33.** Right 5 **35.** Right 5; up 4 For Exercises 37-39, each graph is a straight line containing the given points. **37.** (−1,−2); (2,4) **39.** (3, 1); (−1, 5) **41.** 200 **43.** $\frac{1}{2}$ **45.** $-\frac{5}{3}$ **47.** $-\frac{6}{7}$ **49.** 7 **51.** 3 **53.** −6 **55.** −4 **57.** $T = 0.75a$; 4.50 For Exercises 59-61, each graph is a pair of straight lines. The solution is given for each exercise. **59.** (1, 0) **61.** (−1, −1) For Exercises 63-65, each graph is a region that does not include its linear boundary. The location of the region with respect to the boundary is indicated for each exercise. **63.** Below the line containing (0, 3) and (3, 0) **65.** To the left of the line containing (−3, −5) and (1, 3) **67.** The graph is a broken line connecting these points in order from left to right: (M, 2500); (T, 2400); (W, 3000); (Th, 2200); (F, 1900); (S, 2500); (S, 2000)

Page 436 Chapter Test **1.** (4, 0) **3.** (2, 4) **5.** The graph is a straight line containing (0, −3) and (4, 1) **7.** $\frac{3}{2}$ **9.** 2; −2; the graph is a straight line containing (0, 2) and (−2, 0) **11.** The graph is the region below the line containing (0, 5) and (−3, 2) **13.** The graph is a pair of straight lines. The solution is (3, 9) **15.** $v = 9w$; 45

Page 437 Enrichment **1.** {2, 3, 4, 5} **3.** {3, 4, 5, 6, 9} **5.** {3, 4} **7.** ⊄ **9.** ⊂ **11.** ⊂

Page 438 Additional Practice For Exercises 1-3, start at the origin. Then make the indicated moves for each point. **1.** Right 1; up 1 **3.** Left 2; down 2 For Exercises 5-7, the graph is a straight line containing the given points. **5.** (2, 2); (−3, −3) **7.** (−3, −1); (2, 4) **9.** $\frac{4}{3}$ **11.** $\frac{2}{3}$ **13.** −2; 6; the graph is a straight line containing (0, 6) and (−2, 0) **15.** 3; 9; the graph is a straight line containing (0, 9) and (3, 0) **17.** $xy = 40$ **19.** The graph is a pair of straight lines. The solution is (0, 3) For Exercises 21-23, each graph is a region that does not include its linear boundary. The location of the region with respect to the boundary is indicated for each exercise. **21.** Above the line containing (−4, −1) and (4, 1) **23.** To the right of the line containing (−2, −3) and (1, 3) **25.** $Q = 4, 12, 20$; the graph is a straight line starting at (0, 0) and continuing through (5, 20)

CHAPTER 15

Pages 440-441 Self-Check **1.** 9 **2.** 169 **3.** 324 **4.** 961 **5.** 7 **6.** 3 **7.** −4 **8.** −9

Page 441 Classroom Exercises **1.** 16 **3.** 9; 81 **5.** 4 **7.** 7; 7

Page 441 Written Exercises **1.** 4 **3.** 36 **5.** 100 **7.** 16 **9.** 225 **11.** 900 **13.** 441 **15.** 169

17. 324   **19.** 3600   **21.** 576   **23.** 10,000   **25.** 2
27. 7   **29.** 3   **31.** 6   **33.** 10   **35.** 13   **37.** −3
39. −7   **41.** −6   **43.** 14   **45.** 40   **47.** 60

Pages 442-443   Self-Check   1. rational   **2.** irrational
3. irrational   **4.** rational   **5.** irrational   **6.** rational
7. rational   **8.** irrational

Page 443   Classroom Exercises   1. No   **3.** Yes
5. No   **7.** Yes   **9.** No   **11.** No   **13.** irrational
15. rational   **17.** irrational   **19.** rational   **21.** rational
23. irrational   **25.** rational   **27.** irrational
29. rational   **31.** irrational

Page 444   Written Exercises   1. rational; 25 is a perfect square   **3.** rational; 100 is a perfect square
5. irrational; 18 is not a perfect square   **7.** irrational; 22 is not a perfect square   **9.** irrational; 7 is not a perfect square   **11.** rational; 36 is a perfect square
13. irrational; 112 is not a perfect square   **15.** rational; 1600 is a perfect square   **17.** irrational; 1000 is not a perfect square   **19.** rational; terminating   **21.** irrational; non-terminating, non-repeating   **23.** rational; repeating   **25.** rational; repeating   **27.** irrational; non-terminating, non-repeating   **29.** rational; terminating   **31.** irrational; 19 is not a perfect square
33. rational; terminating   **35.** irrational; non-terminating, non-repeating   **37.** rational; repeating
39. irrational; 61 is not a perfect square   **41.** rational; terminating   **43.** b; c   **45.** c   **47.** d

Pages 445-446   Self-Check   1. 1225   **2.** 8281
3. 4.583   **4.** 9.165   **5.** 29   **6.** 85   **7.** 20   **8.** 99

Page 446   Classroom Exercises   1. 3.742   **3.** 400; 4.472   **5.** 324; 4.243   **7.** 3025; 7.416

Page 446   Written Exercises   1. 169   **3.** 441
5. 1024   **7.** 6561   **9.** 14,400   **11.** 16,900
13. 2.828   **15.** 4.583   **17.** 7.483   **19.** 11
21. 6.708   **23.** 12.207   **25.** 15   **27.** 26   **29.** 39
31. 81   **33.** 60   **35.** 55

Page 448   Special Topic   1. −11.9°C   **3.** −13.7°C

Page 449   Special Topic   1. 60   **3.** 862   **5.** 139

Page 450   Review: Sections 15-1—15-3   1. 16
2. 121   **3.** 361   **4.** 64   **5.** 576   **6.** 900   **7.** 7
8. 2   **9.** 10   **10.** 1   **11.** −6   **12.** −9   **13.** −4
14. 5   **15.** −8   **16.** −12   **17.** 3   **18.** 11   **19.** irrational   **20.** rational   **21.** rational   **22.** irrational
23. rational   **24.** irrational   **25.** irrational   **26.** rational   **27.** rational   **28.** rational   **29.** rational
30. irrational   **31.** 324   **32.** 484   **33.** 12,544
34. 4225   **35.** 5.657   **36.** 8.660   **37.** 11.180
38. 11.790   **39.** 31   **40.** 97   **41.** 130   **42.** 62
43. 29   **44.** 96   **45.** 146   **46.** 71   **47.** 135
48. 149

Page 451   Self-Check   1. 12 ft   **2.** 67 mi

Page 452   Classroom Exercises   1. 3.7 m   **3.** 1.7 km

5. 6 in   **7.** 12 yd   **9.** 60 mi   **11.** 71; 85 mi
13. 98; 118 mi   **15.** 114; 137 mi

Page 452   Written Exercises   1. 4.8 m   **3.** 5.9 cm
5. 11.7 cm   **7.** 6.5 m   **9.** 6 ft   **11.** 11 in   **13.** 18 ft
15. 66 mi.   **17.** 146 mi   **19.** 74 mi   **21.** 122 mi

Page 455   Self-Check   1. 8.6 in   **2.** 10.6 m   **3.** 3 m
4. 12 ft

Page 455   Classroom Exercises   1. 10   **3.** 16   **5.** 35

Page 456   Written Exercises   1. 9 m   **3.** 5.4 m
5. 14.3 ft

Page 458-459   Self-Check   1. 1.5   **2.** 0.42   **3.** 10
4. 1   **5.** 6

Page 459   Classroom Exercises   1. 0.176   **3.** 0.466
5. 1.00   **7.** 6.30   **9.** 365.76

Page 460   Written Exercises   1. 1.00   **3.** 0.67
5. 67 m   **7.** 4.82 mi   **9.** 5 km

Page 461   Calculator Exercises   1. 0.0524078
3. 0.5095254   **5.** 28.636253

Page 461   Review: Sections 15-4—15-6   1. 3.5 m
3. 13 m   **5.** 24 m

Page 462   Chapter Review   1. right triangle   **3.** irrational   **5.** hypotenuse   **7.** 81   **9.** 324   **11.** 1764
13. 4   **15.** 9   **17.** 10   **19.** 11   **21.** −2   **23.** −12
25. rational   **27.** irrational   **29.** rational   **31.** irrational   **33.** rational   **35.** rational   **37.** 324   **39.** 441
41. 12,100   **43.** 8.660   **45.** 10.677   **47.** 9.747
49. 3.5   **51.** 5.2   **53.** 2.8   **55.** 118   **57.** 95
59. 178   **61.** 11.2   **63.** 30   **65.** 1.75   **67.** 0.78
69. 11

Page 464   Chapter Test   1. 25   **3.** 7   **5.** −12
7. rational   **9.** irrational   **11.** 4096   **13.** 9.747
15. 6   **17.** 30   **19.** 6.3 m

Page 465   Enrichment   1. 2.4   **3.** 3.9   **5.** 4.12

Page 466   Additional Practice   1. 49   **3.** 169   **5.** 9
7. rational   **9.** rational   **11.** rational   **13.** 676
15. 2209   **17.** 27   **19.** 13   **21.** 4   **23.** 40
25. 93.25   **27.** 23 m

Page 467   Common Errors   1. Start at the origin. Move 3 units to the left. Then move 6 units down.   **3.** 4
5. 3.16   **7.** 0.75

Page 468   Cumulative Review: Chapters 14-15   1. c
3. b   **5.** a   **7.** d   **9.** d

CHAPTER 16

Pages 470-471   Self-Check   1. $21   **2.** $561; $672

Page 471   Classroom Exercises   1. 40   **3.** 85.5; 85

Page 472   Written Exercises   1. 15,813   **3.** $6.15
5. 26; 26   **7.** 1.1; 0.8   **9.** mode; shows most frequent

Answers to Selected Exercises   565

sales **11.** $8100; $8100 **13.** median and mode

Page 474-475 Self-Check 1. 31-40, 7; 41-50, 3; 51-60, 3; 61-70, 5; 81-90, 5 **2.** The height of the bar for each interval is as given: 21-25: 2 units; 26-30: 4 units; 31-35: 5 units; 36-40: 3 units

Page 476 Classroom Exercises 1. 3; 3; 6; 2; 2

Page 476 Written Exercises 1. 120-219, 5; 220-319, 4; 320-419, 4; 420-519, 1; 520-619, 4; 620-719, 2 For Exercises 3-5, the height of the bar for each interval is as given. **3.** 17-23: 2 units; 24-30: 10 units; 31-37: 6 units; 38-44: 2 units **5.** 15-19: 2 units; 20-24: 3 units; 25-29: 4 units; 30-34: 8 units; 35-39: 3 units

Page 477 Review: Sections 16-1—16-2 1. 88.5; 90.5; 92 **2.** 152; 153.5; 160 **3.** 71; 72; 72 **4.** 21; 21; 19 **5.** The height of the bar for each interval is as given: 400-439: 2 units; 440-479: 1 unit; 480-519: 5 units 520-559: 3 units; 560-599: 1 unit

Page 478 Special Topic 1. 6.1 **3.** 5.3

Pages 479-480 Self-Check 1. $\frac{2}{11}$ **2.** $\frac{1}{11}$ **3.** $\frac{4}{11}$ **4.** $\frac{3}{10}$, or 30% **5.** $\frac{2}{5}$, or 40% **6.** 0, or 0%

Page 480 Classroom Exercises 1. $\frac{3}{5}$, or 60% **3.** 18; 30; $\frac{3}{5}$, or 60% **5.** $\frac{1}{10}$, or 10% **7.** $\frac{2}{5}$, or 40% **9.** $\frac{3}{5}$, or 60% **11.** 0, or 0%

Page 481 Written Exercises 1. $\frac{3}{10}$ **3.** $\frac{1}{2}$ **5.** $\frac{1}{5}$ **7.** $\frac{1}{5}$, or 20% **9.** $\frac{3}{5}$, or 60% **11.** $\frac{4}{5}$, or 80% **13.** $\frac{1}{2}$, or 50% **15.** $\frac{1}{2}$, or 50% **17.** $\frac{3}{8}$, or $37\frac{1}{2}$%

Page 483 Self-Check 1. $\frac{1}{9}$ **2.** $\frac{1}{36}$ **3.** $\frac{1}{12}$ **4.** $\frac{1}{6}$

Page 483 Classroom Exercises 1. 0 **3.** $\frac{1}{18}$ **5.** $\frac{1}{9}$ **7.** $\frac{1}{6}$ **9.** $\frac{1}{9}$ **11.** $\frac{1}{18}$ **13.** $\frac{1}{9}$ **15.** $\frac{5}{36}$ **17.** $\frac{1}{9}$

Page 484 Written Exercises 1. $\frac{1}{18}$ **3.** $\frac{1}{6}$ **5.** $\frac{1}{12}$ **7.** $\frac{5}{36}$ **9.** $\frac{1}{4}$ **11.** $\frac{2}{9}$ **13.** $\frac{1}{4}$ **15.** $\frac{5}{18}$

Page 485 Self-Check 1. $\frac{1}{6}$

Page 486 Classroom Exercises 1. $\frac{1}{3}$ **3.** $\frac{1}{6}$

Page 486 Written Exercises 1. Direct; Coastal; Central; Scenic; Direct; Overland; Scenic **3.** $\frac{1}{6}$ **5.** $\frac{1}{3}$ **7.** $\frac{1}{3}$ **9.** $\frac{1}{8}$ **11.** $\frac{3}{8}$ **13.** $\frac{1}{2}$

Pages 488-489 Self-Check 1. $\frac{11}{42}$ **2.** $\frac{2}{15}$

Page 489 Classroom Exercises 1. $\frac{1}{2}$; $\frac{3}{7}$; $\frac{3}{14}$ **3.** $\frac{1}{2}$; $\frac{7}{15}$; $\frac{7}{30}$

Page 489 Written Exercises 1. $\frac{1}{6}$ **3.** $\frac{3}{20}$ **5.** $\frac{1}{20}$ **7.** $\frac{2}{15}$ **9.** $\frac{1}{45}$ **11.** $\frac{2}{15}$ **13.** $\frac{6}{125}$ **15.** $\frac{1}{125}$ **17.** $\frac{1}{20}$ **19.** $\frac{1}{90}$

Page 490 Calculator Exercises 1. 0.9% **3.** 28.6% **5.** 25.4%

Page 491 Review: Sections 16-3—16-6 1. $\frac{1}{6}$ **2.** $\frac{1}{3}$ **3.** $\frac{2}{3}$ **4.** $\frac{1}{3}$ **5.** $\frac{1}{18}$ **6.** $\frac{1}{6}$ **7.** $\frac{2}{9}$ **8.** $\frac{5}{36}$ **9.** T; HT H; TH; T; TT **10.** $\frac{1}{4}$ **11.** $\frac{1}{4}$ **12.** $\frac{1}{2}$ **13.** $\frac{1}{2}$ **14.** $\frac{5}{36}$ **15.** $\frac{15}{22}$

Page 492 Chapter Review 1. median **3.** 0 **5.** mean **7.** independent events **9.** mode **11.** 237; 212.5; 132 **13.** $\frac{1}{16}$, or $6\frac{1}{4}$% **15.** $\frac{1}{4}$, or 25% **17.** $\frac{2}{25}$ **19.** $\frac{1}{25}$ **21.** $\frac{1}{5}$ **23.** HHH; HHT; HTH; HTT; THH; THT; TTH; TTT **25.** $\frac{1}{8}$; $12\frac{1}{2}$% **27.** $\frac{1}{2}$; 50% **29.** $\frac{1}{15}$; $6\frac{2}{3}$% **31.** $\frac{1}{35}$; $2\frac{6}{7}$% **33.** $\frac{1}{6}$; $16\frac{2}{3}$%

Page 494 Chapter Test 1. 9; 7.5; 7 **3.** $\frac{4}{11}$ **5.** $\frac{7}{11}$ **7.** $\frac{4}{25}$ **9.** $\frac{16}{95}$

Page 495 Enrichment 1. 4%; 32 **3.** 4%; 160 **5.** 1300

Page 496 Additional Practice 1. $4\frac{3}{8}$; $4\frac{1}{2}$; 3 **3.** $\frac{2}{5}$, 40% **5.** $\frac{4}{5}$, 80% **7.** $\frac{7}{36}$ **9.** $\frac{1}{6}$ **11.** HHHH; HHHT; HHTH; HHTT; HTHH; HTHT; HTTH; HTTT; THHH; THHT; THTH; THTT; TTHH; TTHT; TTTH; TTTT **13.** $\frac{28}{45}$

CHAPTER 17

Page 499 Self-Check 1. $\overline{AD}$, $\overline{CD}$; $\overrightarrow{CD}$, $\overrightarrow{BA}$

Page 499 Classroom Exercises 1. $\overrightarrow{TS}$ **3.** angle DEF **5.** $\overline{QR}$ **7.** R **9.** MR; HK **11.** No **13.** HER or REH **15.** same

Page 500 Written Exercises 1. f or l **3.** c **5.** d, g, or k **7.** Answers may vary. Some answers are: $\overleftrightarrow{AB}$; $\overleftrightarrow{FA}$; $\overleftrightarrow{BF}$ **9.** $\overrightarrow{BF}$ **11.** LIGH **13.** $\overleftrightarrow{HE}$; $\overleftrightarrow{BI}$

Pages 501-502 Self-Check 1. 70° **2.** 155° **3.** 110° For Exercises 4-7, teacher should check each student's drawings.

Page 502 Classroom Exercises 1. b **3.** a **5.** a

Page 503 Written Exercises 1. 30°; acute **3.** 95°; obtuse **5.** 155°; obtuse **7.** 25°; acute **9.** 105°; obtuse For Exercises 11-19, teacher should check each student's drawings. **21.** 33° **23.** 27°

Page 506 Self-Check For Exercises 1-3, teacher should check each student's drawings.

Page 506 Classroom Exercises 1. isosceles **3.** scalene **5.** acute **7.** right **9.** $\angle M$ **11.** $\overline{PQ}$ **13.** $\angle T$

Page 507 Written Exercises 1. equilateral **3.** isosceles **5.** SAS; $\overline{DF} \cong \overline{AC}$; $\angle F \cong \angle C$; $\overline{FE} \cong \overline{CB}$ **7.** SSS; $\overline{PQ} \cong \overline{TQ}$; $\overline{PR} \cong \overline{RT}$; $\overline{QR} \cong \overline{QR}$ **9** ASA; $\angle B \cong \angle S$; $\overline{BR} \cong \overline{ST}$; $\angle R \cong \angle T$

Page 509 Self-Check 1. 39 ft

Page 509 Classroom Exercises 1. 17 3. $\frac{1.8}{1.2}$

Page 510 Written Exercises 1. 13 3. 20.3
5. 6.4 m 7. 100 m

Page 511 Calculator Exercises 1. 13 3. 20.3

Page 512 Special Topic 1-3. Angles A and B are supp 5. Angles A and B are not supp 7. 70° 9. 30° 11. 82.1°

Page 513 Review: Sections 17-1–17.4 1. $\overrightarrow{ED}$; $\overleftrightarrow{EG}$; $\overleftrightarrow{EB}$ 2. $\overleftrightarrow{EB}$; $\overleftrightarrow{AF}$ 3. $\angle ACB$; $\angle BCF$; $\angle FCE$; $\angle ECA$ For Exercises 4-8, teacher should check each student's drawings. 9. SAS; $\overline{AC} \cong \overline{RS}$; $\angle A \cong \angle R$; $\overline{AB} \cong \overline{RT}$ 10. SSS; $\overline{XY} \cong \overline{FG}$; $\overline{XZ} \cong \overline{EF}$; $\overline{YZ} \cong \overline{EG}$ 11. ASA; $\angle M \cong \angle W$; $\overline{MQ} \cong \overline{WV}$; $\angle Q \cong \angle V$ 12. 15.4 m

Pages 514-515 Self-Check 1. $\overleftrightarrow{BC}$ intersects $\overleftrightarrow{CD}$ at C 2. $\overleftrightarrow{CD} \perp \overleftrightarrow{AC}$ 3. vertical 4. alternate interior 5. alternate interior 6. vertical

Page 515 Classroom Exercises 1. intersecting; perpendicular 3. intersecting; perpendicular 5. parallel 7. b 9. a 11. c

Page 516 Written Exercises 1. $\overleftrightarrow{BA} \perp \overleftrightarrow{AC}$; $\overleftrightarrow{CD} \perp \overleftrightarrow{AC}$; $\overleftrightarrow{AF} \perp \overleftrightarrow{FD}$; $\overleftrightarrow{CD} \perp \overleftrightarrow{FD}$ 3. $\overleftrightarrow{BE}$ and $\overleftrightarrow{BF}$; $\overleftrightarrow{BE}$ and $\overleftrightarrow{CD}$; $\overleftrightarrow{BE}$ and $\overleftrightarrow{AC}$; $\overleftrightarrow{BE}$ and $\overleftrightarrow{FD}$ 5. vertical 7. alternate interior 9. corresponding 11. alternate exterior

Page 518 Self-Check 1. trapezoid 2. parallelogram; rectangle 3. parallelogram; rectangle

Page 518 Classroom Exercises 1. Yes; Yes 3. No; Yes; Yes; Yes; Yes 5. No; No; Yes; No; Yes

Page 519 Written Exercises 1. 12 3. 120° 5. parallelogram; rectangle 7. parallelogram; rectangle; rhombus; square 9. trapezoid 11. parallelogram; rhombus 13. trapezoid 15. parallelogram; rectangle 17. parallelogram 19. rhombus; parallelogram; rectangle 21. rhombus; square

Page 521 Self-Check 1. 72 cm²

Page 521 Classroom Exercises 1. 24 m²; 40 m²; 30 m²; 94 m² 3. 24 ft²; 12 ft²; 16 ft²; 52 ft²

Page 522 Written Exercises 1. 2255 cm² 3. 54 cm² 5. 512 cm² 7. 47.28 m² 9. 72 ft² 11. 580 ft² 13. $414\frac{2}{3}$ in² 15. 1400 in² 17. 864 cm² 19. $268.80

Page 524 Self-Check 1. 849 cm³

Page 525 Classroom Exercises 1. 170 m³ 3. 18.09 m²; 90 m³ 5. 1540 ft³ 7. 17 ft²; 68 ft³

Page 525 Written Exercises 1. 402 cm² 3. 97,108 cm³ 5. 170 cm³ 7. 661 cm³ 9. 69 in³ 11. 4928 in³ 13. 4505 in³ 15. 1 m³ 17. 629 in³

Pages 527-528 Self-Check 1. 18 m³ 2. 180 ft³ 3. 167 m³ 4. 566 ft³

Page 528 Classroom Exercises 1. 100 ft³ 3. 754 m³

Page 529 Written Exercises 1. 8 m³ 3. 3140 cm³ 5. 300 ft³ 7. 38 ft³ 9. 3.36 m³ 11. 769.3 cm³

Page 530 Review: Sections 17-5–17-9 1. $\overleftrightarrow{NP} \perp \overleftrightarrow{PS}$; $\overleftrightarrow{NP} \perp \overleftrightarrow{NR}$; $\overleftrightarrow{SR} \perp \overleftrightarrow{PS}$; $\overleftrightarrow{SR} \perp \overleftrightarrow{NR}$ 2. $\overleftrightarrow{PN} \parallel \overleftrightarrow{SR}$; $\overleftrightarrow{PS} \parallel \overleftrightarrow{NR}$ 3. 5 4. 90° 5. 120°; 60° 6. 20; 12 7. 132 m² 8. $108\frac{1}{4}$ ft² 9. 603 cm³ 10. 50 in³ 11. 24 m³ 12. 31.5 cm³ 13. 462 m³ 14. 1780 cm³ 15. 2112 ft³ 16. 205 yd³ 17. 2904 in² 18. $29\frac{1}{3}$ in³ 19. 113.04 cm³ 20. 0.032 m³

Page 531 Chapter Review 1. degree 3. line 5. polygon 7. quadrilateral 9. right angle 11. similar triangles 13. acute angle 15. obtuse angle 17. line segment 19. Answers may vary. Some answers are: $\overrightarrow{BH}$; $\overrightarrow{AH}$; $\overrightarrow{AE}$ 21. $\angle ABC$; $\angle CBE$; $\angle EBD$; $\angle DBA$ For Exercises 23-25, teacher should check each student's drawings. 27. isosceles 29. scalene 31. 25 33. alternate interior 35. vertical 37. parallelogram 39. parallelogram; rhombus 41. 1161 m² 43. 1356 m³ 45. 6837 cm³ 47. 4385 m³ 49. 2,592,100 m³

Page 534 Chapter Test 1. $\overrightarrow{GH}$; $\overrightarrow{GB}$; $\overrightarrow{GJ}$ 3. acute 5. $\overleftrightarrow{JG} \perp \overleftrightarrow{HE}$ 7. $\overleftrightarrow{JF} \parallel \overleftrightarrow{BC}$; $\overleftrightarrow{JB} \parallel \overleftrightarrow{FC}$ 9. SSS; $\overline{YZ} \cong \overline{RS}$; $\overline{XY} \cong \overline{ST}$; $\overline{XZ} \cong \overline{RT}$ 11. 16.8 13. 4537.3 m³ 15. 66 in³

Page 535 Enrichment 1. 39 cm³ 3. 310 in³ 5. 4443 cm³ 7. 24,417 m³

Page 536 Additional Practice For Exercises 1-3, answers may vary. 1. Some answers are: $\overrightarrow{DE}$; $\overrightarrow{EF}$; $\overrightarrow{GF}$ 3. Some answers are: $\overrightarrow{DG}$; $\overrightarrow{EF}$; $\overrightarrow{FD}$ 5. acute 7. obtuse 9. obtuse 11. C 13. NC 15. trapezoid 17. parallelogram; rectangle; square 19. 20 m 21. 1538.6 cm³

Page 537 Common Errors 1. 86 3. $\frac{1}{25}$ 5. 12 7. 471 cm³

Page 538 Cumulative Review: Chapters 16-17 1. a 3. b 5. a 7. c 9. b 11. b 13. a 15. d

Answers to Selected Exercises 567

PICTURE CREDITS

Photos not listed below, Blaise Zito Associates, Inc.

Key: (t) top, (b) bottom, (l) left, (r) right, (c) center.

ILLUSTRATORS: Page 134, Tom Powers; all others, Blaise Zito Associates, Inc.

RESEARCH PHOTOS: CHAPTER ONE: Pages 5, Chris Reeberg/DPI; 7, © Marvin E. Newman/Woodfin Camp; 8, Focus on Sports; 10, Focus on Sports; 12, Wil Blanche/DPI; 14, © Craig Aurness/Woodfin Camp; 15, Stephen Kraseman/© Peter Arnold, Inc.; 16, Charlie Ott/DPI; 22, © George Hall/Woodfin Camp. CHAPTER TWO: Pages 34, Katrina Thomas/Photo Researchers, Inc.; 35, Focus on Sports; 37, Focus on Sports; 42, Wil Blanche/DPI; 46, DPI; 54, Focus on Sports; 55, DPI. CHAPTER THREE: Pages 64, Chuck Solo/Focus on Sports; 73, © Randa Bishop/DPI; 85, Gianni Tortoli/Photo Researchers, Inc.; 88, DPI. CHAPTER FOUR: Pages 99, © Shelly Grossman/Woodfin Camp; 107, HBJ Photo. CHAPTER FIVE: Pages 128, Chris Reeberg/DPI; 134, Dorothy Affa/Focus on Sports; 138, H. Armstrong Roberts; 144, © Alastar Black/Focus on Sports; 151, Bob Witt/DPI; 153, NASA. CHAPTER SIX: Pages 164, Weiner Muller/© Peter Arnold, Inc.; 165, Focus on Sports; 167, Focus on Sports; 183, © Larry Smith/DPI. CHAPTER SEVEN: Pages 198, E. Simonsen/H. Armstrong Roberts; 206, © C.B. Jones/Taurus Photos; 209, © David Madison/Focus on Sports; 211, Adolph Rohrer/DPI; 212, Focus on Sports. CHAPTER EIGHT: Pages 223, DPI; 230 top, © Lawrence Fried/The Image Bank; 230 (b), DPI; 231, H. Armstrong Roberts; 237, Focus on Sports; 238 (b), S. Shackman/Monkmeyer; 243, © Robert McElroy/Woodfin Camp. CHAPTER NINE: Pages 254, U.S. Dept. of Energy; 255, Chris Reeberg/DPI; 267, Joel Gordon/DPI; 269, Runk/Schoenberger from Grant Heilman; 287, © Jerry Wachter/Focus on Sports. CHAPTER TEN: Pages 296, DPI; 303, William Kelly/Shostal Associates; 307, © David Burnett/Contact Press Images/Woodfin Camp; 310 (t), © George Gerster/Rapho Division/Photo Researchers, Inc.; 310 (b), © L.L.T. Rhodes/Taurus Photos; 320, © Skyviews Survey, Inc./DPI. CHAPTER ELEVEN: Pages 332, H. Armstrong Roberts; 334, Focus on Sports; 337, Wil Blanche/DPI; 338, © Bill Strode/Woodfin Camp. CHAPTER TWELVE: Pages 359, R. Krubner/H. Armstrong Roberts; 369, © Jonathan Blair/Woodfin Camp. CHAPTER THIRTEEN: Pages 377, DPI; 401, © Adam Woolfitt/Woodfin Camp. CHAPTER FOURTEEN: Pages 410, © John Blaustein/Woodfin Camp; 427, © John Lei/Omni Photo Communications. CHAPTER FIFTEEN: Pages 448, H. Armstrong Roberts. CHAPTER SIXTEEN: Pages 470, National Oceanic and Atmospheric Administration; 474, Focus on Sports. CHAPTER SEVENTEEN: Pages 501, Guiliano Bevilacqua/Focus on Sports; 524, Lizabeth Corlett/DPI; 526, Isaac Geib from Grant Heilman; 527, J.Alex Langley/DPI;533,DPI. CHAPTER OPENERS: CHAPTER ONE: (l & tr), Blaise Zito Associates, Inc.;(br),Charlie Ott/DPI.CHAPTER TWO:(r),DPI;(tl),Blaise Zito Associates,Inc.;(bl),DPI.CHAPTER THREE:(r),Gianni Tortoli/Photo Researchers, Inc.;(tl),HBJ Photo;(bl),Chuck Solo/Focus on Sports.CHAPTER FOUR:Blaise Zito Associates,Inc. CHAPTER FIVE: (l),NASA;(tr),Blaise Zito Associates,Inc.;(br),H.Armstrong Roberts. CHAPTER SIX: (l), Blaise Zito Associates, Inc.; (tr), © Larry Smith/DPI; (br), Blaise Zito Associates, Inc. CHAPTER SEVEN: (r), Blaise Zito Associates, Inc.; (tl), © C.B. Jones/Taurus Photos; (bl), Adolph Rohrer/DPI. CHAPTER EIGHT: Blaise Zito Associates, Inc. CHAPTER NINE: (r), Runk/Schoenberger from Grant Heilman; (tl), Chris Reeberg/DPI; (bl), U.S. Dept. of Energy. CHAPTER TEN: (r), William Kelly/Shostal Associates; (tl), © Skyviews Survey, Inc./DPI; (br), © David Burnett/Contact Press Images/Woodfin Camp. CHAPTER ELEVEN: (l), H. Armstrong Roberts; (tr), Wil Blanche/DPI; bottom right, © Bill Strode/Woodfin Camp. CHAPTER TWELVE: (r), © Jonathan Blair/Woodfin Camp;(tl),and(bl),Blaise Zito Associates, Inc. CHAPTER THIRTREEN: (r),© Adam Woolfitt/Woodfin Camp; (tl), DPI; (bl), Blaise Zito Associates, Inc. CHAPTER FOURTEEN: (l), © John Lei/Omni Photo Communications. CHAPTER FIFTEEN: Blaise Zito Associates, Inc. CHAPTER SIXTEEN: (r), Blaise Zito Associates, Inc.; (tl), National Oceanic and Atmospheric Administration. CHAPTER SEVENTEEN: (l), J. Alex Langley/DPI; (r), Blaise Zito Associates, Inc.